MODERN BIOGEOCHEMISTRY

MODERN BIOGEOCHEMISTRY

VLADIMIR N. BASHKIN

Moscow State University, Russia

**In cooperation with Robert W. Howarth,
Cornell University, USA**

KLUWER ACADEMIC PUBLISHERS

DORDRECHT / BOSTON / LONDON

A C.I.P. Catalogue record for this book is available from the Library of Congress.

ISBN 1-4020-0992-5

Published by Kluwer Academic Publishers,
P.O. Box 17, 3300 AA Dordrecht, The Netherlands.

Sold and distributed in North, Central and South America
by Kluwer Academic Publishers,
101 Philip Drive, Norwell, MA 02061, U.S.A.

In all other countries, sold and distributed
by Kluwer Academic Publishers,
P.O. Box 322, 3300 AH Dordrecht, The Netherlands.

Printed on acid-free paper

Printed in the Netherlands

TABLE OF CONTENTS

PREFACE

Biogeochemistry is becoming an increasingly popular subject in graduate education. Courses in ecology, geography, biology, chemistry, environmental science, public health, and environmental engineering all have to include biogeochemistry in their syllabuses to a greater or lesser extent. Humanity's ever growing impact on the Environment, and the consequent local, regional, and global effect demand a profound understanding of the mechanisms underlying the sustainability of the biosphere and its compounds. The ideas of biogeochemistry about the universality of biogeochemical cycles involving the mass exchange of chemical elements between living organisms and the environment in the Earth's surface appear to be quite productive in this high priority academic and scientific discipline. In biogeochemical cycling the active principles come from biota which global biological and geological activity alter slowly the biosphere's compartments. On the other hand, the environment causes the living organisms evolve.

Biogeochemistry is the study of biological controls on the chemistry of the environment and geochemical regulation of ecological structure and function. Although the term was first used some 75 years ago, roots of this discipline can be traced to the earliest development of the natural sciences, before biology, geology, and chemistry became separate disciplines. Today biogeochemistry serves as a force of reintegration across these fields.

This textbook is aimed at generalizing the modern ideas of biogeochemical development during recent decades. Only a few textbooks are available for undergraduate and graduate students, however, as most books deal mainly with advanced research aspects of the subject. This book aims at supplementing the existing textbooks by providing a modern understanding of biogeochemistry, from evolutionary biogeochemistry to the practical application of biogeochemical ideas as a human biogeochemistry, biogeochemical standards, and biogeochemical technologies.

The reader will start from the history of biogeochemistry, which is important in this science inspite of being relatively young. The names of V. Vernadsky, F. Clark, A. Vinogradov, G. Hutchinson, A. Fersman, V. Goldschmidt, V. Kovalsky, E. Degans, V. Kovda, D. Atkinson, G. Likens, S. Miller, V. Dobrovolsky and many other scientists who carried out the biogeochemical research in various countries have to be known to students. During the last 3.5–3.8 billion years the evolution of the Earth and biogeochemical cycles have been developing in parallel and consequently the ideas of evolutionary biogeochemistry (Chapter 2) provide a key to understanding of the modern atmosphere, biosphere, and hydrosphere. The biogenic depositions which are very important for our civilization, were formed during geological history under the active influence of biogeochemical cycles, and this item is also discussed in this text. The biogeochemical cycling of macro and trace elements is a crucial point of the textbook (Chapters 3 & 4). However, the main attention here is given to the description of natural regularities of global biogeochemical cycling. The conceptual ways of considering the interactions of element cycles such as stoichiometric aspects of nutrient uptake (and nutrient limitation of production), stoichoimetric aspects of nutrient recycling, thermodynamics as applicable to bacterial energetics (and the use of different electron acceptors), etc., are elucidated in Chapter 5. The description of regional biogeochemistry gives readers the possibility of understanding the qualitative and quantitative parameters of different cycles in different places of the Earth (Chapter 6). The text also presents a general understanding of the ideas of biogeochemical mapping. Although this has still not been carried out for the whole planet, some examples are shown, for instance, in North Eurasia. Soil-biogeochemical regionalization is considered as a key for the preliminary mapping on a continental scale (Chapter 7). As a complement to the biogeochemistry of natural ecosystems, environmental biogeochemistry is to show the pollution processes as a disturbance of natural biogeochemical cycles (Chapter 8). Human biogeochemistry is now developing in various countries. This is considered to be very important for understanding human illnesses, human diet, and human adaptation, and the relevant state of the art of human biogeochemistry is shown in Chapter 9. At present ecological standards and norms based on the understanding and simulation of biogeochemical cycles of different elements are forming in many countries. The critical load calculation and mapping are widely used in Europe, North America, and Asia. I think this would be of interest for students as well (Chapter 10).

This text is to a certain extent a summary of both scientific results of various authors and of classes in biogeochemistry, which were given to students by the author during the last 5–10 years in different universities. So we thank the many students of the Universities of Cornell, Moscow, Pushchino, Seoul, and Bangkok, who explored this subject initially without a textbook. The critical discussion and comments during these classes have provided us with the possibility of presenting this book.

Vladimir Bashkin,
Professor,
Moscow State University, Russia

CHAPTER 1

INTRODUCTION

In this chapter readers will find a description of modern biogeochemistry, its challenges and problems. The role of biogeochemistry in sustainable development of the Earth is also a subject of discussion. Readers will become familiar with the history of biogeochemistry, which is important in this science in spite of being relatively young. The researches of V. Vernadsky, A. Vinogradov, G. Hutchinson, A. Fersman, V. Goldschmidt, V. Kovalsky, E. Degans, V. Kovda, D. Atkinson, S. Miller, V. Dobrovolsky and many other scientists will be briefly reviewed.

1. THE BASIC CONCEPTS AND APPROACHES TO THE SUBJECT

Biogeochemistry is the interdisciplinary science of the 21st century that originated in the borderline between biology, geology and chemistry. Biogeochemistry is connected to the role of living organisms in the migration and distribution of chemical elements in the Earth's crust. The term "biogeochemical cycles" is used for qualitative and quantitative understanding of the transport and transformation of substances in the natural and human environment.

The theoretical basis of biogeochemistry is the concept of living matter and the biosphere that were suggested by the prominent Russian scientist Vladimir Vernadsky at the beginning of the 20th century.

The Concept of Living Matter. It is known that all living organisms regardless of their size, morphology, and physiology, have a common feature related to the metabolism of their habitat during the living activity. The mass of these organisms presents a negligibly small portion of the Earth's outer envelope; however, the overall effect of their biogeochemical metabolic activity extends to the global scale. Living organisms uptake chemical elements selectively in accordance with their physiological requirements. Carbon, hydrogen, nitrogen, oxygen, phosphorus, sulfur, silicon, calcium, and iron are the principal chemical elements that living organisms utilize in structural issues, for replication, and for energy-harvesting activity. The same elements are also important components in the ocean, atmosphere and crustal rocks. The uptake of living organisms leads to the biogenic transformation and differentiation of elements in the environment. Furthermore, the gaseous metabolites being released to the atmosphere make its composition gradually different. Liquid metabolites and decay products participate in the acid-base and reduction-oxidation processes in natural waters, which transform soil and the upper geological layers by leaching chemical elements from them. These leaches become involved with water transport and finally participate in transformation of the hydrochemical composition of the World's ocean and the formation of sedimentary rocks.

This influence has continued for about 3.5–3.8 billion years and the specific form of material entity, the living matter, has originated and accumulated during this period. For this reason, living matter plays the most important role in biogeochemical evolution of the outer layer of the Earth (see Chapter 2).

The Concept of Biosphere. The Austrian geologist E. Suess suggested the term "biosphere" in 1875. In accordance with his definition, the biosphere is the outer sphere of the Earth where the living organisms exist. At the beginning of the 20th century Vladimir Vernadsky advanced this definition of the biosphere as an outer shell of the Earth subjected to the biogeochemical activity of the living matter. An essential point to be emphasized is that the biosphere is currently conceived not only as a habitat of living organisms. It is rather as a global system, within which inalienably coexistent are, on the one side, the vast variety of diverse forms of life and their metabolites and, on other side, the inert matter in solid, liquid and gas phases. *V. Vernadsky has determined the biosphere as a unity of the living matter and the outer part of the Earth's globe* (Vernadsky, 1926). Living matter is inconceivable without the biosphere as well as the biosphere makes no sense without living matter.

The importance of the suggested interaction between living matter and inert material of the Earth is related to the exchange of chemical elements between living organisms and their environment. This exchange provides the biogeochemical organization of the biosphere. V. Vernadsky termed these processes, which are in essence geochemical (as natural migrations of chemical elements) and which are effected not by the action of geological factors, but through the community of living organisms, as *biogeochemical processes.* It stands to reason that *biogeochemical processes are the subject matter of biogeochemistry* as well as the application of the research activity of many scientists.

Biogeochemical Cycling as the Universal Feature of the Biosphere. The processes of biogenic mass exchange are cyclic in character, as scientists dealing with biogeochemistry have recognized very early. Life is cyclic by nature. The idea of the perpetuation of life through the inevitable process of death of individuals is inherent in religion and world outlook of all nations. The famous biblical dictum condemning all mortals arisen from the ashes to turn again to dust and ashes has become transformed, by dint of imaginative scientific intellect, into a theory of biogeochemical cycling of chemical elements. These elements are involved in mass exchange and served to maintain, in a compositionally dynamic state, both the living matter and the major components of the biosphere: atmosphere, soil, land, water and the World's ocean (Dobrovolsky, 1994).

Studies carried out during recent years have demonstrated the connection of life cycles of individual organisms and their communities and the universal cycling processes. The latter arises from the geophysical and cosmic events such as rotation of the Earth about its axis and about the Sun, the natural evolution of solar matter, the

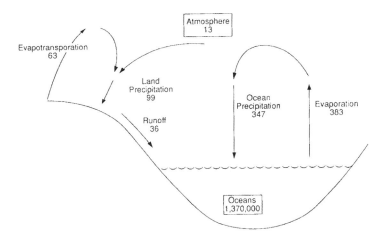

Figure 1. The hydrologic cycle as an example of global cycle. Units are 10^{18} g H$_2$O (burdens) and 10^{18} g H$_2$O/year (fluxes) (Butcher et al., 1992).

displacement of the solar system within the Galaxy, and others. Thus, the spatial and temporal cycles of mass exchange constitute also the dynamic system of the biosphere.

The migration of chemical elements in a thermodynamically open system does by no means lead to chaos. The direction of mass transport is kept under control by a large variety of diverse equilibrium processes. On the other hand, the main feature of the migration cycles of the biosphere, i.e. their openness, depends on this equilibrium.

V. Vernadsky's opinion that the history of most chemical elements that constitute 99.7% of the biosphere mass could be well explained by the concept of biogeochemical cycling. He emphasized that ". . . these cycles are reversible only for a larger part of elements involved, whereas a smaller part of elements is final, that is, the circulatory process is, strictly speaking, irreversible" (V.I. Vernadsky, 1934). The incomplete reversibility of migratory masses and imbalance of migration cycles admit definite limits for the concentration variation of a migratory element, to which the organisms are capable of adapting themselves; simultaneously, this provides for removal of the excess element from a given cycle.

An example of the global biogeochemical cycle is the hydrological cycle, schematically shown in Figure 1.

We can see the amount of water in various reservoirs, the fluxes of water exchange between these reservoirs and the rate of exchange. The World Ocean represents the largest reservoir of water. Water from the ocean is transported into an atmospheric reservoir by the heating of solar radiation. Water from the atmosphere may be returned to the ocean or it may be transported and deposited as precipitation on the terrestrial areas. Furthermore, the water is returned to the ocean both as run off and evapotranspiration.

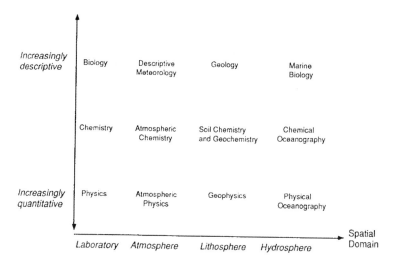

Figure 2. The schematic description of Earth's sciences related to biogeochemistry (Butcher et al., 1992).

   The water cycle is typical of the global biogeochemical cycles, which description is given in more detail in Chapters 3 & 4. The interaction of cycles and stoichiometric aspects of nutrient uptake (and nutrient limitation of production), stoichoimetric aspects of nutrient recycling, thermodynamics as applicable to bacterial energetics (and the use of different electron acceptors), etc. are elucidated in Chapter 5.

Interrelations of Biogeochemistry to Other Scientific Disciplines

Biogeochemistry originated at the beginning of the 20th century as a typical interdisciplinary merger. The appearance of such borderline, synthetic disciplines has provided the impetus for wider scientific research and finally enabled more detailed studies both of separate natural compounds and their interactions. The schematic description of various natural sciences closely related to biogeochemistry is shown in Figure 2.

   The basic natural sciences are chemistry, biology and geology. These disciplines are distinguished according to whether they study complex natural subjects more quantitatively or qualitatively. The border between the qualitative and quantitative area is not, of course, sharp and easily described by disciplinary labels since there are plenty of modern real interdisciplinary sciences such as biophysics, biochemistry, geochemistry and biogeochemistry itself. The spatial domain of various sciences is also very important as it relates to biogeochemistry. This divergence is seen in the development of fields given to the study of global hydrosphere, atmosphere, lithosphere, and pedosphere and is the smallest scale connected with laboratory experiments. Thus, there are the major fields of oceanography, hydrology, atmospheric sciences, geological, land and soil sciences. These fields can be integrated into mathematics, physics, biology, chemistry and geography for a complete understanding of the domain, and in many cases they have developed into the complex fields of atmospheric physics, marine biology, biogeography, geochemistry or soil chemistry.

The laboratory domain or an investigation of the behavior of the system in the test tube or the Petri dish (*in vitro*) may closely simulate the behavior of some natural ecosystem (*in vivo*). For many other ecosystems, however, extrapolation of laboratory results must be taken with care. The laboratory must be included in a description of how we go about studying natural ecosystems for the simplification of many driving forces. At present much experimental work of the natural sciences (biology, chemistry, soil science, biogeochemistry, etc.) occurs in the laboratory. However, the development of advanced techniques for field measurements gives the researcher the possibility to understand biogeochemical processes in natural conditions, like small catchment studies, experimental plot studies in agroecosystems, etc.

2. HISTORICAL DEVELOPMENT OF BIOGEOCHEMISTRY

As has been mentioned above, the basic concepts of biogeochemistry are focused on the assessment of life phenomena and the activity of living matter in the migration and transformation of chemical elements in various spheres of the Earth. Undoubtedly, many scientists have contributed through their selfless labor to the development and acceptance of new ideas in biogeochemistry. However, the truth is also that the most wonderful discoveries and achievements were linked to a few names of remarkable individuals who excelled not only as prominent scientists, but also as brilliant and exceptional persons.

The discovery of oxygen, nitrogen, and carbon dioxide enabled understanding of the chemical composition of the atmosphere in the second part of the 18th century. Scientists from London and Paris were engaged in discussion about the role of these gases in plant physiology. Here we ought to mention the name of the famous French chemist A.L. Lavoisier (1743–1974), one of the founders of modern chemistry. He was one of the first chemists who started the quantitative determination of chemical elements involved in reactions. He studied also the reaction of equivalent exchange of carbon dioxide and oxygen in plants and created the basic ideas of modern carbon geochemistry in the biosphere. Lavoisier provided the evidence that carbon, as the major chemical element of organic matter, is uptaken by plants from the air and consequently carbon is released into the atmosphere as a CO_2 from the decayed plant remains. These findings allowed him to make a conclusion on the universal turnover of elements involved in the interactions of living organisms with the medium. Lavoisier published a treatise "The Turnover of Elements on the Surface of the Terrestrial Globe" where he described the idea of cyclic exchange of chemical elements between three natural components: mineral, vegetable, and animal.

The scientific works of Lavoisier developed the idea that the relationship between organisms and the atmospheric gases is of exceptional importance for the chemistry of life. This problem remained a major concern for researchers through the beginning of the 19th century. In 1841, two prominent French chemists, Jean-Baptist Dumas (1800–1884), one of the founders of organic chemistry, and Jean-Baptist Boussingault (1802–1887), the founder of agrochemistry, suggested a definite formulation of the concept of cyclic turnover of gases in the living organisms-atmosphere system. However, the element exchange between the organisms and the environment is not connected only with gas turnover, in spite of its great importance. The eminent German chemist Justus Liebig (1803–1873) made the next step in the understanding of biogeochemical cycles. He discovered that the chemical elements are supplied to plants via two routes: some of them from air (carbon) and others from soil solution. Liebig carried out numerous determinations of chemical composition in soils, plants and animals as well as their metabolites. Based on vast analytical results, Justus Liebig showed the selectivity of chemical absorption in the soil-plant system. He developed the famous theory of *minimum content of nutrient* that restricted the growth and productivity of plants.

Liebig's works were very important for the development of biogeochemistry. He showed the application of experimental studies to the turnover of chemical elements and outlined the biological cycle of elements for various soil-plant-soil combinations. Liebig discussed the possibility of human control over this process using the application of mineral fertilizers. In his well-known book "Organic Chemistry in its Application to Agriculture" published in 1840 in Germany, he made the first attempts of predicting the effect arising from a disturbed natural mass exchange of certain elements on the human health.

At the turn of the 19th and 20th centuries, the visible frontiers between the traditional natural sciences like chemistry, geology, and biology showed a tendency to carry out new interdisciplinary research and to create new sciences. The genetic soil science that has arisen in the 1880s in Russia was in fact a precursor to the said tendency.

The founder of the given new branch of knowledge was the prominent Russian scientist Vasily V. Dokuchaev (1846–1903). He regarded soil genesis as a combination of many soils forming factors such as geological rocks, biological activity of plants and animals, climatic conditions, relief, ground water, chemical and physical weathering, and others. The soil is made up of diverse components, both biotic (living) and abiotic (nonliving). This differentiates soil from any other simpler natural bodies, like geological rocks. The living edaphic organisms make up the inalienable part of the soil, as well as do the mineral soil particles. Plants, animals, mesofauna and microorganisms form the soil during intricate interactions both among themselves and with mineral geological mass. Climatic, relief and hydrologic conditions influence this interaction. Accordingly, different geographic zones and different landscapes developed various types of soils. Dokuchaev's theory showed the important role of biological activity of living organisms in formation of soil cover.

The other scientific direction that was developed in the 19th century was geochemistry. Geochemistry is a branch of chemistry that focuses its attention not on the chemical species occurring in Nature, but rather on their constitutive elements and proportions thereof. This means that the proportion of its constituent chemical elements can characterize any material object. The geochemical approach allows researchers to compare and correlate diverse natural bodies and processes.

Geochemistry, whose emergence has been predetermined by the general development of both chemistry and geology, progressed in different routes in different countries. In the USA, the study of statistical distribution of chemical elements was the first priority in the second part of the 19th century. In the 1880s, F.W. Clark was concerned with determining the mean concentration for ten major chemical elements in main rock types, natural waters and other media on a basis of available analytical data. He published a number of reports containing replenished and updated results on the subject. His book "The Data of Geochemistry" (1924) was the first substantial characterization of major element distribution over the Earth's crust. At present we use term "clark" for definition of average content of various elements in the Earth's crust.

In Europe, geochemistry was developed as a branch of mineralogy, a science concerned with natural occurrence of chemical compounds and their formation during geological history of the Earth. For instance, at Oslo University the traditionally strong school of mineralogists and chemists whose main interests were related to the distribution and proportion of elements in ores and rocks. The eminent geochemist V.M. Goldschmidt (1888–1947), working with this school, advanced a theory of the global distribution of chemical elements in dependence on their atomic structure. The basic principles of this theory were published by the Chemical Society of London in 1937.

In Russia, V.I. Vernadsky (1863–1945) conducted mineralogical research at Moscow University. In studying minerals genesis, he eventually came to the idea that the natural processes of mineral formation should be evaluated at an atomic level. Subsequently, he advanced the basic principles of geochemistry concerned with the migration of chemical elements, the role of isomorphism in the distribution of elements over the Earth's crust, the forms of occurrence of chemical elements and their dispersal.

To a considerable extent, these principles have highlighted the directions of further development of biogeochemistry. Vladimir Vernadsky belonged to the eminent scientists and thinkers of the 20th century. He was the founder of a number of new scientific disciplines and biogeochemistry is the brightest star in this circle of sciences. The culmination of his scientific creativity was a theory of living matter, which was concentrated on the role of biota in global and planetary geochemistry. Great attention was given also on the biosphere, a unique envelope generated by living matter, within which the Earth is enclosed.

The basic, starting scientific point of Vernadsky's concept of genetic mineralogy, geochemistry and, furthermore, biogeochemistry and biosphere science were Dokuchaev's ideas on close interrelation of all natural components and on the role of living organisms in soil formation. In 1916, V. Vernadsky began the creation of "science of life". The main conceptual principles underlining this theory were the methodology that would enable an objective scientific evaluation of both the living matter and effects arising from its activity. Vernadsky outlined the main routes to a solution of this entirely new problem in 1918–1919 when he organized the first biogeochemical studies in Crimea. In the early 1920s he expounded his ideas in the scientific communities of St-Petersburg and Prague and later in his classes delivered at the Sorbonne University, Paris.

Vladimir Vernadsky believed that "… in order to be able to correctly assess the geochemical role of living matter and to express it in quantitative terms, we must know, firstly, the mean elemental chemical composition of all the organisms of living matter and, secondly, we must know the weight of living matter. The composition and this weight must be correlated to the weight and composition of the medium … within which the terrestrial matter is found" (Vernadsky, 1926).

The Biogeochemical laboratory of the USSR Academy of Sciences was organized in 1928 and V. Vernadsky was the head of this laboratory until his death in 1945. The scientific program of the laboratory on the quantitative determination of living organisms' chemical composition was started in the 1930s. This program has received a new impetus in the second part of the 20th century in many countries of the World.

Methodologically, biogeochemical approaches are closely related to geochemistry, especially to organic geochemistry. These two disciplines are concerned with the distribution of chemical elements in space and time, with the origin and transformation of elements of different occurrence, their migration, accumulation and dissipation. The ideas of V.I. Vernadsky as to the planetary role of living matter have largely contributed to the theory of geochemistry of fossil fuel (organic) deposits or sedimentary ore formation. In this connection the monograph "Organic Geochemistry", edited by M.H. Engel and S.A. Macko (1993) should be noted as a characteristic application of biogeochemical approaches in geochemistry.

Biogeochemistry is also interrelated to other earth sciences, notably those concerned the composition of rocks, minerals, natural waters and gases. The same is true for the biological sciences whose primary concern is the study of the links between organisms and their habitat, such as ecology. In studying ecosystems, a great emphasis is worked out on the mass exchange and element migration in biogeochemical food chains.

The principles of biogeochemistry are of great application in microbiology. The composition of microorganisms and their environment are intimately interdependent. For this reason, the biogeochemical approaches have gained a welcome acceptance by many microbiologists. "Bacterial Biogeochemistry", by T. Fenchel, G.M. King & T.H. Blackburn (1998), is a very good example of applying biogeochemical ideas for estimating bacterial influence on the global biogeochemical cycles and composition of natural waters and atmosphere.

In Russia the principles of biogeochemistry have advantageously been applied to soil science and landscape science. The well-known soil scientist, geochemist and geographer B.B. Polynov (1877–1952) developed a theory of landscape geochemistry. Polynov's followers such as V. Kovda (1904–1988) and V. Dobrovolsky accomplished further progress in biogeochemistry. Victor Kovda's book "Biogeochemistry of Soil Cover", published in1985, described many aspects of soil elements with special attention given to silicon biogeochemistry. Vsevolod Dobrovolsky is the author of "Biogeochemistry of the World's Land", published in 1994. He showed that the bulk of living matter is confined to the World's land. The terrestrial biogeochemical processes are very extensive and play an important role in maintaining the entire biosphere in dynamic equilibrium. Alongside with this, the land is a habitat for human species and the main area of human productive activity. Exactly for this reason, the most profound deformations of natural biogeochemical cycles and mass exchange processes take place on the land. The specific problems of World's land biogeochemistry come to the foreground of major concerns confronting sustainable development of mankind.

By virtue of the fact that the major spheres of productive activity of man (agriculture and industry) are confined to the land, the routes to practical use of biogeochemistry also bear relevance to the processes occurring on the World's land. Until recently, the most effective practical application of biogeochemical approaches has been associated with the so-called biogeochemical method of prospecting for minerals, with effects of environmental levels of chemical elements in ecosystems on human and animal health, and with developing and setting of ecological standards.

Biogeochemical methods of prospecting for minerals include the identification of floristic areas where plants and their vegetable remains and metabolites reveal elevated concentrations of ore-forming materials. The areas of elevated metal concentration in plants and in the upper soil layers present the so-called biogeochemical anomalies and thus are suggestive for the possible occurrence of deep ore deposits. This method started in a geological survey in the former USSR in the 1930s. A.P. Vinogradov, Vernadsky's follower, was head of these researches. N.J. Brundin initiated his experimental studies in Sweden and England. From the 1950s, biogeochemical mineral prospecting has gained a wide acceptance in many other countries such as Canada, Australia, New Zealand, Indonesia, Sri Lanka, Guinea, etc.

A second practically important trend in biogeochemistry is the investigation of the biogeochemical anomalies, i.e. excessive or deficit content of different chemical elements in biogeochemical food chains, and their influence on human and animal health. Russian scientist V. V. Kovalsky and his co-workers (1974) have shown that the productivity of cattle can be correlated to the excess or deficit of boron, cobalt, copper, molybdenum, and selenium in animal feed. J. Webb (1964, 1966) conducted similar studies in the USA, R. Ebens (1973), in the UK and Ireland. The level of trace elements in drinking water and foodstuffs of local production can affect human health. These problems have been studied and discussed in Canada (Warren, 1961), USA (Shakklette, 1970), and Finland (Salmi, 1963). In Russia (former USSR) these biogeochemical approaches to the setting of ecological standards were carried out from the 1950s under the guidance of V. Kovda and N. Zyrin. V. Bashkin et al. (1993) published "Biogeochemical Fundamentals of Ecological Standardization". In this book the problems of ecological standardization were discussed using biogeochemical and physiological approaches.

The famous German biogeochemist Egon T. Degens in his well-known book "Perspectives on Biogeochemistry" (1989) presented the original historical description of various scientific schools in biogeochemistry, especially in studying biogeochemical cycles. There was the first Russian school, and the names of V.I. Vernadsky, A. Fersman, M.M. Kononova, A.I. Oparin, and A.V. Ronov must be mentioned. They first recognized more than 70 years ago the significance of biogeochemical processes in the exogenic cycling of matter. Next in line is the German school and eminent scholars V.M. Goldschmidt, R. Brinkmann, A. Treibs come to mind. Goldschmidt was the first to show that chemical elements have a logic and an affinity for or a version to other elements and feel more at home either in the dead or in the living world. Then the British and U.S. schools followed with famous names of G. Hutchinson, L. Pauling, G.W. Brindley, and J.D. Bernal.

Following the "Big Bang" of organic geochemistry and biogeochemistry in the "roaring" twenties, thirties and forties, the field expanded during the second part of the 20th century at an unbelievable rate. Scholars like Philip H. Abelson, Irving A. Breger, John M. Hart, Stanley L. Miller, Rudolph von Gaerner, Marleis Teichmuller, Alexander Lisitsin, Evgeny Romankevich, Boris Rozanov, Tony Hallam, David R. Atkinson, Geoffrey Eglington, Pierre Albrecht, Bernard Tissotand, Max Blumer, and (we have to add) E.T. Degens stand for the generation of the scientists who advanced the field of biogeochemistry.

The research of these and other biogeochemists will be discussed in more detail in relevant chapters of this textbook.

At present, biogeochemistry is known as a very productive and high-priority academic and scientific discipline, joining many natural sciences like biology, earth sciences and chemistry by definition, mathematics and physics by methods used, and medicine and social sciences as a sphere of application of scientific results and practical conclusions.

With no doubt, biogeochemistry is one of the fastest developing discipline and it attracts great students attention in those Universities where it exists as an academic Program. At present, there are only a few Programs in Biogeochemistry in the USA and that at Cornell University is the best example (Director Prof. R.W. Howarth). However, the annual number of student applications exceeds the possibilities of an enrollment as much as 10:1. Development of new educational opportunities in biogeochemistry programs at the undergraduate, master's, and doctoral levels will gather many high-rank students to select the best of them for education in various universities of the World.

We are sure that the application of biogeochemical fundamentals will present a unique opportunity to create academic and research programs that reflect developing areas as well as traditional disciplines in various natural and social sciences. In keeping with the mission of the universities to provide teaching, research, and public service of the highest quality, we hope this textbook will provide an opportunity to develop biogeochemical research and education activity as one of the vast growing disciplines in the 21st century.

FURTHER READING

Egon T. Degens (1989). Perspectives on Biogeochemistry. Springer-Verlag, 423pp.

Samuel S. Butcher, Robert J. Charlson, Gordon H. Orians and Gordon V. Wolfe, Eds. (1992). Global Biogeochemical Cycles. Academic Press, London et al., 378pp.

Michael H. Engel and Stephen A. Macko, Eds. (1993). Organic Geochemistry. Principles and Applications. Plenum Press, New York and London, 861pp.

Vsevolod V. Dobrovolsky (1994). Biogeochemistry of the World's Land. Mir publishers, Moscow, 362pp.

Tom Fenchel, Gary M. King and T. Henry Blackburn (1998). Bacterial Biogeochemistry. Academic Press, London et al., 307pp.

QUESTIONS AND PROBLEMS

1. What is the subject of biogeochemistry? Give definitions of biogeochemical cycles and biogeochemical structure of the biosphere.

2. Discuss the concept of living matter. Highlight the relationships between this concept and the development of biogeochemistry.

3. Describe the biosphere and its role in sustainable development of the Earth using the modern understanding of biogeochemical cycling parameters. Discuss the contribution of Vladimir Vernadsky to the development of biosphere science.

4. Describe the global biogeochemical cycle. Give examples.

5. Present a review of the historical development of biogeochemistry. Highlight the introduction of chemical and geological researches into understanding of living matter. Discuss the scientific discoveries of such scientists as A.L. Lavoisier, J-B. Dumas, J-B. Boussingault and J. Liebig in the understanding of biogeochemical cycles.

6. Discuss the role of interdisciplinary research in the development of biogeochemistry. Refer to the origin of soil science as a precursor of scientific biogeochemistry.

7. Highlight the role of geochemistry and famous geochemists in the development of biogeochemistry.

8. Highlight the contribution of Vladimir I. Vernadsky as the creator of modern biogeochemistry.

9. Discuss the theoretical and practical application of biogeochemical research. Give examples.

CHAPTER 2

EVOLUTIONARY BIOGEOCHEMISTRY

In this chapter we consider the development and evolution of modern biosphere during its geological history. The reader will find a brief description of the formation of the lithosphere, atmosphere, and hydrosphere and their chemical composition. Creation and evolution of the biogeochemical structure of the biosphere and hydrosphere is also a subject of discussion. Finally, we show the role of biogeochemical cycles in formation of biogenic depositions (oil and gas).

1. INTRODUCTION

Evolutionary biogeochemistry deals with analysis of the origin and evolution of biogeochemical cycles during the geological history of the Earth. Any analysis of the origin and evolution of biogeochemical cycles must address the origin of life itself, which in its many manifestations dominates the cycles of most elements. Addressing this subject, we pose the following questions:

1. How has element synthesis occurred in the Universal?

2. What modes of geochemical cycling existed prior to the origin of life?

3. How did life originate on Earth?

4. How have living organisms qualitatively and quantitatively changed the cycles of individual elements?

5. What geochemical principles have been adapted by or incorporated into metabolism of organisms?

6. How did the biogenic deposition of oil and gas occur?

It is now accepted that physical and chemical factors were exclusively responsible for the distribution of elements 4.5 billion years ago when the Earth was formed. Recent evidence suggests that life originated on Earth over 3.5–3.8 billion years ago. Since that time, biological processes have become increasingly important in redistribution of chemical elements and their compounds.

In subsequent sections of this book, the evolution of biogeochemical cycles is discussed against the background of the questions posed above. The Earth's chemical composition formation presented to the reader mainly through analysis of the origin of elements. The thermodynamics of energy flow systems establish a framework within which solar radiation could have provided energy for the chemical reactions that generated geochemical and biogeochemical cycles in different spheres during evolution of the Earth's biosphere. Within these constraints, life could have originated from a sequence of highly probable geochemical and photochemical reactions. The living organisms thus created then began themselves to change the chemical composition of prebiotic Earth. These changes then would have interrelated with the alteration of the metabolism of living organisms and their adaptation to the different media.

2. ORIGIN OF ELEMENTS

The nucleus of an element consists of two main building blocks: the proton and the neutron. It is the proportions, in which the two nucleons are joined in nucleonic packs termed nuclei which give rise to the various chemical elements and their isotopes. We believe that the essential protons and neutrons needed for the synthesis of chemical elements were generated when the Universe was just a few seconds old. The existing temperature at that time, about 10^{10} K, enabled neutrons to decouple. Further, neutrons and protons react with each other. Calculations showed that among the stable nuclei generated in the first 3 min were deuterium, helium-3, helium-4 and lithium-7. However, due to lack of stable nuclei at mass numbers 5 and 8, the heavier elements could not be created during this enter period.

So, hydrogen and helium were synthesized in fiercely hot plasma during the first few seconds in a mass ratio of about 3 to 1. The remainder of the elements, 90 of them, had their seats of origin in certain stars and their energy levels determined by these stars during their birth, life and death. It seems probable that the elements all evolved from hydrogen, since the proton is stable, while the neutron is not, and hydrogen is the most abundant element, thus assuming a preponoenance of protons. The source of energy in stars has been debated for many years and now the hypothesis of the existence of a kind of "cosmic reactor" in stars is thought to be realistic. Such reactors generate not only energy, but at the same time the full spectrum of chemical elements.

The succession of element formation can best be followed by moving up the temperature scale from a "modest" few million degrees to several billion degrees (Degens, 1989). It is shown schematically in Figure 1.

Examples of reactions proceeding during stellar nucleosynthesis are shown in Table 1. To illustrate the sequence of events, the decay series of uranium-238 (^{238}U) is depicted in this table. Radiogenic nuclides decay by the emission of alpha, beta and gamma radiation or by electron capture into so called daughter nuclides at their half-lives. This half-life ranges from parts of seconds to billions of years.

The time required for the synthesis of all elements in cosmic abundance also ranges from seconds to billions of years. The temperature scales fluctuate over several orders of magnitude (see Figure 1). The chemical composition of the Universe, excepting the matter of "neutron stars" and "black holes", can be presented as follows: of 1,000,000 atoms there are 924,400 hydrogen, 74,000 helium, 830 oxygen,

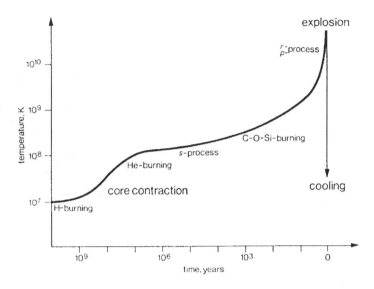

Figure 1. Time scale for the various element syntheses in stars. The highly schematic curve gives the central temperature as a function of time for a star of about one solar mass (Degens, 1989).

470 carbon, 84 nitrogen, 82 neon, 35 magnesium, 33 silicon, 32 iron, 18 sulfur, 8 argon, 3 aluminum, 3 calcium, and the remaining 2 atoms for the rest of the elements (Degens, 1989).

The relative abundance of the more common elements in the solar atmosphere is shown in Figure 2 and Table 2. The remainder of the chemical elements occurs only in trace quantities. By any chemist's measure, the universe is composed mainly of hydrogen and helium, with some impurities of practically all elements.

Thus, the formation of elementary particles and elements follows a logical order, which in turn can be used by scientists to delineate cosmic events and time tags. It is easier to form light elements by fusion of hydrogen, deuterium and helium, as oxygen and carbon are the two most abundant elements after hydrogen and helium.

3. EARTH EVOLUTION

The evolution of the Earth is a long and interesting story — one that is much too complex to be discussed in one chapter. The discussion will follow the essential points in evolution of the main components of prebiotic Earth: lithosphere, atmosphere and hydrosphere. Much more detail can be found in E. Degens (1989), S. Butcher et al. (1992), M.H. Engel and S.A. Macko (1993) and T. Fenchel et al. (1998).

3.1. *Evolution of the Lithosphere*

It is known at present that the solar system has developed from a rotating discoid nebula (stellisk). Temperature and pressure in the inner region were about 1,400 K

Table 1. Decay series of uranium-238 (after Degens, 1989).

Element	Symbol	Half-life	Emission	Original name
Uranium	$_{92}U^{238}$	4.5×10^9 yr.	Alpha	Uranium I
Thorium	$_{90}Th^{234}$	24 days	Beta	Uranium X1
Protactinium	$_{91}Pa^{234}$	1.2 min	Beta	Uranium X2
Uranium	$_{92}U^{234}$	248×10^3 yr	Alpha	Uranium II
Thorium	$_{90}Th^{230}$	80×10^3 yr	Alpha	Thorium
Radium	$_{88}Ra^{226}$	1,622 yr.	Alpha	Radium
Radon	$_{86}Rn^{222}$	3.8 days	Alpha	Radon
Polonium	$_{84}Po^{218}$	3.0 min	Alpha	Radium A
Lead	$_{82}Pb^{214}$	26.8 min	Beta	Radium B
Bismuth	$_{83}Bi^{214}$	19.7 min	Beta	Radium C
Polonium	$_{84}Po^{214}$	1.6×10^{-4} sec	Alpha	Radium C'
Lead	$_{82}Pb^{210}$	21 yr.	Beta	Radium D
Bismuth	$_{83}Bi^{210}$	5.0 days	Beta	Radium E
Polonium	$_{84}Po^{210}$	138.4 days	Alpha	Polonium
Lead	$_{82}Pb^{206}$	Stable	–	Lead

and 10 Pa (10^{-4} atm). In accordance with existing theories, condensation started with calcium, aluminum, titanium oxides and silicates, platinum metals, followed by Fe-Ni metals. This condensation was considered started at 1,360 K. At the temperature 1,200 K the remaining silicates had been condensed in the form of $MgSiO_3$. The bulk of proto-Earth accreted rapidly during the early condensation phase of the solar nebula. This resulted in gross stratification of the proto-Earth with an iron core, surrounded by a mantle of magnesium silicates. The later stages of accretion occurred after the nebula had cooled to a relatively low temperature about 400 K. Resulting equilibrium condensate contained iron, iron sulfide, hydrated magnesium silicates and some volatiles. The mixture of these volatile rich materials with earlier high temperature condensates that had failed to accrete previously, are believed to have accreted upon the Earth over a longer time scale (10^5 to 10^7 years) to produce the upper mantle-crust system (Degens, 1989). This is shown schematically in Figure 3.

It has been estimated that 4 billion years ago at least 50% of the ephemeral global crust had been converted to basin topography. At that time the Earth experienced massive degassing, leading to a proto-atmosphere and proto-hydrosphere. The following steps in the Earth's lithosphere evolution were related to thermal changes. These resulted in convective flow regimes and mantle. Brecciated crustal material and hydrothermally produced iron-rich clays accumulating on the ephemeral oceanic

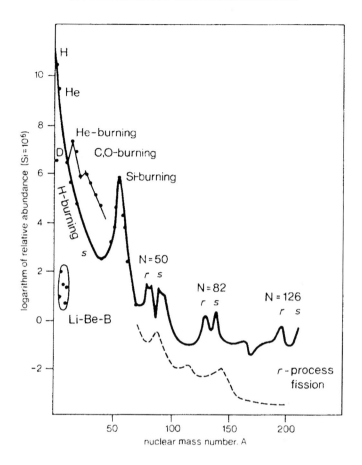

Figure 2. Schematic curve of atomic abundance (relative to Si $= 10^6$ *) as a function of atomic weight (Degens, 1989).*

crust. This crust consisted of alkali-poor basalts and gabbros. It was generated from magma chambers fed by melting of rising fertile peridotite. The schematic sequence of magmatism and cooling is graphed in Figure 4.

The suite of basic to ultrabasic rocks at the Earth's surface can be generated from the magma of different composition. Intense volcanism, vigorous "stirring" by means of polygonal convection cells (Figure 5), and huge impact events inhibited the generation of a substantial continental-like crust prior to 4.0 billion years ago. Small patches, a few hundred kilometers across, might have been presented at that time as the proto-continents.

The way to study the crustal evolution is related to the appearance of the oldest terrestrial rocks of about 3.8 billion years old (see Box 1 and Degens, 1989, for further details).

Table 2. Relative abundance of the more common elements in the solar atmosphere (after Cameron, 1982).

Element	Atomic number	Relative abundance(atoms/10^6 atoms of Si)
H	1	3.18×10^{10}
He	2	2.21×10^9
C	6	1.18×10^7
N	7	3.74×10^6
O	8	2.15×10^7
Ne	10	3.44×10^6
Mg	12	1.06×10^6
Si	14	1.00×10^6
S	16	5.00×10^5
Ar	18	1.17×10^5
Fe	26	8.30×10^5

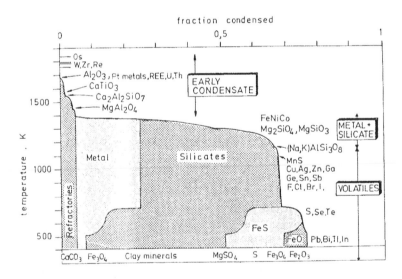

Figure 3. Condensation of solar gas at 10^{-4} atm (10 Pa). Three types of dust condense from a solar gas: refractories, metallic nickel-iron, and magnesium silicates. On cooling, iron reacts with H_2S and H_2O to give FeS and FeO. Further major changes take place below 400 K (Degens, 1989).

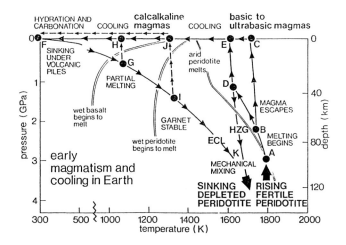

Figure 4. Magnetism and cooling in the early Earth. Depth scale would be expanded for a partly grown Earth (Degens, 1989).

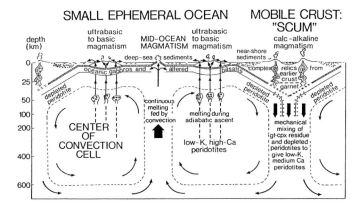

Figure 5. Schematic presentation of polygonal tectonics in the early Earth (Degens, 1989).

Box 1. Earth's Layered Structure (Degens, 1989)

Geologists subdivide our globe into crust, mantle and core. Based on refined geochemical data, the following model of the spherical stratification of the Earth's interior were established:

A — crust, base varying between 10 to 70 km depth;

B — upper mantle, from base of crust to 400 km depth;

C — transition zone, from 400 to 1,000 km depth;

D — lower mantle, from 1,000 to 2,900 km depth;

E — liquid outer core, from base to lower mantle, i.e., to 4,600 km depth;

F — transition region, from 4,600 to 5,150 km depth;

G — solid inner core, from 5,150 to 6,371 km depth.

The thinnest layer is the crust, having an average thickness of only 17 km. It consists of two silicate layers. These layers are *sial*, aluminum-rich material, and *sima*, enriched in mafic elements such as magnesium and iron materials. The continents are mainly composed of *sial*, which lies on a *sima* substratum (Figure 6a). Taking into account that these granites are as old as 3.8 billion years, the *sial* layer was formed quite early.

Those parts of the crust and upper mantle that are rigid are commonly called the *lithosphere*. The layer below the lithosphere (80–100 km) is the *asthenosphere*, which in geophysical terminology is the low velocity layer. It is considered as partially molten and representing the lubricated phase boundary upon which plates move. The lower mantle consists predominantly of iron-magnesium silicates whereas the core is believed to be composed of iron and nickel. Some lighter elements such as sulfur and silicon are also proposed for the last layer. The main structural features of the Earth's layers are shown in Figure 6b schematically, including tectonic structures revealed at the surface of the crust.

It was stated earlier that the primitive crust of the Earth was most likely of basic composition. Massive meteorite bombardments and associated crustal resorption prior to 3.9 billion years ago must have erased the former records. In further crustal evolution at least three cycles can be distinguished. They were laid down between 4.0 and 2.5 billion years ago. The oldest terrestrial rocks from 3.8 billion years ago, gneisses of west Greenland, contain metabasaltic and metasedimentary enclaves having the age of about 4.0 billion years. They can be interpreted as relics of the early crust. So, the first 500 to 700 million years of Earth's geological history have apparently left no stratigraphic records.

Fractionation of chemical elements between a primitive mantle and one that generated the crust can enable an estimate of how much of the primordial matter had to become "digested" and depleted in certain elements to account for all the crustal material. It allows calculation of geochemical trends in three major cycles spaced during 1.5 billion years periods at about 0.5 intervals. The most information has been extracted from geochemical composition of komatiites. Komatiites are chiefly volcanites of effusive or explosive character that formed subaquatically at water depth not exceeding 2 kilometers. They piled up a sialic basement in flow after flow, individually measuring a few decimeters to several tens of meters to yield massive sections 10 to 15 km thick.

So, the global stabilization of continental crust, called *cratonization*, is considered to be at approximately 3.0 to 2.5 billion years ago.

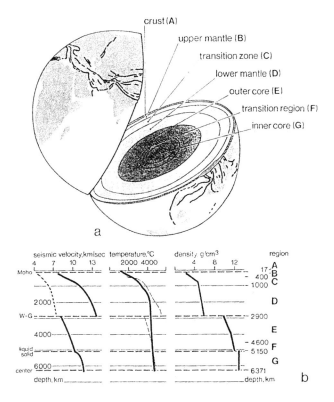

Figure 6. a, b. Structural features of the Earth and related physical properties (Degens, 1989).

3.2. Evolution of the Atmosphere

Subsequent to the differentiation of the lithosphere, i.e., crust and mantle formation, and the cooling that promote formation of the hydrosphere, tectonic cycles and
abiotic weathering are considered to have contributed to temperature regulation of
prebiotic Earth.

Once the main part of accretion had ended, most of the water vapor that was in the
atmosphere would have condensed to form the prebiotic primitive ocean. This led to
an atmosphere dominated by molecular nitrogen, N_2, and carbon compounds, mostly
as carbon dioxide, CO_2, and monoxide, CO (see Box 2). Molecular hydrogen, H_2,
would have been an important trace constituent. Seemingly, an H_2 mixing ratio, i.e.,
mole fraction of $\sim 10^{-3}$, could have been maintained by the balance between the
volcanic outgassing and escape to space. Free molecular oxygen, O_2, would have been
essentially absent. The exception could have related to the high atmosphere where
a thin layer of oxygen formed from photodissociation reactions of carbon dioxide
and water.

Box 2. The Content of CO_2 in the Prebiotic Atmosphere, after Fenchel et al., 1998

Perhaps the most interesting question on the early atmosphere is related to the abundance and oxidation states of carbon. Current models predict a CO_2-rich atmosphere, which would contain a trace amount of methane and a slightly greater amount of carbon monoxide. We can try to explain it on the basis of the following speculations. It is known that even today volcanic activity is a dominant process in releasing CO_2 and it should have been so in the past. It is related to the reasonable suggestion that the oxidation state of the upper mantle was about the same as today. Furthermore, the by-products of water vapor photolysis were enabled to oxidize both CH_4 and CO. The possible reaction pathway could be the following

$$H_2O + h\nu \longrightarrow H + OH$$
$$CH_4 + OH \longrightarrow CH_3 + H_2O$$
$$CH_3 + OH \longrightarrow H_2CO + H_2 \tag{1}$$
$$H_2CO + h\nu \longrightarrow H_2 + CO$$
$$CO + OH \longrightarrow CO_2 + H$$

The hydrogen produced in this reaction chain can escape to space, leaving carbon in its fully oxidized form.

The real values of CO_2 content in the early atmosphere are very uncertain. One way to calculate this value is to estimate the crustal abundance of carbon. This element is mainly stored in carbonate rocks and modern assessments give the value of 10^{11} Tg. This is enough to produce an atmospheric pressure of 60 bars if it had been presented as gaseous CO_2. Even if only one-third of this amount was present in the atmosphere at the moment of accretion, the pressure would be about 20 bars. How long this dense CO_2 atmosphere would have lasted depends on the rate of rock silicate transformation to the carbonates. This rate, in turn, would have depended on the surface temperature on the early Earth and on the amount of continental area exposed to weathering.

Physical models of solar evolution suggest that illumination has increased through time, requiring proportional changes in the Earth's thermal radiation budget for a stable, biologically hospitable climate maintenance. Such changes could result to a large degree from depletion of atmospheric carbon dioxide by weathering of lithospheric calcium silicates according to the following simplified reaction:

$$CaSiO_3 + CO_2 \longrightarrow CaCO_3 + SiO_2 \tag{2}$$

This process might have reduced the content of carbon dioxide in the ancient atmosphere and consequently the degree of thermal trapping (or, in modern words, "greenhouse effect"). It has been postulated that in the early atmosphere this greenhouse effect was related to high content of CO_2. Calcium silicate weathering can be considered as a negative feedback mechanism. It may be related to the fact that the increased greenhouse warming results in higher rates of CO_2 removal and a

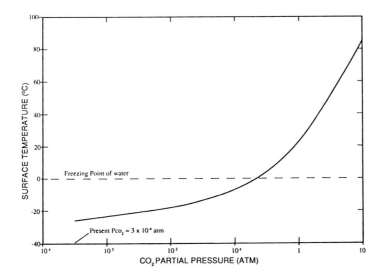

Figure 7. Mean surface temperature of the early Earth as a function of atmospheric CO_2 partial pressure. A 30% reduction in solar luminosity relative to the present-day level is assumed (Engel and Macko, 1993).

compensatory cooling as the greenhouse effect is diminished. It has been argued that a 10- bar CO_2 atmosphere could have persisted for several hundred million years if the early Earth was entirely covered by oceans. Taking into account an estimated 30% decrease in solar luminosity at that time in comparison with today, this 10-bar atmosphere would have produced a mean global temperature about 85 °C (Figure 7).

The existence of such a warm, dense CO_2 atmosphere can be neither demonstrated nor proven by geological evidence, since, as it was mentioned above, the earliest sedimentary rocks are only less than 3.8 billion years old. However, this process can provide a major control for atmospheric CO_2 on a time scale $> 10^7$ years (Figure 8).

Another important environmental parameter of the early Earth was the flux of solar ultraviolet (UV) radiation. At present UV radiation with 200–300 nm length is adsorbed by the stratospheric ozone layer. This layer protects the biota from the harmful influence of dangerous UV radiation. In the early atmosphere the ozone was absent and harmful solar radiation may have reached the Earth's surface without any adsorption and prevented the formation of large, complex molecules for life origin.

A possible explanation of how the UV radiation had been decreased could be presented on a basis of sulfur photochemistry. Sulfur as hydrogen sulfide, H_2S, and sulfur dioxide, SO_2, would have been emitted to the atmosphere from volcanoes. Under intensive solar radiation these gases can be photochemically transformed to different chemical species (Figure 9).

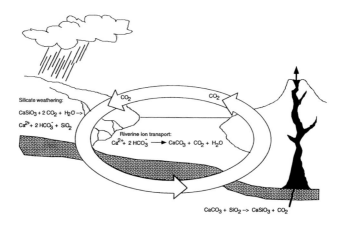

Figure 8. Diagramed representation of interactions among carbon dioxide, continental weathering, carbonate deposition and tectonic activity as controls of atmospheric CO_2 (Fenchel et al., 1998).

Most of the sulfur emitted to the atmosphere should be oxidized to sulfate in reactions with atomic oxygen, O, and hydroxyl radical, OH or dissolved in ocean water as SO_2. However, a substantial fraction of the outgassed sulfur compounds could have been converted into elemental sulfur vapor, consisting of S_8 along with other gaseous sulfur molecules S_n, $2 \leq n \geq 12$. S_8 is a ring molecule, which absorbs well into the near UV. Moreover, these molecules are quite resistant to photolysis. Its UV absorption coefficient is comparable or even higher than ozone (Figure 10).

The concentration of sulfur vapor in the primitive atmosphere would have been limited only by its saturation vapor pressure, which depends strongly on the temperature. As it has been shown above that the temperature was about 85 °C, sulfur vapor could have been presented in sufficient concentrations to shield the Earth's surface from harmful solar UV radiation. Thus, a warm, CO_2-rich ancient atmosphere may have been favorable to early life.

The presence of sulfur gases could have inhibited the photolysis of ammonia, NH_3. Ammonia is considered as a very important compound in any models of prebiotic synthesis of organic molecules. The absorption length of ammonia is below 225 nm. Hence, in the absence of shielding, it would have been rapidly converted to N_2 and H_2. However, both ammonia and its photodissociation reaction product hydrazine, N_2H_2, could have been protected from photolytic reaction by sulfur compounds such as SO_2, H_2S, and especially S_8. The relatively high ammonia concentration in prebiotic atmosphere could have facilitated the origin of life and biogeochemical cycles (Kasting cited by Engel and Macko, 1993).

The evolution and proliferation of life have resulted in an additional set of controls for atmospheric composition on a time scale from $10^0 - > 10^7$ years (Figure 11).

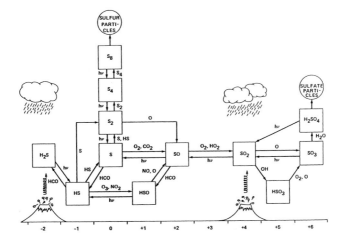

Figure 9. Schematic diagram showing sulfur photochemistry in an anoxic, primitive atmosphere. The numbers at the base of the diagram indicate the oxidation state of the sulfur. Sulfur is emitted to the atmosphere from volcanoes as either SO_2 or H_2S. It is removed by rainout of soluble sulfur gases (SO_2, H_2S, HS, HSO, and H_2SO_4) and by formation of sulfate and elemental sulfur particles (Engel and Macko, 1993).

These parameters are temperature dependent and provide both positive and negative feedbacks on the composition and evolution of the atmosphere. Biological impacts to climate homeostasis involve a number of mechanisms. The reflectance of the Earth's land (albedo) affects the radiation budget. The rate of vegetation colonization on the terrestrial surface depends strongly on the land albedo. The decreasing CO_2 content in the atmosphere is due to biologically enhanced weathering in oxygenated conditions. Moreover, the depletion of CO_2 was followed by a temperature decrease until the average 20 °C that was beneficial for the development of further living organisms. This, in turn, increased the solar illumination over the last 1 billion years.

The Earth's biota has strongly affected all other atmospheric gases such as nitrogen oxides, ozone, and especially oxygen. The only exceptions are the noble gases, i.e., Ar, He, Cr. During the geological history of the Earth, microorganisms have affected atmospheric composition more dramatically than other biological groups, especially on the earlier stages of Biosphere development.

During the early Archaean (> 3.0 billion years ago), the composition of the atmosphere was mostly likely neutral or slightly reducing. The compositional dynamics of air gases were related to the activity of volcanoes and photochemical oxidation. With the advent of methanogenic archea, the atmosphere may have accumulated a significant amount of CH_4. Undoubtedly this increased the greenhouse warming. This climate impact made the global interactions between biosphere and atmosphere very important.

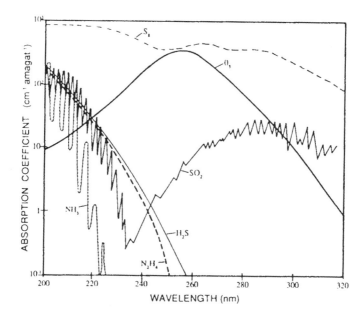

Figure 10. Absorption coefficients of various gases in the near ultraviolet (from Engel and Macko, 1993).

About 3.5 billion years ago the anaerobic photosynthesis started to develop. This and further development of oxygenic photosynthesis began profound changes in the Earth's biosphere. The chemistry and composition of prebiotic atmosphere were altered dramatically due to oxygen production by microorganisms. The main results were related to the formation of hydroxyl radicals, which are responsible for the development of oxidative reactions in atmosphere. This, in turn, has conducted the mechanisms of ozone formation. The ozone layer protected the biota at the Earth, especially in the terrestrial systems and promoted further proliferation of the bio-sphere. Oxygenation of the atmosphere as well as preceding oxygenation of the oceans resulted in the creation of new biogeochemical cycles of elements and the previous primitive anoxic cycles were reorganized significantly by establishing biogeochemical chains and links for complete oxidation of various reduced species, such as ferrous iron and sulfides. The so-called "red beds", geological deposition of iron oxides during the middle Proterozoic, would have been considered as the evidence of these alterations. The evolution of other biogeochemical processes was also the conse-quence of atmosphere oxygenation. We have to mention aerobic methane oxidation, ammonium oxidation to nitrate, and subsequent nitrate reduction to nitrous oxide and molecular nitrogen. Each of the latter processes was related to the bacteria activity. These biogeochemical processes influenced both the composition of the atmosphere and the extent of thermal trapping by greenhouse gases.

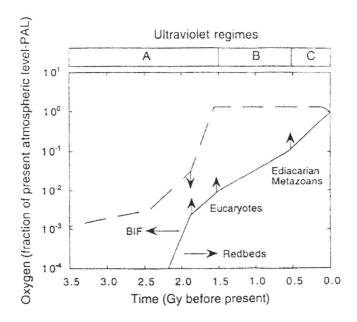

Figure 11. Historical reconstruction of trends in atmospheric oxygen and regimes of UV *at the Earth's surface. Dashed and solid lines represent upper and lower limits for oxygen based on geological and biological constraints; arrows indicate probable directions from oxygen. UV regimes A indicate deep penetration of 210–285 nm wavelengths; B, surface penetration only for 210–285 nm wavelengths; C, surface insulated from 230–290 nm wavelengths. BIF, Banded Iron Formation; PAL, Present Atmospheric Level (Fenchel et al., 1998).*

Microbial activity has also affected the atmospheric concentrations of some other gases, like CO, NO and H_2. The possible changes in atmospheric composition with time are shown in Figure 12.

So, regardless of the specific sequence of cycles in microbial evolution itself, it is clear that microbial activity have changed the chemical composition of atmosphere. Though changes in atmospheric composition have been tremendous, especially to the extent of oxygen content from < 0.001% to > 20.9%, they have typically been very gradual. We have to consider periods of a hundred million years or even a billion years when we are discussing evolution of the atmosphere.

3.3. Evolution of the Hydrosphere

The ocean, as pointed out earlier, was probably formed along with the Earth. It could have been close to the present volume even 4.5 billion years ago. However, its area and chemical composition were quite different from those existing now.

According to Isaac Asimov, the following conditions must be met to create the ocean:

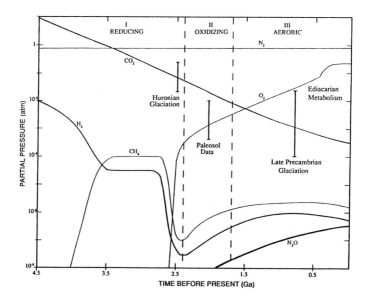

Figure 12. Possible changes in atmospheric composition with time. The two error bars on the CO_2 curve show upper and lower limits on pCO_2 during glacial period calculated with 1-D climate model. The error bar on the O_2 curve shows upper and lower limits on pO_2 between 1.5 billion years and 2.5 billion years, based on a combination of paleosol data and the theoretically derived from pCO_2 curve. The curves for H_2, CH_4, and N_2O are based entirely on theoretical explanations (from Engel and Macko, 1993).

a) a large volume of liquid;

b) elements of the ocean must be plentiful;

c) substance must be a prominent liquid phase at standard temperature and pressure (STP);

d) presence of depressions;

e) Earth gravity must be sufficient to held the ocean.

In the previous subchapters we have noted that there are 4 most abundant elements in the Earth (without inert He and Ne) such as H, O, C, and N. Let's consider the STP values of hydrogen and its chemical combinations. The hydrogen itself melts at $-259.2\,°C$ and boils at $-252.8\,°C$ (1 atm) and these conditions are not met in the Earth surface even 4.5 billion years ago. However, there are different combinations of hydrogen and other most abundant elements: H_2O, NH_3 and CH_4. The physical properties of these compounds under STP are shown in Table 3.

Table 3. Standard temperature values for different physical states of hydrogen chemical species at 1 atm.

Species	Solid	Liquid	Gas
H_2O	$0\,°C$	$100\,°C$	$> 100\,°C$
NH_3	$-77\,°C$	$-44\,°C$	$-33\,°C$
CH_4	$-187\,°C$	$-174\,°C$	$-164\,°C$

Thus, only water is the logical species to qualify for the ocean liquid in the Earth surface. Almost everybody agrees at present that the primordial sea was released from the crust and mantle. The older concept of water origin assumed that hydrothermal sources could readily fill up the ocean during a few billion years. However, the present theory applies the degassing scenario (see also previous section). The modern isotopic data show that this happened before 3.3 billion years ago. Accepting this age, we have to agree that ongoing hydrothermal activities on land and beneath the sea are mainly related to the recycling of ancient water.

Most authors assume that the proto-ocean resembled modern Ocean in that sodium (Na^+) and chloride (Cl^-) were the major ions followed by potassium (K^+), magnesium (Mg^{2+}), calcium (Ca^{2+}) and bicarbonate (HCO_3^-). Sulfate (SO_4^{2-}) was absent since the waters were neutral or slightly acidic. These pH values would have been related to the high CO_2 atmospheric partial pressure at that time (see above 2.3.2). Close to this viewpoint the soda ocean concept was suggested for the explanation of ancient abiotic water chemistry (Box 3.). This concept supposes that sodium carbonate would have been the main chemical species during primordial Ocean.

Box 3. Soda Ocean Concept, after Degens, 1989

This concept is based on two assumptions:

– all of today's water amounting to $1,350 \times 10^9$ Tg was present from the beginning (> 3.3 billion years);

– all crustal carbon had been spontaneously released as CO_2, thus totaling 65.5×10^9 Tg C;

– weathering would greatly have proceeded at this pCO_2 partial pressure (see also 3.2).

It was calculated that all this CO_2 became sequestered via surface weathering in less than 100 million years at the most. Throughout this time, the "runaway greenhouse" threshold was never achieved. Alkalies, most notably sodium and potassium, served as counter ions for the bicarbonate and chloride, whereas calcium and magnesium set free by the weathering precipitated as carbonates due to shift in pH from acid to alkaline conditions. The final chemical product is *the early Sodium Ocean*. It would have resembled the chemical composition of the fourth largest closed lake of the Earth; Lake Van in eastern Anatolia, Turkey, and many other lakes originated in volcanic craters. Liter by liter, these lakes may hold dissolved carbonates in an amount exceeding seawater bicarbonate content by more than 10^3 times before sodium carbonates, Na_2CO_3, precipitate.

For calculation purposes, the assumption was made that sodium is the counter ion for all of the dissolved carbonates in the prebiotic sea. Furthermore, the other presupposition was that the sodium carbonates were the principal salt species, although others may coexist, for instance, sodium bicarbonate, $NaHCO_3$. At the solubility maximum at 35 °C, 320 g Na_2CO_3 dissolves in 1 kg solution corresponding to 470 g Na_2CO_3 per kg water or 53.3 g carbon per kg water. This implies that the primordial Ocean could store any amount of sodium carbonate not becoming supersaturated. The characteristic properties of soda that exist at high pH above 9 acquire alkalinity concentrations of 0.05 to 0.1 meq/L. This means for carbon from 0.5 to 1 g per kg water. Taking into account the enormous pools for carbon and sodium in the lithosphere, such concentrations look quite realistic.

The crucial point of the early Soda Ocean model is the question of further evolution of water chemistry since the modern marine water is dominated by halite. The most plausible explanation based on gradual removal of dissolved Na_2CO_3 from the Ocean is related to

–   accumulation of pore waters in sediments;

–   migration of marine water into the crust;

–   biogenic reduction to organic carbon and subsequent burial in bottom sediments.

After many regressions and transgression of global Ocean during its geological history, the alkalies became part of the continental crust, whereas solid carbonates gave rise to carbonates at depth. At the same time, chlorine, which is present in oceanic geological basalts in small quantities only (20–50 ppm), was leaching and became part of a rising hydrochemical flux to the Ocean. On the basis of geochemical balance calculations, only 0.5 percent of the chloride content in the Ocean is derived from surface weathering of igneous rocks, the rest is crustal leaching. Sinks of hydrochemical chlorine include recent and ancient sediments as well as crystalline rocks derived from granitization and metamorphism of former sediments. Thus, about 25 percent of the chlorine amount present in the Ocean was fixed in sediments in the form of interstitial water, salt depositions and ordinary minerals. Excess of Cl^- in such geological rocks as granites, diorites, syenites and metamorphic rocks above the amount released from the hydrochemical weathering of the oceanic crust yield another 25%. Thus, the total sum of hydrochemical chloride is about 50% above the present marine water chlorinity.

Injection of HCl from the hydrothermal weathering into marine water as well as dissolution of atmospheric SO_2 steadily transformed the ocean's pH value from alkaline to acid. About 1 billion years ago the primitive soda Ocean was transformed into present salt sea. On other side, the changes in pH and possibly Eh have significantly altered the marine environment. This related mainly to carbonate system, which switched from sodium- to calcium- and magnesium-dominated mineral equilibrium about 1 billion years ago (Degens, 1989).

The sulfate system appeared and developed after the development of anaerobic and then oxygenic photosynthesis. Figure 13 shows the scheme for the reorganization of biogeochemical cycles accompanying oxygen increases during Proterozoic-Phanerozoic transition. One can note that in the Proterozoic Ocean slowly sinking organic matter escapes fuel sulfate reduction in a largely anoxic water column. After increases in oxygen, the evolution of biota has led to rapid fecal pellet transport.

The next stages in the evolution of atmospheric oxygen and consequently in the evolution of chemical composition of the Ocean, lasted from approximately 2.4 billion years to 1.7 billion years. During this time, the atmosphere appears to have been oxidizing (based, again, on red beds and detrital mineral deposition, see 3.2), but the deep ocean remained reduced. The evidence for an anoxic deep ocean comes from the occurrence of massive banded iron formations (BIFs), which are considered to require a deep ocean rich in dissolved ferrous iron (Figure 14).

The disappearance of the BIFs at ~1.7 billion years ago signaled the end of the previous state and the beginning of the third and final state of atmospheric oxygen evolution and, accordingly, evolution of Ocean chemistry. Once the deep ocean waters were cleared of ferrous iron, atmospheric O_2 would rapidly have approached its present concentration. The carbon cycle would have been operating more or less as today.

3.4. Prebiotic Earth and Mineral Cycling

We have described the evolution of three spheres existing in the Earth during the early period of its geological history. It is obvious that the specific characteristics of Earth's mineral cycles were very much dependent on our Globe origin and geological evolution. An active mantle and crustal system give the idea that the mineral cycling in the early prebiotic Earth would have been characterized by active exchange of elements between the lithosphere, atmosphere and hydrosphere. The significance of mantle and crustal cycling can be appreciated by comparison of our planets with other planets having seemingly similar space history, such as Mars and Venus. Rapid cooling of Mars due to its small diameter has significantly diminished the tectonic activity in its mass, while on Venus, periodic and apparently catastrophic crust reformation would have promoted elemental cycling but precluded life development simultaneously (Fenchel et al., 1998).

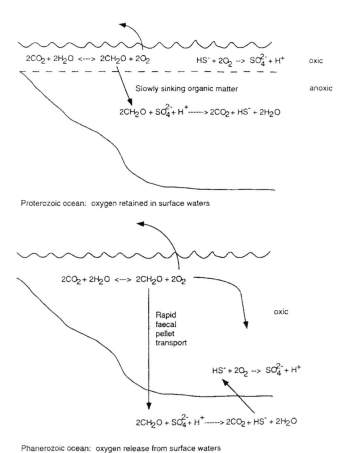

$$2CO_2 + 2H_2O \longleftrightarrow 2CH_2O + 2O_2 \qquad\qquad HS^- + 2O_2 \dashrightarrow SO_4^{2-} + H^+ \qquad \text{oxic}$$

$$\text{Slowly sinking organic matter} \qquad \text{anoxic}$$

$$2CH_2O + SO_4^{2-} + H^+ \dashrightarrow 2CO_2 + HS^- + 2H_2O$$

Proterozoic ocean: oxygen retained in surface waters

$$2CO_2 + 2H_2O \longleftrightarrow 2CH_2O + 2O_2$$

Rapid
faecal
pellet
transport

oxic

$$HS^- + 2O_2 \dashrightarrow SO_4^{2-} + H^+$$

$$2CH_2O + SO_4^{2-} + H^+ \dashrightarrow 2CO_2 + HS^- + 2H_2O$$

Phanerozoic ocean: oxygen release from surface waters

Figure 13. Proposed scheme for "reorganization of biogeochemical cycles" accompanying oxygen increase during the Proterozoic-Phanerozoic transition. Note that in the Proterozoic Ocean (upper panel), slowly sinking organic matter that escapes sulfate reduction in a largely anoxic water column. After increases in oxygen have led to evolution of grazing zooplankton (lower panel), rapid delivery of fecal pellets to the bottom promotes oxygenation of the water column due to decreased organic matter availability and anaerobic processes retreat to sediments following the organic matter supply (Logan et al., 1995).

The geochemical cycling of carbon dioxide between atmosphere, lithosphere and hydrosphere (see Section 3.2) provides an obvious example of abiotic (but temperature-dependent) cycling on the early Earth. These cycles were characterized by changes in chemical status of elements. For instance, CO_2 as a gas has been periodically transforming to carbonate as a solid. However, no accompanying changes in redox status have occurred. Living organisms certainly affect the rates of CO_2 cycling and accelerate them, but it is clear that these cycles occur independently of life.

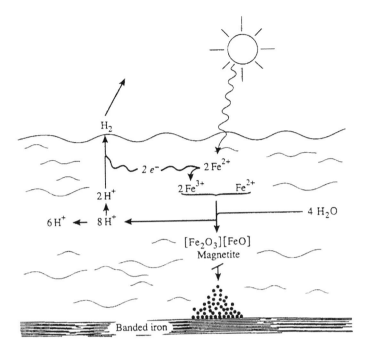

Figure 14. Light-driven hydrogen production and iron oxidation resulting in early banded iron formations (de Duve, 1991).

Other elements, such as halides, various metals and metalloides (iron, manganese, sodium, magnesium, zinc, mercury), nitrogen and phosphorus were also cycled through main Earth's components as a results of volcanism, hydrothermal venting and tectonic shifting of the crust. We can see the summary of prebiotic element cycles in Figure 15.

Important characteristics of the abiotic cycles in early Earth would have included (Fenchel et al., 1998):

–   pressure-driven and thermally-driven changes in the distribution of elements and chemical species among various mineral phases in the crust and mantle;

–   changes over time in the availability of elements and their chemical compounds for the exchange in the lithosphere-atmosphere-hydrosphere system;

–   adsorption, precipitation, aggregation and exchange reactions that differently influence retention and immobilization of elements in hydrosphere, e.g., clay formation in bottom sediments;

–   acid-based weathering of uplifted crustal minerals under the influence of carbon dioxide;

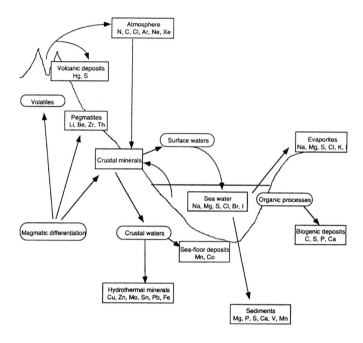

Figure 15. Summary of elemental sources and sinks for selected elements, illustrating lithosphere, atmosphere and hydrosphere reservoirs and basic geochemical processes that mobilized or immobilized elements (Cox, 1995).

–   fluvial and aerial transport of elements and species released by weathering;

–   selective changes in the composition of weathered minerals and transported minerals due to variations in elemental chemistry;

–   abiotic redox transformations.

During the thermally driven differentiation of the Earth into core-mantle-crust, numerous reactions would have produced oxidized forms of iron, sulfur and carbon. These would have contributed to the redox chemistry in the early planet development. Volcanic and hydrothermal emission of sulfur dioxide, SO_2, delivered oxidants to the oceans and atmosphere. Photodissociation of water vapor in the atmosphere have undoubtedly provided a small but significant source of molecular oxygen. Furthermore, UV-driven ferrous iron oxidation could have been coupled to the reduction of a variety of reactants, for instance, CO_2 (Figure 16).

The last seems even more important than the early appearance of a small amount of oxygen. This process provides a reasonable abiological explanation for early banded iron formations (BIFs) and a basis for the evolution of a ferrous iron-based bacterial photosynthesis (Fenchel et al., 1998).

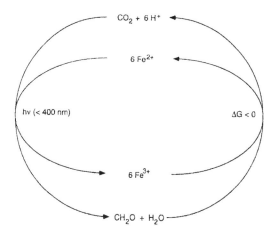

Figure 16. Light-driven iron cycle; ferrous iron is photo-oxidized to produce ferric iron and hydrogen; hydrogen is used as a reductant to form simple organics from carbon dioxide. Simple organics are oxidized to ferric iron completing the cycle; note that ΔG for the oxidation reaction is < 0, thereby providing energy for other processes, such as organic polymerization (de Duve, 1995).

Thus prebiotic Earth was characterized by complex patterns of elemental cycling. The availability of diverse reductants and oxidants, including various forms of organic carbon, would also have supported a range of prebiotic cycling transformations of various elements. Seemingly, these transformations are similar to those known currently, including CO_2 fixation via abiological coupling to pyrite synthesis. During the period of abiotic history of the Earth, contemporary elemental cycling remained dependent on, and subjected to, disturbance by some strictly geochemical processes, such as permanent volcanism, hydrochemical venting, as well tectonic transformations. However, it is evident that the rates of these interrupting processes have decreased over time as Earth's internal heat production has declined (Figure 17A). The relative stabilization of continental lithosphere accompanying the decrease in heat production (see Figure 17B) has undoubtedly had a profound impact on the evolution of living systems and the development of biogeochemical cycling. For example, the colonization of continents by living organisms has drastically changed the rates of geochemical weathering.

4. ORIGIN OF LIFE

The problems related to life's origin in the Earth are central and crucial in any discussion of biogeochemical evolution. Theoretical and experimental analyses of the origins of life occupied a considerable fraction of human thoughts and efforts. Until now it remains the central problem of many sciences, from theosophy to biology.

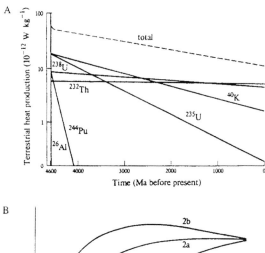

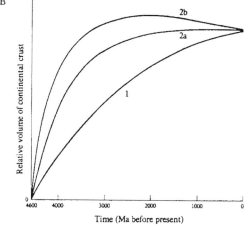

Figure 17. A and B. A. Terrestrial heat production from radionucleide decay; note extremely rapid loss of heat production from decay of ^{26}*Al; this isotope was likely responsible for much of the early heat production after Earth coalesced. B. Three conceptual models for the change in volume of continent crusts over time; model 1 suggests very low early crustal volume and continuous increase while models 2a and 2b suggest a rapid accumulation of near present values, with a stable volume late in model 2a and a slightly decreasing volume from the Proterozoic to the present in model 2b (Fenchel et al., 1998).*

4.1. When Did Life Originate?

At present there are many varieties of theoretical approaches to life origin. A schematic overview of different theories is given in Figure 18.

Progress has quickened during the last 10–15 years on a basis of modern achievements in biochemistry and relative sciences. However, still existing limitations of fossil records, uncertainties in the interpretation of isotopic data, and especially the "dark spot" of the Earth's geological history before 3.8–4.0 billion years ago leave

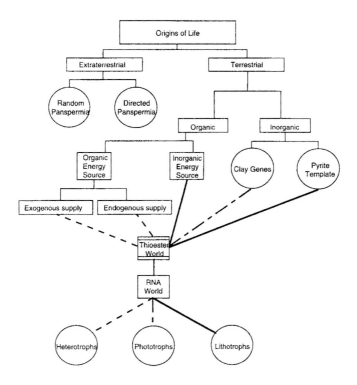

Figure 18. Relationships among various theoretical or conceptual models of the origin of life. Dashed and solid lines of the figure connect mechanisms with putative metabolic type of first organisms. The thioester world is proposed as an intermediate leading to an RNA world (Fenchel et al., 1998).

ample room for alternative and conflicting suggestions. Now the consensus is achieved on a probable course of events, including a likely microbial phylogeny. The other problems, like the compelling evidence for complex early life and the proceeding period when the development would have started, are still the subject of hot scientific and public discussion.

Recent isotopic data from the metamorphic geological rocks of Greenland suggest the evidence of life origin about 3.8 billion years. Some theoretical speculations are possible on the probability of ocean-vaporizing asteroid impacts and this could extend the life origin period to 4.4 billion years. Thus, the period during which the first life originated, differentiated and proliferated was probably very brief (from a geological point of view, of course), perhaps no more than 600 million years and possibly as short as a few tens of million years (Fenchel et al., 1998).

Such a brief geological period of life origin would have clearly sufficed for global proliferation of progenotes or their genote descendants. Recent analyses indicate that substantial differentiation among the bacterial, archean and eukariotic lineages could have occurred during a period of about 200 million years. The geological records also confirm rapid and substantial phylogenic diversification, for instance a variety of microfossil, geochemical and isotopic evidence from the famous and well studied Warrawoona rocks of Australia, which age is about 3.5 billion years. There is a reason to consider that a rapid pace of evolution started from 3.8–3.5 billion years ago (Box 4).

Box 4. Is the Earth Sufficiently Old to Originate Life? (after Fenchel et al., 1998)

There are some theoretical considerations on the period of life's origin on the Earth based on the requirement of a definite time period for protein molecule evolution.

About 100 years ago the famous scientist S. Arrhenius first proposed the idea of panspermia. This idea is based on the assertion that life originated elsewhere, and was transported to Earth. The feasibility of transport was supported by observation of the putative microfossils in a Martian meteorite. Adherents to the concept of panspermia regard any period of before 3.5 billion years as too short for the origin of life. The lack of time for evolution on Earth is supported by calculations of the extremely low probability (4×10^{-83}) of randomly assembling from a pool of 20 different amino acids an enzyme with specific sequence 100 residues in length. Assuming one abiological assembly reaction per second in $1\,cm^3$ of ocean water with a contemporary volume, up to 6×10^{51} years would be required for the acquisition of a single enzyme. Since bacteria contain on the order about 10^3 enzymes or proteins and the age of the Universe is only about 10^{10}, random assembly on Earth is judged impossible.

However, one can put a question, if not on Earth, when and where was life originated and transported to our planet? As it is considered that Earth is one-half to one-third the age of the Universe, then even it is not sufficiently old to originate life.

To resolve this dilemma, some argue for divine creation; some invoke the many n-dimensional Universes. Obviously, these latter explanations are far from scientific approaches. Another view is connected with recognition of the inherent limitations of the time estimates above, and propose that the need for an extraordinarily long "incubation" period is obtained by invoking a plausibly small pool of amino acids, about 10 or so, short catalytic peptides, 10–20 residues, with some tolerance for sequence variability and non-random searches of only portions of the total sequence space due to the existence of chaotic attractors. Under these assumptions, a much slower reaction rate (for instance, 10^5-fold slower) in only a fraction of the ocean is more than sufficient to account for rapid, less then 1 million years evolution of complex, possibly living, organic systems.

4.2. Primordial Soup

The previous discussion of Earth evolution has already demonstrated the feasibility of interactions between carbon, carbon dioxide, and clays. Some modern theories of life origin have been constructed in the suggestions of generations of physiologically active molecules in clay-organic interactions.

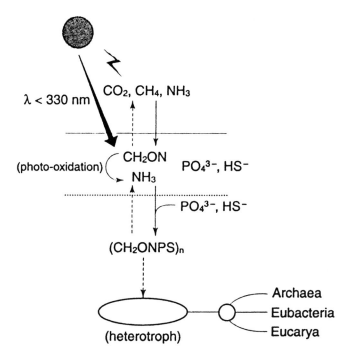

Figure 19. Simplified conceptual representation of the Oparin-Haldane model for origin of life. UV, lighting discharges or other energy sources results in organic matter synthesis from the constituents of a reducing atmosphere. Some photooxidation occurs in the upper layers of a primordial ocean with phosphate and sulfur incorporation and polymerization taking place in lower layers. Evolution leads to a heterotroph initially. Phylogenetic development is unclear (Fenchel et al., 1998)

However, to estimate the significance of these modern theories, we have to mention the previous theories of life origin, which dominated our way of thinking for a long time.

At the beginning of the 20th century the assumption remained undisputed on the preceding chemical evolution to biological evolution and chemical syntheses gave rise to the first living cell. The famous Russian biochemist Alexander Oparin presented the first plausible model on the chemical synthesis of critical biological molecules in the late 1920s. He proposed a reducing proto-atmosphere composed of small inorganic molecules, which synthesized of simple organic molecules such as amino acids or sugars. These products accumulated in the sea, where they interacted, formed polymers, and led to the familiar "primordial organic soup" (Figure 19).

According to A. Oparin, life eventually arose from that substrate. J.B.C. Haldane developed similar theory at nearly the same time.

For several decades, Oparin's model remained uncontradicted, the more so because experiments simulating prebiotic atmospheric conditions and applying electric discharges, ionizing particles, or UV-radiation were able to synthesize amino acids, small peptides, sugars, heterocyclic compounds and other biochemical molecules.

Nevertheless, neither the original Oparin theory nor its more recent variants are especially satisfying accounts of the origins of life. Major deficiencies at present include the inability of this theory to account for the currently accepted phylogenetic and biochemical differentiation of Archaea, Bacteria and Eucarya; inability to account for the presumed thermophilic characteristics of the earliest common ancestor of these groups; and inability to account for the origin of chirality. See also Box 5 for phylogenetic relationships.

Box 5. Classification of Living Organisms (Summons, 1993)

Organisms have traditionally been divided into major taxonomic groups, defined by their physiological and biochemical capacities, their morphology, and their habit (i.e., phenotypic features). Genetic affinities, using data derived from genotypic analysis, at present constitute the main means of assigning relationships (phylogeny). The new information derived from nucleic acid sequences is forcing a reappraisal of many phylogenetic relationships and a revision of classical taxonomy, particularly with respect to microbes. In contrast to the earlier five-kingdom classification, it is now widely accepted that there are three primary kingdoms of organisms (Figure 20).

The simplest cellular organisms, the Prokaryotes, constitute two of the three kingdoms that are the Eubacteria and the Archaebacteria. Prokaryotes are generally very small objects; their typical spheroids are about 1μ m in diameter or less. They live unicellular or in simple colonies. These microbes have little subcellular structure and minimal organization. The other primary kingdom, the Eukariota, is a grouping of larger organisms with typical unicell of $10 \mu m$ in diameter. These are more complex organisms, many of them are mylticellular. Eukaryotes possess the membrane-bound cell nucleus and discrete intracellular organelles such as mitochondria and chloroplasts. These and other morphological characteristics distinguish them from the Prokaryotes. While nucleic acid sequence data confirm the distinctiveness of the genetic material in the Eukaryote nucleus, the mitochondria and chloroplasts have separate genetic material with close affinities to that of the prokaryote group. This gives strong evidence to support the hypothesis that these organisms arose through evolution of earlier symbiotic associations between Prokaryotes and the ancestral Eukaryote as host.

Unicellular or organized colonies of Eukaryotes are termed *protists* while complex and differentiated plants and animals are termed *metaphytes* and *metazoans*, respectively.

Microorganisms grow and multiply generally faster than higher organisms and inhabit a wider range of environments. Microbial microfossils and microbially constructed stromatolites have an unambiguous record extending to at least 3.5 billion years of the 4.6 billion years of Earth's history. We have to add that the earliest part

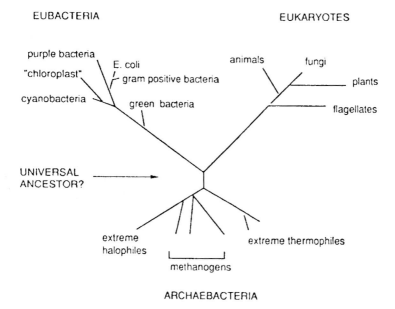

Figure 20. A diagram illustrating phylogenetic relationships between the three kingdoms of organisms. The data used to determine these come from the base sequences of nucleic acids. The tree presently has no root because we have no unambiguous knowledge of the ancestral forms (from Engel and Macko, 1993).

of these records are poorly presented and, accordingly, sometimes cryptic. The oldest preserved microfossils have ages of 3.5 billion years and resemble modern Eubacteria. The oldest recognizable Eukaryotes are in sediments dated approximately 2.1 billion years old. By comparison, the metazoans and metaphytes have an undoubted fossil record of about 0.6 billion years.

Archaebacteria, the third and the most recently recognized primary kingdom of organisms, preferentially inhabit extreme environments and comprise taxa with a variety of quite distinctive biochemistry. They have no presently recognized morphological fossil record and only an uncertain detection through carbon isotopic anomalies in some Archaen and early Proterozoic sediments. Nucleic acid sequence information and difference in membrane composition indicate that the kingdom is at least as old and fundamentally distinct as the Eubacteria and the Eukaryota. Archaebacteria are recognizable through their chemical fossils in some of the oldest, presently known, organic sediments. Archaebacteria include the anaerobic methanogens, which have a capacity to produce methane from CO_2 and other simple carbon compounds and a reductant, such as H_2. Halophylic Archaebacteria are aerobic heterotrophs, which exist only in environments with high salt content. The thermophilic sulfur-dependent Archaebacteria comprise both aerobic and anaerobic orders and can inhabit hot and sulfur rich springs and hydrothermal vents. Sulfide is unstable in the presence of oxygen in modern environments, but would have been an important species in early stages of Earth's history.

4.3. Clays and Life

In the beginning of the 1980s Cairns-Smith proposed that clays produced from igneous rock weathering facilitated biochemical evolution. This model is shown schematically in Figure 21.

It is based on clay minerals or "crystal genes" with replicable defects in their crystal structures. Certain crystal genes increase in abundance through a form of inorganic evolution based on selection of those that promote mutual synthesis and propagation resulting in the formation of "complex vital muds". Light energy drives the synthesis of simple organics, ultimately leading to multi-step complex synthesis. Chirality is introduced at this point, arising primarily from the inherent chirality of clay crystals themselves. Continued evolution of clays and their associated organic synthesis lead gradually to production of structural polymers, including polynucleotides. During further evolution polynucleotides constituted the secondary genetic system resulting in partial control of a subset of the organic synthesis systems, for example peptide synthesis. The evolution of protein enzymes and cellularization with lipid membranes promoted the development of novel mechanisms for organic synthesis and genetic control. It is considered that at this stage, dependence on clay minerals would have been lost. Further evolution would have led to an early ancestral form, so called progenote (Fenchel et al., 1998).

This Cairns-Smith model can be confirmed by a number of its strengths. These are related both to existence of clay varieties and energy sources in the prebiotic Earth. Rapid attenuation of UV by clay in the sediments of shallow lagoons or drying muds minimizes photochemical organic matter oxidation that constraints product accumulation in the water column. The ability of clays to concentrate cations and phosphates facilitates the introduction of metal catalysts into the evolution of enzymatic activity, and phosphorus into biosynthesis and energy metabolism. Furthermore, elemental cycles would have developed simultaneously with the development of clay-based life, thus giving the proto-biogeochemical cycles of various elements.

However, this clay-life theory has many weaknesses as well. The main restriction is related to the unclear abundance of sites suitable for clay evolution. The input of photic energy is also uncertain. There is little information on the kinds of clay that might have existed prior to 3.5 billion years ago. This is very important because the properties of clay would have played a crucial role in organic synthesis and not all clays may be suitable for crystal genes.

This model is also unable to explain microbial phylogeny. For example, the origins of thermophily, sulfur metabolism, and the differentiation of Bacteria and Archaea are not immediately apparent. There are some other specific constraints of clay-life theory and those who are interested can find the details in T. Fenchel et al., 1998 (see Further reading list).

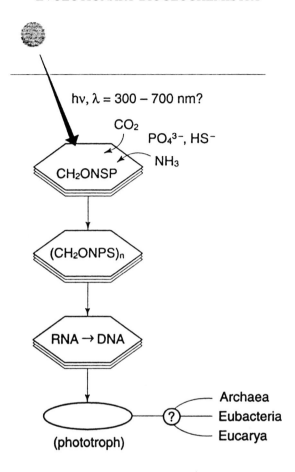

Figure 21. Simplified conceptual representation of the Cairns-Smith clay based model for origin of life. Simple organic synthesis occurs on clay surfaces by a phototransducing system. The clays serve as a "genetic" template that is coupled to organic polymerization. Synthesis of ribose and then ribonucleic acids leads to an RNA world, a transition to DNA as the genetic material and perhaps the origin of a phototroph. Phylogenetic development is unclear (Fenchel et al., 1998).

4.4. Pyrite-Based Life Origin

G. Wächtershäuser has proposed the most biochemical-detailed theory of life origin in the late 1980s. This theory is relatively similar for understanding in spite of its detailed biochemistry. Figure 22 shows the schematic presentation of this theory. We can see that pyrite synthesis provides the driving force for organic synthesis.

Pyrite provides also a surface for stabilizing and protecting reaction of synthesis. These reactions lead ultimately to complex polymerizations and early cellularization by incorporation of pyrite grains.

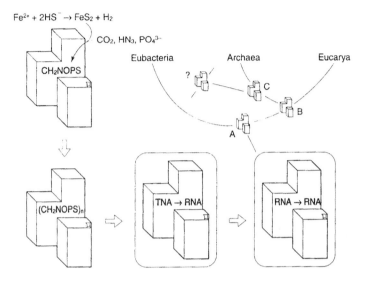

Figure 22. Simplified conceptual representation of Wächtershäuser's pyrite-based model for origin of life. Pyrite synthesis provides the driving force for organic synthesis; pyrite also provides a surface for stabilizing and protecting synthetic reactions, leading ultimately to complex polymerization and early cellularization (incorporating pyrite grains). Development of a "tribonucleic" acid precedes RNA-based genetic systems, which then leads to DNA-based systems (Fenchel et al., 1998).

Thus, pyrite synthesis supplies energy for the corresponding organic synthesis, and resulting pyrite surface serves as an initial 'medium' for the growth and synthesis of organic molecules. The chemical species as ferrous sulfide, carbon dioxide, ammonium, and phosphate would be proposed for involvement in these interactions. Autocatalic cycles based on metabolites that exist now as coenzymes developed rapidly. These cycles were accompanied by both vertical and horizontal extensions of the synthetic systems or, in other words, the spatial growth was enabled. Competition among various produced systems optimized efficiency and kinetics. Semi-cellularization (the cellularization was incomplete, as theory assumed, in the first stage) ensued early with the advent of isoprenoid membranes. Transport of non-ionic substrates through such membranes could be efficient. Complete cellularization followed further, along with incorporation of pyrite crystals within the early organisms. During further development metabolic processes proceeded in pro-cytoplasm. The shifting from pyrite surface was accompanied by the evolution of soluble coenzymes. The corresponding mechanisms of substrate transport across membranes by proteins facilitated this shifting. The evolution of 'tribonucleic acids' allowed the information storage and genetic control. These 'tribonucleic acids', firstly presented as triose-based polymers of purines, were evolved then in syntheses of ribose, ribonucleic acids, pyrimidines, the precursors of tRNA, rRNA and mRNA, and, finally, DNA.

Pyrite's theory includes also the analysis of the differentiation of Bacteria, Archaea and Eucarya (see Figure 22). This shows the resolution of the genetic similarities among domains. Furthermore, it provides the incompatibility between archaeal iso-prenoid lipids and the fatty acid lipids of two other groups of living organisms, Bacteria and Eucarya. The central role of elevated temperature in organic synthesis coupled to pyrite formation also provides a logical explanation for the thermophilic, sulfur-metabolizing character of the putative progenote during early life development.

Unlike two previous theories of life origin, only a few pieces of experimental evidence exist at present to prove the theoretical speculations. However, we have to notice the verification of the basic mechanism of molecular hydrogen generation as a reducing power, furthermore, the amide bond synthesis has been also demonstrated, both at temperatures within the range of hydrothermal vents ($100\,^{\circ}C$). In addition, the evidence for at least sulfide-based amino acid synthesis and polymerization from sim-ple precursors has been shown. The formation of acetic acid and an activated thioester from carbon monoxide, methanethiol and various combinations of ferrous and nickel sulfides has been experimentally proved as well. However, further verification is necessary for the modes and rates of organic synthesis.

In spite of the needs of detailed verification, the pyrite theory has a variety of strengths. We have to mention the following:

a) the ability to explain phylogenetic differentiation,

b) the proposed binding of organisms to pyrite surfaces occurs via phosphate, resulting in much stronger surface associations than are possible for clays,

c) early selection for phosphate esters through strong surface bonding also provides a basis for the central role of this element in intermediary metabolism,

d) the incorporation of metal catalysts (Fe, Ni) into chemical reactions was facili-tated by their availability in the hydrothermal vents,

e) due to chemoautotrophic nature of early pyrite-based life, metabolic reaction pathways were forwarded to synthesis of complex compounds from the relatively simple species, like ribose from surface-bound triose phosphate,

f) finally, pyrite-based life is compatible with an RNA world and may not require an introduction of PNA as an RNA precursor (see Box 6).

There are, of course, visible weaknesses of this theory and the main uncertainties are related to:

a) requirement for hydrothermal systems based on chemical considerations is con-tradictory to instability of biological molecules in the temperature range of these vents, and

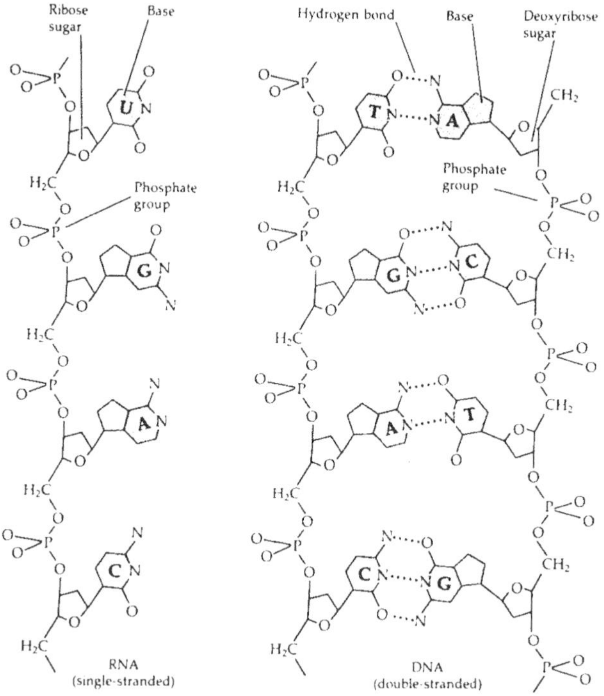

Figure 23. RNA and DNA, the carriers of genetic information. Bases are denoted as: A, *adenine;* T, *thymine;* G, *guanine;* C, *cytosine;* U, *uracil. Note that* RNA *contains* U *where* DNA *contains* T *(from Butcher et al., 1992).*

b) concerns on the sources of phosphate that plays the central role in pyrite-life origin, since modern thermal vents are depleted in phosphate due to its reaction with basalts at high temperatures in the subsurface charging zones, sources of the vents.

Box 6. DNA and RNA (Leeuw and Largeau, 1993)

The biopolymers DNA (deoxyribonucleic acid) and RNA (ribonucleic acid) are present in any living cell and contain or transfer genetic information. They consist of purine (adenine, guanine) and pyrimidine bases (cytosine, uracil, thymine) linked to either 2-deoxy-D-ribose (DNA) or D-ribose (RNA). The sugar units are linked together via phosphates (Figure 23).

Outside living cells, RNA and DNA are biodegraded very easily. Relatively stable bases have been reported for recent sediments. Obviously, minor amounts of DNA, RNA, or their bases can survive to some extent in the bottom sediments and lithosphere.

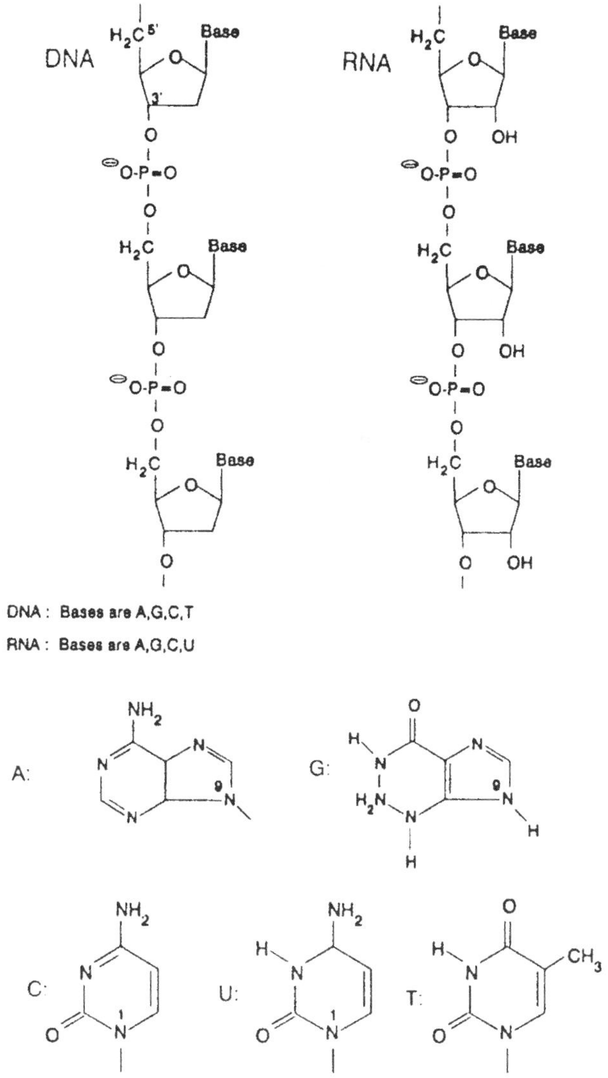

DNA : Bases are A,G,C,T

RNA : Bases are A,G,C,U

A: G:

C: U: T:

Figure 24. Structural units of DNA *and* RNA *(from Butcher et al., 1992).*

4.5. *The "Thioester World"*

In the early 1990s C. de Duve suggested the sulfur-based theory of life origin. The author has described the "thioester world" as a precursor of an RNA world (Figure 25).

The basic aim of this theory is an attempt to avoid the uncertainties of explaining prebiotic mechanism to account for the production of RNA from primordial soup. This is connected with a suggestion that catalytic polypeptides (short sequences of

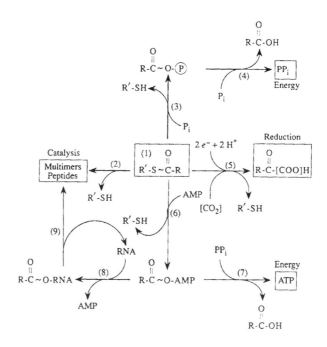

Figure 25. Synopsis of chemical transformations in the thioester world illustrating: (1) a pool of thioesters; (2) polymerization of protoenzymes; (3) generation of high-energy phosphate esters; (4) generation of pyrophosphate, a primordial energy carrier; (5) thioester-based organic synthesis reactions; (6) formation of high-energy adenylate derivatives; (7) production of ATF; (8) generation of acyl-RNA complexes (e.g., amino-charged tRNA); (9) peptide formation (de Duve, 1995).

amino acids) would have arisen spontaneously from thioester derivatives of amino acids. The amino acids are produced either by any variety of prebiotic synthesis in the atmosphere and oceans or are derived from meteoric input. The possible sources of hydrogen and ferric iron are shown in Figure 14 (see above). Further ferric iron reduction in the mechanism of thioester synthesis is shown in Figure 26.

Phosphorolytic attack of thioester by inorganic compounds provides a mechanism for introducing high energetic phosphate esters into metabolism leading to the synthesis of pyrophosphate as the first unit of energy. The synthesis of AMP and ATF followed this step.

The mixture of various catalytic multimers, like thioesters and phosphorylated organic molecules, presents the background for a "protometabolism" in the "thioester world" theory. Protometabolitic reaction pathways rapidly form networks. The latter would have included cyclic mass flows that are stabilized by interactions among metabolites. These pathways could have occurred in associations with pyrite, clays or iron dioxide flocks produced by UV photooxidation.

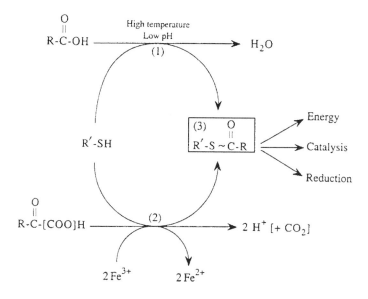

Figure 26. Mechanisms of synthesis of thioesters. (1) High temperature, low pH spontaneous synthesis from thiols and ornic acids; (2) oxidative synthesis from thiols and α-keto organic acids based on ferric iron reduction (de Duve, 1995).

It looks reasonable to hypothesize that globally ubiquitous iron oxides were incorporated into life development from the early evolutionary stage, but subsequent to the synthesis of simple amines and organic acids. The diffusion of sulfide across primitive membranes is also meaningful for the maintenance of the "thioester world". An intracellular iron redox cycle, driven possibly by light and sulfide, could have supported chemosynthesis. Iron respiration and photometabolism could have processed on the external surface of vesicles. This would have contributed to a vectorial transport of thioesters and protons into vesicles and accordingly to the development of the early protonmotive transport system.

We can share the view of some authors (Fenchel et al., 1998) that this theory may serve a plausible explanation of "thioester world" as precursor for the present RNA world. Furthermore, it is attractive in a number of respects too. One has to mention the development of very early, global scale geochemical cycling of carbon, sulfur, and iron. It would present the early interactions between atmosphere and biochemical processes, like thioester-bounded hydrogen production contribution to the formation of reducing species that were critical for any atmospheric organic synthesis.

In spite of these and some other strengths (for instance, linkages between the dynamics of different thioesters and polypeptide synthesis and metabolism of high-energy phosphates), the theory of "thioester world" leaves some basic problems of life origin unanswered. Among these problems, inconsistency with the modern understanding of the microbial phylogeny, the uncertainties related to the catalytic activities for small multimers under only a limited range of available amino acids, unclear explanation of cellularization, the sources of lipids formation and their composition as well as the origin of chirality in amino acids and sugars.

These and other problems pose considerable experimental and even conceptual challenges. Similar problems are typical for other mentioned above theories suggested during the 20th century for the understanding of such a crucial scientific problem as life origin. Seemingly, only unification of experimentally approved components of various theories could ultimately suggest the further development of a coherent, comprehensive framework amenable to unequivocal testing.

5. EVOLUTION OF BIOGEOCHEMICAL CYCLES

We have discussed various theories of life origin in the Earth, however, irrespective of its origin, the advent of life contributed significant alterations to existing prebiotic geochemical cycles. Aside from the colonization of terrestrial surfaces, perhaps the most important changes happened as a result of evolution of oxygenic photosynthesis and respiration.

5.1. Application of Isotopic Analysis for the Geological History of the Biogeochemical Cycles

The passage of the life-forming elements, so-called biophils, through the biogeochemical turnover results very often in stable isotopic fractionations due to equilibrium and kinetic isotope effects. The first effect can be attributed to CO_2/HCO_3^- equilibrium, and the second effect is related, for example, to enzymatic discriminations. Carbon, oxygen, sulfur, nitrogen, and hydrogen all exhibit significant isotopic heterogeneity with and between inorganic and organic reservoirs (Box 7). Isotopic analyses constitute an informative and effective technique for considering different mass exchange reactions during geological time. For instance, the importance and intensity of coupling of the global carbon and sulfur cycles during the Phanerozoic has been proved using recognition of the inverse relationships between the isotopic composition of marine carbonates and sulfates. It has been shown that major faunal extinction events, such as those just prior to the Cambrian, the Permian Terminal Event, would have to be associated with rapid and important disturbance in the functioning of the carbon biogeochemical cycle. This has been supported using isotopic stratigraphy, for instance, the investigation of fine scale carbon isotopic fluctuations in carbonates and organic matter, deposited in the ancient oceans.

Box 7. Isotopic Fractionation

The identification of biogeochemical consequences of carbon, nitrogen, sulfur, oxygen and other biophilic elements incorporation into biomass is recognized as an isotope fractionation or anomaly. When an element exists as a mixture of stable isotopes, as in the case of C, H, S, N, and O, biochemical and chemical transformations

Table 4. Sulfur isotopic composition in various reservoirs.

Species	$\delta^{34}S(\%_oCDT)$
Sedimentary sulfate	$\sim +17 \pm 2$
Sedimentary sulfide	$\sim -18 \pm 6$
Oceanic sulfate S	$\sim +20.0$
Living biosphere	~ 0
Dead biosphere	~ 0
Metabolic processes	Isotopic fractionation
Assimilatory sulfate reduction	~ 0
Dissimilatory sulfate reduction	$+5$–46
Oxidation of sulfate species	$+2$–18

usually take place with a slight preference for one isotope (see for instance Table 4 for sulfur). Most biochemical changes show a preference for the light isotope. This can show up as a measurable isotopic ratio difference between substrate and product. This ratio is expressed in relative terms using the per mil (‰) notation and based on a system of international standards. The standards are generally selected so as to be close to the isotopic composition of the elements in the major reservoir (e.g., in the Ocean) or they were at the time of the Earth's formation. For example, triolite, an iron sulfide mineral from a meteorite (the Canyon Diablo triolite; CDT) is used as an international standard, and all sulfur isotope compositions are expressed relative to the following:

$$\delta^{34}S(\%_oCDT) = \left\{ \left[{}^{34}S/{}^{32}S_{sample} \right] / {}^{34}S/{}^{32}S_{standard}] - 1 \right\} \times 10^3 \qquad (3)$$

Similarly, carbon isotope compositions ($^{13}C/^{12}C$) are expressed relative to that of the Pee Dee Belemnite (PDB) standard.

During the geological history of the Earth, oceanic circulation, tectonism, and climate have strongly influenced the spatial and temporal patterns of organic matter formation and deposition. Since the carbon dioxide present in the ocean-atmosphere system is a major mediator of surface temperature, being a greenhouse gas, it is probable that global climate changes in Earth's history and carbon cycle have been interdependent. Determination of such precise relationships between geochemical and biogeochemical phenomena, such as sedimentary isotopic fluctuations and global "anoxic" events, on one hand, and seemingly simultaneous events of biological evolution, on the other hand, is a great task for many biogeochemists.

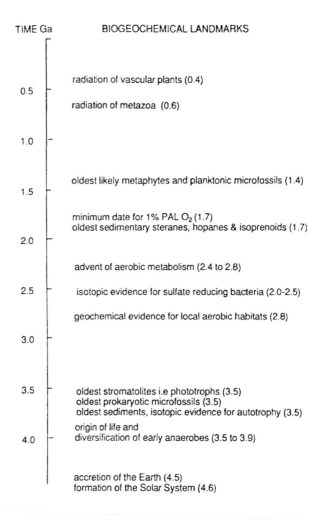

TIME Ga   BIOGEOCHEMICAL LANDMARKS

0.5   radiation of vascular plants (0.4)

           radiation of metazoa (0.6)

1.0

1.5   oldest likely metaphytes and planktonic microfossils (1.4)

           minimum date for 1% PAL O_2 (1.7)
           oldest sedimentary steranes, hopanes & isoprenoids (1.7)
2.0

           advent of aerobic metabolism (2.4 to 2.8)
2.5   isotopic evidence for sulfate reducing bacteria (2.0-2.5)

           geochemical evidence for local aerobic habitats (2.8)

3.0

3.5   oldest stromatolites i.e phototrophs (3.5)
           oldest prokaryotic microfossils (3.5)
           oldest sediments, isotopic evidence for autotrophy (3.5)
           origin of life and
4.0   diversification of early anaerobes (3.5 to 3.9)

           accretion of the Earth (4.5)
           formation of the Solar System (4.6)

Figure 27. Time scale of Earth's history, showing events of biogeochemical significance based on current assessments of fossil evidence (Engel and Macko, 1993).

Biogeochemical cycles had their origins in early life evolution and undoubtedly were interrelated to the stages in this evolution. This is schematically shown in Figure 27.

The evidence of such a relationship can be obtained from the analysis of stromatolites. Stromatolites, i.e., laminated sedimentary structures, were formed by microorganisms, which trapped, bound and cemented marine sediments during the long geological development of our planet.

5.2. Evolution of Oxygen Biogeochemical Cycle

The morphology of the oldest fossil stromatolites (about 3.5 billion years) resembled that of the modern ones constructed by photosynthetic microbes. Isotopic differences between organic and inorganic carbon compounds of the oldest sedimentary rocks are of "similar" magnitude of those found today. On a basis of this and similar information one can conclude that photoautotrophy has approximately the same age as the ancient living organisms, i.e., about 3.5 billion years. Geological data suggests that photosynthesis would have operated from the period of 2.7 billion years ago. At present it is impossible to make a definite conclusion whether the ancient photosynthetic process was based on hydrogen sulfide (anoxygenic; photosystem I), water (oxygenic, photosystems I + II), or some other substrates as an electron donor. Some evidence, however, supports the first type of photosynthesis (Box 8).

Box 8. Early Photosynthesis (Butcher et al., 1992)

Evolution of photosynthesis was one of the most important biological events of the early Earth's history. Anaerobic photosynthetic bacteria, like the modern green and purple sulfur bacteria, would have been the first photosynthetic organisms. There is no remarkable evidence of the development of these organisms before 3.0–3.5 billion years ago, however, the process of anaerobic photosynthesis is thought to have occurred before this time. The following speculations can prove this hypothesis.

The anaerobic photosynthetic bacteria can not produce oxygen in an aerobic atmosphere. They would have produced oxygen carrying out the anaerobic oxidation of the reduced sulfur compounds H_2S and S, with the formation of S and $SO_4{}^{2-}$, correspondingly:

$$2H_2S + CO_2 \longrightarrow (CH_2O) + 2S^0 + H_2O$$
$$2S^0 + 3CO_2 + 5H_2O \longrightarrow 3(CH_2O) + 2SO_4^{2-} + 4H^+$$

(4)

Organic matter produced by this photosynthetic process was represented by CH_2O.

Oxygen was not produced during this type of photosynthesis, unlike oxygenic photosynthesis by other organisms, including cyanobacteria, higher algae and plants. Perhaps, the best evidence that the earlier photosynthesis was carried out by anaerobic bacteria before 3.0–3.5 billion years ago is the discovery of sulfate minerals in deposits of that time. Although small amounts of oxygen from abiotic photolysis of water could have resulted in oxidation of reduced sulfur compounds, the main role would have played by anaerobic photosynthesis in accordance with the reaction mentioned above.

In spite of starting the photosynthesis 2.7 billion years ago (Box 9), the atmospheric and oceanic contents of oxygen in the Archean and Early Proterozoic were much lower than the present atmospheric level (PAL). There is general agreement that oxygen production and accumulation from oxygenic photosynthesis resulted gradually in atmospheric levels capable of sustainable metazoan respiration. The availability of molecular oxygen was an essential factor in numerous biochemical reactions during the life development. This is related especially to biosynthesis of unsaturated fatty acids and sterols in Eukaryots. Molecular oxygen was also a precursor of formation of a protective ozone layer (see Subsection 3.2). Accordingly, the definite concentration of oxygen in the atmosphere has arisen before the living organisms would have developed in surface waters and land surfaces. After achieving this content, the Eukaryotes became widespread in terrestrial surfaces and the biochemistry of Eukaryotes started to play an important role in the carbon cycle.

Box 9. Oxygenic Photosynthesis (Butcher et al., 1992)

Being opposed to anoxygenic photosynthesis (see Box 2.8), oxygenic photosynthesis produces oxygen. This process can be also developed by cyanobacteria but these bacteria are capable of utilizing water instead of hydrogen sulfide and can produce oxygen. This type of photosynthesis is related to all cyanobacteria, algae, and higher plants:

$$H_2O + CO_2 \longrightarrow CH_2O + O_2 \tag{5}$$

The reverse reaction $(CH_2O) + O_2 \rightarrow H_2O + CO_2)$ is termed respiration.

Oxygenic photosynthesis has been dated to before 2.0 billion years ago by some distinctive geological records. The extensive banded iron formation (BIFs) dated 2.0-2.5 billion years ago, can serve as an evidence of the development of this type of photosynthesis. The iron occurs in the form of almost pure ferric oxide and silica, and the formation of BIFs can be explained by the oxidation of ferrous iron in solution with oxygen, produced in large quantities in photosynthesis. The only pre-Cambrian photosynthetic organisms that could carry this out were cyanobacteria. There are extensive fossil records of similar organisms in geological formations of that age.

Here we have to say a few additional words about the role of oxygen in the evolution of Earth's biogeochemical cycling. The paramount importance of oxygen has to be commented on specifically. The chemistry of oxygen suits it uniquely for its biogeochemical role.

Oxygen is the most abundant element at the Earth's surface being behind only hydrogen and helium in the whole Universe. This element occurs in several oxides, and iron, aluminum, and silicon oxides are predominant. The main pool of combined oxygen is in Ocean (about 1.2×10^{12} Tg of oxygen). The atmosphere contains the molecular oxygen (about 1.2×10^9 Tg). It has been estimated that biosphere cycling of oxygen between the atmosphere and the hydrosphere occurs relatively rapidly, with turnover time of about 5,000 and 5,000,000 years, respectively.

Changes in the evolution rates and modes for many elements were influenced by oxygen during Earth's geological history.

5.3. Evolution of the Nitrogen Biogeochemical Cycle

In the anoxic atmosphere the closed anaerobic cycle of nitrogen would have consisted of the following thermodynamically permissible reactions:

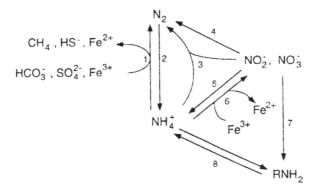

Figure 28. Hypothetical anaerobic nitrogen cycle based on the following thermodynamically permissible reactions: (1) ammonium oxidation to dinitrogen by carbon dioxide, sulfate or ferric iron (no evidence at present, possibly kinetically limited); (2) dinitrogen fixation by various organic and inorganic reductants (known); (3) ammonium oxidation by nitrite or nitrate producing dinitrogen (known); (4) denitrification (known); (5) nitrite or nitrate respiration (known); (6) ferric iron oxidation of ammonium to nitrite or nitrate (no evidence at present); (7) nitrate assimilation (known); (8) ammonium assimilation and dissimilation (known) (Fenchel et al., 1998).

1). Ammonium oxidation to dinitrogen by carbon dioxide, sulfate or ferric iron;

2). Dinitrogen fixation by various organic and inorganic reductants;

3). Ammonium oxidation by nitrite or nitrate producing dinitrogen;

4). Denitrification;

5). Nitrite or nitrate respiration;

6). Ferric iron oxidation of ammonium to nitrite or nitrate;

7). Nitrate assimilation;

8). Ammonium assimilation and dissimilation.

Schematically this reaction pathway is shown in Figure 28.

The first three reactions are considered at present as phylogenetically ancient and known from extant organisms. The remaining redox transformations are thermodynamically possible but still not carried out experimentally, perhaps due to kinetic limitations. In addition, even if such reactions did occur in the early Earth, their continuation would be constrained by electron-acceptor recycling. As an assumption, we can consider the volcanic eruptions of SO_2 as a way to achieve closure, but the continuation via this process is also uncertain. Nitrogen would have been lost from this anoxic atmosphere and accumulated in the oceans.

It is known that the modern closed biogeochemical cycle of nitrogen currently depends on microbial denitrification and two obligatory aerobic steps, related to ammonia and nitrite oxidation (see more details below under discussion of nitrogen biogeochemical cycling in Section 3.3). This closure could be achieved only when the oxygen content in the atmosphere has started to rise. This rise of oxygen ensured closure for the nitrogen cycle by promoting the evolution of ammonia and nitrite oxidizers. The development of denitrifiers would be mentioned as well subsequent to the evolution of respiration.

5.4. Evolution of Carbon and Sulfur Biogeochemical Cycles

Since the terrestrial rock record has hitherto failed to furnish unequivocal evidence of prebiotic organic chemistry, early chemical evolution was almost certainly confined to the "Hadean" era bracketed by the time of Earth's formation ($\sim$ 4.5 billion years ago) and the onset of the sedimentary record about $\sim$ 3.8 billion years ago (Figure 29).

Specifically, the isotopic composition of the carbon compounds in the oldest sediments supports the conclusion that the geochemical cycle of carbon evolved into a biogeochemical cycle by the time of the deposition of these rocks. Thus, a fully developed biogeochemical carbon cycle may date from $\sim$ 3.8 billion years ago. Also, the morphological record of microbial (prokaryotic) life has been shown to proceed over at least 3.5–3.8 billion years of geological history.

Mass-balance calculations, based on isotopic composition and the respective sizes of the various pools of carbon and sulfur demonstrate the coupling of the cycles of these elements during the Phanerozoic with variations of atmospheric oxygen. The records of sulfate-sulfur isotopes rarely appear in geological rocks older than 1.2 billion years. However, even the existing records, which extend each 3.8 billion years, allow geochemists to identify the advent of biogenic sulfide production at about 2.5 billion years ago. The sulfur isotopic composition of diagenetic pyrite is controlled by a number of parameters, including the availability of sulfur source and its isotopic ratio. In the sedimentary depositions of about 2.5 billion years age, the predominance of negative $\delta^{34}S$ (0–50‰ CTD) was measured indicating bacterial sulfate reduction. On the other hand, sulfate shows positive values of $\delta^{34}S$ (+ 10 to + 50‰ CTD) in these depositions. In the geological strata that are older than 2.5 billion years, isotopic composition of both sulfide and sulfate varies about 0‰ CTD, a value thought to be close to that of Earth's primordial sulfur, and reflecting a dominantly magnetic source for all sulfide.

The geochemical balance of carbon changed dramatically during Devonian and Carboniferous times. At those times the main evolutionary radiation of terrestrial vascular plants occurred. These events flowed from the development of various parts of higher plants, such as roots, stems, leaves, and reproductive apparatus enabling the colonization of a new ecological niche. This colonization and accumulation of

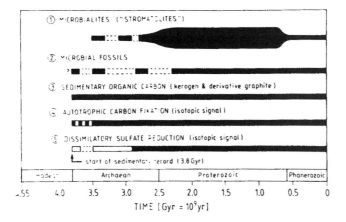

Figure 29. Record of fossil microbial life (1,2) and selected categories of biogeochemical evidence (3–5) within the temporal framework of 4.55 billion years of Earth's history. The Hadean, Archean, and Proterozoic eons are conventionally grouped together as the "Precambrian", with the Hadean (> 3.8 billion years) lacking a sedimentary record and thus being virtually devoid of geological documents. The stromatolite diagram (1) reflects the exuberant proliferation of microbial ecosystems after the establishment of extended stable marine shelves during the Proterozoic as well as their sudden decline following the rise of metazoan grazers at the dawn of the Phanerozoic. The isotopic signature of autotrophic carbon fixation (4) as revealed by sedimentary organic carbon quite uniformly through Earth's history exhibits a well-defined metamorphic overprint for t > 3.5 billion years (broken line). A corresponding isotopic index line for dissimilatory sulfate reduction (5) apparently fades between 2.7 and 2.8 billion years, but it seems most likely that the underlying biologically mediated isotope effect has been camouflaged (stippled signature) by either geochemical or environmental controls that were selectively operative during the Earth's earliest history (from Engel and Macko, 1993).

biomass was connected with a massive deposition of organic carbon. The terrestrial origin of this carbon was determined on the basis of isotopic analyses of both sulfur and carbon.

No remarkable alterations were shown in carbon and sulfur cycles or in atmospheric oxygen concentrations since the Latest Proterozoic. The coupling of carbon and sulfur biogeochemical cycles is supported by systematic covariations in the isotopic records of sedimentary rocks over Phanerozoic time.

The appearance of photosynthetic system II and the consequent rise in atmospheric oxygen content would have also changed significantly the geochemical cycles of ferric iron and hydrogen during the Archean and much of the Proterozoic times.

Peculiarities of Carbon Biogeochemical Evolution.

In addition to carbonate precipitation, the introduction of CO_2-fixing organisms into the surficial environment must have a paramount impact on the terrestrial carbon cycle

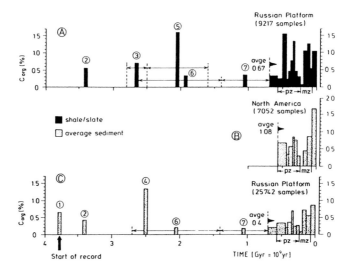

Figure 30. Organic carbon content of shales (A) and average sediments (B, C) through geological time. Note that the data base for t > 0.6 billion years is still deplorably scanty as compared to the younger record, but the scatter of Corg in Precambrian sediments is apparently the same as in Phanerozoic ones. Numbered Precambrian occurrences: (1) Early Archean metasediments from Isua. West Greenland; (2) Swaziland System, South Africa; (3) Archean pelitic sediments, Canadian Shield; (4) Hamersley Group, Australia; (5) Proterozoic pelitic sediments, Canadian Shield; (6) Proterozoic 1-2 of Russian Platform (from Engel and Macko, 1993).

by establishing a second sink for atmospheric CO_2, since carbon is the key biophilic element. Both carbonates (limestone, dolomite, etc) and biologically originated carbon (reduced forms) are presented in sedimentary rocks in abundant quantities. Organic carbon in old depositions presents the fossil residues of primary biological substances. In the average sediment, organic carbon is present in amounts between 0.5 and 0.6%. Carbon exists mostly in the form of kerogen, the acid-insoluble, polycondensed end product of the diagenetic alteration of organic detritus derived from the organisms and the products of their metabolisms (see details below in Section 6.1). With the mass of sedimentary rocks totaling about 2.4×10^{18} tons, the content of organic carbon will be $1.2–1.4 \times 10^{16}$ tons. Carbonate carbon deposits are about four times more abundant, i.e., $\sim 5.0 \times 10^{16}$ tons.

While carbonates were formed by precipitation from aqueous solutions due to equilibrium of the heterogeneous $CO_{2(gas)} - HCO_{3(aq)}^- - CO_{3(solid)}^{2-}$ system, the bulk of terrestrial organic carbon was primarily generated by photosynthesis (see Boxes 8 and 9).

Both carbon species, and specifically organic carbon (Figure 30), can be traced back in sedimentary rocks to the very beginning of the record 3.8 billion years ago.

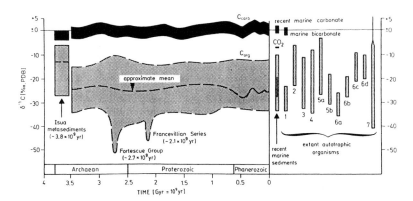

Figure 31. Isotope age functions of organic carbon (C_{org}) over 3.8 billion years of recorded Earth's history as compared with the isotopic composition of their progenitor substances in the present environment (marine bicarbonate and biogenic matter of various parentage; cf. Right panel). Isotopic compositions are given in δ ^{13}C values, indicating either a relative increase (+) or decrease (−) in the $^{13}C/^{12}C$ ratio of the respective substances (in per mil difference) as compared to that of the Pee Dee Belemnite (PDB) standard with $^{13}C/^{12}C$ = 88.99, which defines the zero per mil line on the δ-scale. Note that $\delta^{13}C_{org}$ spread of the extant biomass is basically transcribed into recent marine sediments and the subsequent record back to 3.8 billion years. The envelope shown for fossil organic carbon is an update of the database, comprising the means of some 150 Precambrian kerogen provinces as well as the currently available information on the Phanerozoic record. The negative spikes at 2.7 and 2.1 billion years indicate a large-scale involvement of methane in the formation of the respective kerogen precursors. Extant autotrophs contributing to the contemporary biomass are C3 plants (1), C4 plants (2), crassulacean acid metabolism (CAM) plants (3), eukaryotic algae (4), natural and cultured cyanobacteria (5a, 5b), groups of photosynthetic bacteria other than cyanobacteria (6), and methanogens (7). $\delta^{13}C_{org}$ range in recent marine sediments is based on some 1600 data points (black line) within bar covers > 90% of the database (from Engel and Macko, 1993).

Systematic assays for organic carbon carried out on Phanerozoic rocks as well as observed oscillations of the carbon isotope age function (Figure 31) indicate that organic carbon deposition rates have just moderately oscillated around a mean of perhaps 0.5% in the average sediment over this time span.

Hence, the average content of organic carbon in the oldest sediments does not appear to be basically different from that of geologically younger formations.

Additional proof for the biogenicity of the reduced carbon fractions of common sedimentary rocks, including their oldest occurrences, comes from $^{13}C/^{12}C$ fractionations between the two sedimentary carbon species and has been discussed (Engel and Macko, 1993). It is known that the transformation of inorganic carbon into biological substances entails a marked bias in favor of the light isotope (^{12}C), with the heavy isotope (^{13}C) preferentially retained in the inorganic feeder pool and finally precipitated in the form of carbonate. In terms of conventional δ-notation, the $\delta^{13}C$ values of average biomass are usually between -20–$30‰$ and they are more negative than those of marine carbonate, which is the predominant oxidized carbon form in the Earth (see right panel of Figure 31). The original isotope fractionations between reduced and oxidized carbon as established in the surficial environment was largely "frozen" after the entry of the respective carrier phases into the sedimentary realm. As can be inferred from Figure 31, both the carbon isotope spreads of instant primary producers and those of recent marine bicarbonate and carbonate are virtually superimposable, with minor oscillations, on the record back to 3.5, if not 3.8, billion years. This constitutes evidence that organic carbon and carbonate carbon have been transferred from the surficial exchange reservoir to sedimentary rocks with little change in their isotopic composition.

The narrow spread of the C_{carb} age function suggests that the $\delta^{13}C$ values of the bicarbonate precursor of ancient marine carbonates were closely restricted to the zero per mil line over 3.8 billion years of Earth's history, allowing little departure in either direction. As for the corresponding C_{org} function, it is obvious that the $\delta^{13}C$ range observed in recent marine sediments (see right panel of Figure 31) spans the respective spreads of the $\delta^{13}C$ values of the principal contributors to the contemporary biomass with just the extremes eliminated. This also shows that the influence of a later diagenetic overprint on the primary values is rather limited in extent and, for the most part, gets lost within the broad distribution of the primary values that has consequently come to be projected into the geological past over billions of years. Over the 3.5 billion years of the unmetamorphosed rock record, isotopic fractionations between organic carbon and carbonate carbon thus appear to be demonstrated as being the same as in the present world, with the characteristic differences in $\delta^{13}C$ values of -20–$30‰$ interval and the corresponding enrichment in light isotope in the biogenic phase. There is little doubt that the conspicuous ^{12}C enrichment displayed by the data envelope for fossil organic carbon (see Figure 31) constitutes a coherent signal of autotrophic carbon fixation over 3.8 billion years of Earth's geological history as it ultimately originated with the process that gave rise to the biological precursor materials.

Peculiarities of Sulfur Biogeochemical Evolution

The emergence of dissimilatory sulfate reduction certainly was an important step in early evolution of the biogeochemical cycle of sulfur. As H_2S-based bacterial photosynthesis (see Box 8) surely preceded the H_2O-splitting variant of the photosynthetic process in the evolution of photoautotrophy, sulfate ion must have appeared well before free oxygen. The availability of an oxidized sulfur species was, in turn, a prerequisite for the rise of dissimilatory sulfate reduction, an energy-yielding adaptive reversal of the primary SO_4^{2-} generating photosynthetic process that coupled the reduction of sulfate to hydrogen sulfide with the oxidation of organic substances.

These simplified cyclic transformations driven by solar energy are the following:

$$H_2S + 2HCO_3^- \underset{\text{dissimilation}}{\overset{\text{photosynthesis}}{\rightleftarrows}} 2CH_2O + SO_4^{2-} \tag{6}$$

Since sulfate instead of oxygen serves as an oxidant for organic compounds like CH_2O in the above back reaction, the dissimilatory process may be regarded as a form of anaerobic respiration or sulfate respiration. In the early Earth's history this reaction would have played a crucial role as it would have been responsible for large-scale transformations of sulfide to sulfate in biological mediation of the sulfur cycle.

Bacterial sulfate reduction produces biogeochemically significant amounts of hydrogen sulfide, which $\delta^{34}S$ values are shifted by about 40‰ in the negative range as the substrate marine sulfate. Part of these isotopically light (^{32}S-enriched) sulfides was precipitated as metal sulfides. At present geochemists use these data for estimating the period when the biological sulfate reduction processes would have started.

Utilizing the presently available isotopic evidence (see Figure 32, and corresponding explanation below) for pinning down the emergence of this process, it would appear that a bacteriogenic sulfur isotopic pattern in sedimentary sulfides arose about 2.8 billion years ago.

This is rather astonishing since photosynthesis started at least one billion years earlier (see Subsection 5.2). The only explanation would be the suggestion that the early trends in geochemical isotopic signatures were either suppressed or camouflaged by different geological processes, which are still unknown. Any large scale isotopic reequilibration between marine sulfur and the primordial sulfur reservoirs of the early oceanic trust have possibly restricted the $\delta^{34}S$ values of both sulfide and sulfate to mean not far from zero per mil. This is depicted in the Early Archean record (see 1–5 in Figure 32). Follow the existing data, we can logically suspect that the oldest presumably bacteriogenic sulfur isotope pattern appearing in the record $\sim$ 2.8 billion years ago just give a minimum age for the operation of the dissimilatory sulfate reduction as a biogeochemical and bioenergetic process.

6. ROLE OF BIOGEOCHEMICAL CYCLES IN BIOGENIC DEPOSITION FORMATION

The transformation of geological and geochemical cycles of different elements into biogeochemical cycles had been connected with the formation of various biogenic depositions, such as oil, gas and coal. It is generally believed that kerogen was the major starting material for most oil and gas generations as sediments were subjected to geochemical heating in the subsurface. Kerogen is the complex, high molecular weight, disseminated organic matter (OM) in sediments. Even now kerogen is the most abundant form of organic carbon on Earth. The calculations show that kerogen is about 1000 times more abundant than coal, which formed primarily from terrigenous higher plant residues. Kerogen consists of the altered remains of marine and lacustrine microorganisms, plants, and animals, with various amounts of terrigenous debris in sediments. The structural terrestrial (e.g., woody) portions of kerogen have elemental composition similar to that of coal. So, the main attention will be given here to the oil and gas formation in biogeochemical reaction pathways.

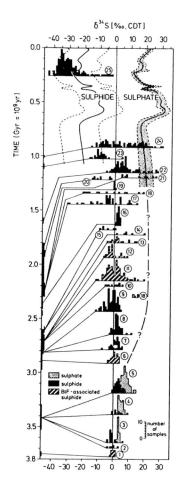

Figure 32. Isotopic evolution of sedimentary sulfide and sulfate over 3.8 billion years of geological history. The microbially generated isotopic differentiation patterns between the two sulfur species are progressively blurred from about t > 1.3 billion years, mainly as a result of the low preservation potential of marine sulfate evaporates that tend to fade from the record with increasing geologic age. Therefore, the oldest isotopic evidence of bacterial sulfate reduction primarily rests on the typical δ^{34}S distribution patterns of bacteriogenic sulfides that are characterized by extended spreads coupled with a "Rayleigh tail" at a positive end, reflecting a reservoir effect (such as exemplified by the Permian Kupferschiefer of Central Europe shown by no. 25). Applying these criteria to the early record, the presumably oldest bacteriogenic sulfide patterns appear around 2.7–2.8 billion years ago [Deer Lake Greenstone Belt, Minnesota (no.8); Michipicoten, Woman River, and Lumby-Finlayson Lakes banded iron formations, Canada (nos. 11–13)]. Note that the δ^{34}S values of the sulfide constituents of the Isua banded iron formations (no. 1) cluster closely around the zero per mil defined by the isotope ratio of the Canyon Diablo triolite (CDT) standard with ^{34}S/^{32}S = 22.22. BIF, banded iron formation (from Engel and Macko, 1993).

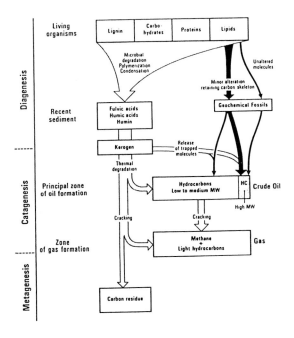

Figure 33. Stages of kerogen transformation and hydrocarbon formation pathways in geological situations. Direct inheritance of low molecule weight compounds is indicated by block arrows (from Engel and Macko, 1993).

6.1. Formation of Biogenic Depositions from Kerogen

During Earth's geological history, kerogen was often depicted as a complex high-molecular weight material formed from random condensation of monomers. These monomers are thought to have been generated by the initial breakdown of polymeric biological precursor molecules during the sediment burial. Other biopolymers, which have been altered in various biogeochemical processes, may participate in kerogen formation as well. These biopolymers have undergone various degrees of alteration prior to and after burial. Since kerogen was formed from the altered residues of different marine and terrestrial organisms, it contains information about geological, geochemical, and biogeochemical history of sediments and their transformation to biogenic deposition.

The evolution of sedimentary organic matter (OM) under the influence of increasing burial depth and related temperature rise is termed maturation and this process is commonly subdivided into three stages, i.e., diagenesis, catagenesis, and metagenesis (Figure 33).

We can determine these stages as follows. Diagenesis is microbial and chemical transformations of sedimentary organic matter at low temperature and it has been started in recently deposited sediments. Catagenesis has involved the formation of

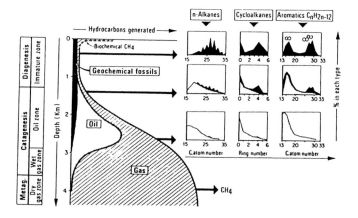

Figure 34. General scheme of hydrocarbon formation as a function of burial of the source rock. The evolution of the hydrocarbon composition for three compound classes is shown schematically in the insets. Depths are only indicative and may vary according to the actual geological situation (from Engel and Macko, 1993).

oil and gas from thermal breakdown of kerogen under increasing temperature during burial of sediments. The subsequent and last stage of the evolution of OM in sedimentary basins at high temperatures termed metagenesis.

Transformation of the initially deposited remains of living organisms started in the water columns and in the upper layers of bottom sediments of ancient seas. Random polymerization and condensation reactions of degraded biopolymers are believed to form the initial geopolymers, which contain humin, fulvic and humic acids. Until now these compounds (very typical for soil humus as well) are not well determined on the molecular level, but they are differentiated on the basis of their solubility in acids and bases.

During oil and gas formation, thermal breakdown of kerogen has been occurring in the catagenic stage (Figure 33). Under the influence of elevated temperatures, usually > 50 °C, and geological time, liquid and gaseous hydrocarbons have been released from kerogen mass. Chemically, it is connected with cleavage of C − C bonds. Other organic compounds containing heteroatoms such as O, S, and N have originated as well. Oil not expelled from source rocks will be cracked to gas if subsidence continue and higher temperatures are reached.

During the ultimate stage, metagenesis, only dry gas methane was stable. It originated from previously formed liquid hydrocarbons (Figure 33) mainly due to high temperature decomposition of kerogen.

Petroleum geologists commonly call the catagenesis range, in which oil is effectively produced from kerogen, 'oil windows'. One can see this in Figure 34 in a bell-form generation curve.

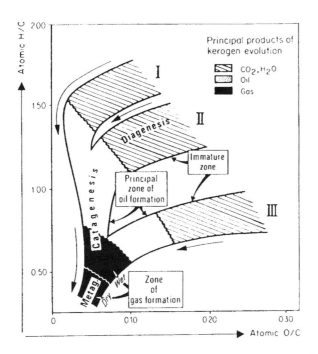

Figure 35. General scheme of kerogen evolution in a van Krevelen diagram (from Engel and Macko, 1993).

Formation of oil from kerogen is a disproportionation reaction related to a hydrogen-rich mobile phase and a hydrogen depleted carbon residue. After crystalline reordering the last H-depleted phase may ultimately form graphite, at the late metagenesis stage. This elemental evolution of kerogen is commonly illustrated in a diagram of H/C versus O/C atomic ratio or a van Krevelen diagram (Figure 35).

In this figure different kerogen types are distinguished due to the different hydrogen and oxygen contents in the precursor materials (types I-II-III). The main evolution pathways, shown in Figure 33, involve an initial decrease in O/C atomic ratio due to the loss of small oxygen-bearing molecules (CO_2, H_2O) during diagenesis. Furthermore, in the stage of early catagenesis the release of bitumen species enriched in heteroatomic compounds have occurred. The significant decrease of the H/C atomic ratio has lasted during catagenesis and metagenesis, until the kerogen was be transformed into an inert carbon residue.

6.2. Geological and Biological Factors of Oil Composition Formation

The depositional environment of oil source rocks, its thermal evolution, and secondary alteration processes are the most important factors determining the composition of crude oil.

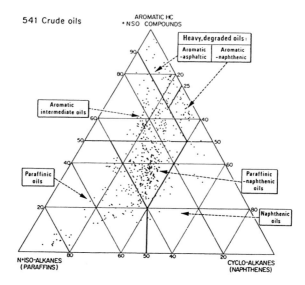

Figure 36. Ternary diagram showing the composition of six classes of crude oils based on the analysis of 541 oils (from Engel and Macko, 1993).

Their bulk properties as well as their chemical composition can characterize crude oils. Distillation of crude oil provides fraction profiles over a certain boiling range. Crude oil as well as its distillation fractions can be described in terms of density, viscosity, refractive index, sulfur content, and other bulk parameters.

Chemically, crude oil can be divided into fractions of different polarities using column or thin-layer liquid chromatography.

The ternary diagram in Figure 36 shows the composition of crude oil samples based on the content of normal plus isoalkenes (paraffiins), cycloalkenes (naphtenes), and aromatic hydrocarbons plus polar, heteroatomic compounds (NSO).

This compound class subdivision provides an optimal spread of the data over the diagram, thus allowing a classification of oil into six groups.

Any composition of crude oil depends on the combination of various factors. Among the environmental factors, those that influence the nature of the organic matter in the source rock and its mineral composition are of primary significance.

Although all hydrocarbon source rocks are deposited under aquatic conditions, they may contain various amounts of land derived organic matter. The biogeochemical terrestrial cycles can determine the contribution of organic matter, particularly in inter-continental basins and in the deltas of large rivers, which may extend far into the open sea. Terrestrial organic matter is usually enriched by such fractions as cellulose and lignin, which are not considered as oil precursors due to their oxidation state. On the contrary, the subordinate lipid fraction together with the biomass of microorganisms incorporated into source rocks yielded crude oils. These oils are rich in aliphatic units originating from waxes, fats, aliphatic biopolymers, etc., which present strait chain and branched alkanes (paraffines). Polycyclic naphthenes, particularly steranes, are present in trace quantities.

Marine organic matter is usually considered as type II kerogen (see Figure 33). This organic matter produces oils of paraffinic-naphthenic or aromatic-intermediate type (Figure 36). The amount of saturated hydrocarbons is moderate, but isoprenoid alkanes and polycyclic alkanes (like steranos from algal steroids), are relatively more abundant than in oils from terrigenous organic matter. Marine kerogen, particularly when it is very rich in sulfur, is particularly suited to release resin and asphaltenes-rich heavy crude oils at a very early stage of catagenesis.

The sulfur content of crude oil shows the close connection to the mineralogical type composition of the source rocks. Typically oil is enriched by sulfur when source rocks include sediments consisting of calcareous (e.g., from dinoflagellates or foraminifera) or siliceous shell fragments (e.g., from diatoms or radiolaria) of decayed planktonic organisms and at the same time containing abundant organic matter. This is related to anoxic conditions, which are required to preserve the organic matter. Under these conditions sulfate-reducing bacteria formed hydrogen sulfide or other reactive inorganic sulfur species. These compounds reacted with organic matter and sulfur has been incorporated into the kerogen and furthermore into oil. Examples of such oils come from onshore and offshore of southern California, northern Caspian onshore and offshore deposits and many carbonate source rocks of the Middle Eastern crude oils.

However, this sulfur would have been removed from oil when clastic rocks contained high amounts of detrital clay minerals abundant in iron. Under these conditions the most H_2S generated by the sulfate-reducing bacteria would have reacted with iron resulting in iron sulfide depositions. Because terrigenous organic matter is commonly deposited together with detrital mineral matter (e.g., in deltas of large rivers), waxy crude oils derived from type III kerogen usually are depleted in sulfur.

A variety of physical, chemical and biogeochemical processes has worked in concert to alter the organic and inorganic composition of particulate and dissolved organic matter as it is transported through the water column (Figure 37).

Several factors may be named as of special importance: (1) the amount and composition of organic matter transformed from land surface or biosynthesized in surface waters; (2) the transport mechanism involved; (3) the biological community structure mediating the transformation mechanisms; and (4) physical characteristics of water, such as the water column depth, redox state, temperature, etc.

Finally, Figure 38 gives an idea on the ages of the oil sources rocks.

An example is presented from the North Slope of Alaska, where the largest producing oil field was discovered in 1968. The petroliferous sedimentary rocks of the North Slope consist of Mississippian through Tertiary sandstone, conglomerates, shales, and carbonates. Two major rock sequences comprise the sedimentary succession. The older Ellesmerian sequence, deposited through the Early Cretaceous, is derived from a postulated northern landmass, whereas the younger Brookian sequence consisted of sedimentary shed from the southerly Brooks Range since the Early Cretaceous. So, the age of various oil deposits is from about 65 to 360 millions years.

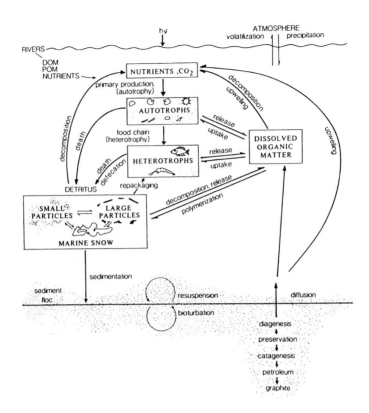

Figure 37. Schematic presentation of organic matter cycle in the ocean. DOM—dissolved organic matter; POM—particulate organic matter (from Engel and Macko, 1993).

FURTHER READING

Samuel S. Butcher, Robert J. Charlson, Gordon H. Orians and Gordon V. Wolfe, Eds. (1992). Global Biogeochemical Cycles. Academic Press, London et al., 9–54.

Egon T. Degens (1989). Perspectives on Biogeochemistry. Springer-Verlag, 20–129, 288–392.

Tom Fenchel, Gary M. King and T. Henty Blackburn (1998). Bacterial Biogeochemistry. Academic Press, London et al., 188–275.

Michael H. Engel and Stephen A. Macko, Eds. (1993). Organic Geochemistry. Principles and Applications. Plenum Press, New York and London, 3–22, 377–396, 611–624, 639–656.

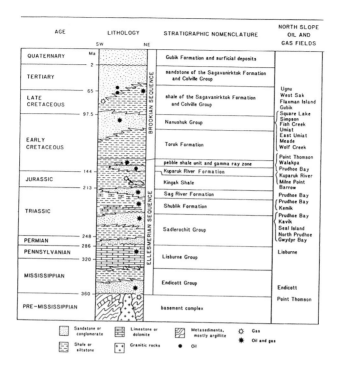

Figure 38. Stratigraphic section above economic basement for the central portion of the North Slope of Alaska. The Brookian and Ellesmerian sequences are separated at the major Lower Cretaceous unconformity. Both commercial and noncommercial oil gas fields are shown at right (from Engel and Macko, 1993).

QUESTIONS AND PROBLEMS

1. Briefly describe the general features of element origin in the solar system. Highlight the processes that produce the origin and distribution of the elements. Give examples, for instance, the decay series of uranium-238.

2. Discuss the earliest development of the Earth crust. What types of geological materials were present at the Earth surface in the Hadean?

3. Present the Earth's layered structure, name the main constituents, give their physical and chemical characteristics.

4. Present the chronology of the lithosphere evolution and highlight the main geological and geochemical changes in the Earth's crust.

5. Discuss the general items and steps in the atmosphere evolution with main attention to the content of CO_2 in the prebiotic atmosphere and its alteration.

6. Give the chemical and geochemical presentation of weathering processes in abiotic Earth and discuss the role of carbon dioxide in this process.

7. The flux of solar ultraviolet (UV) radiation was one of the important environmental parameters on the early Earth. Present your understanding of this phenomenon and describe the main chemical processes in the prebiotic atmosphere.

8. Discuss the greenhouse effect in the ancient atmosphere. What would be the similarities and differences with the modern characteristics of this process?

9. Describe the schematic diagram showing sulfur photochemistry in an anoxic, primitive atmosphere.

10. About 3.5 billion years ago the anaerobic photosynthesis started to develop. This and further development of oxygenic photosynthesis began profound changes in the Earth's biosphere. Discuss the main results related to these natural phenomena in the formation of the modern atmosphere.

11. Outline the main features of hydrosphere evolution. Why was water the main liquid on the Earth?

12. Discuss the Soda conception of ocean evolution. Present a chemical description of the aquatic chemistry during the period when soda was the main component of ocean water.

13. Discuss mineral cycling on prebiotic Earth. Highlight the driving forces of geochemical transformations in that time.

14. Describe the main preconditions of life origin in the Earth. What notions are included in the term "primordial soup"?

15. Oparin's theory has been suggested for the explanation of life origin. Highlight the strengths and weaknesses of his theory.

16. Clay and pyrite theories of life origin have some similarities. Describe these theories from biochemical and phylogenic points of view.

17. Highlight the main biochemical reactions and reaction pathways in the theory of "thioester world". Why could this theory be considered as the most successful? Give your understanding of applicability of various theories for life's origin process.

18. Since the Earth was originally anaerobic, anaerobic microorganisms evolved in the Earth evolution and life origin. Outline the evidence for and implications of an anaerobic origin of life.

19. Describe the evolution of biogeochemical cycles. What was the trigger in the drastic changes of element cycling during early Earth history?

20. Discuss the applicability of isotopic researches for the understanding of biogeochemical evolution. Highlight the record of various isotopes in geological history of the Earth.

21. What are the similarities in the evolution of carbon and sulfur biogeochemical cycling? Give examples of biological processes that possess these cycles.

22. Discuss the main peculiarities of carbon biogeochemical evolution. Present the isotopic record of this evolution in various fossil rocks.

23. Discuss the main peculiarities of sulfur biogeochemical evolution. Present the isotopic record of the sulfur reduction process and its role in evolution of the sulfur cycle.

24. Describe theoretical possibilities of the nitrogen cycle in the anoxic atmosphere. How has this cycle been changed with the appearance of oxygen in the Earth's atmosphere?

25. Kerogen is considered as a main precursor of oil and gas formation. Describe the general stages of transformation and highlight the role of temperature in this process.

26. What chemical alterations have occurred during kerogen transformation? Give an example of a van Krevelen diagram.

27. Discuss geological and biological factors of oil composition formation. Describe the ternary diagram showing the composition of six classes of crude oils and give the main types of chemical compounds.

28. Highlight the origin and sulfur content in crude oil. Why did different oil fields show various sulfur contents?

29. Present your opinion on the age of main oil and gas deposits and discuss the biogeochemical evidence.

CHAPTER 3

BIOGEOCHEMICAL CYCLING
OF MACROELEMENTS

In this chapter we will describe the biogeochemical cycling of macroelements (C, N, P, S, Si, and Ca) in the biosphere. The reader will find the definitions and general regularities of migration, transformation and accumulation of these main elements in natural ecosystems. We consider biogeochemical fluxes and barriers for various elements in terrestrial and aquatic compartments of the global biosphere with relevant estimations of the global biogeochemical cycles.

1. INTRODUCTION TO BIOGEOCHEMICAL CYCLING OF ELEMENTS

As we have seen in Chapter 2, the evolutionary development of the Earth during its geological history was accompanied by the biogeochemical development of the biosphere. The major process at the initial development stages was the planetary differentiation of chemical elements between solid, liquid and gaseous forms with consequent development of biogeochemical cycles of elements in the biosphere.

The major chemical elements of the biosphere as a whole and any living systems are carbon, hydrogen, silicon, oxygen, nitrogen, calcium, phosphorus, iron, and sulfur. They exist in dynamic fluxes between their living and dead organic forms as well as these elements having possibly temporarily coexisted in one or more biological or non-biological reservoirs. The fluxes of each element between pools proceed cyclically by spontaneous chemical and biochemical reactions as well as by biological and geological intervention. These conversions are known as biogeochemical cycles (see also the definition of biogeochemical cycles in Chapter 2). Traditionally, spatial and temporal separations of the cycles have been identified in the gaseous phase (atmosphere), solid phase (lithosphere and pedosphere), and liquid phase (surface and marine waters). Some identification includes *endogenic* (lithospheric or rock) cycles and the *exogenic* cycles as the cycles in the sphere of living organisms.

The major biogeochemical cycles are intimately interconnected and are ultimately powered by solar energy through carbon fixation in photosynthetic processes. Localized exceptions occur in some media, such as hydrothermal environments where the energy for various biochemical processes, like carbon fixation or iron reduction, may be provided by various inorganic reducing agents emanating from rock materials or water. The transformations within each biogeochemical cycle are mainly oxidation

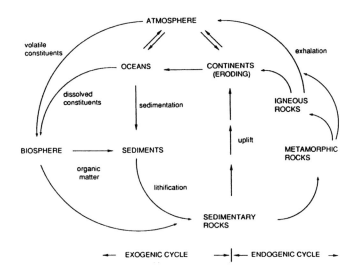

Figure 1. The exogenic and endogenic cycles showing interchanges of matter between the biosphere, oceans, atmosphere and geological rocks. These features are usual in the biogeochemical cycles of different elements (from Engel and Macko, 1993).

and reduction reactions, which give a basis for connection between cycles. The following example shows the exogenic carbon cycle (Figure 1).

In carbon cycles the inorganic pools consist of carbon in its oxidized states of CO_2 gas, in the atmosphere and dissolved in water, and HCO_3^- and CO_3^{2-}, mainly dissolved in surface and marine waters, and as carbonate minerals in geological rocks, soils and sediments. The organic pools comprise carbon in reduced states mainly confined in living organism biomass, in ancient, deeply buried sedimentary organic matter (kerogen, coal, oil, natural gas), and in recently dead organic matter dissolved or suspended in surface waters and oceans and in surficial sediments (Engel and Macko, 1993).

The corresponding example of reservoirs is depicted in Table 1. This Table shows current estimates of the most important carbon reservoirs and averaged residence times for carbon in these pools. The lithosphere is the reservoir for the bulk of global carbon, with about 20% of this being in the form of fossil organic fuels.

Some aspects of the carbon cycle are detailed in Figure 2 and more details are shown in Section 2. One can see that the reduction of 1 mole of CO_2 in oxygenic photosynthetic formation of carbohydrates, and the concomitant splitting of 1 mole of H_2O, releases 1 mole of O_2. This oxygen, in turn, will take part in oxidation processes like respiration. It is known that the content of atmospheric oxygen is about 21%, and to maintain this value the rapid cycling of carbon in the atmosphere, hydrosphere, and biosphere is highly necessary.

Table 1. Carbon reservoirs: residence times and isotopic composition (Modified from Engel and Macko, 1993).

Species	C mass, 10^{12} tons	Residence time (yr.)	$\delta^{13}C$ (‰PDB)
Atmospheric CO_2	0.72	4	~ −7.5
Sedimentary carbonate C	62,400	342,000,000	~ 0
Sedimentary organic C	15,600	342,000,000	~ −24
Oceanic inorganic C	42	385	~ +0.46
Necrotic C	4.0	20–40	−27
Living terrestrial biomass	0.56	16	~ −27
Living marine biomass	0.007	0.1	~ −22

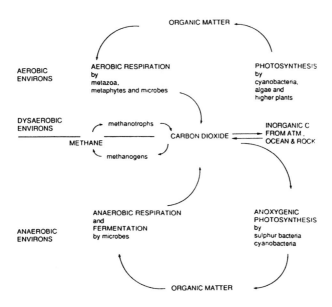

Figure 2. Simplified illustration of the interactions between major players of the carbon cycle (Engel and Macko, 1993).

As has been shown in Chapter 2, oxygen-dependent respiration processes are the most efficient and quantitatively significant in the case of organic matter mineralization. Accordingly, the preservation and accumulation of organic matter in sediments and soils depend strongly on the site of its production. Quantitative data about ecosystem productivity and reservoirs of living biomass are important and these parameters are of great interest for many biogeochemical researchers. It is generally

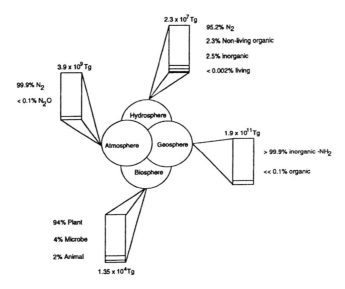

Figure 3. Conceptual illustration of overlapping elemental reservoirs, and distribution of nitrogen within and among reservoirs; data for nitrogen are given in accordance with Fenchel et al, 1998.

agreed that continental ecosystems produce and conserve more organic matter than do the marine ecosystems, despite the area of the oceans that doubles that of the land (Table 1). Correspondingly, most biogeochemical cycles are faster in terrestrial ecosystems than in marine ones.

Thus to give an idea of the biogeochemical cycling of elements in various media, the examples will be given below for different biosphere compartments like air, terrestrial and aquatic ecosystems.

The atmosphere, hydrosphere and lithosphere are the three major reservoirs for all the Earth's elements. The biosphere may be considered as the sphere of their interactions (Figure 3).

We are familiar with the definitions of these reservoirs, but for the aims of this chapter the biosphere includes living organisms and their environments. In addition, the hydrosphere is defined as inclusive of all surface and marine waters and terrestrial ground waters; physically isolated soil water is considered as a component of soil/lithosphere. Analogously, soil gas phases are physically isolated from the atmosphere and also will be treated as a component of the solid sphere.

1.1. Biogeochemical Cycling of Macroelements in the Atmosphere

The biogeochemical history of the atmosphere composition buildup provides a good example of the impact of living organisms on the environment and some aspects of atmosphere evolution are shown in Chapter 2. Recent monitoring indicates that the actual composition of the Earth's gas shell has been the ultimate stage of a long lasting process, in which a major role was assigned to the biogeochemical activity of living matter.

The mass of the atmosphere is about $5.14–5.27 \times 10^{15}$ tons (Walker, 1977; Voitkevich, 1986). The major part of gaseous matter (about 80%) occurs in the troposphere whose upper equatorial boundary reaches as high as about 17 km going down to 8–10 km at the poles. The troposphere is restricted by tropopause, which is characterized by a sharp temperature drop and the absence of water vapors. The active physical, chemical, and biogeochemical interactions of the atmosphere with ocean and land occur mainly in the troposphere. It contains also the bulk of water vapors and air-borne particulate matter. It is also well known that many important biospheric photochemical reactions take place in the troposphere.

Above the troposphere, in the stratosphere and mesosphere, the gas density becomes increasingly more important, and the thermal conditions become subject to complex variations, both temporal and spatial. At a height of 25–35 km, the oxygen molecules are split by the influence of solar UV radiation to form ozone in accordance with Chapman's mechanism, which is responsible for the absorption of 97% of harmful ultraviolet solar radiation. Devoid of this shield, life on the terrestrial surface would have been doomed to extinction.

Up to 80–100 km from the Earth's surface is the ionosphere, a region of highly rarefied and ionized gas molecules. The exosphere, the outermost belt of the gas shield, might extend as far as 1800 km. The losses of the lightest gases, hydrogen and helium, into space from the Earth's atmosphere occur in the exosphere.

Table 2 shows the globally and seasonally averaged composition of the atmosphere.

The Earth's atmosphere composition is to a significant degree determined by the activity of living organisms and is maintained owing to a system of biogeochemical cycles. At present, 99.8% of the gaseous matter of atmosphere is nitrogen, oxygen and argon. The content of water is greatly variable and no average values can be presented. While the concentration of nitrogen and oxygen are nearly invariant, the concentration of other constituents could vary both spatially and temporally.

Atmospheric components which are present in small amounts are water vapor, inert gases and species produced by photochemical reactions and biological processes. The geochemistry of inert gases is of particular interest for the historical re-constructions of the atmosphere. The relative high percentage of argon is associated with ^{40}Ar isotope, which is deriving from decay of the widespread ^{40}K. In contrast, the amount of atmospheric helium is smaller by a factor 1000, as might be expected. This is due to the continuous extra-atmospheric dissipation of this element in the exosphere. The rest of inert gases are present in the amounts in which they were evolved during the entire period of the Earth's existence. An analysis of the isotopic xenon production supports the idea that the gas shell emerged during a very short period of time, which was roughly coincident with the period of Earth's accretion (Dobrovosky, 1994).

The concept of mean residence time (MRT) is useful in the consideration of average atmospheric composition shown in Table 2. For any biogeochemical reservoir, its MRT

Table 2. Chemical composition of the atmosphere.

Constituents/species	Chemical formula	Content (% by volume)	Total mass, 10^9 tons
Total atmosphere			5.27×10^6
Water vapor	H_2O	Variable	0.017×10^6
Dry air		100	3.87×10^6
Molecule nitrogen	N_2	78.08	3.87×10^6
Molecule oxygen	O_2	20.95	1.18×10^6
Ozone	O_3	variable	~ 3.3
Argon	Ar	0.93	6.59×10^4
Carbon dioxide	CO_2	0.032	2.45×10^3
Neon	Ne	1.82×10^{-3}	6.48
Helium	He	5.24×10^{-4}	2.02
Krypton	Kr	1.14×10^{-4}	1.69
Xenon	Xe	1.87×10^{-6}	2.02
Methane	CH_4	1.5×10^{-4}	~ 4.3
Molecule hydrogen	H_2	$\sim 5.0 \times 10^{-5}$	~ 1.8
Nitrous oxide	N_2O	$\sim 3 \times 10^{-5}$	~ 2.3
Carbon monoxide	CO	$\sim 1.2 \times 10^{-5}$	~ 0.59
Ammonia	NH_3/NH_4^+	$\sim 1.0 \times 10^{-6}$	~ 0.03
Nitrogen dioxide	NO_2	1.0×10^{-7}	~ 0.0081
Sulfur dioxide	SO_2	$\sim 2 \times 10^{-8}$	0.0023
Hydrogen sulfide	H_2S	$\sim 2 \times 10^{-8}$	0.0012

value can be calculated as

$$MTR = Mass/flux, \tag{1}$$

where 'Mass' is related to the quantity of any species in the reservoir, and 'Flux' may be either the input or loss from the reservoir. One example of this calculation can be presented for N_2O (Schlesinger, 1997). The average nitrous oxide content in the atmosphere is about 300 ppb. Multiplied by the mass of the atmosphere, we obtain 2.3×10^9 tons for the content of this species in the entire atmosphere. Our best estimates of the sources of N_2O suggest an annual production of at least 20×10^6 ton/yr., giving the mean residence time of over 100 years for nitrous oxide in the atmosphere. During this period, N_2O will be relatively evenly distributed within the global atmosphere. The higher concentrations will be near the strong point sources

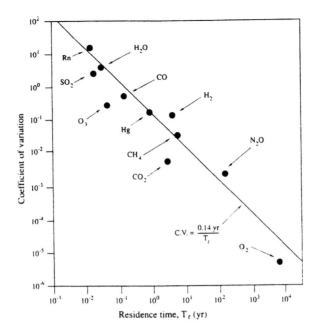

Figure 4. Variability in the concentration of atmospheric gases (expressed as the coefficients of variation in measurements) as a function of their estimated mean residence times in the atmosphere (Schlesinger, 1997).

of pollution. On the other hand, the average volume of water in the atmosphere is equivalent to $\sim$13,000 km^3 at any time, or 25 mm above any point on the Earth's surface. The globally and seasonally average daily precipitation would be about 2.7 mm. Thus, the MRT for H_2O in the atmosphere is

$$MTR = 25\,mm/2.7\,mm\,day^{-1} = 9.3\,days. \tag{2}$$

In comparison with circulation of global and regional atmosphere mass, this is a short time. It means that water vapor content is very variable in time and space.

The MTR values for other atmospheric gases are shown in Figure 4.

The biogeochemical processes that control the atmospheric levels of oxygen and carbon dioxide play the most important role in maintaining normal environmental conditions in the Earth. Free oxygen is a prerequisite for the existence of major forms of life, and carbon dioxide, being the basic building material for photosynthesis, is also the leading factor for greenhouse effect, which determines the thermal and climatic conditions on the surface of our planet.

The major atmospheric species, N_2, O_2, Ar, and CO_2 are relatively unreactive. They show nearly uniform concentrations and relatively long mean residence times. Argon is an inert gas. Biogeochemically, molecular nitrogen is almost inert; with the exception of symbiotic microbes, all living systems can only assimilate nitrogen from bound or fixed forms, like ammonium and nitrates. The atmosphere contains only a small portion of the total O_2 released by photosynthesis through geological time. This value does not depend on the present activity of the photosynthetic process. If we can imagine the instantaneous combustion of all the organic matter that is stored now on land, this would reduce the atmospheric oxygen content by only 0.035% (Schlesinger, 1997). It has been suggested that atmospheric O_2 controls the storage of reduced carbon, not vice versa. However, despite potential reactivity of molecule oxygen (see Chapter 2), these rates of reactions with reduced compounds are sufficiently slow, and O_2 is a relatively stable atmospheric species with mean residence time of about 10,000 years (Figure 4).

In the atmosphere CO_2 is affected by processes that operate at different time scales, including interaction with the silicate cycle (see Chapter 2), dissolution in the oceans, and annual cycles of photosynthesis and respiration (see also Section 3). The relative effect of these processes is described below in the consideration of the whole carbon biogeochemical cycle and environmental aspects of biogeochemistry. Here, it is important to note that carbon dioxide is not reactive with other atmospheric species; its MRT is 3 years (Figure 4). This value is largely determined by exchange with seawater (see Section 2).

As regards other elements indicated in Table 2, their original natural sources are mainly related to volcanoes (H_2S, NH_3, H_2 and many others). The typical example of these gases' composition is given in Table 3.

At present in the troposphere the contents of many minor gases are monitoring in concentrations well in excess of what is predicted under equilibrium geochemistry (Table 4).

In most cases the monitored tropospheric concentrations are maintained by the action of living matter, mainly microbes. Natural biogeochemical cycles of nitrogen and sulfur are driven by biota; however, at present these cycles are complicated by the modern anthropogenic influence (see Chapter 8). Unlike major atmospheric constituents, these gases are highly reactive and correspondingly they show short residence times and variable contents, both spatially and temporally (Figure 4). Oxidation reactions and removal by rainfall drive the losses of these species from the troposphere. Currently the concentration of almost all these constituents is increasing as a result of anthropogenic activity and atmosphere pollution in the global scale.

1.2. Biogeochemical Cycling of Macroelements in Terrestrial Aquatic Ecosystems

Two thirds of the Earth's surface is covered with water. Most of this is ocean, as seen in the Table 5.

As well there is water underground, called 'ground water'. In this section, we will consider the biogeochemistry of macroelements in different natural waters with special emphasis on the general features of various biogeochemical cycles in terrestrial water bodies.

Table 3. Composition of gases of the Great Tol-bachinsky fissure eruption, Kamchatka, Russia (after Dobrovolsky, 1994).

Species	Content (% by volume)	
	Water vapor included	Dehydrated
H_2O	78.56	—
N_2	11.87	55.36
CO_2	4.87	22.71
H_2	3.01	14.04
HCl	0.57	2.66
CO	0.39	1.86
CH_4	0.44	2.05
H_2S	0.16	0.75
NH_3	0.11	0.51
HF	0.06	0.26
Ar	0.06	0.30
SO_2	0.03	0.14
O_2	0.01	0.05
He	0.001	0.005

The interactions between living matter and the hydrosphere is one of the global processes occurring in the biosphere. The living matter of the Earth is inalienably linked to liquid water. Water accounts for 60% of the total mass of terrestrial living organisms. All biochemical reactions and physiological processes proceed in aqueous media. Enormous amounts of water undergo decomposition during the activity of photosystem II of the photosynthetic process (see Chapter 2).

The chemical composition of atmospheric depositions reflects the types and rates of biogeochemical reactions in the troposphere. These chemical compositions change seriously after interactions with humid acids of the soil layer, higher plant metabolites, and soil microbes. Carbon dioxide, the end product of any organic matter decay, is readily soluble in water to yield carbonic acid, H_3CO_3, the dissociation of which produces a slightly acid reaction of a water solution. These factors play the most important role in the dissolving ability of surface waters with respect to the mineral matter of the Earth crust. Simultaneously, the runoff surface waters entrain the particles of soil mineral matter and carry them into streams. Consequently biogeochemical processes affect the composition of both soluble species and suspended matter.

Table 4. Actual and equilibrium partial pressures and contents of N species on the Earth's surface (after Schlesinger, 1997).

Species	Medium	Equilibrium concentrations	Actual contents
NO_3^-	Marine water	$A_{NO_3^-}=10^{+5.7}$	$0\text{--}28\times10^{-6}\,mol/L$
NO_2	Atmosphere	$10^{-3.8}$ ppm	0.001 ppm
NH_3	Atmosphere	$10^{-51.5}$ ppm	0.006–0.020 ppm
N_2O	Atmosphere	$10^{-12.7}$ ppm	0.33 ppm
NO	Atmosphere	$10^{-9.6}$ ppm	0.001 ppm

Table 5. Distribution of the Earth's surface water.

Water type	Mole
Oceans	9.5×10^{19} (> 99%)
Lakes and rivers	1.7×10^{15}
Atmosphere	7.2×10^{14}

To the same extent this is related to the solubility of gases in the natural waters. Many gaseous chemical species are included in biogeochemical cycles in the system atmosphere-natural waters. When any gas equilibrates with a solvent, the amount of gas which dissolves is proportional to the partial pressure of the gas (see Box 1). This statement, which is known as Henry's law, can be written mathematically as follows:

$$K_H = [X, solv]/p(X, g), \tag{3}$$

where X, solv is the mass of chemical species X dissolved in water and X, g is the partial pressure of the gas in the atmosphere in equilibrium with water.

The proportionality constant is an equilibrium constant, or Henry's law constant, usually expressed in mol/L. Table 6 gives the K_H values for some biogeochemically active gases at 25°C.

Box 1. Thermodynamic principles of gas solubility in water (After Bunce, 1994).

Gases become less soluble with increasing temperature. From this variation of K_H with temperature, one can deduce the underlying thermodynamic principles governing the dissolution of gases, namely, that change from the gaseous state to solution is a process for which ΔH^o and ΔS^o are both negative. Enthalpic stabilization accompanies dissolution, but the dissolved state is more ordered than the gas. A rise in temperature thus favors the gaseous state ($-T\Delta S^o$ for dissolution becomes more positive).

Table 6. Examples of the Henry's law constants at 25°C and 1 atm pressure.

Gas	K_H, mol/L/atm
H_2	7.8×10^{-4}
N_2	6.5×10^{-4}
CO_2	3.4×10^{-2}
CO	9×10^{-4}
O_2	41.3×10^{-3}
O_3	1.3×10^{-2}
NO_2	6.4×10^{-3}
HO_2	2.0×10^3
SO_2	1.2
CH_2O	6.3×10^3
NO	1.9×10^{-3}
NO_3	15.0
H_2O_2	7.4×10^4

The low solubility of non-polar gases, such as methane, in water can be also understood using thermodynamic principles. The process $CH_4(g) \rightarrow CH_4(aq)$ is exothermic (ΔH° negative) and the low solubility is due to the enthropic factor. The very negative ΔH° for this process is caused by the intrinsically greater order of a condensed phase compared with the gas phase, and the property of water of ordering itself around the non-polar solute molecule. The latter phenomenon has been likened to forming a miniature iceberg around the solute, thereby greatly reducing the entropy of the water. Please remember that we must consider the whole system, not just the methane or other non-polar gases.

Some other thermodynamic discussion is also shown in Chapter 5, Section 3.

A Stream's waters can be considered as composite solutions containing soluble ions and particulate solid matter. The main forms of the occurrence of chemical elements in stream water are:

i. simple and complex ions;

ii. neutral molecules, mostly existing as a ligand with an inorganic complexing ion; colloidal particles from 0.001 to 0.1 μm, with ions adsorbed at their surface;

iii. finely dispersed, mainly clay mineral particles of 0.5–2 μm; and

iv. large suspended particles from clastic minerals of 2–3 to 10 μm.

In analytical practice the suspended particles are separated by means of membrane filters with various pore diameters. A centrifugation is also of use.

The weight ratio of soluble compounds and solid suspensions in the river stream depends to a significant extent on land use and type of vegetation. The recorded facts provide evidence that in the course of geological history the ratio was subject to multiple variations (Dobrovolsky, 1994). On this basis the French soil scientist Erhart (1956) developed a theory of biorhexistasie. The epochs of biostasie (biological equilibrium) were characterized by a wide spread of stable forest ecosystems, which protected the soil from mechanical erosion and facilitated the involvement of chemical elements in water migration in soluble forms. In the epochs of rhexistasie the biological equilibrium became increasingly impaired owing to the extensive deforestation. As a consequence the denudation and erosion process became more active, causing a rise in concentration of solid suspensions in the runoff.

In a river's waters the most common soluble species are HCO_3^-, SO_4^{2-} and Cl^- anions accounting, respectively, for 48.8%, 10.0% and 5.3% of the sum of the total of soluble compounds. Among the cations calcium (10.8%), magnesium (2.7%) and potassium (1.2%) are typical. The other elements are present in various trace amounts. Most of the water soluble and solid particulate matter elements are biogeochemically active (Table 7).

Perelman (1975) proposed the use of the coefficient of water migration (C_w), which is determined as the ratio of the concentration of the element in the dry residue of a water sample to the concentration of the element in the rock. In estimating the extent of involvement of an element in aqueous migration on a global scale we should know the ratio of the average content of the element in the dry residue of water stream to the abundance of this element in the continental granite layer.

The amount of solid dissolved in natural water varies widely. The values in Table 8 are typical of river and ocean water, although we shall see that river water is quite variable in its mineral content.

Ground water is at least as high in dissolved solids as lake and river water; sometimes it is much higher, with total dissolved solids exceeding 1000 ppm. With exceptions of Ca^{2+} and HCO_3^-, there is a parallel between the average concentrations of ions in fresh and in ocean water. There is relatively more Ca^{2+} and HCO_3^- in river water because some hydrochemical and biogeochemical principles: rivers dissolve ancient rocks containing $CaCO_3$, whereas the oceans precipitate $CaCO_3$ in the form of marine organisms' exoskeletons.

Microbes govern practically all biogeochemical cycles of macroelements in water. However, the quantitative role of water-column bacteria has not been fully appreciated until recently. The presence of suspended bacteria (together with solid suspended matter) has, of course, been long known. The 'microbial loop' describes the current view of the role of bacteria in planktonic food chain or, in other words, the role of bacteria in biogeochemical cycling of organic matter in the water—'microbial loop' (Figure 5).

Table 7. Soluble forms of macroelements in riverine waters and coefficients of their migration (after Dobrovolsky, 1994).

Species	Average concentration		Global stream	Coefficient of
	In water, mg/L	% of total sum of salts	loss, 10^3 ton/yr	water migration, C_w
Cl^-	6.4	5.33	262400	313.0
SO_4^{2-}	12.0	10.00	492000	—
S	3.96	3.30	162360	82.5
HCO_3^-	58.5	48.75	2398500	—
Ca^{2+}	13.0	10.80	533000	4.6
Mg^{2+}	3.3	2.75	135300	2.3
Na^+	4.5	3.75	184500	1.7
K^+	1.5	1.25	61500	—
NO_3^-	1.0	0.83	41000	0.5
SiO_2	13.1	10.9	537100	—
Si	5.7	4.78	233700	0.15

Note: "—" denotes that rock deposits are not important

Table 8. Typical concentrations of ions in rivers and sea water.

Ion	C(river), mol/L	C(ocean), mol/L
HCO_3^-	9.5	0.0023
Ca^{2+}	3.8	0.010
Cl^-	2.2	0.55
Na^+	2.7	0.46
Mg^{2+}	1.7	0.054
SO_4^{2-}	1.2	0.028
K^+	0.59	0.010

The essential fact is that a significant part of primary production in any body of water is not directly consumed by phagotrophs (dead organic matter destructors), but is lost in the form of detritus or dissolved organic matter, which is then included in various biogeochemical chains by bacteria. The resulting bacterial production is in turn consumed by phagotrophs and thus channelled into the classical plankton food chain. The relative importance of the microbial loop varies among aquatic systems

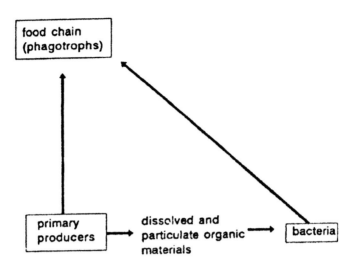

Figure 5. The role of bacteria in biogeochemical cycling of organic matter in the water—'microbial loop' (Fenchel et al, 1998).

and over time. We can draw the conclusion that in general its role is highest under oligotrophic (nutrients-poor) conditions and least in eutrophic (nutrients-rich) ones and during the initial stages of algal blooms, when large algae dominate primary production (Fenchel *et al*, 1998). This process is of great importance for the relationships between different forms of organic matter in the aquatic ecosystems.

The weight ratios between dissolved organics, particulate organic matter and living biota (bacteria, phototrophic plankton, protozoa and zooplankton) in the water column are approximately 100 : 10 : 2. Next to kerogen of sedimentary rocks (see Chapter 2), dissolved organic substances constitute the largest pool of the Earth's organic matter. Preliminary assessments give the value of about $n \times 10^{12}$ tons of dissolved organics for the oceans.

By definition dissolved organic matter (DOM) is a material that passes a 0.2 or 0.1 μm filter. Particles > 1 nm (molecular weight > 10000) are considered colloidal; most studies have found that colloids constitute 10–40% of the total pool of dissolved matter. The chemical analysis of DOM has shown the predominant role of humic acids, up to 40–80% in freshwater, but only 5–25% in seawater. Humic acids are mainly refractile remains of plant polymers. Terrestrial and limnic humic acids derive mainly from the lignin of vascular plant issue and they contain more aromatic groups than do humic acids extracted from seawater. Combined carbohydrates, which to a large degree are resistant to hydrolysis, constitute about 10% and combined amino acids (proteins, peptides) about 1% of total DOM amount. Bacteria can assimilate directly only low molecular weight compounds (monosaccharides and free amino acids).

Table 9. Global river water loss of macroelements (after Dobrovolsky, 1994).

Element	Concentration, mg/L	Annual river loss of suspended forms of elements, 10^6 tons	Total annual sum of dissolved and suspended forms, 10^6 tons	Suspended form, % of total
P	0.51	20.91	21.73	96.2
Ca	11.50	471.5	1004.5	46.9
Mg	5.75	235.75	371.05	63.5
K	6.9	282.9	344.4	82.1
Na	4.6	188.6	373.1	50.4
Si	117.0	4797.0	5030.7	95.4
Al	38.2	1566.2	1569.3	99.8
Fe	23.5	963.0	991.0	97.2

These compounds determine using high sensitive liquid chromatography. The concentrations of individual amino acids and monosaccharides are about 10–50 nm. The concentration ranges of various DOM constituents are presented in Figure 6.

Both inorganic and organic substances form the aquatic particulate matter in rivers. Completely different relationships are found for the chemical elements and their species that migrate as suspensions in stream water. Generally, the mass of suspended matter is 4 times as much as the mass of dissolved solids (Table 9). With an exception of Na and Ca, the transport of macroelements in suspended form is predominant in river water (> 50% from total transport).

1.3. Biogeochemical Cycling of Macroelements in Soils

Soil is a unique natural system. A major property of the soil is the close interrelation of its constitutive living and non-living components. An artificial separation of the soil components destroys it completely as a system and makes the existence of soil impossible. V. I. Vernadsky aptly named the soil a 'biolatent body'.

The soil composition is very complex and depends on many factors. Solid, liquid and gaseous phases are present in any soil. As a physical body, soil can be determined as a polydispersed system, with relatively large particles (over 0.001 mm) and finely dispersed particles (less than 1 μm). The soil consists of both organic and mineral compounds with distinguished mechanical and physicochemical properties. The mineral part, in turn, is very diversified and is made up of clay minerals of parent rocks and various soil-hypergene formations. A vast variety of living organisms is the major specific feature of any soil on the globe.

The soil has evolved as a system with interrelated biological activity of various groups of living organisms. We can mention such groups as

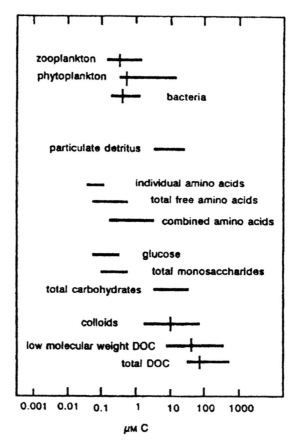

Figure 6. Concentration ranges for different constituents of dissolved organic matter expressed as micromoles of C per liter. Vertical bars represent typical values for offshore seawater (Fenchel et al, 1998).

i.   higher plants with photosynthetic production;

ii. edaphic mesofauna aimed to the destruction of plant decay;

iii. microbes with profound transformation of decay products up to CO_2, H_2O and simple organic compounds.

Various forms of dead organic matter, fragments of durable hypogenic minerals, and numerous products of biogeochemical and geochemical transformations and weathering represent the non-living part of soil.

The provision of photosynthetic organisms with elements of soil nutrition is related to the inner part of soil's biogeochemical structure and is connected to two major soil components. The first of these is the dead organic matter and the second is the soil's

Table 10. Partition of plant nutrients in forest ecosystems of boreal climate, per cent (after Schlesinger, 1997).

Processes	N	P	K	Ca	Mg
Annual total uptake	100	100	100	100	100
Including atmosphere	18	0	1	4	6
Rock weathering	0	13	11	34	37
Soil absorption	25	24	4	0	1
Mineralization*	57	63	84	62	56

*Including throughfall and stemflow

mineral mass. Decomposition of dead organic matter goes on inside the soil system and completes the intra-system biogeochemical cycle by releasing nutrients for plant's uptake. *Decomposition*, in general refers to the breakdown of organic matter as a whole into smaller physical and chemical components. *Mineralization* is a much more specific term that refers to processes which release the elements in simple inorganic form, like carbon dioxide, nitrate, ammonium, orthophosphate, etc.. A variety of soil animals (micro- and mezofauna), including earthworms, mix and fragment the fresh plant residues and corresponding organic matter. However, at present soil fungi and bacteria are confirmed to be mainly responsible for the transformation of organics in the soil's biogeochemical cycling. Soil microbes release extracellular enzymes, which catalyze the mineralization reactions and degrade organic matter. During the course of microbial decomposition, soil humus compounds are synthesized. The interaction of photosynthetic bacteria and heterotrophic organisms provides for a biogeochemical cycling migration of the elements in soil and vegetation.

The dispersed mineral particles are also the source of accessible forms of nutrients for higher plants. These particles have very large specific area per unit volume and contain appreciable amounts of sorbed chemical elements. The absorption protects these species from leaching with rainfall waters filtering through soil profile, but plant roots readily take them up. The finely dispersed soil mineral and organic matter plays a pivotal role in the mechanisms of the biogeochemical cycles, being the most important *biogeochemical barrier* in the turnover of many elements that enlarge the accumulation of nutrients and any other chemical species in the upper soil layers. These layers are the main pools of the macronutrients for plan productivity (Table 10).

The soil micromorphology is also an important parameter in biogeochemical turnover of any species. The aggregation of soil particles facilitates the conservation and promotes the regulation of nutrient supply. The soil pore system favors free gas exchange between soil and atmosphere.

The famous Russian scientist, the founder of genetic soil science, V. V. Dokuchaev, determined the soil as a 'natural historical body'. This reflects the appearance of soil as a product of interaction between rocks, climate, vegetation and biota. Numerous combinations of these factors determine the change of the soil. For this reason each natural landscape exhibits its own specific soil. However, any natural conditions, including soil types, are subject to alteration not only in space, but also in time. Paleogeographic and paleogeochemical data have supported this idea in many parts of the Earth (Perelman and Kasimov, 1999).

The spread of terrestrial living organisms has given rise to the formation of land surface of a specific biogeochemical entity (see Chapter 2), which was absent in the benthic areas of the ocean. The emergence of the soil as a bilatent body was associated with a specific developmental phase of the Earth's living matter (Dobrovolsky, 1994). One of the fundamental attributes of living matter is the tendency to an ever-increasing renewal of living mass and to a maximum occupation of the available space (this law was formulated by V. I. Vernadsky). In the ocean the growth of living matter is restricted within the subsurface oceanic layers and limited by the low content of nutrients in marine waters.

Soil organic matter plays the most important role in the formation of soil humus. On the one hand, humus acts as a source of nitrogen and other nutrients, which are primarily essential for growth of higher plants. These nutrients are released from humus by the activity of soil microbes. Therefore humus is an important parameter of soil fertility and plant productivity. On the other hand, humus acid and their derivatives having high adsorption ability exert an active influence on the migration and accumulation of macroelements in biogeochemical cycles. Humic substances have surface areas as high as $800–900\,m^2/g$, and have adsorptive power of about $1.50–3.00\,meq/g$ ($150–300\,cmol(p^+)/kg$). The composition of soil organic matter is most usefully described in terms of broad solubility classes, collectively referred to as humus substances. They are divided into three main groups on the basis of their solubility in acid and base:

i.   humic acid (HA), soluble in base, but precipitated by acidification of the alkaline extract at pH 2;

ii. fulvic acid (FA), soluble in both alkaline and acid; and

iii. humin, not soluble either base or acid (Kononova, 1966).

Whereas these three humic fractions are structurally similar, they are differentiated by molecule weight and chemical composition (Table 11).

Fulvic acid contains less H, N, and S but more O than HA and humin; the total acidity and ratio of carboxyl to phenolic hydrohyl groups is almost 3 for FA but near 2 for two other fractions.

In the soil, the FA and low molecular weight HA, which are water-soluble, are of particular interest. These humic substances are important in the process of natural weathering of rock materials via complexataion, dissolution, and transport.

Biogeochemical transformation of organic matter in the soil is not related only to the formation of humus. By the action of microbes the process continues to the

Table 11. Analytical characteristics of HA, FA, *and humin from A1 horizon of a haploboroll* (HA *and humin) and from* Bh *horizon of a spodosol* (FA) *(after Butcher et al, 1992).*

Chemical composition	HA	FA	Humin
Elementary composition, per cent of dry, ash-free weight			
C	56.4	50.9	55.4
O	32.9	44.8	33.8
H	5.5	3.3	5.5
N	4.1	0.7	4.6
S	1.1	0.3	0.7
Oxygen-containing groups, meq/g of dry, ash-free weight			
Total acidity	6.6	12.4	5.9
Carboxyl	4.5	9.1	3.9
Total hydroxyl	4.9	6.9	—
Phenolic hydroxyl	2.1	3.3	2.0
Alcoholic hydroxyl	2.8	3.6	—
Total carbonyl	4.4	3.1	4.8
Quinone	2.5	0.6	—
Ketonic carbonyl	1.9	2.5	—
Methoxyl	0.3	0.1	0.1

complete degradation of organic matter yielding CO_2 as the end product. Using the ^{14}C isotope technique the humus renewal in the upper horizon of modern soils takes a period of 300–500 years. In the deeper horizons, the renewal proceeds at a far slower rate, and thus the humic matter has the age of several thousand years. Presumably, this should be attributed to a higher population of microbes in the upper soil layers. The content of humus decreases down the soil profile in parallel with the decreasing population of microbes (Figure 7).

In the context of biogeochemical cycles, the soil is an open system which receives inputs and outputs of many macroelements, such as C, N, Ca, Mg, Si, P, and K. The primary productivity of terrestrial ecosystems basically depend on soil fertility. The fluxes of chemical species from rainfalls pass through soil profiles to river channels and accordingly soil is a regulator of geochemical and biogeochemical migration. Thus, soil occupies a key position and it plays an extremely complex role in any biogeochemical cycle. Only a few examples are given below to show this role.

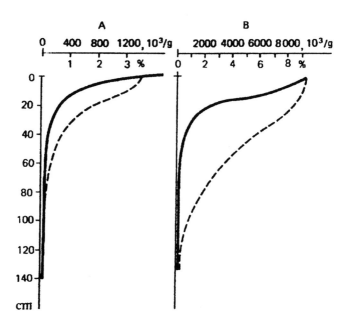

Figure 7. Distribution curves for microbes (solid line) and humus (dashed line) down to soil profiles from Russia. A- Podzol, B- Chernozem (Dobrovolsky, 1994).

The amount of carbon present in soil is closely related to the CO_2 content of the atmosphere. But atmospheric CO_2 is regulated mainly by the ocean rather than by soil (see Section 2). The amount of nitrogen in the soil also does not influence the N pool in the atmosphere because the atmosphere is a huge reservoir regulated mainly by the ocean (Butcher *et al*, 1992). Nevertheless, the soil has a tremendous influence on the nitrate load of the rivers (see Section 3).

Additional elements found in rivers (see above in 1.2) such as Si, Al, Fe, Mg, Ca, K, P, and Na, present as soluble salts or as particles, originate within the soil via weathering of rocks. We could predict that the quality and quantity of this salt loading would depend strongly upon the chemical composition of soils on the river catchments. This is true but not always the case. For instance, the hydrochemistry of rivers from granite bedrocks of Fennoscandia and tropical Africa are similar, whereas the biogeochemistry of soils is very different. This highlights the influence of biogeochemical transformations of macroelements in the atmosphere-soil-plant-water system.

2. BIOGEOCHEMICAL CYCLE OF CARBON

Cyclic processes of exchange of carbon mass are of particular importance for the global biosphere, both in terrestrial and oceanic ecosystems.

This element is distributed in the atmosphere, water and land as follows. According to existing data there are 6160×10^9 tons or 1.4×10^{16} mol of CO_2 in the atmosphere (1680×10^9 tons of C). A major source of atmospheric carbon dioxide is respiration, combustion, and decay, compared with oxygen, whose main source is photosynthesis. In its turn, an important sink of CO_2 is photosynthesis (about 66×10^9 tons/yr or 1.5×10^{15} mol/yr.). Since carbon dioxide is somewhat soluble in water ($K_H = 3.4 \times 10^{-2}$ mol/L/atm), exchange with the global ocean must also be considered. The approximate global balance of atmosphere-ocean water exchange is 7×10^{15} mol/yr. (308×10^9 tons/yr.) being taken up and 6×10^{15} mol/yr. (264×10^9 tons/yr.) being released in different parts of the oceanic ecosystem (Bunce, 1994). The residence time of CO_2 in atmosphere is about 2 years, which makes the atmospheric air quite well mixed with respect to this gas. However, a more recent analysis shows that the terrestrial ecosystems have much stronger sinks of carbon dioxide uptake.

In the global ocean, along with occurrence in living organisms, carbon is present in two major forms: as a constituent of organic matter (in solution and partly in suspension) and as a constituent of exchangeable inorganic ions HCO_3^-, CO_3^{2-}, and CO_2.

$$CO_2(g) \leftrightarrow H_2CO_3(aq) \leftrightarrow -H^+/+H^+ \leftrightarrow HCO_3^-(aq)$$
$$\leftrightarrow -H^+/+H^+ \leftrightarrow CO_3^{2-} \leftrightarrow +Ca^{2+} \leftrightarrow CaCO_3(s)$$

$$(4)$$

The amount of $CO_2(aq)$ in the oceans is sixty times that of $CO_2(g)$ in the Earth's air, suggesting that the oceans might absorb most of the additional carbon dioxide being injected at present into the atmosphere. However, there are some drawbacks restricting this process. First of all, CO_2 uptake into surface oceanic waters (0–100 m) is relatively slow ($t_{1/2} = 1.3$ yrs). Secondly, these surface waters mix with deeper waters very slowly ($t_{1/2} = 35$ yrs). Consequently the surface oceanic waters have the capacity to remove only a fraction of any increase in the anthropogenic CO_2 loading (Figure 8).

The known analytical monitoring data obtained over many years at the Mauna Loa Observatory in Hawaii, a location far from any anthropogenic sources of carbon dioxide pollution, show a pronounced one year cycle of CO_2 content (Figure 9).

One can see the peak about April and then through around October each year. These data indicate that the content of carbon dioxide in the Earth's atmosphere is not perfectly homogeneous. Some explanations would be of interest to understand this figure.

Hawaii is in the Northern Hemisphere where the photosynthetic activity of vegetation is maximal in summer time (May-September). In this period CO_2 is removed from the air a little bit faster than it is added. The reverse situation occurs during the winter. This is a reasonable explanation and accordingly the monitoring stations in the South Hemisphere show the highest concentration of CO_2 in October, and the lowest in April (see http://mlo.hawaii.gov).

A gradual increase in the partial pressure of carbon dioxide over the last decades is clearly pointed out from Figure 9. The value of $p(CO_2)$ was *ca.* 315 ppmv in 1958,

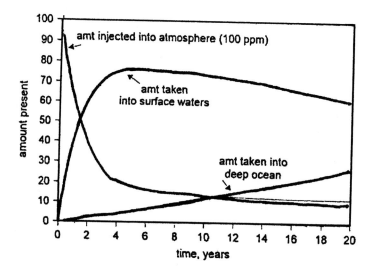

Figure 8. Calculated uptake of CO_2 from atmosphere to the surface and deep oceanic waters (After Bunce, 1994).

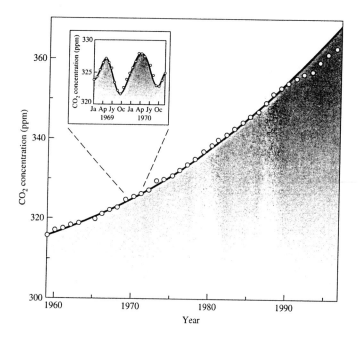

Figure 9. Observations of CO_2 concentration at the Mauna Loa Observatory for the period of 1958–1999.

it had reached 350 ppmv in 1988, and 368 in 1999. Accordingly, this trend can give a doubling of carbon dioxide content in the Earth's atmosphere sometime during the end of the 21st century and this seems to be a reasonable prediction.

Here we should refer to the opinion of some other authors who have argued that increased CO_2 levels in the atmosphere may be a consequence of atmospheric warming, rather than the cause. The statistical analysis led Kuo *et al* (1990) and recently A. Kapitsa (2000) to the conclusion that, although there is a correlation between $p(CO_2)$ and global temperatures, the changes in $p(CO_2)$ appear to lag behind the temperature change by *ca*. 5 months. A possible explanation, if this trend is proved correct, would be that natural climatic variability alters the temperature of the global Ocean, which contains about 90% of total CO_2 mass. In turn, this leads to increase of CO_2 flux from the warmer oceanic water to atmosphere in accordance with the Henry law (see Box 1 for explanation).

2.1. *Turnover of Carbon in the Biosphere*

As it has been pointed out earlier, the terrestrial ecosystems are the main sink of carbon dioxide due to the photosynthesis process. The present bulk of living organisms is confined to land, and their mass (on dry basis) amounts to 1880×10^9 tons. The average carbon concentration in the dry mater of terrestrial vegetation is 46% and, consequently, the carbon mass in the land vegetation is about 865×10^9 tons (after Bolin, 1979 and Dobrovolsky, 1994).

In accordance with Romankevich (1988), the oceanic biomass of photosynthetic organisms contains 1.7×10^9 tons of organic carbon, C_o. In addition, we have to include a large number of consumers. This gives 2.3×10^9 tons of C_o. Totally, the oceanic organic carbon is equal 4.0×10^9 tons or about 0.5% from that in land biomass.

Moreover, a substantial amount of dead organic matter as humus, litterfall and peat is also present in the terrestrial soil cover. The mass of forest litter is close to 200×10^9 tons, mass of peat is around 500×10^9 tons and that of humus, 2400×10^9 tons. Recalculation of this value for organic carbon amounts to 1550×10^9 tons.

However, the greatest amount of carbon in the form of hydrocarbonate, HCO_3^-, $(38600 \times 10^9$ tons) is contained in the ocean, 10 times higher than the total carbon in living matter, atmosphere, and soils.

Thus, in the terrestrial ecosystems the least amount of carbon is monitored in living biomass, followed by dead biomass and atmosphere.

The mass distribution of carbon in the Earth's crust is of interest for understanding of the global biogeochemistry of this element. These values are shown in Table 12. One can see that carbon from carbonates (C_c) is the major form. The C_c/C_o ratio is about 5 for the whole Earth's crust as well as for its main layers (sedimentary, granite, and basalt) and crustal types: continental, sub-continental and oceanic. However, for the latter this ratio is higher.

The sedimentary layer of the Earth's crust is the main carbon reservoir. The C_c and C_o concentrations in the sedimentary layer are by an order of magnitude higher than in granite and basalt layers of lithosphere. The volume of sedimentary shell is about

Table 12. Mass distribution of carbon in the Earth's crust (after Dobrovolsky, 1994).

Earth's compartments	Mass 10^{18} Tons	Average concentration, %			Math 10^{15} tons			Ratio of	
		CO_2	C_c	C_o	CO_2	C_c	C_o	C_c+C_o	C_c/C_o
Total Earth's crust	28.5	1.44	0.38	0.07	409	108	20	128	5.4
Continental type	18.1	1.48	0.40	0.08	267	72	14	86	5.1
Including:									
sedimentary layer	1.8	9.57	2.61	0.50	177	48	9	57	5.3
granite layer	6.8	0.81	0.22	0.05	55	15	3	18	5.0
basalt layer	9.4	0.37	0.10	0.02	35	9.4	1.9	11	5.0
Sub-continental type	4.3	1.37	0.36	0.07	58	16	3	19	5.3
Oceanic type	6.1	1.35	0.36	0.05	82	21	3	24	7.0
Earth's sedimentary shell	2.4	12.4	3.37	0.62	297	81	15	96	5.4
Phanerozoic sedimentary deposits	1.3	15.0	4.08	0.56	194	53	7	60	7.5

0.10 from the crust volume, however, this shell accounts for 75% of both carbonate and organic carbon. Dispersed organic matter contains most of the C_o mass. Localized accumulation of C_o in oil, gas and coal deposits are of secondary importance (see also Chapter 2). It has been estimated that the oil/gas fields amount of 200×10^9 tons of carbon, and the coal deposits contains 600×10^9 tons, totally 800×10^9 tons. This is by three orders of magnitude less than the carbon mass of dispersed organic matter in the sedimentary shell. The general carbon distribution between reservoirs is shown in Table 13.

Thus, there are two major reservoirs of carbon in the Earth: carbonate and organic compounds. It should be stressed that both are of biotic origin. Non-biotic carbonates, for instance, from volcanoes, are the rare exception of the rule. A connecting link between the carbonate and organic species is CO_2, which serves as an essential starting material for both the photosynthesis of organic matter and the microbial formation of carbonates.

Atmospheric CO_2 provides a link between biological, physical, and anthropogenic processes. Carbon is exchanged between atmosphere, the ocean, the terrestrial biosphere, and, more slowly, with sediments and sedimentary rocks. The faster components of the cycle are shown in Figure 10.

Table 13. The major global carbon reservoirs (based on Dobrovolsky, 1994).

Reservoirs	C, 10^9 tons
Atmosphere, CO_2	1680
Global land:	
Vegetable biomass prior to human activity (estimates)	1150
Present natural vegetable biomass	900
Soil cover:	
Forest litterfall	100
Peat	250
Humus	1200
Total	1550
Ocean:	
Photosynthetic organisms	1.7
Consumers	2.3
Soluble and dispersed organic matter	2,100
Hydrocarbonate ions in solution	38,539
Total	40,643
Earth's crust:	
Sedimentary shell, C_o	15,000,000
Sedimentary shell, C_c	81,000,000
Continental granite layer, C_o	4,000,000
Continental granite layer, C_c	18,000,000
Total	118,000,000
Total present global C mass	118,044,773

The component cycles (Figure 10) are simplified and subject to considerable uncertainty (compare with Table 13, for example). In addition, this figure presents average values. The riverine flux, particularly the anthropogenic portion, is currently very poorly qualified and is not shown here. While the surface sediment storage is approximately 150×10^9 tons, the amount of sediment in the bioturbated and potentially active layer is of order 400×10^9 tons. Evidence is accumulating that many of the key fluxes can fluctuate significantly from year to year (for example, in the terrestrial sink and storage). In contrast to the static view conveyed by figures such

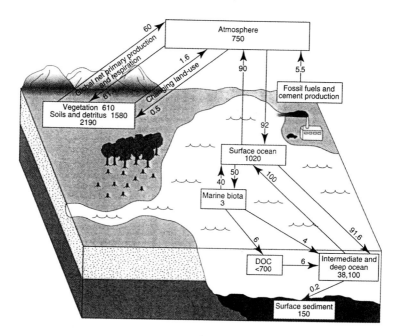

Figure 10. The global carbon cycle, showing the reservoirs (in 10^9 tons per year) relevant to the anthropogenic perturbation as annual averages over the period 1980 to 1989 (Erswaren et al, 1993; Potter et al, 1993; Siegenthaler and Sarmiento, 1993; Schimel et al, 2000).

as this one, the carbon system is clearly dynamic and coupled to the climate system on seasonal, inter-annual and decadal time scale (Keeling and Whorf, 1994; Schimel and Sulzman, 1994).

The biotic activity of organisms is accompanied by the fractionating of the isotopic composition of carbon dioxide. Vladimir Vernadsky (1926) predicted this effect before the first experimental evidence became available. It is known that the carbon of terrestrial reservoirs consists of two stable isotopes ^{12}C and ^{13}C supplemented with small quantities of ^{14}C radioactive species ($T_{1/2} = 5730$ years). For more details on this fractionation see Chapter 2, Box 6.

The carbonate formation and photosynthesis have to be considered as two general processes in the global activity of living matter over geological history of the Earth (see also Box 7 and Chapter 2, Section 5). The C_c-to-C_o mass ratio may specify the 'growth limit' of living matter at sequential stages of Earth's geological history over the period of 3.5–3.8 billion years. This ratio tends to decrease regularly with the last 1.6 billion years. The C_c/C_o ratio was 18 in the sedimentary layers of the Upper Proterozoic period (1600–750 million years); that of the Paleozoic (570–400 million years), 11; of the Mesozoic (235–66 million years), 5.2, and of the Cainozoic (66 million years to the present), 2.9 (Ronov, 1976). The never interrupted increase in the relative content of organic matter in the ancient stream

Table 14. Net primary production of the Earth's major ecosystems (after Rodin et al, 1975).

Global ecosystem zone	Area, 10^6 km^2	Plant mass, 10^9 tons	C-NPP, 10^9 tons
Polar	8.1	13.8	1.3
Coniferous Forests	23.2	439.1	15.2
Temperate	22.5	278.7	18.0
Subtropical	24.3	323.9	34.6
Tropical	55.9	1347.1	102.5
Total land	133.9	2402.1	171.6
Lakes and rivers	2.0	0.04	1.0
Glaciers	13.9	0	0
Total continents	149.3	2402.5	172.6
Oceans	361.0	0.2	60.0
Earth total	510.3	2402.7	232.6

loss provides evidence for a progressively increasing productivity of terrestrial photosynthetic organisms. This provides also the proof for growing importance of global terrestrial ecosystems in fixation of CO_2. Apparently, the increasing productivity of land vegetation would be the major sink of CO_2 under the increasing content of this green-house gas in the atmosphere, however, the role of increasing input of nitrogen, for instance, with atmospheric deposition, has to be considered (see Chapter 5, Section 4 for further details). Moreover, both carbonate formation and the photosynthesis of organic matter share in the common tendency for removal from the atmosphere of CO_2 continually supplied from the mantle. Consequently these processes take part in the global mechanisms for maintaining the present low concentration of carbon dioxide in the Earth's gas shield, which is an essential parameter in the greenhouse effect.

2.2. Carbon Fluxes in Terrestrial Ecosystems

All three CO_2-controlling processes (ocean soaking, photosynthesis and carbonate formation) play an important role in maintaining equilibrium in the biosphere-atmosphere-hydrosphere system. The photosynthetic process is of great importance for living plants and microorganisms. The difference between total photosynthesis and respiration processes is defined as 'net primary production', NPP. The global NPP distribution in the Earth's major ecological zones is shown in Table 14.

Oceans, despite their much larger surface area, contribute much less than half of the global NPP. The reason is related to highly nutrient deficiency in surface waters, which limits the photosynthesis process. Oceanic production is mainly concentrated in coastal zones, especially where upwelling of deep water brings the nutrients (P and N, of major interest) into the surface layer, 0–100 m (see also Chapter 5, Section 2). On land the photosynthetic process is also often limited by nutrient deficit, however, the influence of water storage and low temperature plays a more important role. That is why subtropical and tropical ecosystems contribute much more to global NPP than their proportional share.

The amount of annually decaying organic matter is the subject of speculation. However, some estimates might be done. For instance, in terrestrial ecosystems only, the humus accumulation of carbon in soils is about 70% of the total accumulation of CO_2 in the atmosphere. We may presume therefore that the stable long-lived humic compounds acquire some 30% of carbon annually from the dead organs of plants, and the complete renewal of humus in soils extends over period of 0.3–1.0×10^3 years. The variance depends on the moisture and temperature conditions in the region of question.

Terrestrial biomass is divided into a number of sub-reservoirs with different turnover times. Forest ecosystems contain 90% of all carbon in living matter on land but their NPP is only 60% of the total. About half of the primary production in Forest ecosystems is in the form of twigs, leaves, shrubs, and herbs that only make up 10% of the biomass. Carbon in wood has a turnover time of the order of 50 years, whereas these times for carbon in leaves, flowers, fruits, and rootlets are less than a few years. When plant material becomes detached from the living plant, carbon is moved from phytomass reservoir to litter. *'Litter or litterfall' can refer to a layer of dead plant material on the soil surface.* A litter layer can be a continuous zone without sharp boundaries between the obvious plant structures and a soil layer containing amorphous organic carbon. Decomposing roots are a kind of litter that seldom receives a separate treatment due to difficulties in distinguishing between living and dead roots. Total litter is estimated as 60×10^9 C and total litterfall as 40×10^9 tons C/year (Atjay et al, 1979). The average turnover time for carbon in litter is thus about 1.5 years, although for tropical ecosystems with mean temperature above 30°C, the litter decomposition rate is greater than the supply rate and so storage is impossible. For colder climates, NPP exceeds the rate of decomposition in the soil and organic matter in the form of peat is accumulated. The total global amount of peat might be estimated at 165×10^9 tons C. Average temperature at which there is a balance between production and decomposition is about 25°C.

Humus is a type of organic matter in terrestrial ecosystems that is not readily decomposed and therefore makes up the carbon reservoir with a long turnover time (300–1000 years). Schlesinger (1977) presented an assessment of the various carbon pools for a temperate grassland soil (Figure 11).

The undecomposed litter (4% of the soil carbon) has a turnover time measured in tens of years, and the 22% of the soil carbon in the form of fulvic acids is intermediate with turnover times of hundreds of years. The largest part (74%) of the soil organic carbon (humic acids and humins) also has the longest turnover time (in thousands of years) (Holmen, 1992).

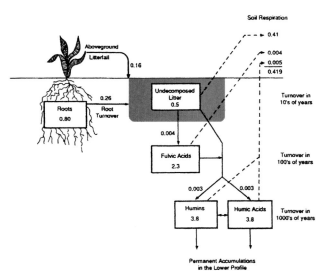

Figure 11. Detrital carbon dynamics for the 0–20 cm layer of chernozem grassland soil. Carbon pool (kgC/m²) and annual transfers (kgC/m²/yr) are shown. Total profile content down to 20 cm is 10.4 kg C/m² (Schlesinger, 1997).

2.3. Comparison of Carbon Biogeochemical Processes in Terrestrial and Aquatic Ecosystems

The synthesis and degradation of organic matter in the ocean are significantly distinct from those in terrestrial ecosystems. The phytoplankton provides for a larger part of photosynthetic organic matter. The dry mass of phytoplankton is three orders of magnitude less than global terrestrial mass, whereas the annual production is only about 3 times smaller. This can be related to the much faster life cycles of plankton organisms in comparison with the terrestrial vegetation.

Let's consider the renewal of terrestrial and oceanic organic matter. The terrestrial biomass might be assessed as $2400–2500 \times 10^9$ tons of dry organic matter and annual production as $170–175 \times 10^9$ tons. These values present a period of 13–15 years for complete renewal of organic matter. In the oceans, the problem is much more complicated. The various authors give 8–10-fold discrepancy in the existing estimates of phytoplankton productivity and biomass. It is estimated also that phytoplankton mass cycle takes 1–2 days to be completed. Taking this into account, we can reasonably consider that the renewal of the total biomass in the global ocean takes about one month. Based on modern assessments, the annual production of photosynthesis varies from $20–30 \times 10^9$ to 100×10^9 tons of organic carbon and the average values are $50–60 \times 10^9$ tons C_0. Furthermore, we can hypothesize that the plankton-synthesized organic matter is almost completely assimilated in subsequent upper food webs. Thus, the organic precipitation would not exceed 0.1×10^9 tons. These calculations present

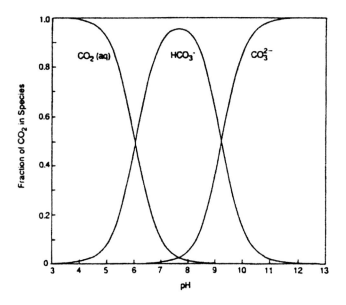

Figure 12. Distribution of dissolved carbon species in seawater as a function of pH *at 15°C and a salinity of 35‰. Average oceanic* pH *is about 8.2 (Holten, 1992).*

the annual uptake of terrestrial and oceanic living organisms of about 440×10^9 tons CO_2 or 120×10^9 tons C_0. Most of this amount recycles into the ocean and atmosphere (Dobrovolsky, 1994).

Carbon Dioxide Interactions in Air-Sea Water System

The interaction between carbon dioxide in the atmosphere and the hydrosphere is the principal factor for understanding large carbon biogeochemical cycles. As it has been mentioned above, the gases of the troposphere and the surface layer of the ocean persist in a state of kinetic equilibrium.

Compared with the atmosphere, where most carbon is presented by CO_2, oceanic carbon is mainly present in four forms: dissolved inorganic carbon (DIC), dissolved organic carbon (DOC), particulate organic carbon (POC), and the marine biota itself.

DIC concentrations have been monitored extensively since the appearance of precise analytical techniques. When CO_2 dissolves in water it may hydrate to form $H_2CO_3(aq)$, which, in turn, dissociates to HCO_3^- and CO_3^{2-}. This process depends on pH and specification is shown in Figure 12.

The conjugate pairs responsible for most of the pH buffer capacity in marine water are HCO_3^-/CO_3^{2-} and $B(OH)_3/B(OH)_4^-$. Although the predominance of HCO_3^- at the oceanic pH of 8.2 actually places the carbonate system close to a pH buffer minimum, its importance is maintained by the high DIC concentration ($\sim 2\,mm$).

Ocean water in contact with the atmosphere will, if the air-sea gas exchange rate is short compared to the mixing time with deeper water, reach equilibrium according to Henry's Law. At the pH of oceanic water around 8.2, most of the DIC is in the form of HCO_3^- and CO_3^{2-} with a very small proportion of H_2CO_3. Although H_2CO_3 changes in proportion to CO_2 (g), the ionic form changes little as a result of various acid-base equilibrium.

From chemical aqueous carbon specification, the alkalinity, Alk, representing the acid-neutralizing capacity of the solution, is given by the following equation:

$$Alk = [OH^-] - [H^+] + [B(OH)_4^-] + [B(OH)_3] + 2[CO_3^{2-}]. \qquad (5)$$

Average DIC and Alk concentrations for the World's Ocean are shown in Figure 13.

With an average DIC of 2.35 mmol/kg sea water and the world oceanic volume of 1370×10^6 km^3, the DIC carbon reservoir is estimated to be 37900×10^9 tons C. The surface waters of the World's ocean contain a minor part of DIC, $\sim 700 \times 10^9$ tons C. However, these waters play an important role in air-deep water exchange (see above).

Oceanic surface water is everywhere supersaturated with respect to the two solid calcium carbonate species calcite and aragonite. Nevertheless, calcium precipitation is exclusively controlled by biological processes, specifically the formation of hard parts (shells, skeletal parts, etc.). The very few existing amounts of spontaneous inorganic precipitation of $CaCO_3(s)$ come from the Bahamas region of the Caribbean.

The detrital rain of carbon-containing particles can be divided into two groups: the hard parts comprised of calcite and aragonite and the soft tissue containing organic carbon. The composition of the soft tissue shows the average ratio of biophils as $P : N : C : Ca : S = 1 : 15 : 131 : 26 : 50$, with $C_c : C_o$ ratio as 1:4. More details of carbon transformation in bottom sediments are presented in Box 2.

Box 2 Heterogenity of carbon cycle in bottom sediments (after Fenchel et al, 1998)

The carbon cycle, in both lake and sea sediments, is predominantly heterotrophic, but there are some aspects of autotrophy. The DOM produced from hydrolysis of fine particulate organic matter (FPOM) is processed by a number of oxidative and fermentative processes in aqueous sediments. The oxidants are O_2, NO_3^-, NH_4^+, Fe^{3+}, SO_4^{2-}, and CO_2 in sequence from the top of the sediment downward. The DOC component of DOM has limited possibilities: it can be oxidized by one of the listed oxidants or it can leave the sediment unoxidized. This is an obvious conclusion, but it has some interesting connotations, for example, the proportion of C oxidized by O_2, etc. and the determination of the factors influencing this proportion. Clearly, the quantity of DOC will determine the depth of O_2 penetration and the extent to which O_2 can participate in C oxidation. An equally important factor is the depth at which DOC is produced by hydrolysis of POM: the deeper the site of POM hydrolysis, the more likely will be the anoxic processing of the soluble products. Another very important factor is the degree to which HS^- is free to diffuse in marine sediments. If HS^- can diffuse to the sediment surface and react with O_2, the depth of O_2 penetration will be greatly reduced.

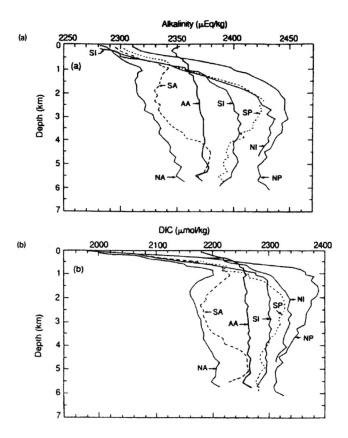

Figure 13. The vertical distribution of alkalinity (a) and dissolved inorganic carbon in the World's Ocean. Ocean regions are shown as NA (North Atlantic), SA (South Atlantic), AA (Antarctic), SI (South Indian), NI (North Indian), SP (South Pacific) and NP (North Pacific) (Holten, 1992).

Simulation modeling of the fate of DOC under different conditions emphasises these simple relationships in marine sediments. Figure 14 shows the rates at which DOC is stipulated to be produced by hydrolysis of POM at proximately 6, 36 and 60 mmol/m^2/day either at the surface of the sediment (TOP), or as a linear gradient from the surface down (LINEAR), or equally to all sediment layers down to 5 cm (MIX). The matrix in Figure 14 represents sediments receiving increasing quantities of POM (arrow down) at increasing frequencies (arrow across).

Both processes, with limits, will increase benthic faunal populations and bioturbation, but sediments receiving excessive organic loading will tend to become anoxic and devoid of macrofauna. This matrix represents the variety of sediment types that might be found in various parts of the World and under various water depth and productivity

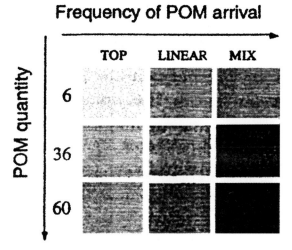

Figure 14. A representation of the addition of particulate organic matter (POM) to sediments. The amount of POM added increases from top to bottom. The frequency of POM addition increases from left to right. The amount of POM hydrolysis per day is 6, 36 or 60 mmol/m²/day. Infrequent addition of POM will not support macrofaunal development and will result in POM hydrolysis at the sediment surface (TOP distribution), whereas more frequent additions will results in increasing macrofaunal populations (LINEAR and MIX distributions) and greater sediment mixing. Dark shading represents the degree of bioturbation (Fenchel et al, 1998).

conditions. These hypothetical types are used as a proxy for actual sediment variety, but if the World's sediments could be categorized in terms of quantity, quality and frequency of POM input, simulation could predict all the important biogeochemical rates and nutrient profiles. More continuous productivity (longer season) would result in a more continuous input than that found in ice-covered seas. The TOP distribution would represent the sediment subject to very little disturbance either by water movement or by bioturbation. At the lower DOC addition rates, this would be equivalent to the shelf sediment with no macrofauna. The absence of macrofauna could be attributed to a single pulse of POM per year, as might occur in the Arctic Ocean. The highest DOC input to the sediment surface will be unrealistic, as macrofauna might be expected to arise in response to this amount of organic input, unless the input occurs as a single pulse. The LINEAR distribution might be found in sediments receiving multiple POM pulses, supporting a moderately active macrofaunal population, which transports a small amount of the fresh POM to 5 cm depth. The MIX distribution is equivalent to a very active macrofaunal population or very efficient mixing by waves or tides.

The estimation of C_c and C_o mass annually eliminated from the biogeochemical cycles in ocean is a very uncertain task. The carbonate-hydrocarbonate system includes the precipitation of calcium carbonate as a deposit:

Atmosphere CO_2

$\uparrow\downarrow$

Surface ocean layer $H_2O \leftrightarrow H_2CO_3 \leftrightarrow H^+ + HCO_3^- \leftrightarrow H^+ + CO_3^{2-} + Ca^{2+}$

$\uparrow\downarrow$

Deep ocean water $CaCO_3$

$$(6)$$

The binding of carbon into carbonates is related to the activity of living organisms. However, the surface runoff of Ca^{2+} ions from the land determines the formation of carbonate deposits to a significant degree. The Ca^{2+} ion stream is roughly 0.53×10^9 tons/year, which can provide for a $CaCO_3$ precipitation rate of 1.33×10^9 tons/year. This would correspond to the loss of 0.57×10^9 tons CO_2, or 0.16×10^9 tons C from the carbonate-hydrocarbonate system.

The surface runoff from the World's land plays an important role in the global carbon mass exchange. The continental runoff supply of HCO_3^- is 2.4×10^9 tons/year, that is, 0.47×10^9 tons/year for carbon. Besides, the stream water contains dissolved organic matter at 6.9 mg/L, which makes up to an annual loss of 0.28×10^9 tons/year. The average carbon concentration of suspended insoluble organic matter in the stream discharge is 5 mg/L, which gives the loss of about 0.2×10^9 tons/year. Most of this mass fails to reach the open ocean and becomes deposited in the shelf and the estuarine delta of rivers. We can see that equal amounts of C_c and C_o (0.5×10^9 tons for each) are annually lost from the World's land surface (Dobrovolsky, 1994).

The formation of carbonates and the accumulation of organic matter are not confined solely to ocean; these processes occur also on the land. The mass of carbonates annually produced in the soils of arid landscapes appears to be high enough (see Chapter 6, Section 4).

2.4. Global Carbon Fluxes

Two large cycles determine global dynamics of carbon mass transport in the biosphere. The first of these is provided for by the assimilation of CO_2 and decomposition of H_2O through photosynthesis of organic matter followed by its degradation to yield CO_2. The second cycle involves the uptake-release of carbon dioxide by natural waters via chemical reactions of CO_2 and H_2O leading to build-up of a carbonate-hydrocarbonate system. The cycles are intimately related to the activity of living matter. The living matter of the biosphere, the global water cycle, and carbonate-hydrocarbonate system regulate the cyclic mass exchange of carbon between atmosphere, land, and ocean. These global carbon fluxes are shown in Table 15.

A specific feature of these two major biogeochemical cycles of carbon is their openness, which is related to the permanent removal of some carbon from the turnover

Table 15. Fluxes of carbon in the biosphere (After Dobrovolsky, 1994).

Fluxes	$C, 10^9$ tons/yr
World's ocean	
Turnover of planktonic photosynthesis organisms	50
CO_2 uptake by ocean	30
CO_2 release by ocean	30
C_o deposited in precipitation	0.08
C_c deposited in precipitation	0.16
World's land	
Biological cycle(photosynthesis-degradation of organic matter)	85
HCO_3^- ion mass exchange between land and troposphere	
Supply to troposphere	0.136
Rainfall washout from troposphere	0.139
Stream loss of:	
DIC	0.47
DOC	0.28
POC	0.20
Transport of oceanic air-borne HCO_3^- ions to land	0.003

as a dead organic matter and carbonates. The carbon burial in the sea deposits is of great importance for biosphere development.

The organic carbon content in the granite crustal layer amounts to 4×10^{15} tons and inorganic (carbonates) amounts to 18×10^{15} tons (Ronov and Yaroshevsky, 1976). This quantity is about 4 times less than in the sedimentary shell. It should be inferred therefore that carbon could not be supplied to the biosphere via hypergenic transformation of the rocks in the granite layer of the lithosphere. Calculations have shown that the atmosphere has served as a reservoir for carbon biogeochemical supply during the last 2 billion years. In the geological epochs of vigorous volcanic eruptions and efflux of enormous masses of lava, the deposition of both carbonate rocks and dispersed organic matter proceeded at a notably high rate (Figure 15). Presumably the laval volume may serve to indicate the extent of erupted volcanic gases.

Over the period of 570 million years, 71.3×10^{15} tons of carbonate-bound carbon and 8.1×10^{15} tons of dispersed organic carbon became buried in sedimentary rocks. Dobrovolsky (1994) has considered that the increased influx of carbon dioxide at the periods of active volcanism was attended by an overall climatic warming up and a reduced temperature contrast between high and low latitudes. Possibly, the wide

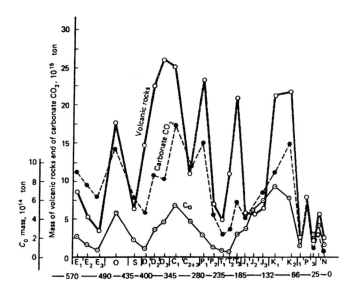

Figure 15. Temporal variations of volcanic rock mass, total carbonate CO_2 pool, and organic carbon pool (C_o) buried in the sedimentary continental layer (after Ronov, 1976).

dispersal of red products of weathering typical for tropic landscapes in the Neogene and their complete disappearance from extratropical territories in the Pleistocene might be associated with the reduced supply of volcanic CO_2 at the end of Alpine tectonogenesis (see Chapter 2).

Dobrodeev *et al* (1976) has suggested that the alteration of glacial and interglacial periods in the Pleistocene was mainly due to fluctuations of CO_2 in the atmosphere. It may be hypothesized that the spread of land ice and the drastic reduction of forest areas with their typically high biomass were favorable to an elevated content of carbon dioxide in the atmosphere and the subsequent climatic warming up. In its turn, the resulting contraction of glacial areas and reforestation was attended by an increased CO_2 uptake from the atmosphere and by its binding to the biomass and soil organic matter. The resulting effect was a gradual cooling and the onset of a new glaciation followed by reduction of forest areas and repetition of the whole cycle.

The role of carbon dioxide in the Earth's historical radiation budget merits modern interest in arising atmospheric CO_2. There are, however, other changes of importance. The atmospheric methane concentration is increasing, probably as a result of increasing cattle population, rice production, and biomass burning. Increasing methane concentrations are important because of the role they plays in stratospheric and tropospheric chemistry. Methane is also important to the radiation budget of our planet.

Analyses of ice cores from Vostok, Antarctica, have provided new data on natural variations of CO_2 and CH_4 levels over the last 220,000 years (Barnola *et al*, 1991;

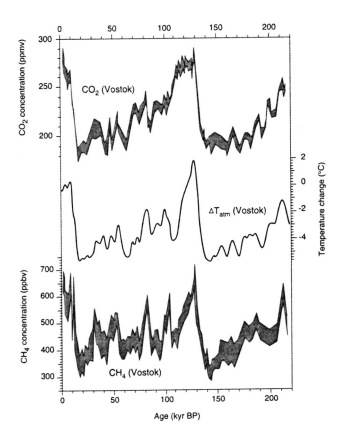

Figure 16. Temperature anomalies and methane and carbon dioxide concentrations over the past 220,000 years as derived from the ice core records at Vostok, Antarctica (Schimel et al, 2000).

Jouzel *et al*, 1993). The records show a marked correlation between Antarctic temperature, as deduced from isotopic composition of the ice and the CO_2/CH_4 profiles (Figure 16).

Clear correlations between CO_2 and global mean temperature are evident in much of the glacial-interglacial palaeo-record. This relationship of CO_2 concentration and temperature may carry forward into the future, possibly causing significant positive climatic feedback on CO_2 fluxes.

3. BIOGEOCHEMICAL CYCLE OF NITROGEN

Nitrogen is an essential element for all forms of life and its biogeochemical cycle is one of the most important in the modern biosphere. It is a structural component of amino acids from which proteins are synthesized. Animal and human tissue (muscle, skin, hair, *etc.*), enzymes and many hormones are composed mainly of proteins.

3.1. Nitrogen Cycling Processes

The biogeochemical cycling of nitrogen has been extensively studied in different ecosystems and the main processes are listed below.

i. *Fixation* is the conversion of atmospheric N_2 to organic N

ii. *Mineralization* is the conversion of organic N to inorganic N

iii. *Nitrification* is the oxidation of NH_4^+ to nitrite (NO_2^-) and nitrate (NO_3^-)

iv. *Denitrification* is the conversion of inorganic N to atmospheric N_2O and N_2

v. *Assimilation* is the conversion of inorganic N to organic N

Nitrogen chemistry and cycling in the environment are quite complex due to the great number of oxidation stages. A large number of chemical, biochemical, geochemical, and biogeochemical transformations of nitrogen are possible, since N is found at valence stages ranging from -3 (in ammonia, NH_3) to $+5$ (in NO_3^-). Microbial species possess the transformations between these stages, and use the energy released by the changes in redox potential to maintain their life processes. Generally these microbial reactions manage the global nitrogen biogeochemical cycling (Figure 17)

The most abundant form of nitrogen at the Earth's surface is molecular nitrogen, N_2, and this form is the least reactive species. Various above-mentioned processes convert atmospheric nitrogen to one of the forms of fixed nitrogen and can be used by living organisms. The reverse process of denitrificaton returns nitrogen to the atmosphere as N_2.

Generally we consider the 0.781 atm of nitrogen in the atmosphere almost as inert filler, because elemental nitrogen is rather unreactive. The known aspects of N biogeochemistry lead us to the conclusion that the atmospheric content of nitrogen, like carbon, is regulated principally by biological processes.

The atmosphere contains about 3.9×10^{15} tons of elemental nitrogen. Biological nitrogen fixation and production of NO in combustion and thunderstorms are the major natural sinks. Finally, NO deposits as HNO_3 in rainwater. In the Haber process N_2 fixes industrially.

$$\text{Combustion: } N_2(g) + O_2(g) + \text{high temperature} \rightarrow 2NO(g)$$
$$\text{Haber process: } N_2(g) + 3H_2(g) + \text{catalyst, } 450° C \rightarrow 2NH_3(g) \tag{7}$$

Terrestrial and aquatic nitrogen in the form of NH_4^+ and NO_3^- are cycled through the biosphere to make proteins and nucleic acids. The processes of decay return the nitrogen to the atmosphere as N_2 and as N_2O by the action of denitrifying bacteria. Almost all the nitrogen fixed by the Haber process is used as fertilizer so that increased

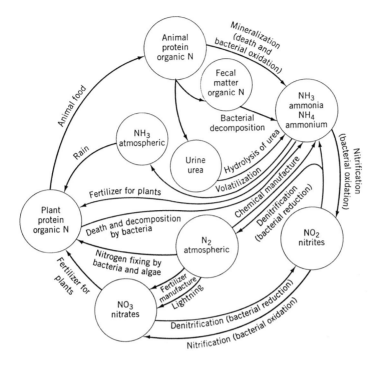

Figure 17. Scheme of nitrogen transformations in the global cycle.

fertilizer uses increase the rate of return nitrogen to the atmosphere through biological denitrification. Indeed, as much as a quarter of all nitrogenous fertilizers applied to crops is denitrified during crop vegetation. This effect is reflected in the gradual increase in the global atmospheric levels of N_2O, which have risen from 0.29 to 0.31 ppmv over the past 25 years. Nitrous oxide, N_2O, is rather unreactive and has a residence time of $\sim$ 20 yr. The "hole-in-pipe" model of nitrous oxide formation is shown in Figure 18.

Most N_2O is formed through biological denitrification in aerobic soils with anaerobic zones or in systems that are alternating between the anaerobic and anaerobic state (see below).

3.2. Main Nitrogen Species

Table 16 lists the most common nitrogen species that exist in nature. This table presents the oxidation stage, the boiling points (b.p.) for each species and its heat of formation $[\Delta H^\circ(f)]$ and free energy of formation $[\Delta G^\circ(f)]$. The same fingers are shown for water for comparison.

Table 16. The list of natural nitrogen species.

Species	Valence	Boiling point °C	$\Delta H^\circ(f)$ kJ/mol, 298 K	$\Delta G^\circ(f)$ kJ/mol, 298 K
$HN_3(g)$	-3	-33	-46	-16.5
$NH_4^+(aq)$	-3		-72	-79
$NH_4Cl(s)$	-3		-201	-203
$CH_3NH_2(g)$	-3		-28	28
$N_2(g)$	0	-196	0	0
$N_2O(g)$	+1	-89	82	104
$NO(g)$	+2	-152	90	87
$HNO_2(g)$	+3		-80	-46
$HNO_2(aq)$	+3		-120	-55
$NO_2(g)$	+4	21	33	51
$N_2O_4(g)$	+4		9	98
$N_2O_5(g)$	+5	11	115	
$HNO_3(g)$	+5	83	-135	-75
$Ca(NO_3)_2(s)$	+5		-900	-720
$HNO_3(aq)$	+5		-200	-108
H_2O		100	-242	-229

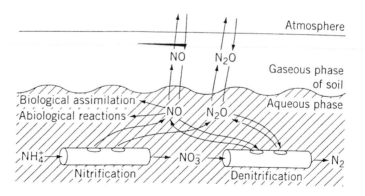

Figure 18. The 'hole-in-pipe' conceptual model of N_2O formation through nitrification and denitrification (Davidson, 1991).

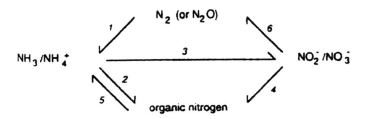

Figure 19. Biological transformation of N species. 1-nitrogen fixation; 2-ammonia assimilation; 3-nitrification; 4-assimilatory nitrate reduction; 5-ammoniafication; 6-denitrification.

3.3. General Characterization of Nitrogen Biogeochemical Cycling Processes

Let us consider the various ways that N is processed by the biosphere. These ways are important for both terrestrial and oceanic nitrogen cycles. They are shown schematically in Figure 19.

All the processes shown in Figure 17 are driven by different microbes. Some of these processes are energy-producing, and some of these occur symbiotically with other organisms. We will start the consideration of biological N transformations from fixation, since this is the only natural process that can bring nitrogen from atmosphere pool to ecosystems.

1. *Biological nitrogen fixation* is any process in which N_2 in the atmosphere reacts to form any N compounds. This process is an enzyme-catalyzing reduction of molecule nitrogen to ammonia (NH_3), ammonium (NH_4^+), and various organic nitrogen forms. In natural processes, biological nitrogen fixation is the ultimate source of nitrogen for all biota. A number of symbiotic and non-symbiotic microbes, both bacteria and algae, have the ability of N fixation. The symbiotic species are quantitatively more significant. There are two major limitations to biological N fixation. The first is related to the required sources of energy to splitting the nitrogen triple bond in N_2, $N \equiv N$. The free energy for formation of NH_3 from N_2 and H_2 ($\Delta G°$) is negative at 25°C (Table 16). However, only some organisms with highly developed catalytic systems are able to fix nitrogen. The second limitation is connected with obligatory anaerobic conditions for N fixation since this is a reductive process. Accordingly, only those organisms that live in anaerobic environment or can create this environment will fix nitrogen.

In terrestrial ecosystems the symbiotic bacteria, particularly strains from the genus *Rhizobium*, play the most important role in N fixation. These bacteria are found on the roots of many leguminous plants (soybeans, clever, chickpeas, etc.). There are other symbiotic diazotrophs (nitrogen-fixing microbes), but till now the Rhizobium has been the most extensively studied.

The cyanobacteria are considered the main N-fixing organisms in both fresh water and marine ecosystems. Some estimates have shown that these species are responsible for up to ~80% of total fixed nitrogen in fresh waters. Cyanobacteria are widespread so that significant nitrogen fixation is possible on local, regional and global scales.

2. *Ammonia assimilation* is the process by which fixed ammonium transforms to the organic form. This process represents significant importance for those organisms that can directly utilize nitrogen from NH_3/NH_4^+. Direct ammonia assimilation yields significant energy saving for those organisms which have an ability to utilize this form of nitrogen. Free ammonium ions do not exist for a long time in any aerobic environments where it is easily nitrified in the nitrification process.

3. *Nitrification* is the oxidation of NH_3/NH_4^+ to nitrite and further to nitrate to gain energy for living microorganisms. Nitrates are the prevalent form of nitrogen in aerobic waters and soils. Nitrification consists of two conjugated stages, which both give an energy yield. The first step is oxidation of ammonium to nitrite and the second one is the subsequent oxidation of formed nitrites to nitrates. We can draw the following typical equations.

$$NH_4^+ + 3/2O_2 \rightarrow NO_2^- + H_2O + 2H^+, \qquad \Delta G° = -290 \text{ kJ/mol}$$
$$NO_2^- + 1/2O_2 \rightarrow NO_3^-, \qquad \Delta G° = -82 \text{ kJ/mol}$$

(8)

The first step in this reaction consequence is processing generally by bacteria genus *Nitrosomonas*, and the second step by *Nitrobacter*. Both these microbes are autotrophic organisms. As carbon sources they utilize CO_2 and gain energy from ammonium oxidation. The role of heterotrophic bacteria, which utilizes organic compounds rather than carbon dioxide, is less significant quantitatively, however, these can perform also the nitrification process. Some intermediates like hydroxylamine (NH_2OH), NO and N_2O were determined in these reactions (Bremner and Blackmer, 1981). The production of N_2O may serve definite environmental problems both with ozone depletion and global warming effects.

Two different pathways can be monitored for nitrates in terrestrial and aquatic ecosystems. The first is related to assimilatory nitrate reduction and the second to denitrification.

4. *Assimilatory nitrate reduction* is simultaneous reduction of nitrate and uptake of N into biomass of any organism. This process might be dominant when reduced nitrogen is in low supply, which is typical in aerobic soil and water column conditions. We can consider this as a primary N input to many microorganisms and plants. Most plants can assimilate both reduced and oxidized forms of nitrogen, even though there is energy cost in first nitrate reduction.

5. *Ammoniafication* is the other major source of reduced nitrogen for living organisms. Ammoniafication can be defined as the breaking down of organic nitrogen compounds, realizing ammonia or ammonium. Decomposition of soil or aquatic organic matter is the typical example of this process and heterotrophic bacteria are

principally responsible for it. During ammoniafication microbes get the carbon source from dead plant or animal biomass and yield the NH_3/NH_4^+ system as additional products. Most of this reduced nitrogen will be conserved in the biological cycle, but a small fraction of it may be volatilized. The more significant source of ammonia is the volatilization during the breakdown of animal excreta and in some regions these values are comparable with losses of nitrogen in the denitrification process.

6. *Denitrification* is the reduction of nitrates to any gaseous N species, generally N_2 and partly N_2O. This is the only process in the biogeochemical cycle of nitrogen, which provides the removal of fixed N compounds to the atmosphere. Under most conditions, N_2 is the prevalent reaction end-product. However, under natural acid reaction of soil or under acidification of terrestrial and especially aquatic ecosystems due to acid deposition, the formation of nitrous oxide can be preferential. Microbes use nitrate as a terminal electron sink (oxidant) in the absence of oxygen (O_2), i.e., in anaerobic soil or water conditions. The overall process is both the oxidizing of an organic compounds and reducing the nitrate. This process leads to energy yielding for approximately 17 genera of facultative anaerobic bacteria. These bacteria can utilize nitrate as an oxidizing agent. The intermediate products of denitrification reactions are nitrite, nitrogen oxide (NO) and nitrous oxide (N_2O). The ratio of $N_2 : N_2O$ in denitrification product is of environmental concern. We can note only that N_2 account for 80–100% of the nitrogen realized or $N_2 : N_2O$ ratio may be equal to 16:1 for global N budget calculations.

It is generally agreed that in terrestrial ecosystems denitrification is a significant source of both N_2 and N_2O under natural and agricultural conditions. See Figure 17 for better understanding of nitrous oxide formation.

Denitrification in Aquatic Ecosystems

The denitrification is still insufficiently quantitatively understood in aquatic ecosystems, especially by comparison to terrestrial ecosystems. There are different views on whether the oceans are a source or sink of nitrous oxide. Various data indicate that the ocean is, on average, supersaturated with respect to N_2O, and that N_2O supersaturations are positively correlated with NO_3^- and negatively correlated with O_2. A number of studies have suggested that denitrification is a major sink for fixed nitrogen in the oceans. Recent studies suggests that denitrification losses in the oceans are on the order of 0.13×10^9 tons/yr (9.2×10^{12} mol N per year) that exceed known oceanic N inputs. More than half of this denitrification (0.067×10^9 tons/yr (4.8×10^{12} mol N per year) takes place in sediments, with the remainder in pelagic oxygen minimum zones (see Box 3 for more details).

Box 3. Estimating denitrification in North Atlantic continental shelf sediments (after Seitzinger & Giblin, 1996)

Although sediments are commonly thought of as sources of inorganic nitrogen to the overlaying water through regeneration, they also serve as a sink for both externally supplied and regenerated nitrogen via denitrification. Sediments are ideal sites for denitrification because they generally are anaerobic a few millimeters below the surface. Nitrates can be supplied to the sediments either directly from the overlaying water (direct denitrification) or through nitrification in the sediments of ammonia released during the decomposition of organic matter (coupled nitrification/denitrification).

Consequently, authors have developed the model of coupled nitrification/denitrification for continental shelf sediments to estimate the spatial distribution of denitrification throughout shelf regions in the North Atlantic basin. Using data from a wide range of continental shelf regions, authors found a linear relationship between denitrification and sediment oxygen uptake (Figure 20).

This relationship was applied to specific continental shelf regions by combining it with a second regression relating sediment oxygen uptake to primary production in the overlaying water. This allows the calculations of denitrification rates as a function of phytoplankton production:

$$DNF_c = 0.019 \times PhytoProd, \qquad (9)$$

where DNF_c equals denitrification coupled to sediment nitrification (mmol N as N_2 per m^2/day) and PhytoProd equals phytoplankton production (mmol C per m^2/day) in the overlaying water.

This relationship suggests that approximately 13% of the N incorporated into phytoplankton in shelf waters is eventually denitrified in the sediments via coupled nitrification/denitrification, assuming a C : N ratio of 6.625 : 1 for phytoplankton. The model calculated denitrification rates compare favorably with rates reported for several shelf regions in the North Atlantic.

The model-predicted average denitrification rate for continental shelf sediments in the North Atlantic Basin is 0.69 mmol N as N_2 per m^2/day. Denitrification rates (per unit area) are highest for the continental shelf region in the western North Atlantic between Cape Hatteras and South Florida and lowest for Hudson Bay, the Baffin Island region, and Greenland (Figure 21).

Within latitudinal belts, average denitrification rates were lowest in the high latitudes, intermediate in the tropics and highest in the mid-latitudes. Although denitrification rates per unit area are lowest in the high latitude, the total N removal by denitrification (53×10^{10} mol N/yr.) is similar to that in the mid-latitudes (60×10^{10} mol N per yr.) due to the large area of continental shelf in the high latitudes. The gulf of St. Lawrence/Grand Banks area and the North Sea are responsible for 75% of the denitrification in the high latitude region. N removal by denitrification in the western North Atlantic (96×10^{10} mol N/yr.) is two times greater than in the eastern North Atlantic (47×10^{10} mol N/yr.). This is primarily due to differences in the area

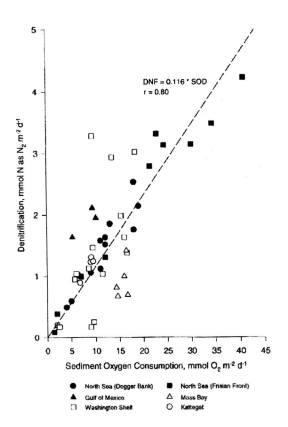

Figure 20. Relationship between denitrification rates coupled to sediment nitrification and sediment oxygen consumption rates in various continental shelf regions. Dashed line indicates regression line (Seitzinger and Giblin, 1996).

of continental shelf in the two regions, as the average denitrification rate per unit area is similar in the western and eastern North Atlantic.

Authors calculate that a total of 143×10^{10} molN/yr. is removed via coupled nitrification/denitrification on the North Atlantic continental shelf. This estimate is expected to underestimate total sediment denitrification because it does not include direct denitrification of nitrate from the overlaying water. The rate of coupled nitrification/denitrification calculated is greater than the N inputs from atmospheric deposition and river discharge combined (Figure 22).

The input of N to the continental shelf from estuarine export plus fluxes from rivers that discharge directly on the shelf was calculated from Nixon *et al* (1996). Atmospheric deposition of $NO_y + NO_x$ (wet plus dry deposition) was calculated using

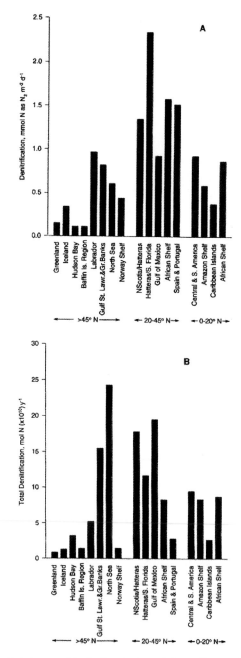

Figure 21. Model predicted rates of denitrification coupled to sediment nitrification for various continental shelf regions in the North Atlantic basin: a) denitrification rates per unit area (mmol N as N_2 per m^2/day), and b) N removal by region (mol $N \times 10^{10}$/yr.); e.g., Nova Scotia/Hattras = 17.8×10^{10} mol N per yr.) (Seitzinger and Giblin, 1996).

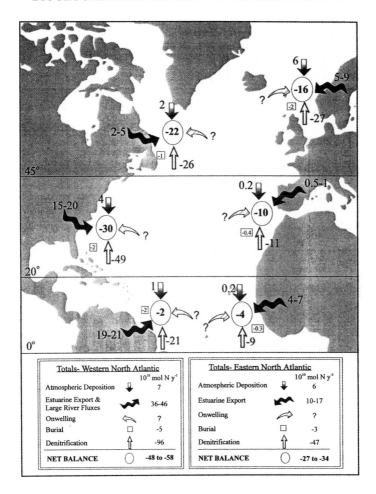

Figure 22. Known sources and sinks of N for continental shelf regions in the western and eastern Atlantic Basin by latitudinal zone (Seitzinger and Giblin, 1996).

data from Nixon *et al* (1996) and Prospero *et al* (1996). Burial of N in shelf sediments exclusive of major river deltas was calculated based on the primary production data and assumption that 0.25% to 1.25% of the primary production with a C : N molar ratio of 10 is buried. The balance of the indicated sources and sinks lies within the closed circles; the mid-point of the range in estuarine export/large river fluxes for a region was used for that calculation. The contribution of N from onwelling of nutrient rich slope water is not known, but is hypothesized to account for the additional N input needed to balance the N budget.

This suggests also that onwelling of nutrient rich slope water is a major source of N for denitrification in shelf regions. For the two regions where inputs to shelf regions from onwelling have been measured (South and Mid-Atlantic Bights), onwelling appears to be able to balance the denitrification rates.

The considered processes relate to N transformation in various chains of the biological cycle. The biological N cycle itself is only part of total global biogeochemical turnover of this element. The overall cycle is the interaction of biotic and abiotic processes.

3.4. Global Nitrogen Cycle

Prior to estimating the global N cycle let us consider the nitrogen reservoirs in the biosphere (Table 17).

The effect of human activity on the global cycling of nitrogen is great, and, furthermore, the rate of change in the pattern of use is much greater even for many other elements (Galloway et al, 1995). The single largest global change in the nitrogen cycle comes from increased reliance on synthetic inorganic fertilizers, which accounts for more than half of the human alteration of the nitrogen cycle (Vitousek et al, 1997). The process for making inorganic nitrogen fertilizer was invented during World War I, but was not widely used until the 1950s. The rate of use increased steadily until the late 1980s, when the collapse of the former Soviet Union led to great disruptions in agriculture and fertilizer use in Russia and much of eastern Europe. These disruptions resulted in a slight decline in global nitrogen fertilizer use for a few years (Matson et al, 1997). By 1995, the global use of inorganic nitrogen fertilizer was again growing rapidly, with much of the growth driven by use in China (Figure 23).

Use as of 1996 was approximately 83×10^6 ton N per year. Approximately half of the inorganic nitrogen fertilizer that has ever been used on Earth has been applied during the last 15 years.

Production of nitrogen fertilizer is the largest process whereby human activity mobilizes nitrogen globally (Box 4).

Box 4. The Fate of Nitrogen Fertilizer in North America (After Howarth et al, 2002)

When nitrogen fertilizer is applied to a field, it can move through a variety of flow paths to surface and grounds waters (Figure 24).

Some of the fertilizer leaches directly to groundwater and surface waters, with the range varying from 3% to 80% of the fertilizer applied, depending upon soil characteristics, climate, and crop type. On average, for North America some 20% is leached directly to surface waters (NRC, 1993; Howarth et al, 1996). Some fertilizer is volatilized directly to the atmosphere; in the United States, this averages 2% of the application, but the value is higher in tropical countries and also in countries that use more ammonium-based fertilizers, such as China. Much of the nitrogen from fertilizer is incorporated into crops and is removed from the field in the crops when they are harvested, which is of course the objective of the farmer. A recent NRC report

Table 17. Nitrogen reservoirs in the biosphere, 10^9 tons N.

Reservoir	Soderlund & Svensson, 1976	McElroy, 1976	Stedman & Shetter, 1983	Dobrovolsky, 1994	Mackenzie, 1998
Crustal					
Rocks				165000	
Sediments	400000	600000		600000	
Coal dep.	120				
Terrestrial					
Soil:					
Organic	300	60		110	
Inorganic		10			
Biomass:					
Plants	11–14	10		25	
Animals	0.2				
Oceanic					
Dissolved N_2	22000			20000	
Dissolved N_2O	20				
NO_3^-	570			685	
NH_4^+	7				
NO_2^-	0.5				
Total inorganic		600			
Organic matter:					
Dissolved	530				
Particulate	3–24				
Total organic		200		300	
Biomass:					
Plants	0.3		0.2		
Animals	0.17			0.32	
Total biomass		0.8			
Atmospheric Gaseous					
N_2	3900000	4000000	3900000	3900000	
N_2O	1.3	1.1	1.4		
NH_3	0.0009		0.0003		
NO_x	0.001–0.004		0.0002		
NO_y					
Aerosols:					
NH_4^+	0.0018		0.0006		
NO_3^-	0.0005		0.0002		
Total N(-III)					
$NO_x + NO_y$					
Total reactive, excludes $N_2 + N_2O$		0.003			
Total organics	0.001				
Total N in biosphere					5.0

(NRC, 1993) suggests that on average 65% of the nitrogen applied to croplands in the United States is thus harvested, although other estimates are somewhat lower. By difference, on average somewhere around 13% or so of the nitrogen applied must be building up in soils or denitrified to nitrogen gas.

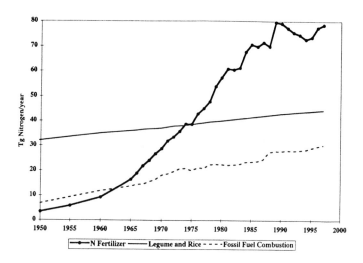

Figure 23. Temporal trends (1950–1997) in the anthropogenic production of reactive nitrogen (Tg N/yr.=106 N tons/yr.) from fertilizer; fossil fuel combustion and legume and rice cultivation (Galloway, 1998).

Since much of the nitrogen is harvested in crops, it is important to trace its eventual fate. The majority of the nitrogen is fed to animals (an amount equivalent to 45% of the amount of fertilizer originally applied, if 65% of the nitrogen is actually harvested in crops). Some of the nitrogen is directly consumed by humans eating vegetable crops—in North America perhaps 10 percent of the amount of nitrogen originally applied to the fields. By comparison perhaps 10% of the amount of nitrogen originally applied to fields is lost during food processing, being placed in landfills or released to surface waters from food-processing plants.

Of the nitrogen that is consumed by animals, much is volatilized from animal wastes to the atmosphere as ammonia. In North America, this volatilization is roughly one-third of the nitrogen fed to animals, or 15% of the amount of nitrogen originally placed on the fields. This ammonia is deposited back onto the landscape, often near the source of volatilization but some of it first travels for long distances through the atmosphere (Holland *et al*, 1999). Some of the nitrogen in animals is consumed by humans, an amount roughly equivalent to 10% of the amount of nitrogen fed to the animals, or 4% of the nitrogen originally applied to fields. By comparison, the rest of the nitrogen—over 25% of the amount of nitrogen originally applied to the fields—is in animals wastes that are building up somewhere in the environment. Most of this may be leached to surface waters.

Of the nitrogen consumed by humans, either through vegetable crops or meat, some is released through wastewater treatment plants and from septic tanks. In North America, this is an amount equivalent to approximately 5% of the amount of nitrogen

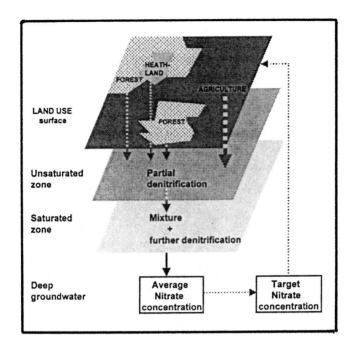

Figure 24. Schematic view of nitrate leaching to groundwater (Joosten et al, 1998).

originally applied to fields. By comparison, the rest of the nitrogen is placed as food wastes in landfills or is denitrified to nitrogen in wastewater treatment plants and septic tanks.

The conclusion is that fertilizer leaching from fields is only a portion of the nitrogen that potentially reaches estuaries and coastal waters. Probably of greater importance for North America as a whole is the nitrogen that is volatilized to the atmosphere or released to surface waters from animal wastes and landfills. Since food is often shipped over long distances in the United States, the environmental effect of the nitrogen can occur well away from the original site of application of the fertilizer.

However, other human-controlled processes, such as combustion of fossil fuels and production of nitrogen-fixing crops in agriculture, convert atmospheric nitrogen into biologically available forms of nitrogen (see Figure 21). Overall, human fixation of nitrogen (including production of fertilizer, combustion of fossil fuel, and production of nitrogen-fixing agricultural crops) increased globally some two to three fold between 1960 to 1990 and continues to grow (Galloway, 1998). By the mid

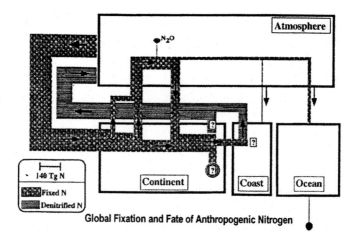

Figure 25. Schematic of global N cycle for 1990 (Galloway, 1998).

1990s, human activities made new nitrogen available at a rate of some 140×10^6 ton N per year, matching the natural rate of biological nitrogen fixation on all the land surfaces of the world (Vitousek *et al.*, 1997; Cleveland *et al*, 1999). Thus, the rate at which humans have altered nitrogen availability globally far exceeds the rate at which humans have altered the global carbon cycle (Figure 25).

The human alteration of nutrient cycles is not uniform over the Earth, and the greatest changes are concentrated in the areas of greatest population density and greatest agricultural production. Some regions of the world have seen very little change in the flux of either nitrogen or phosphorus to the coast (Howarth *et al*, 1996), while in other places the change has been tremendous. Human activity is estimated to have increased nitrogen inputs to the coastal waters of the northeastern United States generally and to Chesapeake Bay specifically by some six- to eight-fold (Howarth *et al*, 1996). Atmospheric deposition of nitrogen has increased even more than this in the northeast (Holland *et al*, 1999). The time trends in human perturbation of nutrient cycles can also vary among regions. For example, while the global use of inorganic nitrogen fertilizer continues to increase, the use of nitrogen fertilizer began in 1960 in the United States, but has increased relatively little since 1985 (Figure 26).

Note, however, that the use of nitrogen fertilizer in the United States in the next century may again increase to support greater exports of food to developing countries. Countries such as China have been largely self sufficient in food production for the past two decades, in part because of increased use of nitrogen fertilizer. The use of fertilizer in China is now very high—almost 10-fold greater than in the United States—and further increases in fertilizer use are less likely to lead to huge increases in food production, as they have in the past. Therefore, if China's population continues to grow it may once again be forced to import food from the United States and other

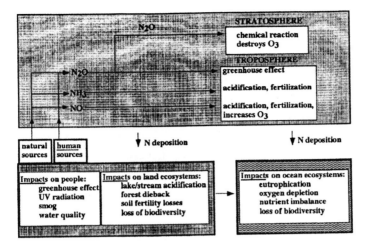

Figure 26. Schematic of the environmental consequences of anthropogenic N mobilization (Galloway, 1998).

developed countries, leading to more use of nitrogen fertilizer here. More discussion of environmental biogeochemistry of nitrogen in regional scale (North America and East Asia) is presented in Chapter 8, Section 1).

4. BIOGEOCHEMICAL CYCLE OF PHOSPHORUS

4.1. Phosphorus Forms in the Biosphere

The content of phosphorus in the Earth's crust is about 0.1%, however P plays a very important role in the biosphere. This element was originated as a melted-out component of the crustal matter. Its further evolutionary history in the biosphere was complicated and in many details poorly understood. In basalts the P concentration is 0.14% and in granites twice as low. There are about 200 phosphorus-containing minerals. However, due to low crustal abundance of this element, they are not rock-forming materials. The overall phosphorus mass in the granite layer of the lithosphere is 63300×10^9 tons. The most abundant phosphate mineral is apatite, which accounts for more than 95% of all P in the Earth's crust. The basic composition of this mineral is listed in Table 18.

Since apatites serve as a row material for P fertilizer production, the impurities of heavy metals are the source of great pollution of agroecosystems. In general, the major phosphorite deposits are of marine origin and occur as sedimentary beds ranging from a few centimeters to tens of meters in thickness. The biogenic matter produced in water column settles to the sediment surface and decomposes, releasing PO_4^{3-} to the sea water and pore water.

Table 18. The chemical composition of apatites.

General formula	X ions	Possible substitutes for Ca^{2+}	Possible substitutes for $PO_4{}^{3-}$
$Ca_{10}(PO_4)_6X_2$	F^- = Fluorapatite, OH^- = Hydroxylapatite, Cl^- = Chlorapatite	Na^+, K^+, Ag^+, Mn^{2+}, Sr^{2+}, Mg^{2+}, Zn^{2+}, Cd^{2+}, Ba^{2+}, Sc^{2+}, Y^{2+}, Bi^{3+}, U^{4+} and rare earth elements	$CO_3{}^{2-}$, $SO_4{}^{2-}$, $CrO_4{}^{2-}$, $AsO_3{}^{2-}$, $VO_4{}^{3-}$, $F \cdot CO_3{}^{2-}$, $OH \cdot CO_3{}^{2-}$, $SiO_4{}^{4-}$

Table 19. Content of phosphorus in biotic and abiotic Earth's matter.

Biosphere compartment	Content, %	Mass, 10^9 tons
Dry matter of terrestrial ecosystems	0.2	5.0
Dry matter of oceanic ecosystems	1.1	0.035
Soil organic matter	0.15	4.65
Total		9.685

Phosphorus is one of the most important elements on the Earth. It participates in or controls many of the biogeochemical processes occurring in the biosphere. The important role of phosphorus in the biosphere is owed to its vital role in protein synthesis. The exothermal reaction of adenosine triphosphate with photosynthesized hydrocarbons provides the subsequent biochemical reactions with energy (see structure and functions of DNA and RNA in Chapter 2, Box 5). The N : P ratio in plant issues is within 8–15 (Bazilevich, 1974; Romankevich, 1982, Vitoushek and Howarth, 1991). Almost in all biogeochemical systems P was found as a deficient element limiting the productivity of ecosystems.

The average P concentration in the dry matter of terrestrial ecosystems is about 0.2%, whereas that in oceanic ecosystems is about 1.1%. A significant part of this element has been retained in soil humus biogeochemical barrier, where the P content is about 0.15%. The total mass of P in biotic and abiotic matter is about 9.7×10^9 tons (Table 19).

A great number of phosphorus species is found in fresh and marine waters. Phosphate, $PO_4{}^{3-}$, is the fully dissociated anion of triprotic phosphoric acid, H_3PO_4.

$$H_3PO_4 \leftrightarrow H^+ + H_2PO_4{}^{3-} \leftrightarrow 2H^+ + HPO_4{}^{2-} \leftrightarrow 3H^+ + PO_4{}^{3-}. \qquad (10)$$

The dissociation constants for these equilibria for fresh and marine waters are shown in Table 20.

Table 20. Dissociation constants (pK) of phosphoric acid at 25°C.

Reaction	Fresh water	Marine water
$H_3PO_4 \leftrightarrow H^+ + H_2PO_4^{3-}$	2.2	1.6
$H_2PO_4^{3-} \leftrightarrow H^+ + HPO_4^{2-}$	7.2	6.1
$HPO_4^{2-} \leftrightarrow H^+ + PO_4^{3-}$	12.3	8.6

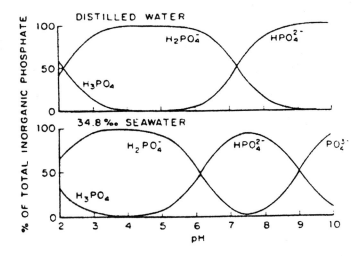

Figure 27. Distribution of phosphoric acid species as a function of pH in fresh and sea waters (Atlas, 1975).

The speciation of various P species in fresh and seawaters is shown in Figure 27 for the pH range of 2–10.

The typical pH of surface waters is between 6 and 8, and $H_2PO_4^{3-}$ is the dominant P species. In predominant seawater pH of 8, HPO_4^{2-} is prevalent.

The condensed phosphates or polyphosphates are another important class of inorganic phosphates. In these compounds, two or more phosphate groups bond together via $P - O - P$ bonds to form chains or in some cases cyclic species. In soils and waters polyphosphates generally account for only a small part of the total P content. However, these species are very reactive and in many places they are responsible for anthropogenic pollution of natural water, for instance, by detergents. The polyphosphates are formed also in reactions between orthophosphates of mineral fertilizers and soil organic matter and can be leached to surface waters (Kudeyarova and Bashkin, 1984; Kudeyarova, 1996).

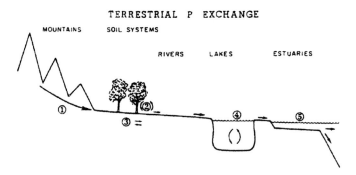

Figure 28. Schematic representation of the transport of P *through the terrestrial system. The dominant processes considered in this description are: (1) mechanical and chemical weathering of rocks, (2) incorporation of P into terrestrial biomass and its return to the soil system through decomposition, (3) exchange reactions between groundwater and soil particles, (4) cycling in freshwater lakes, and (5) transport through the estuaries to the oceans of both particulate and dissolved* P *(Jahnke, 1992).*

4.2. Fluxes and Pools of Phosphorus in the Biosphere

The biogeochemical cycle of phosphorus on local, regional or global scale is unique among the cycles of macroelements since it has no significant gaseous component. This is related to the soil redox potential, which is generally too high to produce the phosphine gas, PH_3. Unlike transfers in the N cycle, the biogeochemistry of P is not driven by microbes. Nearly all the phosphorus in cycling processes is originally derived from weathering P-containing minerals. These cycling processes include the transfer of P within the terrestrial ecosystems and various transformations in fresh and marine aquatic ecosystems (Box 5).

Box 5. Conceptual models of sub-global phosphorus cycles (after Jahnke, 1992)

Conceptual ideas behind simulation of P cycling are related to the construction of models for freshwater terrestrial ecosystems and a generalized oceanic system and understanding the restrictions of its application.

In a general way, the overall movement of phosphorus on the continents can be considered as the constant water erosion of rock and transport of P in both particulate and dissolved forms with surface runoff to river channels and further to the oceans. The intermediate transformations are connected with uptake of P as a nutrient by biota and interactions between river waters and bottom sediments. The majority (up to 90%) of eroded P remains trapped in the mineral lattices of the particulate matter and will reach the estuaries and ocean without entering the biological cycle. The smallest soluble part of eroded phosphorus is readily available to enter the biological cycle (Figure 28).

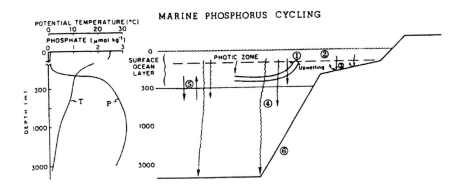

Figure 29. Schematic profiles of potential temperature and phosphate in the open ocean and representation of processes that control the P distribution. The dominant processes considered are (1) upwelling of nutrient-rich waters, (2) biological productivity and the sinking of the biogenic particles produced, (30 regeneration of nutrients by the decomposition of organic matter in surface waters and shallow sediments, (4) decomposition of particles below the main thermocline, (5) slow exchange between surface and deep waters, (6) incorporation of P into the bottom sediments (Jahnke, 1992).

Dissolved P in ground and subsurface waters may be taken by plants and microbes. When organisms die, the organic P species decompose and return to the mineral turnover. Inorganic chemical reactions in the soil-water system also greatly determine the geochemical mobility of phosphorus. These reactions include the dissolution or precipitation of P-containing minerals or the adsorption and desorption onto and from mineral surfaces. The inorganic mobility of P is pH dependent. In alkaline soils apatite is less dissolved than in acid soils and the aluminum phosphates are the main forms in the latter environment. Formation of complexes between phosphate and iron and aluminum oxyhydroxides and clays during adsorption reactions may limit the solubility and migration of this species. Thus, these links of P biogeochemical cycle influence its behavior on the way towards the ocean.

Lakes also constitute an important component of the terrestrial phosphorus system. The degree of trophic stage of water body and its productivity in many cases directly depend on the P content in the water. In the water column, in summer, warming of the surface layers produces strong stratification, which restrict exchange between the lighter, warm surface water and the colder, denser deep water. During photosynthesis, the dissolved P in the photic zone is incorporated into plants and is eventually transported below the thermocline on sinking particles. The upward motion of this P is very slow due to water column stratification. The P depletion of water pool leads to low P content and finally limits the biological productivity of the aquatic ecosystem.

Over the main part of the ocean area, the vertical distribution of dissolved phosphates is similar to the shape of that shown for lake waters (Figure 29).

Unlike a temperate lake, stratification in the ocean does not completely break down in the winter and only a few nutrients come up due to exchange between deep and surface water layers. More important sources of P to the photic zone (surface oceanic layer) are the major upwelling regions. Under oceanic currents and wind influence the open oceanic waters exchange with waters from the continental margins accompanied by the corresponding input of nutrients.

Once in the photic zone, P is readily incorporated into biogenic particles via the photosynthetic activity of plankton and begins to sink. The majority of these particles decompose in the surface waters or in shallow layers of sediments and the P is recycled directly back into the photic zone to be incorporated into biological particles. A small part of these particles escapes the surface layer and sinks into the deep ocean. Most of these particles eventually decompose and the cycle is repeated. Only a very small faction of these particles may incorporate into sediment without decomposition.

The additional forms in which P is removed from marine waters are the burial of P adsorbed onto the $CaCO_3$ surface and the formation of apatite in continental margin sediments. In lesser degree, this is related to the adsorption of P onto iron oxyhydroxide phases and decomposition of fish debnis.

Turnover of P in the aquatic ecosystems is expressed by significant values, because of its importance for living aquatic organisms (Box 6).

Box 6. Export of phosphorus from agricultural systems in USA (after Howarth et al, 2002)

Several surveys of U.S. watersheds have clearly shown that phosphorus loss in runoff increases as the portion of the watershed under forest decreases and agriculture increases. In general, forested watersheds conserve phosphorus, with phosphorus input in dust and in rainfall usually exceeding outputs in stream flow. Surface runoff from forests, grasslands, and other noncultivated soils carries little sediment, so phosphorus fluxes are low and the export that occurs is generally dominated by dissolved phosphorus. This loss of phosphorus from forested land tends to be similar to that found in subsurface or dissolved base flow from agricultural land. The cultivation of land in agriculture greatly increases erosion and with it the export of particle-bound phosphorus. Typically, particulate fluxes constitute 60 to 90% of phosphorus exported from most cultivated land. In the eastern United States, conversion of land from forests to agriculture between 1700 and 1900 resulted in a 10-fold increase in soil erosion and a presumed similar increase in phosphorus export to coastal waters, even without any addition of phosphorus fertilizer. The soil-bound phosphorus includes both inorganic phosphorus associated with soil particles and phosphorus bound in organic material eroded during flow events. Some of the sediment-bound phosphorus is not readily available, but much of it can be a long-term source of phosphorus for aquatic biota.

Increases in phosphorus export from agricultural landscapes have been measured after the application of phosphorus. Phosphorus losses are influenced by the rate, time, and method of phosphorus application, form of fertilizer or manure applied, amount and time of rainfall after application, and land cover. These losses are often small from the standpoint of farmers (generally less than $200\,kg\,P\,km^{-2}$) and represent a minor proportion of fertilizer or manure phosphorus applied (generally less than 5%). Thus, these losses are not of economic importance to farmers in terms of irreplaceable fertility. However, they can contribute to eutrophication of downstream aquatic ecosystems.

While phosphorus export from agricultural systems is usually dominated by surface runoff, important exceptions occur in sandy, acid organic, or peaty soils that have low phosphorus adsorption capacities and in soils where the preferential flow of water can occur rapidly through macropores. Soils that allow substantial subsurface exports of dissolved phosphorus are common on parts of the Atlantic coastal plain and Florida, and are thus important to consider in the management of coastal eutrophication in these regions.

Although there exists a good understanding of the chemistry of phosphorus in soil-water systems, the hydrologic pathways linking spatially variable phosphorus sources, sinks, temporary storages, and transport processes in landscapes are less well understood. This information is critical to the development of effective management programs that address the reduction of phosphorus export from agricultural watersheds.

Runoff production in many watersheds in humid climates is controlled by the variable source area concept of watershed hydrology. Here, surface runoff is usually generated only from limited source areas in a watershed. These source areas vary over time, expanding and contracting rapidly during a storm as a function of precipitation, temperature, soils, topography, ground water, and moisture status over the watershed. Surface runoff from these areas is limited by soil-water storage rather than infiltration capacity. This situation usually results from high water tables or soil moisture contents in near-stream areas.

The boundaries of surface runoff-producing areas will be dynamic both in and between rainfalls. During a rainfall, area boundaries will migrate upslope as rainwater input increases. In dry summer months, the runoff-producing area will be closer to the stream than during wetter winter months, when the boundaries expand away from the stream channel.

Soil structure, geologic strata, and topography influence the location and movement of variable source areas of surface runoff in a watershed. Fragipans or other layers, such as clay pans of distinct permeability changes, can determine when and where perched water tables occur. Shale or sandstone strata also influence soil moisture content and location of saturated zones. For example, water will perch on less permeable layers in the subsurface profile and become evident as surface flow or springs at specific locations in a watershed. Converging topography in vertical or horizontal planes, slope breaks, and hill slope depressions or spurs, also influence variable source area hydrology in watersheds. Net precipitation (precipitation minus evapotranspiration) governs watershed discharge and thus total phosphorus loads to surface waters. This should be taken into account when comparing the load estimates from different regions. It is also one reason why there seems to be more concern with phosphorus in humid regions than in more arid regions.

In watersheds where surface runoff is limited by infiltration rate rather than soil-water storage capacity, areas of the watershed can alternate between sources and sinks of surface flow. This again will be a function of soil properties, rainfall intensity and duration, and antecedent moisture condition. As surface runoff is the main mechanism by which phosphorus is exported from most watersheds, it is clear that, if surface runoff does not occur, phosphorus export can be small. Thus, consideration of hydrologic pathways and variable source areas is critical to a more detailed understanding of phosphorus export from agricultural watersheds.

At present, the uncertainty of quantitative estimates of the total marine and freshwater biomass is very high, mainly due to a dynamic picture of oceanic productivity. As a tentative estimate, we can accept the calculations based on the C_o : P rate in the oceanic biomass as 100 : 1.

The solubility of orthophosphates in fresh and marine waters is low, 0.04 and 0.088 μg/L. Despite low solubility, P is retained for a long time in the ocean waters, being involved in small biological cycles between water and aquatic organisms.

The sedimentary rocks store 1.3×10^{15} tons of P and the overall amount of this element in sedimentary shell and in the granite layer of the continental crust is estimated at 6.33×10^{15} tons (Dobrovolsky, 1994).

Therefore, only about 17% of the total phosphorus mass, found in the granite layer of the lithosphere, would have to be lost during biosphere development. The distribution of P in various biospheric compartments is shown in Table 21.

We can estimate that 0.345×10^9 tons of P is annually included into the biological cycle in terrestrial ecosystems and 1.21×10^9 tons in the biological cycle of marine ecosystems. The major values of biological turnover in the ocean are probably related to the more profound P deficit in seawaters than in soil humus.

Surface runoff drives the transport of phosphorus to the ocean. The annual supply of ionic P to oceanic ecosystems might be 1×10^6 tons, with averaged content of orthophosphate in the riverine water of 0.04 μg/L (Meybeck, 1982). Given the 5% content of P in dispersed and partly soluble organic matter, the transport of P in organic form from terrestrial to oceanic ecosystems would be about 2×10^6 tons. The content of P in suspended matter was assessed to be equal to 510 μg/L (Gordeev, 1983). Therefore, the amount of phosphorus, transported as a suspended particulate material, is estimated as 21×10^6 tons, being in marked excess over the mass of soluble P: 88% of river transport is in particulate form and only 12% in dissolved form. The summary of annual global phosphorus fluxes between land and ocean is given in Table 22.

Table 21. Phosphorus pools in the Earth.

Pools	P mass, 10^9 tons
Earth's crust:	
Continental granite layer	6,330,000
Sedimentary shell	1,311,000
Ocean:	
Biomass of photosynthetic organisms	0.04
Dissolved inorganic mater	15.0
Land:	
Land vegetation	5.0
Organic matter of pedosphere	4.7
Total	7,641,024.74

Table 22. Major fluxes of global phosphorus cycle (after Dobrovolsky, 1994 and Howarth et al, 1995).

Sub-cycles	Fluxes, 10^6 tons/yr.
Ocean	
Photosynthetic organisms	1,210
Land	
Photosynthetic organisms	345
Riverine discharge:	
Soluble inorganic species	1
Dispersed organic matter	2
Particulate matter	21
Total flux	1,579

The biogeochemical cycle of phosphorus is shown in more detail in Figure 30.

Human activity has an enormous influence on the global cycling of nutrients, especially on the movement of nutrients to estuaries and other coastal waters. For phosphorus, global fluxes are dominated by the essentially one way flow of phosphorus carried in eroded materials and wastewater from the land to the oceans, where it is ultimately buried in ocean sediments. The size of this flux is currently estimated at 22×10^6 ton per year. Prior to increased human agricultural and industrial activity,

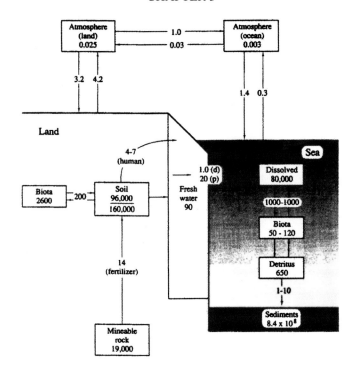

Figure 30. The global phosphorus cycle. Pools are in million tons (10^6 tons) and annual fluxes in million tons per year, 10^6 tons/yr. (Schlesinger, 1991).

the flow is estimated to have been around 8×10^6 ton per year (Howarth *et al*, 1995). Thus, current human activities cause an extra 14×10^6 tons of phosphorus annually to flow into the ocean sediment sink each year, or approximately the same as the amount of phosphorus fertilizer (16×10^6 ton per year) applied to agricultural land each year.

The global biogeochemical cycle of P is almost open. This differentiates phosphorus from carbon and nitrogen, which natural biogeochemical cycles are, on the contrary, almost closed.

5. BIOGEOCHEMICAL CYCLE OF SULFUR

5.1. Sulfur in the Earth

Sulfur is a typical representative of the group of active outgassed elements. Sulfur dioxide (SO_2) and hydrogen sulfide (H_2S) are of common occurrence among the gaseous sulfur compounds evolved with volcanic gases.

There are nine known isotopes of S of which five are stable: ^{32}S - 95.0%, ^{33}S - 0.76%, ^{34}S - 4.22 and ^{36}S - 0.014% of crustal abundance.

Table 23. Examples of natural sulfur compounds in the Earth biosphere.

Oxidation state	Form of existence					
	Gas	Aerosol	Aqueous	Soil	Mineral	Biotic
+6	SO_3	H_2SO_4	SO_4^{2-}	$CaSO_4$	$CaSO_4 \cdot H_2O$	
		HSO_4^-	HSO_4^-		$MgSO_4$	
		SO_4^{2-}	$CH_3SO_3^-$			
		$(NH_4)_2SO_4$				
		Na_2SO_4				
		CH_3SO_3H				
+4	SO_2	$SO_2 \cdot H_2O$	HSO_3^-	SO_3^{2-}		
			$SO_2 \cdot H_2O$			
			$HCNO \cdot S_2O$			
			SO_3^{2-}			
+2			$S_2O_3^{2-}$			
0	$CH_3SOCH_3^+$			S_8		
−1	RSSR		RSSR	SS^{2-}	FeS_2	Methionine
−2	H_2S		H_2S	S^{2-}	S^{2-}	$CH_3S(CH_2)_2$
						$CHNH_2$
	RSH		HS^-	HS^-	HgS	Cysteine
	OCS		S^{2-}	MS	CuS_2	$HSCH_2CH$
						NH_2COOH
	CS_2		RS^-			Dicysteine

Sulfur is found in valence stages ranging from $+6$ in sulfate anion (SO_4^{2-}) to -2 in sulfides (S^{-2}). Naturally sulfur exits in several oxidation states, and its participation in oxidation-reduction reactions has important geochemical and biogeochemical consequences (Table 23).

The data presented in Table 23 demonstrates the existence of sulfur in solid, liquid, and gaseous forms and in living organisms. Figure 31 shows the total Earth's key reservoirs and the approximate content of S in each.

In any given time the vast majority of S is in the lithosphere. However, the main transport of this element occurs in the atmosphere, hydrosphere, and biosphere. The role of the biosphere often involves reactions that result in the movement of sulfur from one reservoir to another.

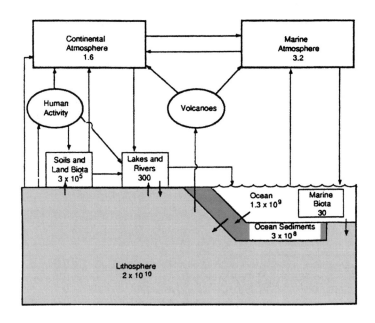

Figure 31. Major sulfur reservoirs (Charlson et al, 1992).

5.2. Sulfur in the Biosphere

In the biosphere, the global sulfur mass transport is not associated only with the migration of gaseous compounds. It includes also the transport of water-soluble sulfurous species in underground and surface waters. The formation of water-soluble S compounds is associated with the hypergenic transformation of insoluble rock sulfides into easily soluble sulfates. The metal sulfides (MeS) in the hypergenesis zone undergone hydrolysis and oxidation, with the resultant formation of sulfuric acid (H_2SO_4), insoluble hydroxides of iron (Fe^{3+}), manganese (Mn^{4+}), and soluble sulfates (SO_4^{2-}). The microbiological formation of SO_4^{2-} occurs in soils.

Sulfates are water-soluble and transported with the riverine discharge to the ocean. The SO_4^{2-} content in marine water is 2.7 g/L. Taking into account for the whole volume of the Ocean, 1.2×10^{15} tons of S or 3.7×10^{15} tons of sulfates has been accumulated during Earth's geological history in the oceanic waters.

Sulfur is the vital structural part of proteins and essential component of living matter. The concentration of sulfur in plant tissues is relatively small, 0.34% of dry biomass (Bowen, 1966). Sulfur concentration is much higher in animals and microbes due to higher content of proteins themselves. The C : S ratio in protein is about 16, in carbohydrates 80, in terrestrial plants over 200, and in animals about 70. The average S content in the oceanic living matter is 1.20% of the dry weight. In marine plant C : S ratio is around 50.

The whole oceanic consumption consists of 0.05×10^9 tons, oceanic photosynthetic organisms, 0.07×10^9 tons, and terrestrial biomass, 8.5×10^9 tons. Accounting for the 0.5% of humus dry weight, the pool of S in the soil global layer is 15.5×10^9 tons (Dobrovolsky, 1994).

Consequently, an intensive uptake of sulfur occurs in hydrosphere, lithosphere, and biosphere. This is the main reason that the content of gaseous sulfur species in the atmosphere is rather small and even in polluted air does not exceed 2–3 ppmv. In unpolluted atmosphere the concentration of most S compounds is at ppbv levels, despite the intense sulfur outgassing from the Earth's interior. The atmospheric content of the major gaseous S species, either SO_2 or H_2S, is highly variable and is influenced by both natural and anthropogenic factors. The role of anthropogenic sulfur emission in acid rain chemistry will be discussed in Chapter 10. The influence of natural parameters, microbiological activity in particular, is described in Box 7.

Box 7. Microbial sulfur transformation in the system hydrogen sulfide - sulfate - dimethyl sulfide (after Fenchel et al, 1998)

Hydrogen sulfide (H_2S) production and consumption quantitatively dominate microbial sulfur transformation in marine ecosystems. However, in spite of relatively high rates of sulfate reduction in marine environments, gaseous fluxes of H_2S are relatively small. This emission constraining is related to:

i. rapid oxidation by aerobic and photosynthetic bacteria in conspicuous mats and biofilms at sea water surface, and

ii. reaction of sulfide with metals both dissolved in seawater and in aerosols.

The iron is especially important. In freshwater ecosystems, fluxes of hydrogen sulfide are also relatively small owing to the lack of sufficient sulfate as a substrate for dissimilatory reduction and to the relatively greater incorporation of the available sulfur into biomass. However, the release of hydrogen sulfide is significant from wetlands. In addition, H_2S emission from plant canopy occurs when S plant uptake is in excess of biosynthetic demands. The latter process may account for as much as 40% of total natural S emission.

The list of S species, which are important for exchange between biosphere and atmosphere, includes also carbonyl sulfide (COS), carbon disulfide (CS_2) and various methylated species (methanethiol, dimethyl sulfide and dimethyl disulfide). Dimethyl sulfide (DMS) dominates biogenic S emission from marine and freshwater ecosystems.

With the exception of DMS and direct emission of H_2S from sediments, soils and plants, the various sulfur gases originate primarily from microbial decomposition of sulfur-containing organic matter (Figure 32).

In this microbial cycling of organic sulfur, the methanethiol is opposed to dimethyl sulfide. DMS arise from hydrolysis of dimethylsulfoniopropionate (DMSP) by plant and animal activity as well as microbial activity in marine ecosystems (Figure 33).

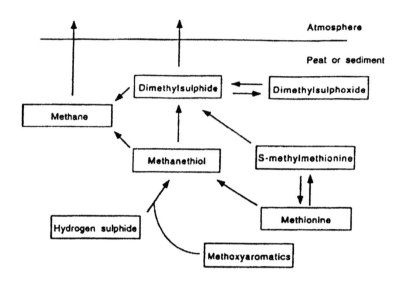

Figure 32. Scheme for organic sulfur cycling in a freshwater peatland (Fenchel et al, 1998).

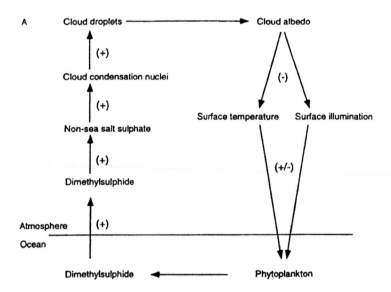

Figure 33. A. Relationships between dimethyl sulfide (DMS), cloud condensation nuclei and climate control; + and − signs indicate the effect of the preceding parameter on the downstream parameter. DMSO, dimethyl sulfoxide (Fenchel et al, 1998).

Typical organic precursors for COS, CS_2 and the methylated sulfur gases include methionine and cysteine from proteins and isothiocyanates and thiocyanates from plant secondary metabolites. Methanethiol and DMS are also formed in anoxic freshwater sediments from reactions based on H_2S and various methyl donors, for example, methoxylated aromatic acids, such as syringic acid from lignins. The rates of DMS emission per unit area are similar for both the oceans and *Sphagnum*-dominated wetlands. Only the area of this peat land limited the relative importance of the latter source.

Sulfide methylation reactions couple dissimilatory sulfate reduction to DMS production and determine the rates of DMS emission in freshwater wetlands. This process involves acetogenic bacteria, some of which degrade aromatic acids to acetone. In soils, freshwater, and marine ecosystems a wide diversity of other anaerobic and aerobic bacteria can contribute to sulfur gas production. In addition, diverse aerobes (e.g. methylotrophs and sulfate oxidizers) and anaerobes (e.g. methanogenes) consume S gas, thereby regulating fluxes in the atmosphere-biosphere system.

5.3. Global Fluxes and Pools of Sulfur

We can presume that there are three major reasons which specify the biogeochemical cycle of sulfur:

1. Microbial transformation of sulfur gases play an important role in S exchange between atmosphere and biosphere;

2. Sulfurous gases transform easily to readily soluble sulfates, which are a predominate form of riverine S migration;

3. Formation of hydrogen sulfide in wetlands leads to the reactions between H_2S and soluble metals, especially iron, with subsequent deposition of insoluble metal sulfides.

These reasons are connected with S biogeochemical fluxes and pools in biosphere, atmosphere and hydrosphere. However, the main reservoirs are related to lithosphere. According to Ronov (1976) and Dobrovolsky (1994), the average concentration of sulfide sulfur in sedimentary shell is 0.183% and the total amount of sulfur is 9.3×10^{15} tons. In addition, the granite crustal layer contains 8.6×10^{15} tons of S. Totally in the Earth's crust there is around 94% of the global S mass (Table 24).

Living matter plays a very important role in the global biogeochemical fluxes of sulfur. As we have seen above, the microbes are the major living organisms in S cycling. Here we will try to assess the quantitative fluxes of sulfur gases in its biogeochemical turnover. According to Dobrovolsky (1994), the annual supply of microbial produced sulfur from ocean to atmosphere is around 48×10^6 tons/yr. Part of this amount (about 25×10^6 tons/yr.) is scavenged into water drops, and the rest is finally oxidized to sulfate and washed out by precipitation. In terrestrial ecosystems, soil bacteria release in S gaseous form about 58×10^6 tons/yr., of which

Table 24. Sulfur reservoirs in biosphere (After Dobrovoslky, 1994).

Reservoirs	S, 10^9 tons
Earth's crust:	
Sedimentary shell	
Sulfide sulfur	4,100,000
Sulfate sulfur	5,200,000
Continental granite layer	
Sulfide sulfur	5,300,000
Sulfate sulfur	3,300,000
Ocean:	
Soluble inorganic ions	1,200,000
Biomass of consumers	0.09
Biomass of photosynthetic organisms	0.02
Atmosphere	0.0014
Terrestrial biosphere	
Vegetation	8.5
Soil organic matter	15.5
Total	19,100,024.1114

15×10^6 tons/yr. is taken by vegetation, and 43×10^6 tons/yr. is oxidized in the troposphere to be washed out with precipitation.

We have described already the isotope fractionation effects in Chapter 2. Here we would like to recall that the living cell components and extracellular metabolites are, as a rule, enriched in light isotope. From four known stable S isotopes (see above this section), two are of practical interest, ^{32}S and ^{34}S. The relative abundance of these two isotopes may be estimated by the following:

$$\delta^{34}S = \left\{\left[^{34}S/^{32}S(\text{sample})\right] \Big/ \left[^{34}S/^{32}S(\text{reference})\right] - 1\right\} \cdot 1000‰. \qquad (11)$$

As a reference, the triolite $^{34}S/^{32}S$ ratio equal to 0.0450045 from Canon Diablo meteorite has been accepted. Any proportion of two S isotopes in natural compounds may be expressed in terms of $\delta^{34}S$ quantity, whose value is positive (+ sign) or negative (− sign) relating the reference value.

Solid and gaseous biogenous sulfur products are enriched in the light isotope. For instance, the content of ^{32}S isotope is increasing in the reduced H_2S whereas

^{34}S is increasing in the $SO_4{}^{2-}$ during the process of biological sulfate reduction by *Desulfovibrio desulphuricans bacteria*. Since sulfates are accumulated in the ocean, the values of $\delta^{34}S$ in marine waters and evaporates are, respectively, $+20$, and $+17$, whereas the values of $\delta^{34}S$ for the sedimentary rocks, markedly abundant in biogenuos iron sulfide, is -12. The sulfur of hydrogen sulfide as produced by sulfate-reducing microbes, has the value of $\delta^{34}S$ reaching down to -43.

The formation of microbe-generated S species in the global ocean is still uncertain. Some estimates deviate from 6×10^6 tons/yr. (Lein *et al*, 1988) up to 48×10^6 tons/yr. (Zehnder *et al*, 1980) for sulfur release rate from the ocean surface. The uncertainty relates mainly to unknown rates of H_2S oxidation in the troposphere that is coming back to the ocean with atmospheric precipitation.

The average amount of annual precipitation over the global ocean surface is around 411×10^{15} L with mean content of mineral salts of $10 \, mg/L$. This gives the value of dissolved salts equal to 4.1×10^9 tons/yr. We can also assume that additionally about 20% from this sum would enter the ocean with dry deposition, i.e., 0.8×10^9 tons/yr. Totally, it would be 4.9×10^9 tons/yr., of which S accounts for 0.29×10^9 tons/yr. It seems reasonable that about 10% of this sum might be transported onto Earth's land. Therefore, about 0.31×10^9 tons/yr. is supplied to the atmosphere over the global ocean; of this, 0.83×10^9 tons/yr. is precipitated into the ocean as sulfates and $\sim 0.03 \times 10^9$ tons S per year is transported to the terrestrial ecosystems.

The average content of $S - SO_4{}^{2-}$ in the deposition over the Earth's land is $\sim 1.7 \, mg/L$ and the annual amount of precipitation is 62×10^{15} L. This presents 0.103×10^9 tons/yr. of sulfur as wet deposition and additionally 0.02×10^9 tons/yr. as dry deposition, totally 0.123×10^9 tons S per year. Accounting for ocean input ($\sim 0.03 \times 10^9$ tons S per year), annually 0.126×10^9 tons of sulfur is deposited in terrestrial ecosystems.

According to different estimates, the river runoff to the ocean changes from 0.104×10^9 tons S/yr. (Ivanov, 1983) up to 0.162×10^9 tons S/yr. (Dobrovolsky, 1994). A significant amount of various sulfur compounds are input through hydrothermal sources, up to $\sim 130 \times 10^9$ tons/yr. The volcanic sulfur emission to the continental atmosphere is estimated as 0.001×10^9 tons annually and a similar number applied to for oceanic atmosphere (Fried, 1973).

Along with the important role of the chemolithotrophic bacteria in the global sulfur mass exchange, the contribution from the photosynthetic bacteria is also substantial. Accounting for the productivity of terrestrial ecosystems and the mean content of sulfur in the dry biomass (0.34%), we can estimate that $\sim 0.6 \times 10^9$ tons/yr. of sulfur is cycling in terrestrial biogeochemical turnover and $\sim 1.3 \times 10^9$ tons/yr in the marine biogeochemical S cycle, since the sulfur content in dry mass of marine photosynthetic organisms is higher than in terrestrial ones (1.2).

These estimates of sulfur fluxes and similar assessments made by other researchers are shown in Table 25.

Depending on the estimates of various authors, the total annual turnover of sulfur varies from 0.291×10^9 up to 2.846×10^9 tons/yr.

Table 25. Annual global biogeochemical fluxes of sulfur, 10^6 *tons S /yr. (Modified from Charlson et al, 1992).*

Flux	Description	Kellag et al, 1973	Friend, 1973	Granat et al, 1976	Moller, 1984	Ivanov, 1983 Natural	Ivanov, 1983 Anthrop.	Charlson et al, 1992	Dobrovolsky, 1994
Continental part of cycle									
F1a	Emission to the atmosphere from fossil fuel burning and metal smelting	50	65	65	75	—	113	—	110
F1b	Effluents from chemical industry and mining	—	—	—	—	—	29	—	60
F1c	Input from soils to rivers	—	26	—	—	—	28	8	—
F2	Aeolian emission (dust)	—	—	0.2	—	20	—	20	—
F3a	Volcanic emission to continental atmosphere	1	1	1.5	1	14	—	9	—
F4	Biogenic gases (land)	—	58	5	35	18	—	18	210
F5	Gravitational settling of large (aeolian) particle to land	—	—	—	—	12	—	12	—
F6	Washout and dry deposition of gases and fine particles to land	111	121	71	—	25	47	46	230
F7	Transport to oceanic atmosphere	5	8	18	—	35	66	13	—
F8	Transport from oceanic atmosphere	4	4	17	—	20	—	24	22
F9	Weathering to soil	—	42	66	—	114	—	26	—
F10	Weathering to river	—	—	—	—	—	—	93	—
F11	Burial of sulfur in sediments from continental water bodies	—	—	—	—	—	—	35	—

Table 25. Annual global biogeochemical fluxes of sulfur, 10^6 *tons S/yr. (Modified from Charlson et al, 1992) (continued).*

Flux	Description	Kellag et al, 1973	Friend, 1973	Granat et al, 1976	Moller, 1984	Ivanov, 1983 Natural	Anthrop.	Charlson et al, 1992	Dobrovolsky, 1994
Continental part of cycle (continued)									
F12	River runoff to oceans	—	136	122	—	104	104	104	164
2. Oceanic part of the cycle (see also F7, F8, and F12 above)									
F3b	Volcanic emission to marine atmosphere	1	1	1.5	1	14	—	19	—
F13	Aeolian emission (seasalts)	47	44	44	175	140	—	140	—
F14	Biogenic gases (marine)	—	48	27	35	23	—	39	310
F15	Washout and dry deposition of gases and particles to oceans	72	96	73	—	258	—	187	290
F16	Burial of sulfur in oceanic sediments	—	—	—	—	139	—	69	130
F17	Deposition of marine sulfate via thermal vent reactions at mid-ocean ridges	—	—	—	—	—	—	43	—
F18	Lithification of marine sediments	—	—	—	—	—	—	69	—
F19	Turnover of planktonic photosynthetic organisms	—	—	—	—	—	—	—	1320
F1–F18	Total turnover of S in biogeochemical cycle	291	650	528	322	936	387	974	2846

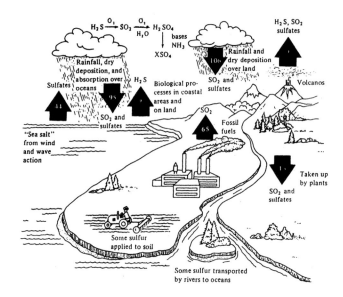

Figure 34. Modern estimates of global sulfur cycle.

The global biogeochemical cycle of sulfur is shown in Figure 34.

The global sulfur cycle is open testifying to the removal of excessive S mass from the biogeochemical cycling in the form of metal sulfides or calcium and magnesium sulfates. This maintains the concentration of sulfur gases in the atmosphere at very low levels. The local increase of SO_2 concentrations and corresponding acid deposition problems due to anthropogenic pollution will be discussed in Chapter 10.

6. BIOGEOCHEMICAL CYCLE OF SILICON

6.1. Silicon in the Earth

Silicon (Si) is the second (after oxygen) element in the Earth's crust. It was actively accumulated in the lithosphere during geological history and Earth's evolution. According to Vinogradov (1962), the Si content in the upper mantle is about 19%, in basalts, 24%, and in granites, 32.3%. The silicon tetroxide tetrahedron, made up of a silicon cation tightly liganded to four oxygen anions, this is a major structural unit of crystalline matter of the Earth's crust. In the granite layer of the lithosphere, SiO_2 accounts for 63.08%, or 2427.5×10^{15} tons of silicon.

6.2. Migration and Accumulation of Silicon Compounds in Soil-Water Systems

Silicon is permanently entering soil solution and ground waters as a result of hydrolysis of dispersed alumosilicate and dissolution of various silicon minerals, like quartz,

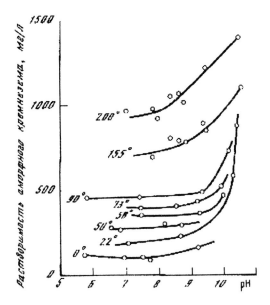

Figure 35. The influence of pH *and temperature on the solubility of amorphous silicon.*

opal, and chalzedon. The role of decomposition of dead organic matter containing Si is also important. The enrichment of water solutions that leaches different silicon minerals varies from < 5 to > 20 mg/L SiO_2. In the maximum extent this process occurs during the weathering of fresh volcano deposits. This type of chemical weathering is characteristic for the soils of Japan, Indonesia, Philippines, Chili, Peru, and the Russian Kamchatka peninsula. For instance, the annual fluxes of soluble silicon in surface waters of Kamchatka peninsula is $3–13$ tons/km^2 (Sokolov, 1967). The predominant forms of silicon species in soil solution and natural waters are ionic and molecular solutions of orthosilicic acid, H_4SiO_4, and metasilicic acid, H_2SiO_3. The other chemical species are silicon-organic compounds like polymers, colloidal solutions of $Si(OH)_4$ and fine suspension gels with common formula $nSiO_2 \cdot mH_2O$. The polymers of silicon acid have high molecular weight, from 1000 to 70000.

The solubility and migration of SiO_2 is sharply increasing with increase of pH and temperature. In alkaline soil solution of Solonets, the solubility of amorphous silicon may be up to 100–200 mg/L. The growth of temperature increases also the solubility (Figure 35).

The results shown in Figure 35 explain why the thermal waters are enriched by soluble silicon, up to 500–1000 mg/L. The solubility of silicon gels is also increasing under the influence of sodium salts, like NaCl, Na_2SO_3 and especially $NaHCO_3$ and Na_2CO_3. For instance, soil solution from Solonets, which contains soda, enriches by silicon, up to 500 mg/L. In the contrary, the presence of sulfate, bicarbonate and carbonate of calcium and magnesium decreases sharply the Si solubility and induces the deposition of relevant alkaline earth salts or formation of SiO_2 films on soil particle surfaces.

Similar deposition effects are related to transpiration or freezing of soil solutions. Sometimes, these processes are accompanied with the formation of silicon layers up to 4–5 m thickness, as in the Mississippi River valley.

The Si entering biogeochemical food webs leads also to formation of biogenous depositions. Diatomic algae, radiolaria, some plants (bamboo) may bioaccumulate the soluble silicon with subsequent transformation in amorphous silicon compounds, like opal. However, this process seems reversible and opal may be de-polymerized back to colloidal and soluble species of silicon (Kovda, 1984).

6.3. Biogeochemical Migration of Silicon in Arid Tropical Ecosystems

About 40–60 years ago the significant geochemical and biogeochemical migration of silicon was shown in arid ecosystems. The results have been analyzed in relevant reviews (Stephens, 1981, Kovda, 1984). The comparison of these data with recent research allows us to conclude that chemical weathering, especially in arid tropical conditions of Australia and Africa, releases a significant amount of ionic and colloidal-dissolved Si species from geological rocks to natural waters. The presence of natural waters with a wide spectrum of redox potential and chemical composition stimulates the biogeochemical migration of silicon. Simultaneously, other processes, like transpiration, chemical interaction with calcium and magnesium, lead to the restriction of migration and accumulation of secondary silicon minerals in soils. In accordance with Stephens (1981), in Central Australia these processes were responsible for the formation of vast areas with a predominance of silcrite and opal silicon minerals (Figure 36).

The Arid Steppe and Desert ecosystems (for instance, Savanna ecosystems) of Africa and South America present also favorable conditions for the formation and accumulation of secondary silicon minerals. The bioaccumulation of Si in plants and formation of secondary Si minerals was monitored in the Egyptian deserts (Kovda, 1984). The most characteristic examples are from the Great African Rift depressions. This area has had a very complicated geological history during the last Tertiary and Quarternary times, including strong volcanic activity, lava fluxes, volcanic ash depositions, thermal ground water formations, tectonic depressions and upwelling, pluvial periods of significant hydromorphism and aridization changes. The practical absence of geochemical runoff in this northeastern part of Africa, the ancient geological history and intensive physical and chemical weathering processes, affected the migration and differentiation of many chemical species. This has led to the formation and accumulation of soda, migration and accumulation of silicon species, and formation of silicon Solonchaks and silcrites. The tropical weathering of relatively fresh magma rocks, volcanic lava and ash in Africa is the permanent source of soluble silicon and carbonate leached to different types of natural waters. The spatial migration of these chemical species is forwarded from mountains (Kilimanjaro, 3700 m above sea level) and hill plateaus (1500–2000 m above sea level) and volcanoes to inland depressions

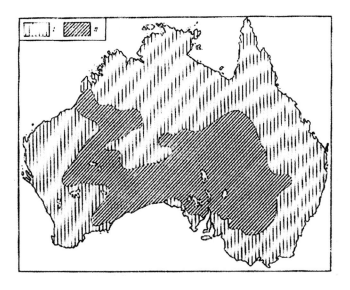

Figure 36. The spatial distribution of laterite and silcrite in Australia (Stephens, 1981).

and lakes (200–800 m above sea level). These geochemical fluxes are complicated by outputs of thermal water sources with increasing content of carbonate and silicon species.

These factors lead to the formation of Si-enriched soils and alluvial deposits in river deltas with layers of secondary quartz and opal.

Biogeochemical migration and transformation of silicon in Kenya was monitored also in Amboseli swampy plain with Meadow Steppe and Solonchak soils. The fluxes of groundwater from neighboring Kilimanjaro Mountains form the swamps and lakes. The strong transpiration of these ground waters initiates the biogeochemical and geochemical formation of secondary minerals in various soils of this plain. The soil mass is undergoing an intensive accumulation of silicon species and carbonate (Table 26).

We can see that the accumulation of both water soluble and amorphous silicon occurs in the upper 25 cm of soil profile, whereas the accumulation of carbonate is found up to 50 cm.

6.4. Biogeochemical Migration of Silicon in Wet Boreal and Tropical Ecosystems

Wet Boreal Ecosystems

The modern biogeochemical accumulation of silicon species was monitored in various parts of northern Eurasia, like Island Greenland, European (Karelia and Kola peninsula) and the Asian part of Russia (Chukotka peninsula). Enlarged concentrations of silicon (15–30 mg/L) were found in surface waters, especially in lakes. The source

Table 26. Biogeochemical accumulation of silicon species and carbonate in Meadow Steppe soil profile of Ambiseli plain, Kenya (Kovda, 1984).

Soil layer, cm	Horizon	Chemical composition			
		pH	Water soluble SiO_2, ppp	KOH soluble SiO_2, %	Calcium carbonate, %
0–4	Upper salt crust	8.2	81	28.3	30.2
5–25	Pore hardpan	9.0	204	37.2	28.8
25–50+	Concrete hardpan	8.2	20	0.2	26.8

of silicon is chemical weathering of geological rocks. The deep diatomic and trepel layers (1–5 m) are widespread in bottom depositions of lakes.

A similar modern accumulation of silicon, as opal mineral, was monitored in Boreal Peat ecosystems in Belarussia and in floodplains of various rivers in the Central Russian Plain. The biogeochemical mechanism of this silicon species formation is connected with the evapotranspiration of ground waters enriched in silicon by plants and deposition of amorphous silicon (Kovda, 1984).

On the other hand, the abundant silicon powder formation was monitored in Podsoluvisols and Phaerozems at 0.5–2.0 m depth in Forest Steppe Ecosystems of Central Russian Plain, South Siberia, Amur River valley, and Manjury region of China. In this case, the most reasonable explanation is connected with the deposition of silicon from glacial melting waters after the glaciation period.

Periodic soil freezing leads also to the Si deposition from soil and ground waters as a powder, SiO_2 sand and opal-like materials. The physical mechanism of Si deposits formation due to winter freezing is similar to that of summer evaporation. Diatomic algae seem to accelerate this process. The chemical composition of Si-containing deposits is shown in Table 27.

The secondary fine-crystal and amorphous quartz material is deposited around living fine roots of trees, forming 'silicon powder' as the spots, nets or bunch at the depth of 20–80–100 cm (Figure 37)

The accumulation of silicon in the upper 40 cm of Podsoluvisols from Boreal Forest ecosystems achieves up to 55 kg/ha and in 0–200 cm layer, up to 80–100 kg/ha of SiO_2. On the contrary, the accumulation of iron, aluminum and titanium oxides was calculated as much as a few kg per ha (Kovda, 1984).

Wet Tropical Ecosystems

In the lower terraces of the Amazon river the development of soils enriched in silicon species have been monitored during the 1960's and 1970's. Similar soils have been

Table 27. Chemical composition of opal deposits and crusts from soils in Amur-Manjury River watershed, Russia-China, % (after Kovda, 1984).

Samples	5% KOH extraction		
	SiO_2	Al_2O_3	Fe_2O_3
Crust materials from Brunozems of Amur river basin	50.2	0.6	0.2
Silicon powder in Meadow Steppe soils of Amur river basin	39.3	1.7	0.1
Silicon powder in Floodplain soils of Amur river basin	4–10	No	No
Opal formation in Meadow Steppe soil of Manjury river watershed	22.03	No	3.53

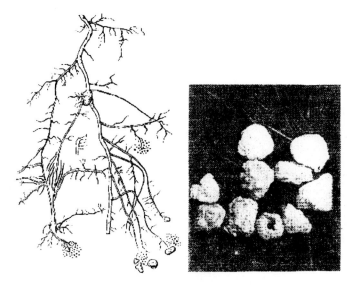

Figure 37. Formation of secondary quartz material in rhizosphere of spruce roots (Nazarov, 1983).

described in Indonesia in 1970's (Driessen *et al*, 1976). The geochemical catena of these soils is shown in Figure 38.

The most interesting biogeochemical phenomenon connected with Si, Al and S species formation was monitored in delta-marine soils of Mangrove ecosystems. At pH 2–3 of soil solution, even silicon minerals become soluble and the content of SiO_2 reaches 5–8 mg/L. In soil and ground waters of salty soils of Mangrove ecosystems of delta low plains of Senegal (West Africa), SiO_2 concentrations are 20–50 mg/L with dry seasonal maximum up to 90–100 mg/L. The evaporation of these solutions is accompanied with deposition of amorphous silicon gel, which is crystallized further. The long term duration of this process has led to the formation of

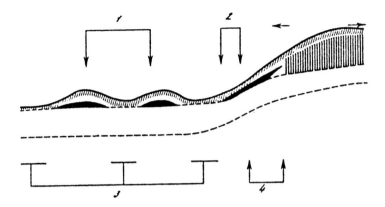

Figure 38. Geochemical profile of Si-enriched soils in Indonesia. 1-hydromorphic Podzol; 2-Deep Wet Humic soils; 3-Deep Tropical Podzols; 4-Hydromorphic Podzols; dash line-groundwater level; horizontal arrows-degree of Si enrichment (After Driessen et al, 1976).

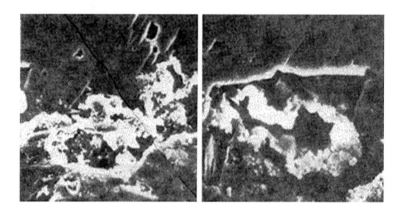

Figure 39. Secondary amorphous accumulation of SiO_2 *in hydromorphic soils in Mangroves ecosystems of Senegal, West Africa (Kovda, 1984).*

Si-enriched A1 - A2 soil horizon, underlined by accumulation of R_2O_3 (Al_2O_3 and Fe_2O_3) oxides and hydrogenic clays, deposited in the lower part of capillar ground waters (Figure 39).

The modern clearcutting of Tropical Forest ecosystems leads to soil erosion and destruction of silicon enriched horizons. The deeper and more ancient soil layers with Fe and Al high contents, so called laterite and hardpan, become common for many hilly tropical and subtropical agricultural areas.

6.5. *The Importance of Silicon in Coastal Ecosystems*

Although nitrogen is the element primarily controlling eutrophication in estuaries and coastal seas, and phosphorus is the element primarily controlling eutrophication in lakes, other elements can have a major influence on the community structure of aquatic ecosystems and can influence the nature of the response to eutrophication. A key element in this regard is silica (silicon), an element required by diatoms. The availability of silica in a water body has little or no influence on the overall rate of primary production, but when silica is abundant, diatoms are one of the major components of the phytoplankton. When silica is in low supply other classes of algae dominate the phytoplankton composition.

Inputs of biologically available silica to aquatic systems come largely from weathering of soils and sediments. The major human influence on silica delivery to coastal marine systems is to decrease it, as eutrophication in upstream ecosystems tends to trap silica before it reaches the coast. Thus the concentration of silicate in the Mississippi River water entering the Gulf of Mexico decreased by 50 percent from the 1950s to the 1980s, a time during which nitrogen and phosphorus fluxes and concentrations increased.

Eutrophication in a system can further decrease silica availability as it is deposited with diatoms and stored in bottom sediments, as demonstrated in the Baltic Sea. The decrease in silica availability, particularly if accompanied by increases in nitrogen, may encourage the formation of some blooms of harmful algae as competition with diatoms is decreased (NRC, 1993). For all cases where long term data sets are available on silica availability in coastal waters, a decrease in silica availability relative to nitrogen or phosphorus has been correlated with an increase in harmful algal blooms.

Decreasing silica availability and the consequent lower abundances of diatoms also lowers the sedimentation of organic matter into bottom waters, and thereby have a partially mitigating influence on low-oxygen events associated with eutrophication. In many coastal systems there may, however, still be sufficient silica to fuel diatom blooms during the critical spring bloom period when the majority of sedimentation often occurs. Further, eutrophication can lead to other complex shifts in trophic structure that might either increase or decrease the sedimentation of organic carbon.

6.6. *Global Pools and Fluxes of Silicon*

The average Si content in terrestrial biomass is 0.5% (on dry mass basis) that corresponds to 12.5×10^9 tons. In marine plankton the average content is 5% and relevant pool is 0.17×10^9 tons. The annual biological Si cycle in the global terrestrial ecosystem takes about 0.9×10^9 tons/yr.

The biological cycle of silicon in the global ocean is driven by diatomaceous and radiolarian plankton algae. The exact estimate of this cycle is uncertain at present. Dobrovolsky (1994) has assumed that using the 5% Si content in plankton, the biological turnover might be 5.5×10^9 tons/yr. In the soil humus and other types of dead organic matter, Si average content is about 1%, which means 31×10^9 tons on the global scale.

Table 28. Major fluxes of global silicon cycle (After Dobrovolsky, 1994).

Sub-cycles	Si, 10^9 tons/yr.
Ocean	
Photosynthetic organisms	5.5
Land	
Photosynthetic organisms	12.5
Dead organic mater	31
Riverine discharge	
Soluble inorganic species	0.2
Particulate matter	4.8
Wind transport	0.47

Neutral hydroxide $Si(OH)_4$ is predominant in the natural water, the content of anion $Si(OH)_3O^-$ is in a lesser degree. The continental river water discharge is responsible for 0.2×10^9 tons of soluble silicon species. The mass of Si compounds in the ocean is 4, 110×10^9 tons, and the residence time of Si in the marine waters is 20,550 years. The transport of silicon from terrestrial to oceanic ecosystems is not counterbalanced by the reverse transport. In addition to the soluble species, the content of silicon in river particulate matter is about $120 \ \mu g/L$. This gives the elemental transport of 4.8×10^9 tons/yr. The total estimate of river water fluxes from the global land area to the ocean is 5.0×10^9 tons/yr. Aeolian migration of silicon is responsible for 0.47×10^9 tons per year. It means the annual global land losses (river and wind fluxes) are 5.47×10^9 tons (Dobrovolsky, 1994).

The soluble Si species in the global ocean account for less than 0.001% of the mass of this element in sedimentary rocks. According to Ronov (1976), the mass of sedimentary rocks contains 44.03% of SiO_2, which corresponds to 493.6×10^{15} tons of silicon. The total pool of this element in the granite layer and sedimentary shell amounts to 2.918×10^{18} tons (Table 28).

Modern calculations show that during the Earth's geological history about 17% of the total mass of silicon has been release via geochemical and biogeochemical cycling.

7. BIOGEOCHEMICAL CYCLE OF CALCIUM

7.1. Calcium in the Earth

Calcium (Ca) is ranked among the major elements of the Earth's crust; its content is 3.6%. The Ca content decreases with depth of the lithosphere from 5.8% in the basalt layer until 2.7% in granite. The calcium minerals were formed at the earlier stages of magma crystallization (see Chapter 2, Section 2). The high crust content of Ca provides the occurrence of 385 mineral species, of which a half accounts for deep-layer silicates. However, owing to its large size, Ca^{2+} cation is incapable of entering the structure of hypergenic silicates.

Ronov (1976) estimated the average CaO content in sedimentary layer of 15.91%, and in granite layer, of 2.71%. Accordingly, the calcium reservoir in sedimentary shell is 272.8×10^{15} tons, and in the granite pool is 222.8×10^{15} tons. The weathering and metamorphosis of deep-layer silicates is accompanied by the formation of clay minerals with release of calcium available for plant and microbial uptake.

7.2. Calcium Pools and Fluxes in the Biosphere

Calcium plays an important role in the physiology of living organisms. This element is involved in the carbon and nitrogen metabolism in plants and it is essential for building up the inner and outer skeleton of animals. Calcium takes part in a great variety of physiological processes. For example, it regulates blood coagulation. The average content of Ca in plant issues is from 0.9% (Bazilevich, 1974) up to 1.8% (Boven, 1966). Accordingly, the calcium reservoir in the terrestrial plant biomass is between $22.5–45 \times 10^9$ tons. The oceanic photosynthetic organisms are depleted by Ca content in their tissues and the corresponding reservoir is 0.034×10^9 tons, being three orders of magnitude less than in the land biomass. Using a modern estimate of 0.5% for Ca content in dead organic terrestrial matter, this pool is 15×10^9 tons. Thus, the values of calcium pools in dead and living biomass of terrestrial ecosystems are similar.

In the oceanic soluble organic matter a reasonable average Ca content is 0.5%, and it represents 20×10^9 tons in the corresponding reservoir.

The water soluble inorganic calcium compounds, most commonly bicarbonate, $Ca(HCO_3)_2$, leaches permanently to river waters and finally migrates to the ocean. This bicarbonate forms in the reactions of calcium carbonate with carbonic acid.

7.3. Solubility of Calcium Species in Natural Waters

Calcium carbonate has intrinsically low solubility in water, $K_{sp} = 6 \times 10^{-9} \, (mol/L)^2$ at 25°C. However, because the carbonate anion is basic, $CaCO_3$ becomes increasingly soluble as the pH drops. It is known that even unpolluted rain water has pH 5.6 and the rainwater is slightly acidic. The reaction may be summarized as

$$CaCO_3(s) + H_2CO_3(aq) \rightarrow Ca(HCO_3)_2(aq). \qquad (12)$$

This reaction is responsible for formation, over thousands of years, of caves and gorges in limestone areas, as CO_2-laden rainwater very slowly dissolves the rock. The process is slow because the equilibrium constant is small. This constant can be estimated as

$$CaCO_3(s) \leftrightarrow Ca^{2+}(aq) + CO_3{}^{2-}(aq) \qquad K_1 = K_{sp} \text{ for } CaCO_3$$

$$H^+(aq) + CO_3{}^{2-}(aq) \leftrightarrow HCO_3{}^-(aq) \qquad K_2 = 1/K_a \text{ for } HCO_3{}^-$$

$$H_2CO_3(aq) \leftrightarrow H^+(aq) + HCO_3{}^-(aq) \qquad K_3 = K_a \text{ for } H_2CO_3 \qquad (13)$$

$$\overline{CaCO_3(s) + H_2CO_3(aq) \leftrightarrow Ca(HCO_3)_2(aq) \qquad K_4 = K_1 \times K_2 \times K_3}$$

where $K_4 = (6.0 \times 10^{-9})(1/4.8 \times 10^{-11})(4.2 \times 10^{-11}) = 5.3 \times 10^{-5} \, (mol/L)^2$.

Due to the dynamic equilibrium between the atmospheric carbon dioxide and the oceanic bicarbonate and carbonate anions, the greatest amount of soluble calcium cations is contained in the ocean. This mass is four orders of magnitude higher than the total mass of bound calcium in living and dead matter of both terrestrial and aquatic organisms. The average calcium content in the seawater is 408 mg/L, and the overall pool is 559×10^{12} tons.

The actual calcium concentration in the marine waters is about 30 times higher than in the riverine waters. This is owed to the limited solubility of calcium carbonate and, especially, related to the extensive calcium uptake by planktonic organisms followed by calcium deposition as pellets. These processes have facilitated the vast accumulation of calcium as a component of massive layers of limestone, dolomite, marl, calcareous clay, and other Ca-containing rocks (see Box 8).

Box 8. Why calcium carbonate does not spontaneously precipitate? (After Bunce, 1994)

An interesting paradox about seawater is that calcium carbonate does not spontaneously precipitate, but neither do sea shells on the beach dissolve. This suggests that the oceans are not far from equilibrium with respect to the system $CaCO_3(s)/Ca^{2+}(aq)/CO_3{}^{2-}(aq)$. However, this precipitation is not confirmed by calculations unless considerable care is taken.

At 15°C, K_{sp} for $CaCO_3$ is $6.0 \times 10^{-9} \, (mol/L)^2$. From the ionic concentration (Table 29) we can write:

$$Q_{sp} = c(Ca^{2+}, aq) \times c(CO_3{}^{2-}, aq) = (0.010) \times (2.7 \times 10^{-4})$$
$$= 2.7 \times 10^{-6} \, (mol/L)^2. \qquad (14)$$

Hence $Q_{sp} \gg K_{sp}$, and according to this calculation, the oceans are vastly supersaturated with respect to calcium carbonate. We would therefore expect $CaCO_3$ to precipitate spontaneously. It turns out that there are two factors which have been overlooked in this simple calculation, namely ionic strength and ion complexation. We will consider these processes.

Ionic Strength

According to chemical laws, equilibrium constants have to be written in terms of activities rather than concentrations. Activity is the effective concentration, by definition.

Table 29. Ions in seawater.

Ion	Concentration, mol/L	Input from rivers, 10^{10} mol/yr.	Residence time, 10^6 yr.
Ca^{2+}	0.010	1220	1
Mg^{2+}	0.054	550	10
Na^+	0.46	900	70
K^+	0.010	190	7
Cl^-	0.55	720	100
$SO_4{}^{2-}$	0.028	380	10
$HCO_3{}^-$	0.0023	3200	0.1

For low concentrations, we can usually approximate by writing concentration where we real mean activity. Since:

$$\text{Activity} = \text{activity coefficient} \times \text{concentration} = a \times c. \qquad (15)$$

The approximation is equivalent to saying that the activity coefficient is unity. The approximation fails in a solution of high ionic strength such as seawater. Activity coefficients appropriate for seawater are 0.26 for Ca^{2+} and 0.20 for $CO_3{}^{2-}$. Now the reaction quotient looks like this:

$$Q_{sp} = a(Ca^{2+}, aq) \times a(CO_3{}^{2-}, aq) = (0.010 \times 0.27) \times (2.7 \times 10^{-4} \times 0.20)$$
$$= 1.4 \times 10^{-7} \text{ (no units, activities are dimensionless).} \qquad (16)$$

This represents an improvement of nearly an order of magnitude, but we would still predict that the solution is supersaturated with respect to $CaCO_3$.

Complexation

The ionic strength/activity coefficient effects arise because at high ionic strength a particular ion (say, Ca^{2+}) is not completely free in solution; it will be surrounded by ions of opposite charge. This is a generalized effect in that the identities of these counter ions are not important. Besides this general effect, the ion in question can associate with specific counter ions to form recognizable chemical species whose concentrations can be measured. For Ca^{2+}, these include $(CaSO_4)^0$ and $(CaHCO_3)^+$. These tight ion pairs are observable chemical species. Note that $(CaSO_4)^0$, for example, is an aqueous entity, different from $CaSO_4(s)$. As an aside, the existence of these complexes is the reason that simple K_{sp} calculations frequently underestimate the true solubility of ionic complexes. Calculations have shown that in seawater, Ca^{2+} and $CO_3{}^{2-}$ are speciated as shown in the Table 30.

Table 30. Complexation of species in seawater.

Species	K_{assoc}, L mol^{-1}	% of total
Calcium, total concentration: 0.010 mol/L		
Ca^{2+}(aq)	—	91
$(CaSO_4)^0$	2×10^2	8
$(CaHCO_3)^+$	2×10^1	1
Hydrocarbonate, total concentration: 0.0023 mol/L		
HCO_3^-(aq)	—	75
$CaHCO_3^+$	2×10^1	3
$MgHCO_3^+$	1×10^1	12
$(NaHCO_3)^0$	2	10
Carbonate, total concentration: 0.00030 mol/L		
CO_3^{2-}(aq)	—	10
$(MgCO_3)^0$	3×10^3	64
$(CaCO_3)^0$	3×10^3	7
$(NaCO_3)^0$	3×10^1	19

We can now put the finishing touches to our Q_{sp} calculations.

$$Q_{sp} - a(Ca^{2+}, aq) \times a(CO_3^{2-}, aq)$$
$$= (0.010 \times 0.27 \times 0.91) \times (2.7 \times 10^4 \times 0.20 \times 0.10) \qquad (17)$$
$$= 1.3 \times 10^{-8} \text{ (no units, activities are dimensionless).}$$

We are now within a factor of two to the value of K_{sp}, and have shown $CaCO_3$ in seawater to be close to saturation, as found experimentally.

For the global calcium fluxes, the biological cycle and the aqueous migration of ions with the land-ocean system are of primary importance. About $1.5–3.1 \times 10^9$ tons of Ca per is annually involved in the biological cycle in terrestrial ecosystems. In the oceanic ecosystems, this value is about 1.1×10^9 tons/yr.

Table 31. Major fluxes of global calcium cycle.

Sub-cycles	Ca, 10^9 tons/yr
Ocean	
Photosynthetic organisms	0.034
Land	
Photosynthetic organisms	22.5–45.0
Dead organic mater	15
Riverine discharge	
Soluble inorganic species	0.484
Particulate matter	0.471
Wind transport	0.048
Deposition	
Over ocean	0.19
Over land	0.20
Ocean-land transport	0.02

7.4. Overall Global Biogeochemical Fluxes of Calcium

The annual continental river discharge of Ca^{2+} ions to the ocean is 0.48×10^9 tons. The similar value is discharged in suspended form, 0.47×10^9 tons/yr. In addition, 0.048×10^9 tons/yr. is windblown off the land into the ocean. The average calcium content in the precipitation over the ocean is 0.36 mg/L (Savenko, 1976). Therefore, the total annual deposition flux of Ca over the global ocean is 0.16×10^9 tons, including about 20% of dry deposition, 0.03×10^9 tons/yr. (Dobrovolsky, 1994). The total amount of Ca entering the atmosphere from the global ocean is about 0.20×10^9 tons annually. About 0.02×10^9 tons/yr. is transported to terrestrial ecosystems, and the rest is returned to the ocean (Table 31).

The average calcium content in deposition over land is about 3 mg/L, and the annual flux to the global terrestrial ecosystem is 0.34×10^9 tons. Thus, the overall annual calcium fluxes in the land-atmosphere system is 0.41×10^9 tons per year.

We can see that the major fluxes are related to biogeochemical soil-plant exchange in terrestrial ecosystems. These fluxes are 1–2 orders of magnitude higher than those fluxes in oceanic ecosystems and the deposition fluxes from the land to the ocean and back. The averaged global biogeochemical fluxes of calcium in the atmosphere-land-ocean system are $38.0–60.5 \times 10^9$ tons per year and these deviations are related mainly to uncertainties in estimating Ca contents in land photosynthetic organisms.

FURTHER READING

Vicror A. Kovda, 1984. Biogeochemistry of soil cover. Moscow: Nauka Publishing House, 159–180.

Samuel S. Butcher, Robert J. Charlson, Gordon H. Orians and Gordon V. Wolfe, Eds. (1992). Global Biogeochemical Cycles. Academic Press, London *et al*, 239–316.

Beldrich Moldan and Jiri Cherny, Eds. (1994). Biogeochemistry of Small Catchments. John Wiley and Sons, 229–299.

Vsevolod Dobrovolsky (1994). Biogeochemistry of the World's Land. Mir Publishers, Moscow/CRC Press, Boca Raton-Ann Arbor-Tokyo-London, 105–205.

Robert W. Howarth, Ed. (1996). Nitrogen Cycling in the North Atlantic Ocean and its Watersheds. Kluwer Academic Publishers, Dordrecht-Boston-London, 304pp.

William H. Schlesinger (1991/1997). Biogeochemistry. An Analysis of Global Changes. Academic Press, 308–348.

T. Fenchel, G. M. King and T. H. Blackborn (1998). Bacterial Biogeochemistry. Academic Press, London *et al*, 43–165.

Vladimir Bashkin and Soon-Ung Park, Eds. (1998). Acid Deposition and Ecosystem Sensitivity in East Asia. NovaScience Publisher, USA, 95–122, 229–268.

QUESTIONS AND PROBLEMS

1. Provide a simple explanation for the reason why the major biogeochemical cycles are intimately interconnected.

2. Discuss the exogenic and endogenic cycles showing interchanges of matter between the biosphere, oceans, atmosphere and geological rocks. Why are these features usual to the biogeochemical cycles of different elements?

3. Discuss the role of biogeochemical activity of living matter in the formation of chemical composition of modern atmosphere.

4. What is the concept of mean residence time and why is it useful in consideration of average atmospheric composition?

5. Describe the differences between the chemical composition of natural unpolluted atmosphere and that of volcano eruption. Explain the possible role of biogeochemical processes in air composition changes.

6. Characterize the interactions between living matter and the hydrosphere as one of the global processes occurring in the biosphere. Give some examples.

7. Discuss the possible role of biogeochemical processes in solubility of gases in natural waters. Give some characteristic examples regarding the Henry's law constants.

8. What are the main chemical species in natural waters? What biogeochemical processes are connected with these aquatic chemical species? Give some examples for C, S and Cl.

9. Discuss the role of microbial communities in governing biogeochemical cycles of macroelements in water.

10. What soil processes are the most important in biogeochemical cycling in terrestrial ecosystems? Present the characteristic element and discuss the role of soil in its turnover.

11. Describe the role of soil organic matter in biogeochemical processes. Why can the upper humus layer be considered as biogeochemical barrier?

12. Discuss the nature of physical and chemical weathering in soil and soil forming geological rocks. Make some connections with biogeochemical turnover of elements in any simulated ecosystem.

13. Highlight the main characteristic features of the natural biogeochemical cycle of carbon. Explain why the ocean can not fully soak up the excess of carbon dioxide in the atmosphere.

14. Compare the main pools and fluxes of carbon in terrestrial and aquatic ecosystems. Explain the relevant role of living matter in these ecosystems.

15. What would be the consequences of man-made disturbance of the natural carbon biogeochemical cycle? Choose an example and present a biogeochemical explanation.

16. Describe the main links of the nitrogen biogeochemical food web in natural environment. Highlight the role of microbes in biogeochemical turnover of this element.

17. Compare the significance of denitrification in various ecosystems. Give the characteristic reactions of this process and describe the quantitative role of denitrification in terrestrial and marine environments.

18. Table 17 depicts data on the global biogeochemical cycle of nitrogen calculated by various authors. Describe these results and explain the possible reasons of data deviations.

19. Discuss the role of phosphorus in the biosphere and its ratios with other essential nutrients. Why may this element be the most important in productivity limitation?

20. Characterize the global biogeochemical cycle of phosphorus. Note the similarity and differences with natural biogeochemical cycles of carbon and nitrogen.

21. Present examples of natural sulfur compounds and describe their role in biogeochemical cycling of this element. Highlight the application of sulfur isotopic composition for quantitative estimates of sulfur turnover in the biosphere.

22. Microbial sulfur transformations in the system hydrogen sulfide-sulfate-dimethyl sulfide play the most important role. Compare the role of these transformations in terrestrial and aquatic ecosystems.

23. Compare the historical and modern data of the global sulfur biogeochemical cycle. Explain the relevant importance of natural and anthropogenic sources and sinks of this element in various compartments of the biosphere.

24. Discuss the biogeochemical migration of silicon species and compare this migration in arid and wet ecosystems. Explain the geochemical and biogeochemical mechanisms of this process.

25. Explain the role of living organisms in the global biogeochemical cycle of silicon. Compare the terrestrial and oceanic ecosystems.

26. Estimate the solubility of calcium species in natural waters. Rewrite the relevant chemical reactions and calculate the corresponding constants.

27. Why do calcium carbonate not spontaneously precipitate in seawaters? Present relevant explanations.

28. Discuss the global biogeochemical cycle of calcium. Draw your attention to the most important fluxes and pools. Explain the importance of calcium biogeochemical fluxes in terrestrial ecosystems for the global budget of this element.

CHAPTER 4

BIOGEOCHEMICAL CYCLING
OF TRACE ELEMENTS

In accordance with modern classification, the trace elements are those with content in the Earth (clarks) about $1 \times 10^{-4}\%$ or less. Many trace elements are important elements in the biogeochemistry of any terrestrial or aquatic ecosystem and some of them are required micro-nutrients for plant, human, or animal life (Figure 1).

At very low concentrations or availability living organisms may indicate a deficiency of certain trace elements, which serve as nutrients (Kovalsky, 1984; Simkins and Taylor, 1989). Anthropogenic activity, such as fossil fuel combustion, transportation, industrial processes and mining, have greatly altered the biogeochemical cycles of many trace elements and enhanced their bioavailability (Driscoll *et al*, 1994). At elevated concentrations or availability some trace metals may be toxic, for instance, V, Cr, Mn, Co, Ni, Cu, Zn; a similar toxicity may be shown by B, Mo, Se and As as well. These metals and non-metals have known biochemical and physiological functions in living organisms' metabolism and they are active migrants in biogeochemical food webs.

However, there is a set of trace elements with unknown physiological or biochemical functions and their accumulation in biogeochemical cycles leads to acute or chronic poisoning of both plant and animal organisms. Amongst them, the most dangerous are Pb, Cd, U, and Hg (see Figure 1). The latter will be considered in Chapter 8 "Environmental biogeochemistry". In this chapter the main attention will be given to trace essential elements, like copper, selenium, boron, molybdenum, and zinc.

1. BIOGEOCHEMISTRY OF COPPER

Amongst various trace elements, copper plays the most important essential role. Copper is the main constituent of many oxidation-reduction ferments, in which Cu forms a stable complex with a specific protein. Copper is known to be metabolically essential for virtually all organisms. It displays the well-known trace metal properties, being essential to growth at low concentrations and toxic at high concentrations. The requirement for copper stems from its inclusion in several proteins, in which it is always coordinated with N, S, or O ligands. At the other extreme, copper's toxic properties have been exploited to reduce growth of unwanted organisms such as algae and fungus in soil and water bodies. For instance, the Cu-containing ferment plastociannin being the most abundant in plant chloroplasts takes part in photosynthetic reactions.

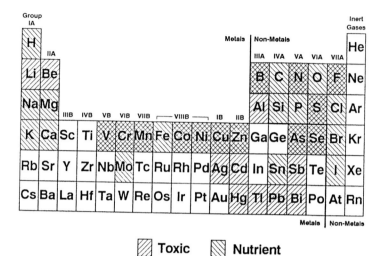

Figure 1. The periodic table showing the essential elements for plant or animal life with an indication of their toxicity. The division between nonmetals and metals is given (possible, arbitrary division).

The deficit of this element leads to delay in crop maceration, as well as to different anemia, disease of bone systems and endemic ataxia of animal organisms. However, the excessive uptake and accumulation of copper in animals is accompanied by hemolytic jaundice, kidney disease, and by plant chlorosis.

1.1. Copper Speciation

The local and global biogeochemical cycles of copper have been extensively studied. Copper biogeochemistry has been discussed and reviewed frequently in recent years (see for example, Benjamin and Honeyman, 1992). The researches during the last half of the twentieth century provide considerable insight into the flux rates among various reservoirs in terrestrial and aquatic ecosystems. The speciation of copper in natural waters has also been well studied. In recent years a rough consensus seems to have been reached regarding the dominant copper species in various environments. For instance, the predominant inorganic species are Cu^{2+}, $CuCO_3$ and $CuOH^+$ and among solid phases regulating aqueous concentration of copper are CuS, $CuFeS_2$, $Cu_2CO_3(OH)_2$, $Cu_3(CO_3)_2(OH)_2$, $Cu(OH)_2$ and CuO (after Morel, 1983). However, the exact values of the stability constants of some important copper complexes are still somewhat uncertain (see below). Moreover, some questions remain regarding the gross inventories of copper in various environmental reservoirs. In this section, only an outline of the problem will be provided, with emphasis on how the speciation of copper controls its biogeochemical cycling.

Copper exists in crustal rocks at concentrations ranging from about 10 to a few hundred ppm(m), and 70 ppm(m) is an average value. More than 20 copper-containing minerals have been identified with different oxidation states of this metal (0, +I, and +II). The chalcopyrite ($CuFeS_2$) is the most common and others include sulfides, hydroxides, and carbonates (see above). Copper is also found in relatively high content (> 0.5%) in deep-sea ferromanganese nodules. Solids containing oxidized anions (carbonates, sulfates, hydroxides, and oxides) are the dominant forms of Cu in airborne particles and about 50% of this copper is soluble. Since the solubility is strongly dependent on pH values, acid or alkaline precipitation and acidification/alkalinization of surface waters may have a significant effect on the forms and fate of copper compounds in waters and soil solutions.

Copper in solution, as in solids, can carry either a +1 or +2 charge. The divalent form is the prevalent one in natural waters as the most stable one in oxygenated conditions. The migration and transport of Cu-containing compounds are practically dependent on the valence, since they form complexes with different ligands. Cuprous (Cu^+) ions react strongly with formation of Cl^--bonds. This is the most important for marine water where the chlorine ions are abundant. In contrast, cupric ions (Cu^{2+}) form strong complexes with carbonate, phosphate, hydroxide and ammonia and with many organic molecules. For instance, dissolved Cu is associated primarily with humic or fulvic acids, in spite of rather low concentrations of these organic species. It has been calculated that in sea waters with a total humic acid concentration of 10^{-6} mol as carbon, organic complexes could account for 47% of total dissolved copper (Turner *et al*, 1981). Most non-organically bound Cu^{2+} is complexed with inorganic carbon. In freshwater most copper (> 90%) is bound with organic humic and fulvic acids since the content of these acids is much higher than in marine environments. In general, organic complexation is predominant in any surface waters containing more than 30 mg/L of organic carbon. Only in extremely pristine waters, hydroxide and carbonate complexes may dominate.

Taking into account the physicochemical similarities between complexation and adsorption reaction mechanisms, one can conclude that divalent copper ions are the most strongly sorbing of the various trace metals. This species can sorb onto both inorganic and organic solids. Sorption of copper onto oxides and clays has been investigated extensively (see Box 1).

Box 1. Copper and zinc sorption in soil (after Pampura, 1997)

In accordance with modern theories, the main mechanism of metal sorption in various soils is the heterogenic surface sorption of metals by soil particles. In the sorption process, the various metals react with different functional groups of various soil compounds, like clay minerals, organic matter, oxides of Fe, Mn, Al, Si, etc. Various mechanisms of Me sorption in soil and soil solution are connected with various forms of their existence in soil such as exchangeable, specifically sorbed, occluded by oxides and hydroxides, bounded with organic matter and included into interlayer spaces of silicon minerals. In this study Tessier *et al* (1979)'s method has been

Table 1. The method of subsequent metal extraction from soil sample (Tessier et al, 1976)

Extractant	Extraction duration, h	Extracted metal forms
1 M MgCl$_2$, pH7	1	Exchangeable and surface sorbed, Me$_{ex}$
1M CH$_3$COONa + CH$_3$COOH, pH 5	5	Carbonate bounded, Mecarb
0.04MNH$_2$OH · HCl in 5% CH$_3$COONH$_4$	8	Bounded with Fe and Mn oxides, Me$_{(Fe+Mn)ox}$
H$_2$O$_2$/HNO$_3$, pH 2 and then CH$_3$COONH$_4$	5.5	Organic matter bounded, Me$_{org}$
Total destruction in HF/HClO$_4$ mixture	24	Residual, included in mineral crystal grid, Me$_{res}$

applied to estimate the sorption of copper and zinc in Typical Chernozem, Russia. The description of this method is shown in Table 1.

In initial Chernozem, the copper (30 ppm) was distributed as follows:

$$Cu_{res}(92\%) \gg Cu_{org}(7\%) \gg Cu_{(Fe+Mn)ox}(0.3\%) \quad (1)$$

Other forms (exchangeable and carbonate bounded) were not detected. The distribution of zinc (76 ppm) was different:

$$Zn_{res}(59\%) > Zn_{(Fe+Mn)ox}(33\%) > Zn_{ex}(7\%) > Zn_{org}(0.6\%) > Zn_{carb}(0.3\%). \quad (2)$$

The total sorption was 1200 ppm for Zn and 750 ppm for Cu after experimental treatment of Chernozem samples with relevant metal salts. The illustration of metal distribution is shown in Figure 2.

We can see that after treatment, the relative content of Me$_{res}$ was sharply decreased. The predominant sorption of Zn was connected with Zn$_{ex}$ and Zn$_{carb}$ forms, whereas the predominant copper sorption was in Cu$_{org}$ form. The role of Me$_{(Fe+Mn)ox}$ form was significant for both studied metals. In general, increasing amount of sorbed metal was connected with the increase of less bounded forms.

This sorption, especially in the soil system, is very important both from points of view of biogeochemical cycling and plant availability of Cu from these complexes.

The strength of the copper-organic bond and the well established dual potential of copper as a biological stimulant or inhibitor has led to many studies of the effect of copper speciation on plant growth. The general conclusion from these researches is related to the more biological activity of hydrated Cu^{2+} than that of chelated and some organically or inorganically complexed forms. The toxicity of copper does not correlate with the total content of this metal or even total soluble copper in a

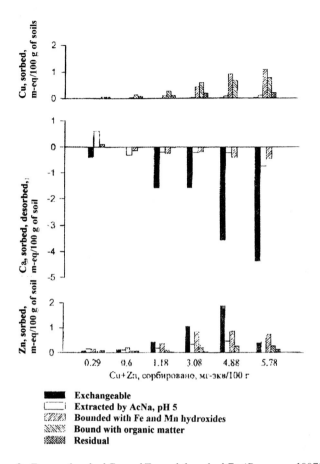

Figure 2. Forms of sorbed Cu and Zn and desorbed Ca (Pampura, 1997).

system. However, until now uncertainty remains regarding the toxicity of certain inorganic complexes (particularly the hydroxyl and carbonate complexes) and the roles of alkalinity of soil or alkalinity and hardness of natural waters in modifying Cu toxicity (Box 2).

Box 2. Biogeochemical monitoring of copper in soils and crops of agroecosystems of Southern Russia (after Zakrutkin and Shishkina, 1997)

Let us consider a case study monitoring of Cu in the soils of the Rostov region, southern Russia. The soil-geochemical monitoring was carried on the scale of 1:500,000 and included about 2500 soil and more than 2000 crop samples. The following crops were sampled: winter and spring wheat, barley, ray, corn, rice, beans, alfalfa, Sudan grass, tomato, cabbage, apple, grape and sunflower.

Table 2. Content of total copper in soil samples, ppm

Soil types (Russian classification)*	Limits	Mean
South Chernozem	18–146	42.7
North Pre-Azov Chernozem	18–120	32.6
Pre-Caucasian Chernozem	19–290	40.8
Kastanozem	18–210	39.3

* Various suborders of Mollisol in USDA classification

The contents of copper in soils vary significantly (Table 2).

The minimum contents are in sandy soils of river floodplains. The maximum values (70–290 ppm) monitor in the Pre-Caucasian Chernozem. The zones of elevated concentrations are locally distributed and afforded to both natural and anthropogenic factors. The first factor is related to the biogeochemical barrier in the upper soil humus horizon, with increasing content of organic matter, and the second one is related to the pollution due to application of Cu-containing fungicides in orchards and vineyards.

The content of mobile copper form (extracted by acetate-ammonium buffer with pH 4.8) in all soil types varies from 0.05 to 0.41 ppm, being less than 1% from total content. However, it increases sharply in polluted soils of orchards and vineyards until 2.4–12.5 ppm. Some increase occurs also in rice soils, up to 0.26–0.94 ppm. It is clear that this increase is connected with application of irrigation and fungicides. The comparison of Cu content with maximum permissible levels for soils (55 ppm for total and 5 ppm for mobile forms) allowed the researchers to conclude that about 5% of agroecosystem soils are polluted by copper.

The content of copper in various crops is shown in Table 3.

Taking into account that the crop Cu content of 2 ppm characterizes the physiological deficit, one can note the insufficient availability of copper for corn, rice and in less degree for wheat and barley. This should require the application of copper fertilizers.

In general, the background content of copper in soils of southern Russia is less than known geochemical clark values and those for crops are similar with the background numbers (Cabata-Pendias, 1989).

1.2. Global Cycle of Copper

The global cycling of copper has been reviewed by Nriagu (1979) and shown in Figure 3.

The flux of copper is dominated by transport in rivers. Copper reaching the oceans by atmospheric transport is of the same order of magnitude as that by three strictly anthropogenic sources: direct discharge of wastes and sludge into the oceans, discharge of domestic and industrial wastes into rivers, which eventually discharge into the ocean, and the use of copper-containing anti-fouling paints. Each of these sources

Table 3. Copper content in crops of southern Russian ecosystems, ppm per dry mass

Crops	Limits	Mean
Cereals		
Wheat	1.9–8.9	4.4
Barley	1.3–8.0	4.5
Corn	1.9–6.1	3.6
Ray	3.5–18.0	7.3
Rice	1.3–2.5	1.7
Bean	3.6–11.4	6.8
Vegetables and fruits		
Tomato	5.9–10.8	7.6
Cabbage	2.0–5.6	3.9
Grape	14.0–18.9	9.6
Apple	1.7–7.0	3.4
Fodder grasses		
Alfalfa	2.0–7.0	3.7
Sudan grass	2.1–6.4	3.7
Oil crops		
Sunflower	2.0–14.0	6.1

is several hundred times less than the Cu flux with riverine waters as a part of natural biogeochemical cycle. The air transport occurs mainly with particles and most of this (> 90%) is injected to the atmosphere as a result of anthropogenic activities (smelting, fossil fuel burning, fungicide application).

Despite the relatively small part of anthropogenic copper in its global biogeochemical cycle, atmospheric transport is responsible for transboundary long range air pollution and impact in many regions of the World, especially in Europe and North America.

2. BIOGEOCHEMISTRY OF ZINC

The zinc concentration in the Earth's crust tends to increase on passing from the upper mantle ($3 \times 10^{-3}\%$) toward the basalt ($1.3 \times 10^{-2}\%$) and further on to the granite layer ($6 \times 10^{-2}\%$). Substantial masses of zinc are found in the post-magnetic geological formations. The zinc-lead ore deposits have accumulated over 20×10^6 tons of Zn. This enormous quantity accounts for a mere 0.001% of the zinc mass occurring in a dispersed state in the uppermost granite crustal layer of 1 km depth.

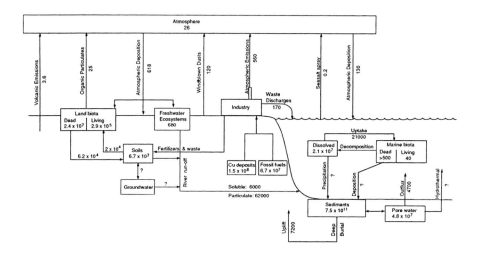

Figure 3. The global copper cycle. The units are 10^2 tons Cu (pools) and 10^2 tons Cu/year (fluxes) (Benjamin and Honeyman, 1992).

2.1. Zinc in Biosphere

Similarly to copper, zinc is an important micro-nutrient. This element is the constituent of more than 200 ferments, which play a significant role in metabolism of proteins, carbohydrates, lipids and nucleic acids. Being one of the essential trace elements, it takes part in synthesis of ribonucleic acids and chlorophill. The Zn-containing enzyme carboanhydrase presents in the erythrocytes. Zinc plays a very important role in the gonadal activity of animals and is involved in the mechanisms providing for the resistance of plants to frost and drought (Kabata-Pendias and Pendias, 1989). Zinc is a physiologically important element and accordingly, in animals and humans the deficiency of this trace metal is accompanied by a delay in growth and development, decrease of immunity, disturbance of skin, bone and blood formation, and nervous diseases. The chlorosis and insufficient growth of leaves are the most characteristic symptoms of zinc deficit in plant species.

Table 4. Content of total zinc in soil samples, ppm.

Soil types (Russian classification)*	Limits	Mean
South Chernozem	32–140	69.1
North Pre-Azov Chernozem	31–140	62.2
Pre-Caucasian Chernozem	31–130	59.9
Kastanozem	24–150	64.8

* Various suborders of Mollisol in USDA classification.

However, the excessive intake and accumulation of zinc leads to toxic effects on all living organisms. This is accompanied with oxygen system depression and anemia. The excessive content of zinc in human organisms induces vomiting, nausea, pneumonia and fibrosis of pulmonary system.

All vegetation species accumulate zinc very extensively. The global mean coefficient of biogeochemical uptake (C_b), defined as the ratio of the average concentration of a metal in the annual production of terrestrial vegetation to the metal abundance in the granite crustal layer, is equal to 12, whereas for lead, for instance, it is slightly higher than 1.

The zinc concentration in terrestrial plants is widely varied depending on the pedo-geochemical and biogeochemical conditions. Plants are known that grow in the areas with abnormally high zinc soil concentrations (zinc-enriched biogeochemical provinces) and which Zn content reaches 10 or even 17% by dry ash weight. These plants are called "galmei flora". At the same time the available data provide information on the relatively small variations of Zn contents in many systematic botanical groups (see Box 3).

Box 3. Biogeochemical monitoring of zinc in soils and crops of agroecosystems of Southern Russia (after Zakrutkin and Shishkina, 1997)

Similarly to copper (see Box 2) the content of zinc in soils varies remarkably, being from 24 to 150 ppm, with <70 ppm for 80% of the monitored area. There are a few zones with excessive content of this trace element, mainly in watersheds of Don-Sal-Manych rivers, where the maximum concentrations exceed the 100 ppm level. The background contents of zinc are similar for different soil types (Table 4).

The averaged values are similar to those determined for chermozemic soils of Kursk (50–54 ppm), and Voronezh (57–76 ppm) regions, for Ukrainian chernozem (42–56 ppm). The clark values of zinc content in the chernozemic and kastanozemic soils of the former USSR (Russia, Kazakhstan, Ukraine, Moldova) are equal to 62 and 52.3 ppm, correspondingly.

The biogeochemical accumulation of this metal occurs in the upper humus horizon. The content of mobile zinc form (acetate-ammonium buffer with pH 4.8) is less than 1 ppm, mainly in the limits of 0.12–0.90 ppm, and this content is known as

Table 5. Zinc content in crops of southern Russian ecosystems, ppm per dry mass.

Crops	Limits	Mean
Cereals		
Wheat	10.0–60.0	27.0
Barley	10.0–54.0	30.3
Corn	18.0–46.0	30.0
Ray	20.0–45.0	30.0
Rice	14.0–18.9	15.8
Peas	17.3–50.5	31.8
Vegetables and fruits		
Tomato	8.7–19.4	13.3
Cabbage	11.0–17.4	12.7
Grape	1.1–8.6	3.6
Apple	1.9–45.9	8.5
Fodder grasses		
Alfalfa	14.0–50.0	28.9
Sudan grass	12.0–63.0	29.8
Oil crops		
Sunflower	12.0–46.0	26.0

deficient even for plants with weak Zn accumulation. The relatively enriched soils are monitored only in orchards and vineyards, where the average content of mobile Zn is increasing up to 2.1 ppm. These values are far away from Russian hygienic standards (23 ppm for mobile form and 150 ppm of total content). The whole area of soils with exceedances of these standards is less than 1%.

The background contents of zinc in various cereal crops are similar and these values are between 27.0 and 30.7 ppm (Table 5). The exceptions are rice and peas.

Taking into account that deficient physiological norm for crops is 10–20 ppm Zn, one can note the pronounced deficit of zinc for rice.

In general, the background content of Zinc in most crops of southern Russia is similar to average global content, 30 ppm, of this trace metal in vegetation (Dobrovolsky, 1994). The content of zinc is not exceeding the average global values for any crops and plants (Cabata-Pendias and Pendias, 1989). However, a deficient content of zinc was monitored for 15% of all wheat and barley samples, 21% of oil flower samples, 14% of fodder grasses, vegetables and fruits. The order of zinc accumulation in various crops is as follows: grape (1.0)→ apple (2.4) → cabbage (3.5) → tomato (3.7) → rice (4.4) → sunflower (7.2) → wheat, barley (7.5) → alfalfa (8.0) → corn, ray (8.3) → bean (8.5 ppm). This is less than maximum permissible levels, which were set for cereals as 50, for sunflower as 30 and vegetables and fruits as 10 ppm of wet mass.

2.2. Biogeochemical Fluxes of Zinc

Zinc in plant-soil system

In the widespread natural plant species of the USA, the zinc concentration falls within a range of 320–640 ppm by dry ash weight (Connor and Shacklette, 1975; Ebens and Shacklette, 1982), whereas the most typical values for plants from the Russian Ural area are 150–750 ppm. According to Brooks (1983), the average zinc content in plants is 50 ppm on a dry mass basis or about 1000 ppm by dry ash. In accordance with different estimates, the reasonable average value for global vegetation is 600 ppm by dry ash weight, or 30 ppm for dry phytomass weight, or 12 ppm for life weight of plants (Dobrovolsky, 1994).

On the basis of this estimate, the overall Zn mass in terrestrial ecosystems might be equal to 75×10^6 tons, and annual growth uptake of this trace metal is about 5.2×10^6 tons/year. The similar annual amount of zinc returns to the pedosphere. Most Zn-containing plant organic matter is easily degradable and after mineralization this metal is removed from plant residues. The average content of zinc in peat and forest litterfall is about 20 ppm by dry weight and in soil humus, 4–5 times higher. Accordingly, the Zn content in peat and forest litterfall is about 14×10^6 tons and that in soil humus is about 200×10^6 tons. The total Zn content in humus horizon of North Eurasian soils varies from 20 to 80 ppm. In the European part of Russia this average value is about 50 ppm and it is very similar to that for USA soils, 48 ppm (Dobrovolsky, 1994; Shacklette and Boerngen, 1984).

From 40 to 60 % of the total soil zinc is complexed to the organic matter and sorbed on iron hydroxide films (see Box 1). In spite of being a very tiny fraction of total soil Zn content, water-soluble species plays a very important role in biogeochemical migration fluxes. The global coefficient of aqueous migration (C_a) for zinc is greater than 3, whereas for other metals, like Pb, it is only 0.5. The average global riverine Zn content is about 20 ppb. This gives the river transport in the World scale as much as 0.82×10^6 tons/year, and the mass transport of suspended forms accounts for 87% of the total riverine flux.

Zinc fluxes in troposphere

Zinc exchange between plant canopy and ground level troposphere is also responsible for the input of this trace metal to the air pool. One square meter of the canopy may annually exude up to 9 kg of Zn as a terpene component (Beaufort et al, 1975). During bacterial biometallization in sub-aquatic coastal areas, the formation and corresponding volatilization of organozinc species take place however, at present there is no quantitative parameterization of these processes in regional or global scale.

Table 6. Zinc content in the Earth's sedimentary rocks (after Turekian et al, 1961).

Sedimentary rock	Zn content, ppm	Zinc mass, 10^{12} tons
Clay and agrillaceous slate	95	108.0
Sandstone	16	6.9
Carbonate	20	14.2
Total	—	129.1

In the background atmosphere over land area, the zinc concentration is between 2 and 70 ng/m^3 (Ostromogilsky *et al*, 1981). A certain amount of zinc is supplied to the atmosphere in mineral dust. The average Zn content in soil is about 50 ppm (see box 3). This gives the estimate of Zn air transport of 0.3×10^6 tons/year, of which about 0.1×10^6 tons can be precipitated over the ocean and the remaining can be deposited over the global land (Dobrovolsky, 1994).

The background Zn content in the precipitation over various regions varies significantly. For instance, the smallest concentrations are in polar and high mountain areas with low particle air content. In the Antarctic snow, the Zn content is about 0.03–0.04 μg/L, however, in the more dust polluted snow samples from Greenland, these values are till 0.3-0.4 μg/L (Herron *et al*, 1977). The highest content of zinc in the snow cover of the Spitzbergen islands (Arctic Ocean) is as much as 31 μg/L, and in high mountain areas of Central Eurasia these values vary from 20 to 50 μg/L.

The monitored rainfall Zn content over background areas is between 10 and 40 μg/L, and in polluted areas these values are much higher. Giving the average content as 20 μg/L, Dobrovolsky (1994) estimated the global deposition flux as much as 2.3×10^6 tons/year.

Volcanic eruptions supply 0.2×10^6 tons of Zn annually to the atmosphere from the inner part of the Earth's lithosphere.

Zinc pools in geological rocks

The zinc content in different sedimentary rocks is shown in Table 6.

We can see that the total mass of Zn in sedimentary rocks is 129.1×10^{12} tons and this is only 30% from the corresponding mass of this element in crustal continental block (419×10^{12} tons). Thus, the accumulation of zinc in sedimentary rocks during the Earth's geological history is equal to 23% from the total zinc mass in the Earth's crust. This mass is in excess of the percentage of elemental mass release (17–19%) by hypergenic metamorphosis of the granite layer. Presumably, a certain additional quantity of this trace metal has been supplied to the biosphere as a result of outgassing processes like volcanic eruption.

Zinc fluxes in the oceanic ecosystems

Water soluble inorganic zinc species are predominant in the global ocean waters. It has been calculated that 90% of suspended and 35% of soluble riverine zinc are deposited at the ocean-land interface (Lisitsin *et al*, 1983). It leads to the annual accumulation of 0.6×10^6 tons of Zn in suspension and about 0.5×10^6 tons in soluble form in the pelagic part of the ocean. The average concentration of soluble Zn in the global marine waters is about 5 μg/L, with relevant amount in ocean water of $6,800 \times 10^6$ tons. The suspended Zn amount is not estimated quantitatively yet.

The existing estimates of Zn content in the oceanic photosynthetic organisms vary from 38 to 850 ppm by dry weight, with average value of 50 ppm (Demina *et al*, 1983). This provides the amount of this trace metal in oceanic photosynthetic organisms as much as 0.17×10^6 tons. This is 440 times less than in terrestrial living matter. However, we have to take into account the high rate of biomass renewal related to a high bioproductivity of the ocean. The simple calculations show that during the year, about 30×10^6 tons of zinc becomes assimilated by the oceanic photosynthetic organisms in multiple cycles of assimilation and dissimilation of plankton biomass. Depending on the region, from 4 to 50% of soluble Zn may be in form of organozinc species.

Over the ocean area the zinc concentration in air ranges widely from 0.05 to $60 \, \text{ng/m}^3$ and the average value of Zn air content over the pelagic regions of the ocean is 7.8 μg/L (Bezborodov and Eremeyev, 1984). The concentration of zinc in oceanic rainwater is about 6 μg/L, and this presents the mass of 2.5×10^6 tons/year. This is a three-fold excess over the riverine discharge of soluble zinc to the global ocean.

The major source of zinc in the troposphere over the ocean is the microbial zinc methylation in marine water (Craig, 1980). Additionally, about 1×10^6 tons/year is related to dust-borne zinc transported from the land areas. In turn, 0.26×10^6 tons of air-borne zinc is transported annually from the ocean troposphere to the troposphere over terrestrial ecosystems and deposited there.

2.3. Global Biogeochemical Fluxes and Pools of Zinc

V. Dobrovolsky in 1994 (Table 7) estimated the global distribution of zinc in the biosphere.

Thus the biogeochemistry of zinc in the biosphere is connected with its main reservoir in the Earth's crust, which is the source of this trace metal for all other links of its biogeochemical cycle in the biosphere.

3. BIOGEOCHEMISTRY OF SELENIUM

Selenium is a chalkophilic chemical trace element. It belongs to group VI of Mendeleev's periodic table, sub-group of oxygen and is similar to sulfur by its chemical properties (Figure 1). Chemical species include the seleneous and selenic acids, selenides, selenites and organoselenides.

Table 7. Zinc mass distribution in various reservoirs of the global biosphere.

Reservoir	Zn mass, 10^6 tons
Global oceanic ecosystem	
Soluble species	6,800
Photosynthetic organisms	0.17
Lower troposphere	0.0028
Global terrestrial ecosystem	
Vegetation	75
Forest litterfall and peat	14
Lower troposphere	0.0005–0.005
Earth's crust	
Sedimentary layer	134,000,000
Granite layer	490,000,000

3.1. Selenium in the Biosphere

The average content of selenium in soil forming rocks (clark) is about 0.3 ppm. The elevated contents of Se occur in some sedimentary rocks (sandstone, shale), volcanic sulfur, sulfides and phosphorites. The contents of this trace element in most soil types are between 0.01 and 1.0 ppm. Chernozems, Kastanozems, Serozems and Floodplain soils are relatively more enriched, 0.3–1.0 ppm, and Podsoluvisols and Arenosols are relatively depleted, 0.05–0.2 ppm.

The content of this element in most living organisms is from 0.01 to 1.0 ppm by dry weight. Selenium is concentrated in some microbial species, fungus, marine organisms and some terrestrial species, like leguminous, cruciferous, and multifloverous, accumulating up to 1 ppm of Se by dry weight. The accumulation of selenium in fodder grasses depends on plant species and soil types. This element is bioconcentrated in ecosystems with weakly alkaline soils. The availability of Se decreases in acid Podzol and Peat soils.

Microbes play an important role in Se migration. They can reduce selenites to metallic selenium and oxidize the selenides. Living organisms can synthesize selenomethionin, selenocistine and form the methylselenium species from inorganic compounds. The following chemical organic species have been determined in plants with concentrated selenium: selenocistionin, selenohomocistein, methylselenomethionin, and selenosinigrin.

The human and animal Se daily requirement is 50–100 ppm by dry weight of food or fodder. The physiological role of selenium in animals and humans is connected

Table 8. The comparison of Se *content in various fodder crops and animal blood with development of diseases.*

Se content, ppb		Disease development
Fodder crops	Animal blood	
3.8–13.0	2–10	No
<3.8	1–2	Myopatia
> 100.0	> 20	Chronic toxicity

with antioxidant properties, stimulation of the phosphorilation process, and decarboxilation of tricarbonic acids. This metal improves the perception of light by the eye's retina. Selenium is a constituent of the important ferment glatutionperoxidase.

3.2. Selenium Enriched Biogeochemical Food Webs

The spatial heterogeneity of selenium content in rocks, various soils, and especially the peculiarities of its translocation from soils to plants, lead to the formation of so-called selenium biogeochemical provinces (see Chapter 7 'Biogeochemical mapping'). These provinces are characterized by a deficient or excessive concentration of this trace metal in all links of biogeochemical food webs. Selenium deficiency in fodder crops is related to the less than 30 ppm content of this element (Table 8). This leads to myopatia (white colored animal tissues), necrotic degeneration of kidney, exude diathesis. The addition of sodium selenite is used for prevention of these diseases.

There are many regions in the World with deficient content of selenium in biogeochemical food webs. In the Eurasian area these regions are in the Baltic Sea states (Estonia, Latvia), the Taiga Forest belt of the European part of Russia, Central and East Siberia (see the map of biogeochemical regionalization of North Eurasia, Chapter 7). As a rule the soils of these areas are formed on deep weathered granite and metamorphic sandstone with a low content of Se, as well as on fluvio-glacial deposition of the tertiary ice formation. Furthermore, the weak crop uptake of selenium is monitored in floodplain and peat soils and this is accompanied also by animal myopatia.

The excessive accumulation of selenium, more than 1 ppm, in crops and grasses is related to soils formed on volcanic ash and sedimentary rocks such as shale and sandstone. This leads to chronic selenium toxicity (Table 7). The acute toxicity is at Se content > 25 ppm by dry weight.

3.3. Case Studies

San Joaquin Valley, California

The characteristic example of Se-enriched regions is the San Joaquin Valley in California, USA (Figure 4).

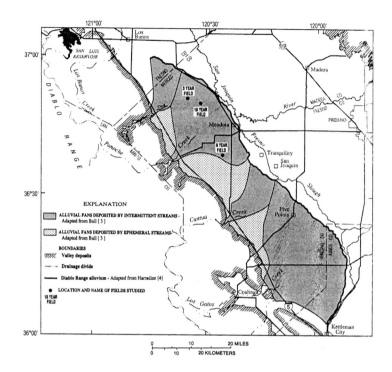

Figure 4. Location and geographic features of the western San Joaquin Valley (Deverel et al, 1994).

The irrigation systems were constructed in this valley in 1930–1960s. It was found afterward that irrigation drainage water from parts of the San Joaquin Valley contains levels of selenium and other trace elements that have been implicated in bird deformations in the Kesterson Reservoir (Tanji *et al*, 1986). Depending on location and season, the drainage water contains 100–1400 μg/L Se, predominantly as selenate (SeO_4^{-2}, Se-IV), the most soluble form, whereas the California State Water Resources Control Board has recommended an interim maximum mean monthly selenium concentration of 2–5 μg/L in receiving waters and wetlands.

Selenium concentrations in drainage and groundwater underlying agricultural areas of the western San Joaquin Valley result from complex interactions amongst irrigated agricultural practices and physical and chemical processes. Although irrigated agriculture in the western valley began in the late 1800s, most of the area remained unirrigated until the 1930s and 1940s. The total area of land irrigated by pumped groundwater increased more than threefold in 1924 and continued to increase through 1955. Surface water imported from northern California replaced groundwater for irrigation from the early 1950s and in the late 1960s and further increased the amount of irrigated acreage.

Irrigation using surface and ground water in this region has substantially altered the physical and chemical nature of the groundwater flow system and finally the total biogeochemical food web.

The deposits of the San Joaquin Valley are derived from the Sierra Nevada on the East and the Coast Range on the West. The Sierra Nevada, a fault block that dips southwestward, is composed of igneous and metamorphic rock of mostly pre-Tertiary age. The Diablo Range of the California Coast Range, which borders the study area to the west, consists of an exposed core assemblage of Cretaceous and Upper Jurassic age and overlain by and juxtaposed with marine deposits of Cretaceous age and marine and continental deposits of Tertiary age. The alluvial fan deposits of the western valley are derived from the Diablo Range. The fans deposited by ephemeral streams are smaller than and enriched by the four major alluvial fans deposited by intermittent streams (Little Panoche, Panoche, Cantua, and Los Gatos Creeks) as shown in Figure 4 (Deverel *et al*, 1994).

Under natural conditions, groundwater discharge was primarily due to an infiltration of water from intermitted streams in the upper parts of the alluvial fans. This was changed during the irrigation period. Pumping of groundwater increased the depth of the water table, and irrigation with pumped groundwater caused downward displacement of soil salts. Replacement of groundwater with surface water imported from northern California in the 1960s caused hydraulic pressures to increase and the water table to rise, creating a need for drainage in low lying areas of the valley. Soil salts, including selenium species, were further displaced in the valley through places where irrigation water generally had not been applied previously. Application of irrigation water caused leaching of soil selenium salts and increased selenium concentrations in the groundwater. Subsequent evaporation of shallow groundwater increased selenium concentrations in the valley through and lower alluvial fans. Drainage systems were installed to prevent evaporation and to leach accumulated salts (Figure 5).

Concentrations of selenium in shallow groundwater generally were less than $20 \mu g/L$ in the middle alluvial fan deposits (Figure 6).

In the lower fan areas Se concentrations were monitored up to $400 \mu g/L$. Similar historical distribution of soil Se content and shallow groundwater content indicate that dissolved selenium species were leached from saline soils by irrigation water. The drainage discharge of shallow groundwater and subsurface irrigation water was accompanied by increasing accumulation of Se in Kesterson Reservoir.

Aquatic birds nesting at Kesterson Reservoir in 1983 were found to have high rates of embryo deformities and mortality. Beginning in 1984, adult birds were also found dead in unusually high numbers. Through a series of field and laboratory studies, these effects were attributed to the exceptionally high concentrations of selenium in the biogeochemical food web of the birds.

Selenium in fodder crops of the USA

The spatial distribution of selenium content in fodder crops of the USA area is shown in Figure 7.

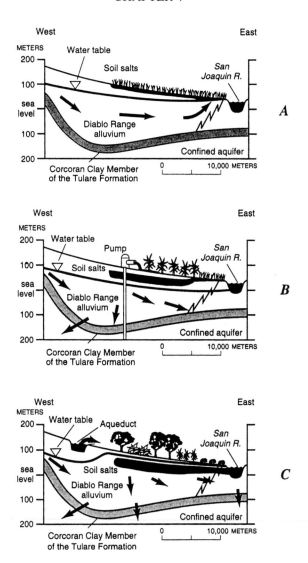

Figure 5. Geohydrologic sections through Panoche Creek alluvial fan illustrating the evolution of groundwater flow system and the concentration of selenium in these waters in the western San Joaquin Valley. Arrows indicate direction of flow. (A) Shallow distribution of soil selenium salts and primary horizontal direction of groundwater flow between recharge areas in the upper part of the fan and discharge areas along the San Joaquin River during pre-irrigation time. (B) Changes in groundwater flow direction and distribution of soil salts from the 1930s through the 1960s. (C) Discontinuation of pumping in the late 1960s caused a rise in the water table. Irrigation of low-lying areas and continued irrigation of middle and upper fan areas caused further downward displacement of soil selenium-containing salts and increasing their content in ground- and drainage waters (Deverel et al, 1994).

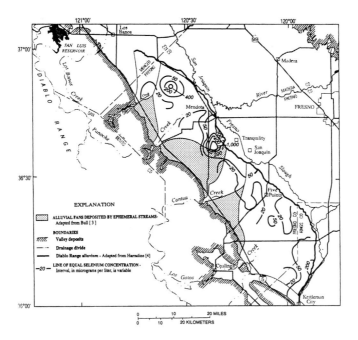

Figure 6. Concentrations of selenium in shallow groundwater in the middle alluvial fan deposits (Deverel et al, 1994).

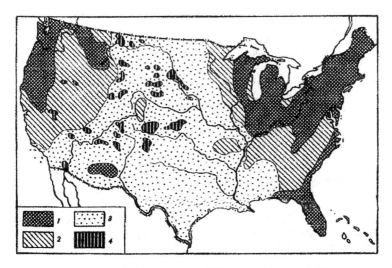

Figure 7. Content of Se *in fodder crops in USA, ppm: 1 low, <0.05; 2—intermediate; 3—sufficient, > 0.1 ppm, and 4—high, up to 5000 and more (Beston and Martone, 1976).*

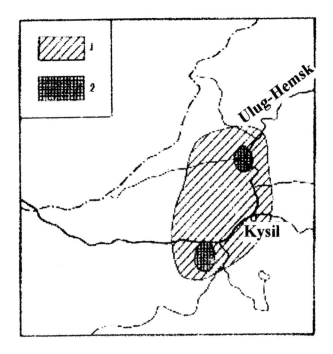

Figure 8. Biogeochemical sub-region and provinces enriched by selenium. 1—sub-region with Se content in soil from 0.2 to 0.5 ppm and in plant species from 0.08 to 0.5 ppm; 2—Ulug-Hemsk and Turan-Uluks biogeochemical provinces with Se content in soil as much as 0.3–6.0 ppm and in plant species from 0.1 to 13.1 ppm by dry weight (Ermakov, 1993).

Plants exhibit genetic differences in Se uptake when grown on seleniferous soil. Some plants accumulate surprisingly low levels of Se. For example, white clover (*Trifolium repens*), buffalograss (*Buchloe dactiloides*) and grama (*Bouteloua spp.*) are poor accumulators of selenium. On the other hand, Se-rich plants like *Brassica spp.* (mustard, cabbage, broccoli, and sunflower) and other crucifers are good concentrators of this element.

Northern Eurasia

In the Asian part of Russia the biogeochemical sub-region with excessive content of Se in different biogeochemical food webs has been monitored. The excessive Se content is connected with high concentration of this metal in local sedimentary rocks of the Tuva administrative region (South of Central Siberia). This sub-region in the Ulug-Hemsk and Turan-Uluks depressions and two corresponding biogeochemical provinces were described (Figure 8).

The first province is shown in Figure 9 and it has been studied extensively (Ermakov, 1993).

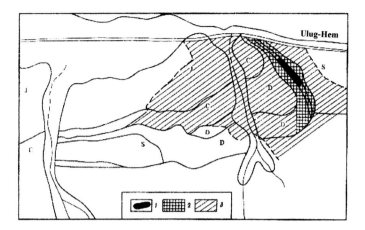

Figure 9. The map of Ulug-Hemsk biogeochemical province with high Se concentrations in biogeochemical food web. Se concentration, ppm: 1- in soil 2–4, in plant 0.7–13.1; 2- in soil 0.7–1.0, in plant 0.4–6.0; 3- in soil 0.4–0.7, in plant 0.1–2.4 (Ermakov, 1993).

This province occupies the central part of the Baryk valley. The geological composition includes Carbon sediments over the Devonian rocks. Selenium was accumulated in the Middle Devonian pink-gray sandstone up to 20 ppm. This has led to the formation of soils enriched by Se up to 6.0 ppm with corresponding enrichment of plant species of *Cruciferae, Leguminae* and various multiflorous botanic families. These species are *Alyssum lenese Adams*, Se accumulation up to 13.1 ppm by dry weight, *Artemisia glauca Pall*, Se accumulation up to 6.0 ppm by dry weight, etc. Biogeochemical researches have shown that only microbial communities have adapted to this Se enrichment. The plant species indicate the chlorosis and necrosis of leaves. Various physiological abnormalities have been monitored in sheep, like hoof deformation, baldness, hypochromic anemia and increasing activity of phosphatase in different organs. The content of Se in various organs and tissues of sheep is 2.5 times higher in the Ulug-Hemsk biogeochemical province in comparison with other studied sites of the Tuva administrative region.

The average concentration of selenium in various biogeochemical food webs is shown in Table 9.

Similar regions were monitored in other sites of Russia, especially in the South Ural mountains, where the elevated contents of selenium in soils and natural waters are coincided with increasing rates of corresponding animal diseases. The analogous biogeochemical provinces have been also monitored in Uzbekistan.

There are physiological standards for diagnosis of both Se deficit and excess. The standard content of Se in blood samples is 4–10 ppb, and in kidney, 10-20 ppb. Under Se deficit this content decreases till 1–3 ppb, and under excessive intake, it increases up to 40–100 ppb (Ermakov, 1993).

Table 9. Se concentrations in biogeochemical food webs of the Tuva biogeochemical sub-region and selenium biogeochemical province (After Ermakov, 1993)

Links of biogeochemical food web	Se content in enriched biogeochemical areas, ppm	
	Tuva sub-region	Ulug-Hemsk province
Geological rock	2.64 ± 0.43	0.29 ± 0.02
Soils	0.82 ± 0.09	0.18 ± 0.02
Natural waters, ppb	3.45 ± 0.85	0.47 ± 0.26
Plants	2.18 ± 0.52	0.38 ± 0.02

Table 10. The average relative selenium contents in biogeochemical food webs of different biogeochemical sub-regions in Eurasia (per cent to the physiological standard values).

			Biogeochemical food webs				
Rocks	Soils	Fodder crops	Winter wheat	Sheep wool	Sheep blood	Activity of sheep glutathionperoxidase ferment	
Chita administration region, Russia, C_b—0.08							
50	60	15	20	30	25	25	
North-Latvian moraine plain, Latvia, C_b —0.2–0.08							
300	70–200	50	60	30	40	40	
Fergana valley, Uzbekistan, C_b—1.50							
200	100	450	160	130	180	130	

Footnote: C_b, coefficient of biogeochemical plant uptake, is the ratio between Se content in plant and soils.

A comparison of different selenium biogeochemical regions is shown in Table 10.

We can see that Se deficit is often monitored in the biogeochemical food webs in the Chita region, Russia, and Latvia and the Se enrichment in the Fergana valley of Uzbekistan. The Se deficit is also connected with the lower values of C_b in the first two regions in comparison with the latter, where selenium is more mobile due to the alkaline reaction of soils (see above). The lower biogeochemical mobility of selenium in the south Siberia (Chita region) is related to the low level of this trace metal in rocks and in Latvia it is connected with low mobility of Se in predominant local acid sand and peat soils.

Selenium in China ecosystems

At the end of the 1960s, biogeochemical studies of selenium were initiated in China to determine the causes of two endemic human diseases, Keshan disease and

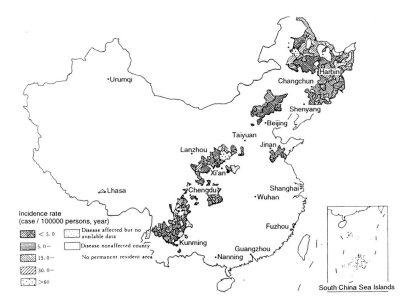

Figure 10. Distribution map of annual average incidences of Keshan disease (acute and sub-acute) in China (Tan, 1989).

Kachin-Beck disease. The former is an endemic cardiomyopathy, and the latter is an endemic osteoarthrosis (Tan *et al*, 1994).

The distribution of both endemic diseases has been found to relate to selenium content in the soils. The two diseases are distributed mainly in a distinct wide belt, usually referred to as the disease belt, running from the northeast to southeast of China and located in the middle transition belt from the southern coast to the northwest inland region (Figure 10 and 11).

The belt is mainly represented by Temperate Forest ecosystems on forest-steppe soils (Brow Earth). The analyses of selenium content in various links of the biogeochemical food web (rock, water, soils, grains, hair, etc.) has shown that these two diseases are always located in low-selenium biogeochemical sub-regions of the biosphere.

The following biogeochemical mapping of Se content in the China ecosystems has been suggested.

1. The low-Se ecosystems occurs mainly in and near the temperate forest and forest steppe landscapes as an axis in China, and the relatively high Se content in the ecosystems usually appear in the typical humid tropical and subtropical landscapes and typical temperate desert and steppe landscapes.

2. In juvenile soil landscapes, Se from parent materials is a very important factor controlling the biogeochemical food web in the whole ecosystem.

Figure 11. Distribution map of annual average incidences of Kaschin-Beck in China (Tan, 1989).

3. In some mountain districts or elevated areas, the distribution of the low-Se ecosystems is also associated with vertical distribution of such mountain landscapes as mountain forest, forest steppe, and meadow steppe ones.

4. Relatively high Se contents are in the ecosystems of large accumulation plains, such as the Songliao, Weine, and Hua Bei plains, compared with the above located landscapes of similar area.

These four principles of mapping are shown in Figure 12 and represent the low Se belt running from northwest to southeast, whereas two relatively high Se belts are flanking it on both sides, to the southeast and northeast.

3.4. *Global Selenium Cycle*
The overview of the global cycle of selenium was presented by Nriagu (1989) for principal biogeochemical concentrations and pools (Table 11).

An understanding of compartmental concentrations has only preliminary use in considering the global Se cycle, because it is the fluxes, driven by both natural (biogenic) and human (anthropogenic) forces, which still have many uncertainties in estimates.

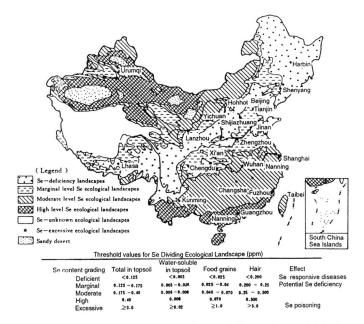

Threshold values for Se Dividing Ecological Landscape (ppm)

Se content grading	Total in topsoil	Water-soluble in topsoil	Food grains	Hair	Effect
Deficient	<0.125	<0.003	<0.025	<0.200	Se responsive diseases
Marginal	0.125 – 0.175	0.003 – 0.006	0.025 – 0.04	0.200 – 0.25	Potential Se deficiency
Moderate	0.175 – 0.40	0.006 – 0.006	0.040 – 0.070	0.25 – 0.500	
High	0.40	0.006	0.070	0.500	
Excessive	≥3.0	≥0.02	≥1.0	>3.0	Se poisoning

Figure 12. Selenium biogeochemical map of China (Tan et al, 1994).

4. BIOGEOCHEMISTRY OF BORON

Sources of boron (B) input into the biosphere are volcanic eruptions, weathering of geological rocks, marine salts, borates, and marine waters with relatively high content of this trace element. As a result of biogeochemical activity boron is redistributed in various compounds of the biosphere, like biogeochemical food webs: rock → soil → plant → animal → human.

4.1. Boron Enriched Biogeochemical Food Webs in Arid Ecosystems

Similarly to selenium, geochemical B mobility is higher in alkaline conditions. This migration also increases in arid conditions and this can lead to the excessive accumulation of this element in biogeochemical food webs. The following climate and geochemical conditions are favorable for boron accumulation in arid regions. Climate aridity, related to the 4–5-fold excess of transpiration over precipitation, is accompanied by small surface and ground runoff. The abundance of salt lakes enhances the transpiration concentration of various elements, including boron. The high mineralization rates of plant residues favor high rates of biogeochemical B cycling. The hydrochemical parameters of natural water bodies in arid conditions stimulate also the accumulation of boron owed to salt saturation, low content of organic matter and high alkalinity. The predominance of clay, salty and boron enriched rocks in neutral and alkaline geochemical conditions, favors leaching of this

Table 11. Global averaged concentrations and pool of selenium (after Nriagu, 1989)

Reservoir	Average Se content, ppm	Pool, ton x 10^6
Lithosphere, down to 45 km	0.05	2.8×10^6
Soils, down to 1.0 m	0.4	1.3×10^6
Soil organic matter	0.2	0.32
Fossil fuel deposits		
Coal	3.4	34
Oil shale	2.3	1.1×10^2
Crude oil	0.2	0.046
Terrestrial biomass		
Plants	0.05	0.06
Animals	0.15	0.00015
Forest litter	0.08	0.0096
Oceans		
Dissolved, surface layer	30 ppt	0.84
Dissolved, deep layer	95 ppt	1.3×10^2
Suspended particulate	3	0.21
Sediment pore water	0.3 ppb	99
Rivers		
Dissolved	60 ppt	0.002
Suspended	0.8	2.4
Shallow groundwater	0.2 ppb	0.0008
Polar ice	20 ppt	0.4

element from geological rocks and its accumulation in the soil profile at sorption, transpiration, and chemical barriers. The biogenic accumulation of boron is also widespread.

In arid conditions a soil content of boron > 30 ppm leads to boron toxicity for plants and animals. We can consider this value as the maximum permissible concentration (standard). Physiological and morphological alterations of plant growth are typical forms of boron toxicity. The animal toxicity is related to the decrease of resistance to different diseases and the enhancement of endemic enteritis. These biological effects are characteristic for about 1–20% of the population owing to the heterogeneity of individuals and different adaptation to boron excess.

As an example of a boron biogeochemical sub-region and corresponding B enriched provinces, we can consider the Central Asian arid steppes and deserts in neighboring areas of Kazakhstan and Uzbekistan.

The predominant landscapes are dry steppe, semi-desert and desert, mainly calcium-sodium, gypsum and salt-enriched types. The well expressed forms of meso- and micro-relief, such as saucer-like and outstretched depressions, temporary water channels, and micro-hills, as typical plain relief and the shallow ground water level are accompanied with the contrast moisture redistribution, distinguish differentiation of soil and vegetation types. This is finally accompanied with an extensive accumulation of boron in salty depressions due to the transpiration biogeochemical barrier.

Now we will consider the content of boron in various links of biogeochemical food web.

Boron in the rocks

The soil forming rocks in this region are soft Tertiary and Quaternary sedimentary rocks, very often salty, with carbonate, and predominantly heavy granulometric composition (clay and loam-mergel depositions). The micas are responsible for the main part of boron in clay, sandstone and sand and the boron minerals represent the minor part (Table 12).

The content of boron in different sedimentary rocks of this region is from 3 to 600 ppm, mainly 100–300 ppm. This exceeds by 2–3 times the average content for the given rocks, 100 ppm. It is related to the occurrence of Perm B-containing geological depositions in these arid regions. The heavier the geological rocks the higher content of boron. For example, the Tertiary clays of Kazakhstan contain on average 126 ppm of boron, the modern clays contain 77.1 ppm, and B concentration in sands is 19.4 ppm only.

Boron in the soils

The regional arid soils have a relatively high B content and the mosaic spots of these soils are characterized by very high boron accumulation (Table 13).

We can see that Chernozems in the north part of the Aktubinsk administration region, Kazakhstan, contains 52 ± 4.2 ppm B that exceeds significantly the average content of this element in these soils in the European part of Russia, 31.3 ppm (Figure 13).

The most enriched soils are Solonchaks, Solonets and desert Xerosols of Kazakhstan and irrigated Fluvisols and Fluvisol-Histosol soil complexes of Uzbekistan (> 70 ppm in average). The minimal content of this element is in primitive Arenosols, less than 34–36 ppm, and even < 18 ppm.

Table 12. Boron content in geological rocks of Central Asian arid regions, ppm.

Rock	B content	Region
Tertiary clays	126	Kazakhstan
Modern depositions		
Clay - heavy clay-loam	77.1	Kazakhstan
Clay loam	51.6	Kazakhstan
Light clay-loam	32.5	Kazakhstan
Sands	19.4	Kazakhstan
Marine depositions		
Subbentonite clays	59	Uzbekistan
Clays	36	Uzbekistan
Alevrolite	35	Uzbekistan
Sands	22	Uzbekistan
Gravelite	19	Uzbekistan
Clays		
Typical marine types	36	Uzbekistan
Laguna-marine types	65	Uzbekistan
Continental salt basins	97	Uzbekistan
Limestone		
Marine origin	25	Uzbekistan
Laguna origin	31	Uzbekistan

Boron in surface waters

The boron content in natural waters depends on the hydrochemical composition and the content of total soluble solids (TSS). There is a direct relationship between content of total water-soluble solids and B content (Table 14).

We can see that at TSS content of natural water up to 5000 ppm, the boron content is 1 ppm, at TSS content up to 50,000 ppm, B content is 15 ppm. The increasing content of boron in the Aktubisk administration region, Kazakhstan, was monitored in 37.2% of surface water bodies and 55.5% of wells. The average content of boron in the large rivers (Amu-Daria and Zerafshan) is 0.15 ppm, and it is connected with the formation of water chemical composition outside of these arid regions.

Table 13. Boron content in soils

Soils (FAO classification)	B content, ppm		Region
	Limits	Average	
South Chernozem	18–100	52 ± 4.2	Kazakhstan
Kastanozem	1.1–88	47.2	Kazakhstan
Light Kastanozem	10–82	35.4 ± 3.05	Kazakhstan
Xerosol	17–64	32.4 ± 2.7	Kazakhstan
Xerosol	8–3560.0	299.3	Russia
Solonetz	8–318.2	101.3	Russia
Solonetz	14–98	46.9 ± 3.4	Kazakhstan
Solonchak	20.0–400.0	114.7	Russia
Greyzem	71–93	76.7	Uzbekistan
Fluvisol	19–55	35.1	Uzbekistan
Fluvisol irrigated	21–61	43 ± 4.7	Uzbekistan
Fluvisol-Histosol complex	37–98	65.0	Uzbekistan
Solonchak-Fluvisol	110–170	145.8	Uzbekistan

Many wells with increasing B content in ground water are the sources of drinking water for humans and animals. Moreover, in the end of summer, the excessive B intake is due to the increasing content of this element in the well waters.

Boron in plants

The content of boron in plant species is connected with the spatial distribution of this element in soils. The major enrichment is characteristic for the plant species growing on Solonchaks. The B content is from 50 up to 260 ppm by dry weight. The high content of boron in meadow species of salty floodplains is 30–140 ppm. The other steppe grass species associations, like grass-wormwood, wormwood-soddy-cereal and multi-grassy — cereal associations, accumulate less boron, on average 2.4 ppm by dry weight. The grassy communities are depleted of boron, the maximum accumulation is up to 13–19 ppm. The Graminea family is characterized by minimal accumulation of this element, whereas the lamb's-quarters may accumulate up to 89–260 ppm, and crucifers up to 300 ppm.

The various plant species uptake non-equivalent amounts of boron from soils. An unlimited uptake, proportional to soil content, is common for *Artemisia maritima*,

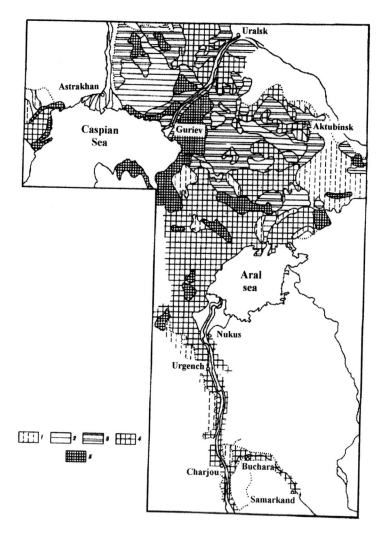

Figure 13. Mapping of boron content in arid soils, ppm. 1<20; 2- 20–30; 3- 30–40; 4- 40–70; 5- > 70 (Kovalsky, 1978).

Limonium suffriticosum Ktze, Salicornia herbacea L., Anabasis aphyllia L., and Kochia prostrata Schat. For instance, the same *Artemisia maritima* species growing in enriched soil accumulated 36.8 ppm whereas, in less enriched soil the B content was only 12.4 ppm by dry weight. The maximum biogeochemical accumulation was characteristic for halophytic plants. At the Ustyurt plateau, the B content in *Salicornia herbacea L.* species was 108 ppm. There are some species, so called 'boron concentrators', which accumulate this element in increasing amounts. Salrosa and Anabasis

Table 14. Boron content in natural waters of arid regions

Region	Type of natural waters	TTS, ppm	B content, ppm limits	Average
Caspian low plain	Ural river	390–630	0.03–0.12	0.09
	Uil river	3,030–4,560	0.11–0.48	0.25
	Wells	9,500–12,400	0.2–5.0	1.5
	Salt lakes	95,600–122,000	0.9–108	19.5
	Dossor lake	87,700	7.7–13.0	10.5
Ustyurt plateau	Sam lake	1,500	—	0.35
	Kosbulak lake	4,340	—	0.65
	Barsakelmes lake (brine)	92,710	—	13.0
	Dengiz-Kul lake (brine)	98,700	—	40.0
Alluvial plain of Amu-Daria river delta	Amu-Daria river	400–760	0.03–0.14	0.09
River Zerafshan valley	Zerafshan river	400–830	0.04–40	0.20

species are typical examples. On the other hand, grasses do not concentrate boron and its content in *Stipa pratenses* was only 9.5 ppm.

Microbes play an important role in biogeochemical boron migration in arid regions. In these conditions, from 0.018 up to 0.125% of B content in soils were found in microbial biomass. The content of boron in microbial plasma depends directly upon B content in soil and the adaptation to increasing content is very high.

Boron in animals

The high content of boron in fodder species leads to increasing intake of this element by animals. The sheep intake of boron with fodder grasses (grass-wormwood association) on the pasture of Calcic Kastanozem soil was increasing by 4-fold in comparison with the control non-boron region, up to 86 mg per day by the living weight. The similar values for other plant species associations were increasing from 10 (Desert-Steppe Xerosol, wormwood steppe association) up to 40 times (Desert-Steppe Xerosol, wormwood-halophyte association) in comparison with the control. This high accumulation is accompanied by endemic enteritis. This diseased is characterized by decreasing activity of protealytic ferments of pancreas and gastrointestinal tract, decreasing exudation of boron and increasing losses of copper, zinc and nitrogen from the animal organisms. The boron enrichment in diseased animals is up to 5 times in comparison with the control region with background content of this element, up to 154 ppm against 24 ppm. The B accumulation occurs in liver, skeleton muscles and spleen. The percentage of affected animals was up to 16 % and the death rates were ca. 40-45% of the illness number.

5. BIOGEOCHEMISTRY OF MOLYBDENUM

The molybdenum (Mo) is the one of the most important trace elements, being essential for plants and animals. The most important biogeochemical role of this metal is related to the process of biological fixation of molecule nitrogen (N_2) from the atmosphere, since Mo is the obligatory constituent of the relevant biochemical ferment at N-fixing leguminous species.

Both Mo deficiencies in crops and toxicity in foraging animals have been reported. Deficiency of molybdenum is possible at levels ≤ 0.1 ppm, while toxic effects are observed in cattle feeding on plants with Mo levels > 10 ppm by dry weight. The average Mo contents in soils range from 0.1 to 40 ppm.

5.1. Mo-Enriched Biogeochemical Food Web

Similar to boron, the consideration of biogeochemistry of molybdenum will be carried out on the example of a Mo-enriched area of Armenia, a mountain country in the south Caucasus. In the high-mountain part of this country, two biogeochemical provinces have been monitored: the Ankawan and the Kanzharan provinces (Figures 14 and 15).

The Ankawan biogeochemical province is in the northern wing of the Mischan-Arzakansk anticline, and the Kanzharan biogeochemical province is in the Zangesur metallogenic geochemical province. The geological rocks of the Anwakan province are enriched in molybdenum, and those of the Kanzharan province are enriched in molybdenum and copper. Molybdenum migrates from the rocks due to oxidation processes and its accumulation and dissipation occurs in soil, water, microbes, plants and animals. All of these components are links of the biogeochemical food webs and we can consider the biogeochemistry of this trace metal on a basis of Mo content in each media. Since the biogeochemical cycle of molybdenum coincides with the relevant cycle of copper, we will consider these cycles in both provinces.

Molybdenum and copper in soils

For understanding the biological role of Mo, its ratio to Cu should be determined in various biogeochemical food webs. The soil is the primary component of these Mo and Cu biogeochemical food webs. The Cu:Mo ratio in soils of the Ankawan province is different from this ratio in soils of the Kanzharan province. For instance, the average Cu:Mo ratio in the first province is 1:1.4, being as much as 1:1.6 in Mountain-Meadow soils, 1:1 in Mountain-Forest Cambisols, and 1:1.4 in Histosols. In the latter province, this ratio is 1:0.09, being 1:0.1 in Kastanozems and 1:0.07

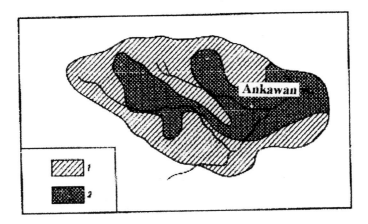

Figure 14. Schematic map of the Ankawan biogeochemical province with increasing content of molybdenum. 1—background area, soil Mo *content is 3.1 ppm, soil* Cu *content is 34 ppm, plant* Mo *content is 0.61 ppm, plant* Cu *content is 4.0 ppm, the Cu:Mo ratio in soil is 1:0.09, in plant is 1:0.15; 2—biogeochemical province, soil* Mo *content is 69 ppm, soil* Cu *content is 50 ppm, plant* Mo *content is 26 ppm, plant* Cu *content 5.7 ppm, the Cu:Mo ratio in soil is 1: 1.4, in plant is 1:4.6 (Kovalsky, 1981)*

in Mountain-Forest Phaerozems. However, this ratio in background Russian steppe Chernozem is 1:0.09, in Non-Chernozemic soils of Taiga-Forest ecosystems is 1:0.19, and in the soils of Arid Steppe, Semi-Desert, and Desert ecosystems is averaged to 1:0.43. Thus, the direct influence of molybdenum on human and animal organisms is monitored only in the Ankawan biogeochemical province, where the Mo:Cu ratio is about 1.5 times higher than in any other regions. The high content of this trace metal determines the peculiarities of its biogeochemical behavior.

Molybdenum and copper in waters

The average content of Mo in surface waters is usually less than 1.0 ppb (μg/L). However, this content is 25–35 ppb in the Ankawan biogeochemical province, i.e., 25–35 times higher. On the other hand, the concentration of copper in surface waters of the given province, 20 ppb, is similar to the average value of this metal. The shallow ground waters are also enriched in Mo and its content is 12 ppb. The enrichment of deep ground waters achieves 890 ppb. These values are higher than those are in background sites. The Mo:Cu ratio is also increasing in surface and ground water in comparison with these values in rocks and soils.

Molybdenum and copper in microbial plasma

The Mo requirement of nitrogen-fixing microbe, *Azotobacter chroococcum, sp. 6*, extracted from Mo-enriched soil of the Ankawan province is higher than that of the

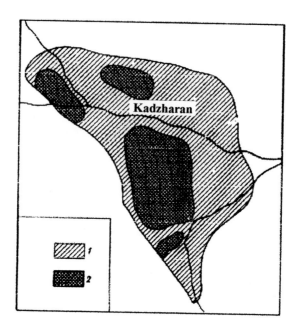

Figure 15. Schematic map of the Kadzharan biogeochemical province with high content of copper and molybdenum. 1—background area, soil Mo *content is* > 4.5 *ppm, soil* Cu *content is 52 ppm, plant* Mo *content is* > 1.2 *ppm, plant* Cu *content is 9.2 ppm, the* Cu:Mo *ratio in soil is 1:0.09, in plant is 1:0.2; 2—biogeochemical province, soil* Mo *content is* > 110 *ppm, soil* Cu *content is* > 1200 *ppm, plant* Mo *content is* > 29 *ppm, plant* Cu *content is* > 170 *ppm, the* Cu:Mo *ratio in soil is 1:0.09, in plant is 1:0.17 (Kovalsky, 1981).*

similar strain from background soils. The microbes from the Ankawan soil have been adapted to the wide range of Mo concentrations conserving the ability for biological fixation of nitrogen.

Molybdenum and copper in plants
In the Ankawan biogeochemical province the content of molybdenum in plants is from 2.6 to 64 ppm, copper from 5.1 to 15 ppm by dry weight. The average Mo content is 26 and Cu 5.7 ppm, i.e. Cu:Mo ratio is as 1:4.6. In the Kadzharan biogeochemical province the content of molybdenum in plants is from 9.7 to 55 ppm, copper from 41 to 350 ppm by dry weight. The average Mo content is 29 and Cu 170 ppm, i.e., Cu:Mo ratio is 1:0.17. The average value of this ratio in plants of standard Chernozem (Kursk region, Russia) is as 1:0.15. This means that in the Kadzharan biogeochemical province the ratio between copper and molybdenum in plant tissues is similar to the standard one. These high contents of Cu and Mo are accompanied with the morphological alterations. For instance, these morphological alterations were found for *Papaver commutatum* with large black pigmentation of petals, when the Cu content was about

210 ppm or 0.021% by dry weight. No morphological alterations and relevant plant diseases were found in the Ankawan biogeochemical province.

5.2. Biochemical and Physiological Response to High Content of Molybdenum in Biogeochemical Food Webs

In the Ankawan biogeochemical province, where Mo content in soil, plant and fodder exceeds significantly the relevant Cu content, different diseases of stomach-intestinal tract and wool quality worsening were monitored. These symptoms are similar to those characteristics for molybdenum toxicity (molybdenosis).

This disease was not monitored in the Kadzharan biogeochemical province, where the pasture grasses are enriched by copper in comparison with molybdenum. The Cu excess protected the animals from high Mo intake.

The alteration of purine exchange in animal and human organisms

In Ankawan biogeochemical province with permanent high content of molybdenum in sheep foraging, the activity of xantinoxidase ferment, containing Mo, is permanently high. This activity is higher on 106% in liver, on 65% in kidney, on 170% in the intestinal tract, and on 206% in milk in comparison with the background region. In the Kadzharan biogeochemical province with relatively high content of copper in forage, the similar increase of ferment activity was determined only on 34, 17, 37, and 49%, correspondingly. This indicates the antagonistic role of copper to molybdenum in animal organisms. In spite of significant increase of the xantinoxidase ferment activity, the accumulation of uric acid was not found in the blood and tissues of animals, since this acid is transformed to allantoin, which is exuded from the body with partial exudation of non-oxidized uric acid.

It has been found that in the Ankawan biogeochemical province, the human daily intake of Mo is 10–15 mg and that of Cu is 5–10 mg, whereas in control regions these values are 1–2 and 10–15 mg for Mo and Cu, correspondingly. Molybdenum is accumulated in the animal blood and tissues and in the human blood and this Mo quantity is significantly correlated with the content of uric acid in blood and activity of xantinoxidase ferment. Statistically significant differences were shown between healthy and ill people both for the Ankawan province and in comparison with non-molybdenum control region population. The uric acid remains in the human organism since it is not subjected to further ferment oxidation. In such cases, this acid is deposited in the tissues of joints, inducing the symptoms of gout disease. The high content of molybdenum in the biogeochemical food webs is stimulating the xantinoxidase ferment synthesis and uric acid formation. In turn, it stimulates this endemic disease. An important role in disease pathogenesis belongs to molybdenum and xantinoxidase ferment and accordingly this gout-like human disease can be called "endemic molybdenum gout". The appearance of this disease is correlated with the duration of human habitation in the Mo-enriched biogeochemical province. It has been shown that the content of uric acid was doubled in human blood when people were living there more than 5 years in comparison with less than 1 year. It was concluded that the alterations of purine exchange are permanent. The monitoring of humans has shown that about 30% of the adult population were ill.

The effects of natural biogeochemical Mo enrichment has been significantly increased due to the development of molybdenum ore exploration and enhancement of natural biogeochemical migration of this trace element during the 1970s.

FURTHER READING

Vsevolod V. Dobrovolsky (1994). Biogeochemistry of the World's Land. Mir Publishers, Moscow/CRC Press, Boca Raton-Ann Arbor-Tokyo-London, 212–215.

William T. Frankenberger and Sally Benson, Eds. (1994). Selenium in the Environment, Marcel Dekker, New York-Basel-Hong Kong, 456 pp.

QUESTIONS AND PROBLEMS

1. Characterize the essential trace elements and describe their position in the periodic table of chemical elements.

2. Discuss the role of chemical speciation in biogeochemistry of trace metals. Give examples.

3. Describe the sorption of trace metals in soils and discuss the forms of sorbed metals in natural and polluted soils.

4. Characterize the background values of trace metals in soils and plants. Determine the clark values.

5. Present the definition of coefficient of biogeochemical uptake. Discuss the significance of these coefficients for understanding the trace metal migration in biogeochemical food webs.

6. Discuss biogeochemical fluxes and pools of copper and the global mass balance of this trace metal.

7. Characterize the biogeochemical role of zinc in the biosphere. Discuss why this trace metal is considered as the most physiologically important for living organisms.

8. Describe the biogeochemical fluxes of zinc in various soil-plant systems of the World. Present the quantitative estimates of these fluxes.

9. Discuss tropospheric transport and deposition of zinc. Compare the deposition rates in land and oceanic areas.

10. Present a comparative assessment of biogeochemical cycles of zinc in terrestrial and oceanic ecosystems.

11. Make a general conclusion on the global biogeochemical fluxes and pools of zinc in the Earth. Highlight the role of living matter in biogeochemical migration of this element.

12. Discuss the role of selenium in the biosphere. What are the physiological indicators of deficient and excessive content of selenium in biogeochemical food webs?

13. Present examples of Se toxicity. Discuss the alterations of food webs in these regions. How can we eliminate selenium toxicity in animals and humans?

14. Explain the processes of geochemical mobility of boron. Compare the migration of this trace element in acid and alkaline soils.

15. Discuss the physiological role of boron in living organisms. What are the consequences of excessive accumulation of boron in biogeochemical food webs?

16. Characterize the synergetic and antagonistic role of copper and molybdenum in biogeochemical food webs. Give relevant examples.

CHAPTER 5

INTERACTIONS OF BIOGEOCHEMICAL CYCLES

In Chapters 3 and 4 we have considered the qualitative and quantitative parameters of biogeochemical cycles of individual macro and microelements. Carbon, nitrogen, phosphorus, sulfur, calcium, and silicon are the principal considered essential chemical macroelements that living organisms utilize in structural tissues. Furthermore, for replication and energy-harvesting activities, zinc, copper, molybdenum, selenium, and boron are also very important in spite of being minor nutrients. These elements are significant in the oceans, atmosphere and crustal rocks. The physiological processes in living organisms combined with chemical, physical and geological forces, continually redistribute these and other chemical species between living and nonliving reservoirs in processes that have been defined as the major and minor biogeochemical cycles.

These biogeochemical cycles are intimately inter-connected and are ultimately powered by energy from the Sun via photosynthetic carbon fixation. Localized exceptions occur such as, for example, hydrothermal environments, where the energy for carbon fixation and biomaterial synthesis may be provided by inorganic reducing species emanating from rocks. The transformations within each biogeochemical cycle are reduction and oxidation reactions, each providing a basis for connection between biogeochemical cycles.

Thus, the purpose of this chapter is to consider the interactions of biogeochemical cycles of various essential elements. The main attention will be given to (i) the stoichiometric aspects of nutrient uptake and nutrient limitations of living matter production, (ii) stoichiometric problems of nutrient recycling, and (iii) thermodynamics of bacterial energenic process.

1. STOICHIOMETRIC ASPECTS OF NUTRIENT UPTAKE AND NUTRIENT LIMITATIONS OF LIVING MATTER PRODUCTION

The various essential chemical nutrients are often determined in predicted proportions in living organisms. It has been found, for instance, that the C:N ratio in microbial plasma is between 4 to 10, in soil humus 6 to 35, and in forest biomass about 160. At the global level, we can consequently calculate how much nitrogen should be involved in biogeochemical cycling to provide the annual net primary productivity (NPP) of terrestrial ecosystems at this ratio for as much as 60×10^9 tons of carbon per year. Simple calculations show that at least 0.4×10^9 tons of nitrogen should be supplied through biogeochemical cycling in global terrestrial ecosystems. Furthermore, as we have seen already, nitrogen and phosphorus are the most deficient elements, and their availability is the limiting factor of living matter productivity.

Reduction of one mole of carbon dioxide in oxygenic photosynthesis formation of carbohydrate, and concomitant splitting of one mole of water, releases one mole of oxygen. This, in turn, will be available for respiratory (oxidation) processes. Organisms fixing carbon by photosynthesis also require a continual supply of nutrients such as N, P, K, Ca, Mg, etc, thus forming links in the biogeochemical food web and interactions between cycles of various nutrients. The availability of these elements in various terrestrial and aquatic ecosystems influences the heterogeneity of NPP values and spatial aspects of ecosystem productivity.

The nitrogen biogeochemical cycle is similar to the carbon cycle and characterized by a rapid recycling within relatively small reservoirs in various compartments of the biosphere. Even smaller pools of nitrogen are in living organisms. Much larger and slower turnover rates are typical for interactions between nitrogen reservoirs in atmosphere and ocean bottom sediments, or between atmosphere and soil organic matter (see Chapter 3, Sections 2 and 3). The atomic carbon-to-nitrogen ratio in living biomass ranges from about 7 (marine microbes) to more than 100 for woody terrestrial materials. In dead organic matter, this ratio increases gradually during burial and diagenesis. This is accompanied with nitrogen release and recycling.

Phosphorus is even less abundant in biota than nitrogen. For marine organic matter, the atomic ratio of carbon-to-phosphorus is 106:1 (Redfield *et al*, 1963). Fossil fuels and fossil organic matter on the whole are depleted in phosphorus, consistent with rapid and efficient recycling of organic phosphorus species. The slowly cycling lithospheric phosphorus reservoir contains $0.8-1.1\times10^9$ tons of P (see Chapter 3, Section 4). Deficiency or availability of this nutrient is considered more likely to affect the productivity of many terrestrial and marine ecosystems than that of nitrogen.

As we have seen earlier, the biogeochemical carbon and sulfur cycles are tightly interrelated. The atomic C:S ratio in living biomass is 200:1. However, fossil organic matter, particularly that deposited in certain ancient marine environments, contains significant amounts of organic sulfur, up to 10% w/w and atomic C:S ratio as much as 1:1. This is far in excess of the normal sulfur content in living organisms, ranging from 0.5 to 1.5% by dry weight.

The biogeochemical sulfur cycle is one of the most important biogeochemical phenomena despite the relatively low content of this element in biota and the fact that sulfur compounds themselves are a quantitatively minor, albeit essential, component. Certain classes of bacteria, the sulfide-oxidizing bacteria and phototrophic sulfur bacteria, continually transform vast quantities of sulfur from sulfide to sulfate in energy transfer processes (see Chapter 3, Section 5). Other groups of bacteria can utilize sulfate as an oxidant for respiration in the process of assimilatory sulfate reduction. These processes are not only for biogeochemical S cycle but also for interplay between many essential nutrients and their biogeochemical cycling in ecosystems.

Many elements, in addition to the above mentioned chemical species, are minor, in quantitative terms, but nevertheless essential constituents of living organisms. Some of them, like Cu, Zn, Co, B, Se, Mo, have been described in Chapter 4. Among others we must account Na, F, Cl, Br, I, Fe, Mn, Cr, and V. Distribution of these elements is affected, in various extent, by biochemical and physiological processes. Of these elements, iron is a particularly significant chemical element. In living organisms, Fe is a constituent of a number of enzymes in cells, especially those related with respiration. We have seen already (see Chapter 2, Section 5) that *Thiobacillus ferrooxidans* and some other microbes can derive energy from the oxidation of Fe^{2+}. Furthermore, deposition of sedimentary pyrite and organic matter were the two principal reduced partners, which balanced atmospheric oxygen during biosphere evolution. These processes are important at present as well.

In summary, biogeochemical cycles of many essential nutrients are redox processes whereby chemical elements in the Earth's crust are transferred in food webs. Fluxes between various reservoirs are mediated by living organisms and microbes play the most important role both in the rates of individual flexes and in flux interactions. Examples of various inorganic processes with intimate interactions of many chemical species are the precipitation of minerals, chemical weathering, physical erosion, oceanic and atmospheric circulation. Sun energy trapped in photosynthetic process is stored as organic carbon. The biogeochemical recycling of this energy leads to the redistribution and consumption by various living organisms in biogeochemical food chains.

The main biochemical reactions coupled with energy transport and incorporation of chemical elements into living matter are shown in Table 1.

1.1. Interactions of biogeochemical cycles in terrestrial ecosystems

Since most terrestrial ecosystems are deficient in nitrogen, we can expect enhancement of photosynthetic processes and an increase in NPP with increasing nitrogen input. The rate of photosynthesis per unit of leaf nitrogen is one measure of nutrient use efficiency (NUE). Subtle variations in the slope of the relationships in Figure 1 reflect differences in NUE during photosynthesis among plants grown in different ecosystems (Schlezinger, 1991).

This relationship is very important for modeling the possible greenhouse effects due to accumulation of carbon dioxide in the troposphere and the possible role of simultaneous increasing nitrogen supply due to air pollution by nitrogen oxides. It is reasonable that this artificial fertilization of natural ecosystems, forest ecosystems especially, will lead to increasing photosynthesis, increasing NPP and increasing sequestration of both C and P in plant biomass. The only open question is on the P deficiency, which may limit all the processes mentioned above.

The typical ratios between essential nutrients in the plant biomass of the World's terrestrial ecosystems are shown in Table 2.

Taking into account the limiting role of phosphorus in many terrestrial ecosystems (see Chapter 3, Section 4), the ratio between various elements is calculated

Table 1. Major biochemical pathways for C, N and S incorporation into biomass (after Engel and Macko, 1993)

Element source[a]	Substrate/product[b]	Electron donor	Type of organisms
CO_2	RuBP/3-PGA	H_2O[c]	Cyanobacteria,[d] algae,[d] C3 plants,[d] CAM plants[d]
		$H_2, H_2S, S, S_2O_3^{2-}$, organic C	Purple photosynthtic bacteria[d]
		$H_2, H_2S, S, S_2O_3^{2-}$, NH_3, NO_2^-, Fe^{2+}	Chemoautotrophic bacteria
CO_2	PEP/OAA	H_2O[e]	C3 plants (dark),[d] C4 plants,[d] CAM plants(dark)[d]
		Organic C	Anaerobic bacteria
CO_2	AcCoA/pyruvate	$H_2, H_2S, S, S_2O_3^{2-}$	Green photosynthetic bacteria[d]
		H_2	Methanogens
CO_2	CO_2/AcCoA	$H_2, H_2S, S, S_2O_3^{2-}$	Green photosynthetic bacteria[d], anaerobic bacteria
CH_4	$O_2 \rightarrow HCHO$		Methanotrophs[g]
NH_4^+	Glu/Gln		Photosynthetic organisms and bacteria[h,I]
NH_4^+	OAA/Glu		Photosynthetic organisms[j]
SO_4^{2-}	ATP/Cys		Photosynthetic organisms, fungi and bacteria[k]
SO_4^{2-}	Org-C $\rightarrow S^{2-}+CO_2$		Sulfate-reducing bacteria[l]

Note:
[a] Reduce and oxidizing forms.
[b] Abbreviations: RuBP, Ribulose 1,5-bephosphate; 3-PGA, 3-phosphoglyceric acid; PEP, phosphoenolpyruvate: OAA, oxaloacetic acid; AcCoA, acetyl coenzyme A; Glu, glutamic acid; Gln, glutamine; ATP, adenosine triphosphate,; Cys, cysteine.
[c] Calvin-Benson or three-carbon pathway catalyzed by ribulose biphospahate carboxilase oxygenase (Rubisco)
[d] Photosynthetically driven reactions
[e] Hatch-Slack or four-carbon pathway catalyzed by phosphoenolpyruvate carboxylase.
[f] Reverse or reductive tricarboxylic acid (TCA) cycle.
[g] Initial oxidation step followed by incorporation of formaldehyde.
[h] Glutamate synthase (GS-GOGAT) pathway.
[i] Assimilation of NH_3 may be preceded by reduction of N_2 catalyzed by nitrogenase of NO_3/NO_2 catalyzed by nitrate and nitrite reductases.
[j] Glutamate dehydrogenase (GDH) pathway.
[k] Asssimilatory sulfate reduction via adenosine phosphosulfate (and phosphoadenylyl sulfate in prokaryotes and fungi), thiosulfate, and sulfide
[l] Dissimilatory sulfate reduction coupled to oxidation of organic acid, and particularly acetate and lactate in sulfate-reducing bacteria (SRBs): this S^{2-} utilized in further reactions but not fixed into biomass.

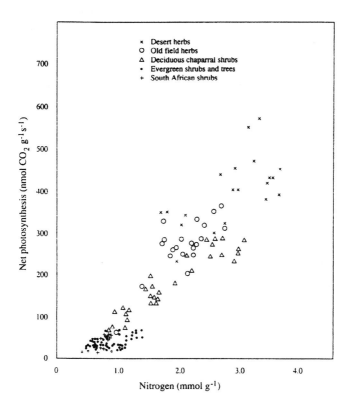

Figure 1. Relationships between net photosynthesis and leaf nitrogen content among 21 species from different ecosystems (Field and Mooney, 1986).

assuming P content as a unit. In this case, the averaged ratio of main essential elements may be drawn as C:N:S:P=400:9:1.4:1 for plant biomass of the global terrestrial ecosystems.

Vitousek *et al* (1988) have compiled results indicating the proportions of carbon and major essential nutrients in various Forest ecosystems of the World (Table 3).

The nutrient ratios vary insignificantly, i.e., 143-165 for C:N, 1246–1383 for C:P, and 8.40–8.80 for N:P ratio for the total plant biomass. The content of nutrients in leaf tissues is higher and C:N and C:P ratios are correspondingly smaller. Thus, we should remember that nutrient ratios increase with time as the vegetation species become increasingly dominated by structural tissues with lower concentration of essential species (Schlezinger, 1991).

Generally, in humus the ratio of C:N:P:S is close to 140:10:1.3:1.3 (Stevenson, 1986). As a result of its high nutrient content, humus plays a role of biogeochemical barrier in soil profile and dominates the storage of biogeochemical species in most ecosystems. For instance, in Temperate Forest ecosystems, the aboveground biomass

Table 2. Accumulation of major elements and their ratios in plant biomass of the World's terrestrial ecosystems (Modified from Bolin, 1966, and Dobrovolsky, 1994).

Elements	Annual growth uptake, 10^6 tons/year	Element ratios*
C	60,000	400
N	3,450	9
Ca	3,105	8
K	2,415	6
Si	862	2
S	586	1.4
Mg	552	1.4
P	397	1
Cl	345	0.9
Na	207	0.5
Al	86	0.2
Fe	34	0.1

*Note: the P content is 1 by definition.

contains only 4–8 % of the total pool of nitrogen accumulated in the whole ecosystem. Slightly higher percentages are monitored in Tropical Forest ecosystems, since the rates of biogeochemical cycling are very high in these ecosystems and the litterfall and dead organic matter mineralize very rapidly, during a month or so.

1.2. Eutrophication of natural waters

In natural waters, phosphorus is usually the limiting element. This means that phosphorus limits the algal growth, but not, for instance, the supply of nitrogen or carbon. There are many monitoring results showing the relationship between input of various nutrients into surface waters, both freshwater and marine waters, and development of eutrophication processes, or accelerated aging of lakes. The uptake of nutrients into biomass occurs in the approximate ratio C:N:P = 100:15:1. However, phosphorus concentrations in natural waters are usually so much lower than those of nitrogen and carbon that phosphorus can be a limiting nutrient, even though only 0.01 times as much phosphorus as carbon is needed for algal growth.

Table 3. Mass ratio of major elements in various Forest ecosystems of the World (Vitousek et al,1988).

Forest ecosystem	C:N	C:P	N:P
Northern and Subalpine Conifer	143	1246	8.71
Temperate Broad Leafed Deciduous	165	1384	8.40
Temperate Conifers	158	1345	8.53
Temperate Broad leafed evergreen	159	1383	8.73
Tropical and Subtropical Closed	161	1394	8.65
Tropical and Subtropical Woodland and Savanna	147	1290	8.80

Table 4. Stoichiometric ratios of major nutrients in different crustacean grazers.

Crustacean grazer	C:N:P ratio
Acanthodiaptomus pacificus	240:48:1
Bosmina longirostris	151:26:1
Daphnia similis	80:14:1

In practice the measured content of P species in fresh and marine waters is higher than dissolved carbon species. This apparent inconsistency arises from sources of nutrient supply. We have seen (Chapter 3, Sections 1 and 2), that carbonate can be re-supplied by atmospheric CO_2. A similar explanation applies to nitrogen since this element can be fixed by blue-green algae with corresponding increasing of dissolved nitrogen species in waters.

Thus we can see that the interactions of biogeochemical cycles in natural waters and relevant ratios between nutrients in various aquatic living organisms determine the productivity of freshwater and marine ecosystems.

A.C. Redfield *et al* in 1963 first demonstrated a characteristic elemental ratio between N, P and C in marine phytoplankton, showing that this left a "fingerprint" on the cycling of major elements throughout oceanic and indeed biosphere biogeochemical food webs. This ratio was then shown to be relatively reliable in marine waters, whereas algal nutrient ratios in fact vary widely in fresh waters. Since then, there has also been considerable research work indicating that the nutrient stoichiometry of grazers is often very divergent from the typical "Redfield ratio" (Elser, 1999). Ratios for three typical crustacean grazers are given in Table 4 as illustration.

Equally, many species have looked at grazer nutrient recycling, generally for N in marine waters and for P in freshwaters, but rarely were both nutrients taken into account or their relative recycling rates looked at.

Interactions of biogeochemical cycles in microbial mats

Microbial mats are communities in surface water ecosystems where bacteria and bacterial processes dominate. In microbial mats dissolved nutrients and metabolites are transformed by one-dimensional (vertical) molecular diffusion. The distinction between microbial mats and biofilms is not sharp. By definition (Fenchel *et al*, 1998), microbial mats are typically stratified vertically with respect to different functional types of bacteria. Microbial mats are thicker (often several millimeters) than biofilms. In microbial mats various types of filamentous prokaryotes are the most conspicuous part and they are responsible for the mechanical coherence of the mat. The mechanical stability of microbial mats is reinforced by the bacterial excretion of mucous polymers, producing a gelatinous matrix.

Requirements for the development of microbial mats include a sufficient energy supply and conditions that more or less exclude eukaryotic activity, especially grazing and mechanical disturbance (bioturbation). Microbial mats are widely distributed in spice of the somewhat special conditions required for their formation and integrity. In most places they are transient or seasonal phenomena of limited extension. They grow at most a few millimeters in thickness.

There are different types of microbial mats, based on colorless sulfur bacteria, purple sulfur bacteria, iron bacteria and cyanobacteria. The most studied mats are those represented by filamentous cyanobacteria. They are widely distributed in protected intertidal sand flats where periodic desiccation discourages colonization by marine invertebrates. The characteristic, sharply defined, colored bands of such mats (green in the top, red or purple a few millimeters down and beneath this black) were recorded as early as the middle of the ninetieth century. Such shallow waters or intertidal mats have now been recorded in many places of the World. These mats are all ephemeral or seasonal, and only in some specific conditions of tropical and subtropical lagoons can they form permanently.

The major processes of cyanobacterial mats are shown in Figures 2 and 3.

Cyanobacteria dominate oxygenic photosynthesis. It has been shown that almost all the activity in terms of energy flow, and both photosynthetic and heterotrophic element cycling, occurs in the surface layer, which thickness is less than 1 cm. The mat metabolism is driven by the rapid turnover of cyanobacterial photosynthesis. In the light, cyanobacterial mats are net producers of oxygen and accumulators of reduced carbon and sulfur species. During nighttime, the mat consumes oxygen. Accordingly, carbon dioxide is taken up during the day and released during the night. Most sulfur, carbon and oxygen is recycled within the mats implying relatively closed systems.

The nitrogen cycle is probably also largely internal. However, cyanobacteria, anoxygenic phototrophs and other anaerobic or microaerobic bacteria can fix nitrogen from the atmosphere. This N fixation compensates for the losses of nitrogen due to

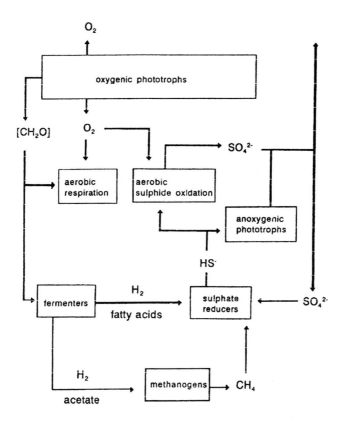

Figure 2. The principal cycling of C, O and S of cyanobacterial mats (Fenchel et al, 1998).

denitrification. The N cycle varies diurnally due to vertical migration of the oxic-anoxic interface and the fact that N fixation is largely confined to darkness when conditions close to the surface are micro-oxic or anaerobic (Fenchel *et al*, 1998).

Biogeochemical mechanisms of nitrogen limitation in coastal marine ecosystems
Coastal waters lie over the inner continental shelf, typically within 4.8 km of shore (i.e., within the boundary of the terrestrial sea) and less enclosed and more saline than estuaries. In addition, oceanic processes affect coastal waters much more than estuaries, with various oceanic phenomena (e.g., waves, tidal action, long-shore currents, coastal upwelling of bottom waters, eddies, and riptides) strongly influencing the movement of water and materials along the inner shelf. Because of the direct link to the open ocean, various nutrients and/or pollutants are dispersed and diluted more readily than in estuaries or lakes, where trapping of substances is of overwhelming importance.

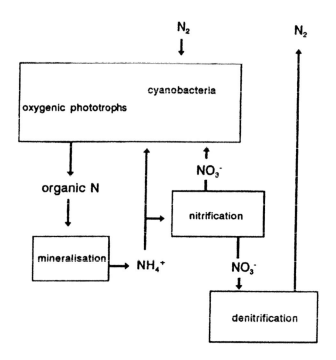

Figure 3. The nitrogen cycle of cyanobacterial mats (Fenchel et al, 1998).

What ecological or biogeochemical mechanisms can lead to nitrogen control of eutrophication in most coastal marine systems and to phosphorus control in so many freshwater lakes? This question was reviewed by Howarth (1988) and Vitousek and Howarth (1991). Here we summarize and update those reviews. Whether primary production by phytoplankton is nitrogen or phosphorus limited is a function of the relative availability of nitrogen and phosphorus in the water. As we have shown above, phytoplankton require approximately 16 moles of nitrogen for every mole of phosphorus they assimilate. If the ratio of available nitrogen to available phosphorus is less than 16:1, primary production will tend to be nitrogen limited. If the ratio is higher, production will tend to be phosphorus limited.

The relative availability of nitrogen and phosphorus to the phytoplankton is determined by three factors (Figure 4):

– the ratio of nitrogen to phosphorus in inputs to the ecosystem;

– preferential storage, recycling, or loss of one of these nutrients in the ecosystem; and

– the amount of biological nitrogen fixation.

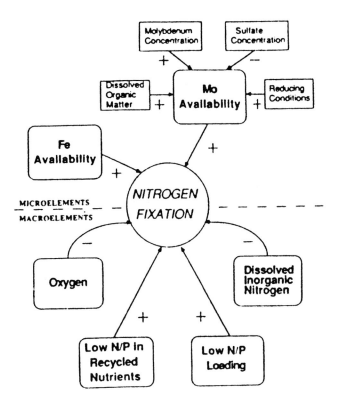

Figure 4. Summary model of biogeochemical controls of nitrogen fixation in water columns, sediments, and wetlands. Symbols represent positive (+) and negative (−) effects (Howarth et al, 1988).

For each of these factors, there are reasons why nitrogen limitation tends to be more prevalent in coastal marine ecosystems than in lakes. For instance, lakes receive nutrient inputs from upstream terrestrial ecosystems and from the atmosphere, while estuaries and coastal marine systems receive nutrients from these sources as well as from neighboring oceanic water masses. For estuaries such as those along the northeastern coast of the United States, the ocean-water inputs of nutrients tend to have a nitrogen:phosphorus ratio well below the Redfield ratio due to denitrification on the continental shelves (Nixon *et al.* 1995, 1996). Thus, given similar nutrient inputs from land, estuaries are likely to be more nitrogen limited than are lakes.

Another factor to consider is that the ratio of nitrogen to phosphorus in nutrient inputs from land will tend to reflect the extent of human activity in the landscape. As the landscape changes from one dominated by forests to one dominated by agriculture and then industry, total nutrient fluxes from land increase for both nitrogen and phosphorus, but the change is often greater for phosphorus and so the nitrogen:phosphorus ratio tends to fall (Billen *et al.* 1991; Howarth *et al.* 1996). This, too, influences why nitrogen limitation is of primary importance in estuaries (NRC 1993). The occurrence of phosphorus limitation in the Apalachicola estuary, for instance, may be the result of the relative low level of human activity in most of the watershed. This suggests that there is a tendency for estuaries to become more nitrogen limited as they become more affected by humans and as nutrient inputs increase overall.

The biogeochemical processes active in an aquatic ecosystem affect the availability of nutrients to phytoplankton in that particular system. Of these processes, the sediment processes of denitrification and phosphate adsorption are the dominant forces that affect the relative importance of nitrogen or phosphorus limitation on an annual or greater time scale. Other processes, such as preferential storage of phosphorus in zooplankton (Sterner *et al.* 1992), act only over a relatively short period. Denitrification is often a major sink for nitrogen in aquatic ecosystems, and it tends to drive systems toward nitrogen limitation unless counterbalanced by other processes such as phosphorus adsorption and storage. The overall magnitude of denitrification tends to be greater in estuaries than in freshwater ecosystems, but this may simply be a result of greater nitrogen fluxes through estuaries. When expressed as a percentage of the nitrogen input to the system lost through denitrification, there appears to be relatively little difference between estuaries and freshwater ecosystems (Nixon *et al.* 1996). That is, available evidence indicates that denitrification tends to drive both coastal marine and freshwater ecosystems toward nitrogen limitation, with no greater tendency in estuaries. In fact, the tendency toward nitrogen limitation—based on this process alone—might be greater in lakes, since lakes generally have a longer water residence time, and the percent of nitrogen loss through denitrification is greater in ecosystems having a longer water residence time (Nixon *et al.* 1996; Howarth *et al.* 1996).

A sediment process counteracting the influence of denitrification on nutrient limitation is phosphorus adsorption. Sediments potentially can absorb and store large quantities of phosphorus, making the phosphorus unavailable to phytoplankton and tending to drive the system toward phosphorus limitation. This process is variable among ecosystems. At one extreme, little or no phosphorus is adsorbed by the sediments of Narragansett Bay and virtually all of the phosphate produced during decomposition in the sediments is released back to the water column (Nixon *et al.* 1980). This, in combination with nitrogen lost through denitrification (Nixon *et al.* 1980; Howarth 1988a), is a major reason that Narragansett Bay is nitrogen limited (Figure 5).

Caraco *et al.* (1989, 1990) suggested that lake sediments have a greater tendency to adsorb and store phosphorus than do estuarine sediments; if this were true, this differential process would make phosphorus limitation more likely in lakes than in estuaries. However, the generality of a difference in phosphorus retention between lakes and coastal marine sediments has yet to be established. It is also important to note that eutrophication may lead to less denitrification since the coupled processes of nitrification and denitrification are disrupted in anoxic waters.

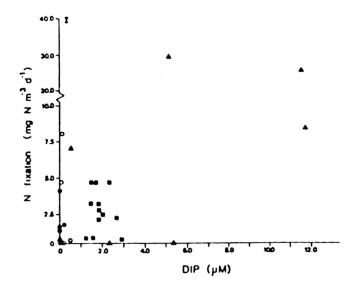

Figure 5. Average rate of planktonic nitrogen fixation plotted vs. average concentration of dissolved inorganic phosphorus (DIP) for a variety of freshwater and marine ecosystems. Freshwater ecosystems—solid sysmbols; marine ecosystems—hollow symbols (Howarth et al, 1988).

Among estuaries, the ability of sediments to adsorb phosphorus is variable, with little or no adsorption occurring in systems such as Narragansett Bay and almost complete adsorption of inorganic phosphate in some other systems, such as those along the coast of the Netherlands. Chesapeake Bay sediments show an intermediate behavior, with some of the inorganic phosphorus released during sediment decomposition being adsorbed and some released to the overlying water. The reasons for this difference in behavior among systems are not well understood. However, there is some indication that the ability of coastal marine sediments—both in tropical and in temperate systems—to adsorb and store phosphorus decreases as an ecosystem becomes more eutrophic, at least until they become extremely hypereutrophic, as in the case of some of the estuaries in the Netherlands. For temperate systems, the lessened ability to sorb phosphate as a system causes more eutrophic results from decreased amounts of oxidized iron and more iron sulfides in the sediments; for tropical carbonate systems, the rate of sorption of phosphate decreases as the phosphorus content of the sediment increases. These changes result in an increase in phosphorus availability in eutrophic systems, intensifying nitrogen limitation and encouraging the growth of algae and other organisms (including some heterotrophic organisms, such as the heterotrophic life stages of *Pfisteria*) with high phosphorus requirements.

The process of nitrogen fixation clearly has different affects on nutrient limitation in freshwater lakes and coastal marine ecosystems. If a lake of moderate productivity is driven toward nitrogen limitation, blooms of heterocystic, nitrogen-fixing cyanobacteria ("blue-green algae") occur, and these tend to fix enough nitrogen to alleviate the nitrogen shortage. Primary productivity of the lake remains limited by phosphorus. This was demonstrated experimentally in whole-lake experiments at the Experimental Lakes Area (USA), where a lake was fertilized with a constant amount of phosphorus over several years. For the first several years, the lake also received relatively high levels of nitrogen fertilizer, so that the ratio of nitrogen:phosphorus of the fertilization treatment was above the Redfield ratio of 16:1 (by moles). Under these conditions, no nitrogen fixation occurred in the lake. The regime was then altered so that the lake received the same amount of phosphorus, but the nitrogen input was decreased so that the nitrogen:phosphorus ratio of the inputs was below the Redfield ratio. Nitrogen-fixing organisms quickly appeared and made up the nitrogen deficit (Flett *et al.*, 1980). This response is a major reason that nitrogen limitation is so prevalent in mesotrophic and eutrophic lakes.

Estuaries and eutrophic coastal waters provide a striking contrast with this behavior. With only a few exceptions anywhere in the world, nitrogen fixation by planktonic, heterocystic cyanobacteria is immeasurably low in mesotrophic and eutrophic coastal marine systems, even when they are quite nitrogen limited (Doremus, 1982; Fogg, 1987; Howarth *et al.*, 1988b; Paerl, 1990; Howarth and Marino, 1990, 1998). This major difference in the behavior between lakes and estuaries allows nitrogen limitation to continue in estuaries (Howarth, 1988; Vitousek and Howarth, 1991).

Much research has been directed at the question of why nitrogen fixation by planktonic organisms differs between lakes and coastal marine ecosystems, with much of this directed at single-factor controls, such as short residence times, turbulence, limitation by iron, limitation by molybdenum, or limitation by phosphorus (Paerl, 1985; Howarth and Cole, 1985; Howarth *et al.*, 1988a; Howarth *et al.*, 1999). A growing concensus has developed, however, that nitrogen fixation in marine systems—estuaries, coastal seas, as well as oceanic waters—probably is regulated by complex interactions of chemical, biotic, and physical factors (Howarth *et al.*, 1999). With regard to estuaries and coastal seas, recent evidence indicates that a combination of slow growth rates caused by low availabilities of trace metals required for nitrogen fixation (iron and/or molybdenum) and grazing by zooplankton and benthic animals combine to exclude nitrogen-fixing heterocystic cyanobacteria (Figure 6).

Nitrogen fixation by planktonic cyanobacteria does occur in a few coastal marine ecosystems, notably the Baltic Sea and the Peel-Harvey inlet in Australia (Howarth and Marino, 1998). In the Baltic, rates of nitrogen fixation are not sufficient to fully alleviate nitrogen limitation (Granéli *et al.*, 1990; Elmgren and Larsson, 1997). The reason that nitrogen fixation occurs in the Baltic but not in most other estuaries and seas remains disputed (Hellström, 1996; Howarth and Marino, 1998), but a model based on the interplay of trace metal availability and grazing as controls on nitrogen fixation correctly predicts that nitrogen fixation would occur in the Baltic but not in most estuaries (Howarth *et al.*, 1999); this model result is driven by the greater

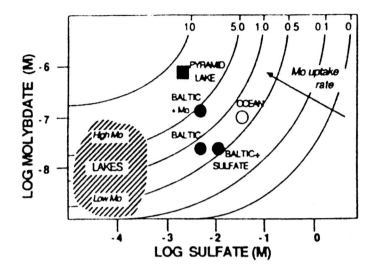

Figure 6. Summary results of a simple Michaelis-Menton model of molybdenum uptake by plankton including the effect of sulfate inhibition. Isolines are in units of pmol Mo (μ g Chl) 'h' and are shown as a function of the sulfate and molybdate concentrations. The cloud labelled "lakes" represents the range of sulfate and molybdate concentrations for most freshwater lakes. "Low-Mo" refers to molybdenum-deficient, ultra-oligotrophic lakes such as Castle Lake and a variety of New-Zealand lakes. "High-Mo" refers to molybdenum-rich, eutrophic lakes such as Lake Donk. Pyramis Lake is saline. Data for Baltic seawater include points for experimental additions of molybdenum ("Baltic+Mo") and of sulfate ("Baltic+sulfate") (Howarth et al., 1988).

availability of trace metals at the low salinity of the Baltic compared to most estuaries. The reason why nitrogen fixation occurs in the Peel-Harvey (Lindahl and Wallstrom, 1985; Huber, 1986) and also a similar estuary in Tasmania (Jones *et al.*, 1994) remains unknown. One hypothesis is that this is a result of extreme eutrophication, which has driven these systems anoxic, increasing trace metal availability and lowering grazing by animals (Howarth and Marino, 1998; Howarth *et al.*, 1999). Nitrogen fixation in both estuaries has only begun in the recent past, and only as they became extremely eutrophic.

A long tradition of thought by oceanographic scientists has held that phosphorus is the long-term regulator of primary production in the oceans as a whole (Howarth *et al.*, 1995). In this view, nitrogen limitation can occur in oceanic surface waters, but this is a transient effect that is made up for by nitrogen fixation over geological time scales. Recently, this concept with a simple, 6-variable model of nutrient cycling and primary production in the world's oceans. The basic concept is appealing, if as yet unproven, in that it explains the strong correlation of dissolved nitrogen and phosphorus compounds over depth profiles in the oceans.

Based on this conceptual view of the interaction of nitrogen and phosphorus over geological time scales, concluded that coastal eutrophication is largely a phosphorus problem, and that "removal of nitrates in the river supply should lead to increased nitrogen fixation, no significant effects on final nitrate concentrations, and no significant effect on eutrophication." The Committee on Causes and Management of Coastal Eutrophication disagrees most strongly. While nitrogen fixation in oceanic waters may alleviate nitrogen deficits over tens of thousands of years, nitrogen fixation simply does not occur in most estuaries and coastal seas and does not alleviate nitrogen shortages. Therefore, decreasing nitrogen inputs to estuaries will not in general lead to increased nitrogen fixation. This model operates on geological time scales for oceans on the whole, a time scale not applicable to estuaries, coastal seas, and continental shelves where water residence time varies from less than one day to at most a few years. Nitrogen fixation does occur in the Baltic Sea, yet even there the water residence time is on the scale of a few decades, thousands-fold shorter than the time scale of response by nitrogen fixation in Tyrrell's model. While debate continues as to whether or not nitrogen fixation completely alleviates nitrogen shortages in the Baltic, much evidence shows that it does not and that much of the Baltic Sea remains nitrogen limited (Granéli *et al.*, 1990; Elmgren and Larsson, 1997; Hellström, 1998; Howarth and Marino, 1998).

Iron is another element that can affect the community composition of phytoplankton. Greater availability of iron may encourage some harmful algal blooms. In some oceanic waters away from shore, iron availability appears to be a major control on rates of primary production. However, there is no evidence that iron limits primary production in estuaries and coastal seas (although it may partially limit nitrogen-fixing cyanobacteria in estuaries) (Howarth and Marino, 1998; Howarth *et al.*, 1999). Although iron concentrations are lower in estuaries than in freshwater lakes, concentrations in estuaries and coastal seas are far greater than in oceanic waters (Marino *et al.*, 1990; Schlesinger, 1997). The solubility of iron in seawater and estuarine waters is low, and complexation with organic matter is critical to keeping iron in solution and maintaining its biological availability. Eutrophication tends to increase the amount of dissolved organic matter in water, and therefore may act to increase iron availability. Furthermore, hypoxia and anoxia accompanying eutrophication may enhance iron availability in the water column due to iron release from sediments as the reducing intensity increases (NRC, 1993).

2. STOICHIOMETRIC ASPECTS OF NUTRIENT RECYCLING

Nutrient recycling plays a very important role in all terrestrial and aquatic ecosystems. The rate of this recycling depends on many biogeochemical features of ecosystems and climate factors. However, the principal factors are related to stoichiometric aspects.

2.1. *Stoichiometric aspects of nutrient recycling in terrestrial ecosystems*

Despite new inputs of nutrients due to atmospheric deposition and chemical weathering, the major part of biogeochemical nutrient fluxes in terrestrial ecosystems is originating from the decomposition of dead material in the litterfall and soil (see Chapter 3).

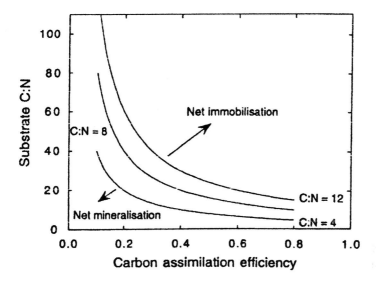

Figure 7. Relationship between net mineralization and immobilization of nitrogen as a function of substrate C:N ratio and microbial assimilation efficiency (F) for three different biomass nitrogen levels equivalent to C:N ratios of 4, 6, and 12 (After Soderlund and Rosswell, 1982).

The composition of organic inputs is as important as size in determining rates of microbial activity and differences in biogeochemical cycling. A complex microcrystalline cellulose base with various intercalated and distinct hemicelluloses, pectins and lignins, all of which contain very little organic nitrogen, forms the bulk of terrestrial plant matter. The typical ratio of C:N in plant residues is more than 100 (see Table 3).

This typical high C:N ratio for terrestrial plant organic matter have a major impact on the nitrogen and carbon biogeochemical cycles in ecosystems (Fenchel *et al*, 1998). A number of empirical and theoretical analyses have shown the strong relationships between mineralization, assimilation and organic mater decomposition in biogeochemical cycles of carbon and nitrogen. The nitrogen cycle processes are quantitatively connected with C:N ratio (Figure 7).

In particular, C:N ratios <10–15 stimulate mineralization of organic matter with release of mineral nitrogen, whereas C:N ratios >30 decrease mineralization and increase assimilation instead with the balance between these processes of uptake and release dependent on the nitrogen content in microbial biomass. Non-symbiotic N fixation can interfere with these processes and ameliorate nitrogen deficiency.

Bacteria and fungi have high contents of nitrogen and phosphorus in their biomass. However, the C:N and C:P ratio in soil microbe biomass is rather constant over a broad range of values (Schlezinger, 1991). This is shown for C:P ratio in Figure 8.

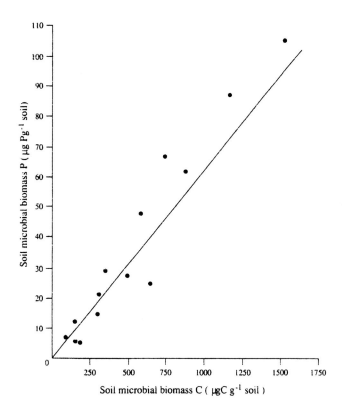

Figure 8. Relationship between the phosphorus and carbon contained in microbial biomass of 14 soils (Brookes et al, 1984).

However, in experimental trials with decomposition of fresh litter in special litterbags, these C:N and C:P ratios decline as decomposition of organic matter proceeds. These may be connected to an increase in the relative part of microbial biomass in litterbags in comparison with the remaining plant materials and remineralization of this biomass (Table 5).

We can note that C:N and C:P ratios decrease. This may be related to the retention of nitrogen and phosphorus in microbial biomass, while carbon is lost as CO_2. The increase of C:Ca and C:K ratios indicate that Ca and K are lost more rapidly than carbon (Schlesinger, 1991).

More recent data of R.Kuperman (1999) on interaction of biogeochemical cycles during litter decomposition have supported this trend. White oak (Quercus alba L.) leaf litter decomposition rates and patterns of N, S and P immobilization and release in decomposing litter were quantified in Oak-Hickory Forest ecosystems in the Ohio river valley for a long term (several decades) bulk atmospheric deposition gradient

Table 5. Ratios of nutrient elements to carbon in the litter of Scots pine (Pinus silvestris) during decomposition in modeled conditions (after Staaf and Berg, 1982).

Duration	Ratios of C to different essential nutrients						
	C:N	C:P	C:K	C:Ca	C:Mg	C:S	C:Mn
Needle litter							
Initial	134	2630	705	79	1350	1210	330
After incubation of:							
1 yr.	85	1330	735	101	1870	864	576
2 yr.	66	912	867	107	2360	ND	800
3 yr.	53	948	1970	132	1710	ND	1110
4 yr.	46	869	1360	104	704	496	988
5 yr.	41	656	591	231	1600	497	1120
Fungal biomass							
Scots Pine Forest	12	64	64	ND	ND	ND	ND

from Illinois, Indiana to Ohio (USA). Historical data on deposition pattern between 1956-1985 showed that 30-yr cumulative annual total (wet+dry) S − SO_4^{2-} deposition were 19.9, 20.8 and 23.9 kg/ha/yr. and total amounts of nitrogen deposition were 5.99, 6.57 and 7.50 kg/ha/yr. in Illinois, Indiana and Ohio, respectively.

We can see that annual decay rates increased along the gradient of increasing atmospheric N and S deposition from Illinois to Ohio as well as with time of field trials (Figure 9).

This increase in decomposition rates suggests a possible simulation effect of higher N availability. The addition of exogenous nitrogen to the soil-litter subsystem of the nitrogen biogeochemical cycle alters considerably the rate of decomposition. The differences in available exogenous N pools across the gradient affect decomposition rates by lowering C-to-N ratios available to decomposers during litter decay (Table 6).

The patterns of net S fluxes in decomposing litter also reflect differences in the S availability across the deposition gradient. There was a net immobilization of sulfur in the decomposing litter at the Illinois site after 6 and 13 months suggesting an unmet microbial demand for this element in a low deposition site. In contrast, litter in both Indiana and Ohio sites showed a net S mineralization throughout the study indicating that sulfur availability exceeded microbial requirements at these sites. Additions of nitrogen were also shown to increase soil organic S mineralization.

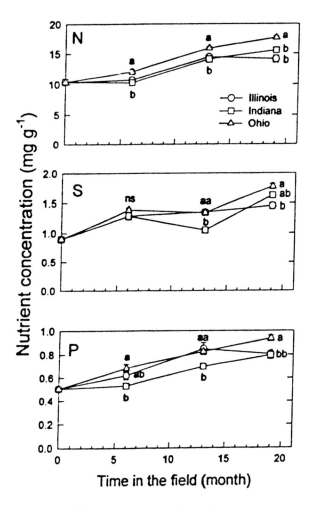

Figure 9. Concentrations of nitrogen, sulfur and phosphorus in white oak litter during 19-month litterbag study along a tri-state atmospheric deposition gradient from Illinois to Ohio. Values are mean s ± 1 SE (n=20 for 6 and 13 months; n=4 for 19 months). Different lowercase letters indicate a significance difference between study sites (p<0.05; Fisher's least significant difference test) (Kuperman, 1999).

Patterns of phosphorus fluxes during decomposition were also variable (Figure 9). Immobilization of phosphorus occurred only in the Illinois site after 13 and 19 months. In the Indiana and Ohio sites P was mineralized throughout the study. High amounts of the atmospheric nitrogen deposition lead to a relative storage of phosphorus in plant residue for microbial decomposers, resulting in a stronger control of litter decay chemistry variables.

Table 6. Changes in carbon-to-element ratios during the decomposition of white oak litter from Oak-Hickory Forest ecosystems along an atmospheric deposition gradient from Illinois to Ohio (After Kuperman, 1999).

Site	Mean long-term deposition, kg/ha/yr.		Time, months	Carbon-to-element ratio		
	Sulfate-S	N		C:N	C:S	C:P
Illinois	19.90	5.99	0	43.2	494.4	879.9
			6	43.0	349.6	761.3
			13	31.2	334.4	553.9
			19	31.7	308.1	554.7
Indiana	20.80	6.57	0	43.2	494.4	879.9
			6	43.8	354.6	872.4
			13	31.9	431.0	649.1
			19	28.7	274.6	561.9
Ohio	23.90	7.50	0	43.2	494.4	879.9
			6	37.8	323.9	689.7
			13	28.1	341.1	543.4
			19	25.2	251.2	475.2

The comparison of deposition chemistry variables was made with other possible factors, which might affect the interaction of biogeochemical cycles of C, N, S, and P, like the climate patterns (annual precipitation and mean temperature). There was no correlation between these climate parameters and decomposition rates. This examination shows that atmospheric deposition chemistry is the most important factor affecting litter decomposition rates and carbon-to-element ratios at study sites along the gradient.

2.2. *Stoichiometric aspects of nutrient recycling in aquatic ecosystems*

We have considered already that phosphate is a limiting nutrient in most freshwater and probably very often in marine water ecosystems as well. The P cycling is thus of great importance for understanding biogeochemical interactions inside aquatic ecosystems. Phosphorus always occurs as phosphate, in its many inorganic and organic molecules. It occurs in cells mostly in polynucleotides, which are hydrolyzed quite rapidly. The phosphates are easily mineralized from the hydrolytic products. Cell detritus loses P preferentially and particulate organic matter, POM, reaching the sediment often has a high C:P ratio. Preferential phosphorus stripping continues in the sediment, but this does not automatically lead to the rapid efflux of P from sediment to water and to phytoplankton (Fenchel *et al*, 1998). Phosphate is retained in oxidized ferric complexes at the sediment surface and is released only when the sediment becomes reduced or when phosphate forms the complexes with organic molecules. The control of phosphate fluxes is partly biological, as it is microbial oxygen uptake that creates the conditions for phosphate recycling in the photic zone.

In the 1980s papers began to appear regarding the food web in aquatic ecosystems, which suggested different cycling rates between nitrogen and phosphorus. It has been shown that the grazers *Daphnia* had a much lower P cycling rate when algal C:P ratios were low. The shifting of limiting nutrient from phosphorus to nitrogen was shown when the dominant grazer changed from *Daphnia* to copepods, as a result of fish population modification. *Daphnia* is known to have a 2–3 times lower body N:P ratio than copepods (Elser, 1999).

Model simulations have shown us how grazer N:P ratios affect the relative rate of release of these nutrients back into water by the grazers, thus, in turn, influencing the nutrient regime experiences by algae. It has been suggested that the rate of P recycling by grazers would fall with increasing C:P ratios. It is clear from experimental and model results that for a C:P ratio >370 *Daphnia* no longer cycles soluble P back into water. Phosphorus limitation of *Daphnia* development is likely in natural environment. Algal ratios in many natural freshwater ecosystems will be lower than P:C ratios for most grazer species. This would significantly limit biogeochemical cycling of P in the food web of aquatic ecosystems.

Static models have been extended to look at the dynamic feedback that changing nutrient cycling would have on algal nutrient ratio, and thus again on grazer ratios and cycling. In case of algal P:N ratios below grazer needs, this results in relatively lower P cycling, reducing P availability for algae, and thus further reducing algal N:P ratios.

Anderson (1997) concluded from this simulation that aquatic ecosystems could move between two distinct stable states. The different food web levels can adapt with nutrient elements being cycled indefinitely scarcer, increasing pressure on grazers until they are driven to extinction. Movement between these states depends on given properties of the ecosystem's trophic dynamics (grazing, ingestion and assimilation rates). When Andersen extended his model to include nutrients held in particles, detritus and bacteria caused the model to predict dynamic chaos. The capacity of grazers to differentiate between such particles and algae will significantly affect the real behavior of the ecosystems.

There is experimental evidence of differential release of nutrients by zooplankton. Firstly, evidence shows that grazers maintain nearly constant body nutrient ratios despite variations in food ratios. Secondly, other experiments show that grazer nutrient release ratios are not constant, but vary considerably in relation to food nutrient ratios. Statistical analysis shows also that zooplankton with relatively higher body ratios tend to cycle less phosphorus, resulting in lower P:N ratios in water.

Some evidence indicates that differential nutrient cycling can indeed affect algal growth, but this would appear to be affected by other factors as well. In Castle Lake, California, USA, for example, Esler (1999) showed that the relative severity of N to P limitation of algae was correlated to estimated grazer N:P ratios in two out of three study years.

One interesting application of grazer stoichiometry effects may be on blue-green algae. Grazing by Daphnia or other high P:N ratio grazers, which cycle nitrogen relatively faster than P, may be to counter risks of blue-green development by decreasing relative availability of phosphorus.

3. THERMODYNAMICS OF BACTERIAL ENERGETICS

Predicting which links of biogeochemical processes driven by microbes, predominate under given natural conditions requires understanding of the energetics of dissimilatory metabolism. There are two aspects of the problem to discuss:

a) considerations based on chemical thermodynamics, and

b) kinetic constraints of chemical reaction.

Kinetic constraints suggest that certain processes, which are possibly based on thermodynamics, do not occur spontaneously since a high activation energy is needed.

Thus, thermodynamic considerations alone would imply that oxidation of N_2 with O_2 could provide a possible way of making a living for bacteria. However, the $N{\equiv}N$ bond is strong so the process requires considerable activation energy and is therefore not realized (Fenchel et al, 1998).

A brief discussion of thermodynamics of bacterial energetics is possible on a basis of equilibrium considerations. These are for several reasons only approximate, but they do provide a heuristic insight with respect to the distribution of different types of bacterial metabolism in biogeochemical food webs. Much more detail can be found in T. Fenchel et al (1998).

The standard free energy ($\Delta G^{0'}$) of a given process can be calculated from the free energy of formation according to:

$$\Delta G^{0'} = \sum \Delta G_f^{0'} (\text{products}) - \sum \Delta G_f^{0'} (\text{reactions}). \qquad (1)$$

Calculations of standard free energy changes approximate the energetics of particular metabolic reactions. Table 7 shows the free energy of formation from elements $\Delta G_f^{0'}$.

Table 7. The free energy of formation from elements $\Delta G_f^{0'}$. For the most stable form of the elements $\Delta G_f^{0'} = 0$. Data are after Fenchel et al (1998).

Substance	State*	$\Delta G_f^{0'}$ (kJ mol^{-1})
H^+	Aq	40.01 (pH7)
H_2O	L	-237.57
CO_2	Aq	-394.90
CH_4	G	-50.82
Methanol	Aq	-175.56
Ethanol	Aq	-181.84
NH_4^+	Aq	-79.61
NO	Aq	+86.83
NO_2^-	G	-37.29
NO_3^-	Aq	-111.45
N_2O	G	+104.33
HS^-	Aq	-12.06
SO_3^{2-}	Aq	-486.04
SO_4^{2-}	Aq	-745.82
Fe^{2+}	Aq	-85.05
Fe^{3+}	Aq	-10.47

Note: Aq-aqueous solution; L-liquid; G-gas phase.

Based on values in Table 7, the free energy change of hydrogen oxidation with four different electron acceptors can be calculated as (expressed as kJ per mole H_2 oxidized):

$$2H_2 + O_2 \longrightarrow 2H_2O; \qquad \Delta G^{0'} = -238\,\text{kJ}$$
$$5H_2 + NO_3^- + 2H^+ \longrightarrow N_2 + 6H_2O; \qquad \Delta G^{0'} = -224\,\text{kJ}$$
$$4H_2 + SO_4^{2-} + H^+ \longrightarrow 4H_2O + HS^-; \qquad \Delta G^{0'} = -38\,\text{kJ} \tag{2}$$
$$4H_2 + CO_2 \longrightarrow CH_4 + 2H_2O; \qquad \Delta G^{0'} = -33\,\text{kJ}$$

Since H^+ is often a reactant or product in biogeochemical reactions, ΔG^0 is usually modified by the addition of the Gf for H^+ at a concentration of 10^{-7} M, or

Table 8. Standard potentials (E'_0) at pH 7 for some redox pairs. Data are from Fenchel et al (1998).

Redox pairs	E'_0 (V)
Some cellular e$^-$ transfer systems	
Cytochrome a ox/red	+0.38
Cytochrome c_1 ox/red	+0.23
Some important organic redox pair	
Pyruvate/lactate	-0.19
CO_2/acetate	-0.29
CO_2/pyruvate	-0.31
CO_2/formate	-0.43
Some important inorganic redox reactions	
O_2/H_2O	+0.82
Fe^{3+}/Fe^{2+}	+0.77
NO_3^-/N_2	+0.75
MnO_2/$MnCO_3$	+0.52
NO_3^-/NO_2^-	+0.43
NO_3^-/NH_4^+	+0.38
SO_4^{2-}/SH^-	-0.22
CO_2/CH_4	-0.24
S^0/HS^-	-0.27
H_2O/H_2	-0.41

the physiologically common pH of 7. In this case the resulting term is referred to as $\Delta G^{0'}$. The ΔG^0 for many reactions of biogeochemical interest can be calculated from Table 8 using compilations of ΔG^0 that are available for numerous organic and inorganic species.

The information provided in Figure 10 is largely equivalent to that presented in Table 7, but clearly indicates which processes are thermodynamically possible.

This graph can also be considered as a biosphere model: oxygenic photosynthesis creates the chemical potential, and chemical equilibrium (mineralization) is restored through a number of redox processes carried out in a variety of organisms. Figure 10 also presents a simplified model of Earth's biosphere. The driving force of oxygenic photosynthesis, which creates the potential energy constituted by free oxygen together

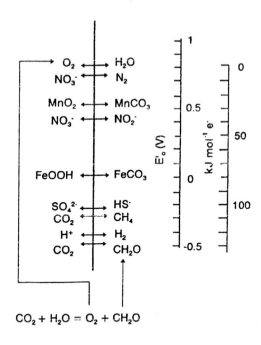

Figure 10. The standard redox potentials (pH 7) of some important redox couples and the free energy changes of processes involving two redox couples (i.e. respiratory processes or H_2/CO_2 methanogenesis) (Fenchel et al, 1998).

with reduced organic material ($[CH_2O]$). Part of this energy of organic matter will be released through mineralization, but most will be released via oxidation-reduction process (respiration) involving external electron acceptors. As long as oxygen is available, oxidative phosphorylation will be responsible for mineralization. When oxygen is depleted, the energetically less favorable nitrate reduction will occur followed by the reduction of oxidized iron and manganese. Thereafter, sulfate reduction will predominate as an electron acceptor. After depletion of sulfate, H_2/CO_2 methanogenesis takes over. This redox sequence describes and explains the temporal succession of the degradation of organic matter and the spatial distribution of processes in general forms (Fenchel *et al*, 1998).

Figure 10 also shows that reduced products of anaerobic mineralization processes like H_2, CH_4, HS^-, NH_4^+, are ultimately oxidized by other electron acceptors. Eventually, everything is in principle oxidized by O_2 so that chemical equilibrium is restored through the concerted action of many different types of bacteria.

We can conclude that the knowledge of these thermodynamic processes allows us to understand the quantitative parameters of interaction of various biogeochemical cycles in terrestrial and aquatic ecosystems. These parameters can be also used in different biogeochemical models.

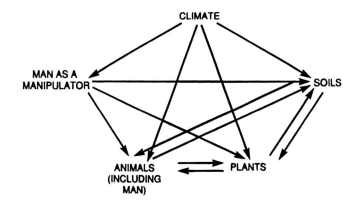

Figure 11. An ecosystem as an integrated complex of living and nonliving compounds (Van Dyne, 1995).

4. BIOGEOCHEMICAL MODELING

4.1. General principles

It is known that all biogeochemical cycles are proceeding inside of various natural ecosystems or their anthropogenically modified analogs. In 1935 A. Tansley introduced the term ecosystem as the system resulting from the integration of all living and nonliving factors of the environment. The term "eco" implies environment; the tem "system" implies an interacting, interdependent complex (Figure 11).

Each component is influenced by the others, with the possible exception of microclimate. At present human beings are on the verge of exerting meaningful influence over the whole ecosystem structure, transforming it from natural into anthropogenic form.

We can see that the ecosystem as a unit is a complex level of organization. It contains both abiotic and biotic components, which integration occurs by cycling migration of energy and chemical elements known as biogeochemical cycling.

So, biogeochemical models have been developed as a variety of general ecosystem models. Let us consider, accordingly, these models and after that make a bridge to so-called biogeochemical models themselves.

Like any models, ecosystem models are mathematical abstractions of a real world situations. In this process some real situation is abstracted into a mathematical model or a mathematical system (Figure 12).

Mathematical analysis has become increasingly important in providing advances in ecosystem behavior and metabolism, including biogeochemical cycling of energy and elements during the last 40–50 years after introduction or powerful computer tools. This allowed the development of special methods of analyzing and studying complex systems in the biosphere.

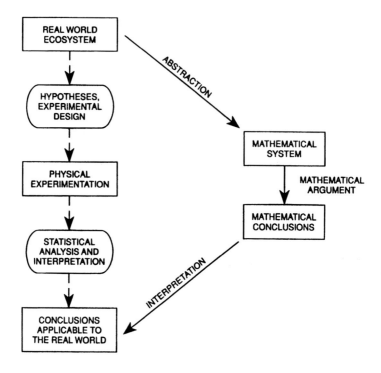

Figure 12. Two ways of experimenting with ecosystems. One involves the conventional process of formulating hypotheses, designing and conducting experiments, and analysis and interpretation of results. The second involves the abstraction of the system into a model, application of mathematical argument, and interpretation of mathematical conclusions (Van Dyne, 1995).

Now we will consider the characterization of ecosystems from thermodynamic theory. The structure of the thermodynamic theory of ecosystems, which aims at developing a general method for the analysis and modeling of these systems, is shown in Figure 13.

Ecosystems are open systems. Their boundaries are permeable, permitting energy and matter to cross them. Effects of environmental constraints and influences on the system play an important role in the regulation and maintenance of the system's spatio-temporal as well as trophic organization and functioning. Indeed, ecosystems operate outside the realm of classical thermodynamics. Biological, chemical, and some physical processes inside of ecosystems are nonlinear. Stationary states of ecosystems are non-equilibrium states far from thermostatic equilibrium. In the course of time, entropy does not tend to a maximum value, or entropy production to a minimum. Entropy decreases when the order of organization and structure of the ecosystem increases. Entropy production is counterbalanced by export of entropy out of the system.

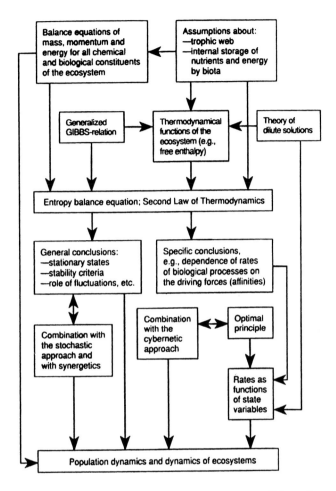

Figure 13. Structure of the thermodynamic theory of ecosystems (Mauersberger, 1995).

Based upon the formalism provided by irreversible or non-equilibrium thermo-dynamics, the ecosystem like any persistent biological structure, may be viewed as a configuration of matter and energy that persists in far-from-equilibrium states by dissipating biologically elaborated, free-energy gradients, thereby forming organic structures out of inorganic elements mobilized from the physicochemical environ-ment. Thus, the ecosystem may be viewed as a functional, biogeochemical system (Figure 14); energy in solar radiation is converted into chemical bond energy by autotrophic photosynthesis.

The chemical bond energy represents a free-energy gradient that is dissipated in order to build persistent, high-energy organic structures out of low-energy inorganic

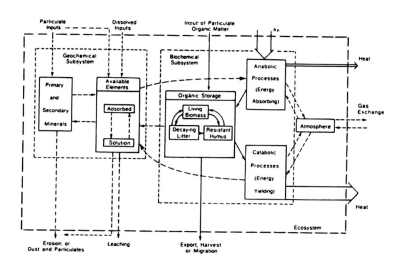

Figure 14. Conceptual model of the ecosystem as a biogeochemical system. Boxes in the diagram represent important storage pools of elements within ecosystems or processes involving the synthesis and degradation of high-energy organic structures out of low-energy organic compounds. Double arrows depict the flow of energy through the system, solid single arrows represent transfer of elements bound in organic forms, and dashed arrows symbolize transfers of inorganic elements in solid, solution, or gaseous forms. The biochemical subsystem is seen to depend on constant energy input and dissipation and on continual exchanges of elements with the atmosphere and the geochemical subsystem. The entire ecosystem is thus represented thermodynamically as an open, dissipative structure, exchanging both energy and matter with the surrounding biosphere (Waide, 1995).

compounds acquired from the surrounding geochemical matrix. Organic structures are decomposed, with energy being dissipated as heat and contained elements returned to the geochemical matrix in a state of low chemical potential (Weide, 1995).

Hence, it is appropriate to view biogeochemical cycles both as a necessary consequence of energy dissipation at the level of ecosystems and as allowing or facilitating a certain level of energy dissipation governed by physicochemical constraints on biological element mobilization and recycling. Thus, the ecosystem may be conceptualized macroscopically as an open dissipative structure and as a persistent organic configuration maintained in far-from-equilibrium states by coupled levels of energy dissipation and biogeochemical cycling.

Therefore, these very general principles should be used when the ecosystem, and in particular cases biogeochemical models, are developed. The set of biogeochemical models was reviewed by Jorgensen *et al* (1995). We will consider only some examples of biogeochemical models to familiarize the reader with typical approaches to the simulation of biogeochemical cycles.

4.2. Models

Mathematical model of global ecological processes (MMGEP)

The purpose of this biogeochemical model is the parameterization of the hierarchy of the biogeochemical, biocenotic and hydrological processes in the biosphere. It gives a prognosis of the consequences of the anthropogenic impacts on these processes (Krapivin, 1993). MMGEP describes the interaction of the atmosphere with the land and ocean ecosystems. It comprises blocks describing biogeochemical cycles of carbon, nitrogen, sulfur, phosphorus and oxygen; global hydrological balance in liquid, gaseous and solid phases; productivity of soil-plant formation with 30 types defined; photosynthesis in ocean ecosystems taking into account its depth and surface inhomogeneity; demographic processes and anthropogenic changes. The model is designed to be connected to a global climate model.

MMGEP includes the blocks describing the function of the world ocean pelagic, arctic and shelf ecosystems. The world ocean is divided in four parts: Arctic, Pacific, Atlantic and Indian Oceans.

The spatial structure of the MMGEP model is determined by the databases available. Spatial inhomogeneity is provided for by the various forms of space discretization. The basic type of spatial discretization of the Earth surface is a uniform geographic grid with arbitrary latitude and longitude steps.

MMGEP has about 50 forcing functions, up to 70 state variables and more than 400 parameters. The inputs of MMGEP include spatial initial data on the parameterized processes. The model has been tested in more than 30 case studies.

Multi-element limitation (MEL) model

Responses of terrestrial ecosystems to increased atmospheric carbon dioxide concentration are of great current interest. Part of this debate relates to the role of limitation by various nutrients, like nitrogen, phosphorus and other vital nutrients on growth and biomass accumulation in terrestrial vegetation. Because vegetation must maintain a nutritional balance, it is expected that nutrient limitation will constrain responses to increased CO_2 level.

From the other side, increasing nitrogen deposition is also a problem of environmental concern in many areas of Europe, North America and Asia (see Chapters 8 and 10). Thus this increase in CO_2 and N in the atmosphere could lead to various global environmental problems, both negative (global warming effect and acidification) and positive (additional sequestration of both elements in plant biomass).

However, responses are also tied to the ability of vegetation to 'acclimate' to an increased CO_2 level by increasing nutrient concentration in biomass or by increasing effort to acquire limiting soil elements, i.e., by increasing biomass, carbohydrate, or enzyme allocation (Bastetter et al, 1997). In this context, 'allocation' includes a broad range of processes, acting at several different time scales that result in compensatory responses to nutritional imbalance in vegetation. These processes include various links of biogeochemical cycling in ecosystems and are generally related to tissue chemistry, morphology, community composition, and genetic adaptation.

In addition to limitation and acclimation in vegetation, soil processes play an important role in ecosystem responses to the increasing input of both carbon dioxide and nitrogen. Here soil characteristics, responsible for the available nutrient pools, are of great importance. Even under increasing input of nitrogen with atmospheric deposition, nutrient concentrations can be depleted by increased uptake, especially in long term period.

Recycling of nutrients further complicates the nature of limitation. Short term depletion of soil nutrient concentration by vegetation can eventually feed back, through litter production and mineralization processes.

We can see that either in local (ecosystem) or in global scale, the experimental description of these processes is very complicated and restricted by current knowledge of biogeochemical cycling structure. Simulation seems a good tool to make a theoretical overview of many possible processes in both current and future teams.

The Multiple-Element Limitation (MEL) biogeochemical model is intended to be a theoretical tool to investigate interactions between the cycles of any two element resources whose storage in biomass will elicit compensatory responses in vegetation (Bastetter et al, 1997).

In this model, the biogeochemical cycles of elements are coupled to one another through vegetation and soil microbial processes that serve to maintain elemental ratios in biomass within an approximate, nutritionally balanced range (Figure 15).

The important feature of the MEL model is the 'acclimation' of vegetation to maintain a nutritional balance when faced with changes in resource availability, V_C, V_N. By "acclimation" authors (Bastetter et al, 1997) mean any process that results in a compensatory redistribution of uptake effort in response to a deviation from optimal element ratios in biomass. These processes include not only physiological and morphological responses of individual plants, such as changes in enzyme concentrations in tissue and root:shoot ratios, but also genetic adaptation of populations and competitive replacement of species by other species with more favorable distribution of uptake efforts. This aspect is essentially important for the time scale of response.

The parameterization of this model has been applied to a 350-yr old forest in Hubbard Brook Experimental Forest, New Hampshire, USA.

The output of this MEL model is to understand the mechanisms by which ecosystems might sequester C under elevated CO_2 and N concentrations in the atmosphere. From the perspective of coupled biogeochemical cycles of carbon and nitrogen, sequestration can be only achieved in combination with one or more of the following changes in ecosystem priorities:

(1) increased C:N ratios of ecosystem components;

(2) a redistribution of N from components with low C:N ratios (e.g., soils) to components with high C:N ratios (e.g., woody vegetation); or

(3) an increase in total ecosystem nitrogen.

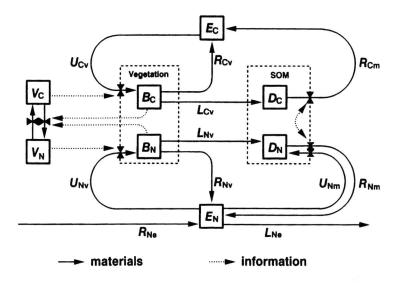

Figure 15. The Multiple-Element Limitation Model applied to C and N cycles in a terrestrial ecosystem. The elements are cycled among vegetation, soil organic matter (SOM), and inorganic pools and are linked through the vegetation and microbial processes. Solid arrows indicate material fluxes, and dotted arrows indicate the transfer of information used to calculate those fluxes. B_C, B_N—*content of carbon and nitrogen in biomass;* D_C, D_N—*content of carbon and nitrogen in soil detritus;* E_C, E_N—*atmospheric input of carbon and nitrogen;* L_{Cv}, L_{Nv}—*litter loss from vegetation;* L_{Ne}—*leaching of inorganic nitrogen;* R_{Ne}—*external replenishment of nitrogen;* R_{Cm}, R_{Nm}—*gross mineralization of carbon and nitrogen in soil;* R_{Cv}, R_{Nv}—*inorganic release of carbon and nitrogen from vegetation;* U_{Nm}, U_{Nm}—*gross microbial uptake of inorganic carbon and nitrogen;* U_{Cv}, U_{Nv}—*uptake of inorganic carbon and nitrogen by vegetation;* V_C, V_N—*uptake effort allocated to elements by vegetation (Bastetter et al, 1997).*

Application of this model shows that the carbon storage potential of forested ecosystems is particularly higher if the nitrogen cycle is augmented by increased N deposition rates and enhanced fixation. The preliminary analyses with the MEL model indicate also that warming due to greenhouse effects has the potential to increase carbon storage in N-limited forests. The mechanism of storage is a net movement of nitrogen from soils (low C:N) to vegetation (high C:N). Increased temperature stimulates both decomposition and mineralization of nitrogen from organic pools. Faster rates of decomposition release CO_2 to the atmosphere. However, this loss of CO_2 is more than made up for by the faster rates of vegetation growth resulting from the N fertilization effect of increased mineralization. Preliminary simulations of the Hubbard Brook ecosystem (with a reasonable assumption of the open N cycle) indicate that about twice as much carbon is stored when CO_2 concentration is doubled, if temperature is also increased by 5 °C (Bastetter *et al*, 1997).

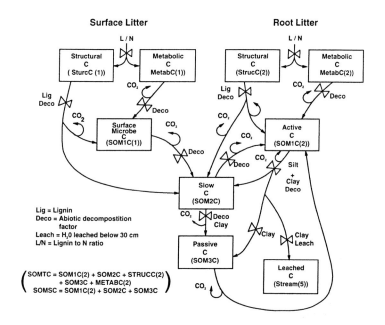

Figure 16. General structure of the Century model (Hall et al, 2000).

Century model

The Century biogeochemical model was developed to study the impact of climate and atmospheric perturbations on soil organic matter and ecosystem dynamics. Century is a general model of plant-soil interactions that incorporates simplified representations of key processes relating to carbon assimilation and turnover (Figure 16).

Century simulates the dynamics of carbon and nitrogen for different plant-soil systems. Plant production in grasslands is a function of soil temperature and available water, limited by nutrient availability and a self-shading factor. The model includes the impact of fire and grazing on grassland ecosystems (Holland *et al*, 1992, Parton *et al*, 1993). This model can be linked to a soil organic matter sub-model, which simulates the flow of carbon and nutrients through the different inorganic and organic pools in the soil, running on a monthly time step.

Macrophyte ecosystem model

This model has been developed by Kemp *et al* (1995) and the conceptual diagram is shown in Figure 17.

This model shows the aquatic ecosystem where phytoplankton, epiflora, macrophytes, and benthic microalgae all compete for limited availability of light and nutrients. Competition for the light occurs through direct shading, while nutrient competition involves two separate sources of nitrogen (water column and sediment

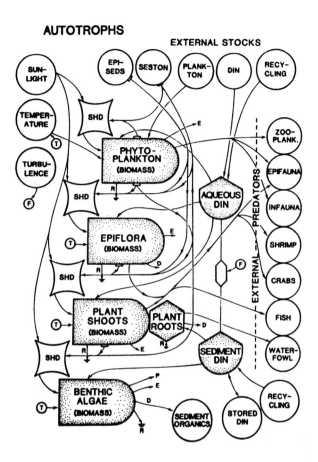

Figure 17. Conceptual scheme of autotroph model of aquatic ecosystem depicting interactions among four autotrophic groups (phytoplankton, epiflora, macrophyte plants, and benthic algae) that compete for limited availabilities of light and dissolved inorganic nitrogen (DIN). Sunlight reaching each autotroph is reduced by shading (SHD) associated with the autotrophs them-selves, along with seston and epiphitic sediments (Epi-Seds) on macrophyte leaves. External forcing functions are represented by circles; interactions are incated by lines with arrows; state variables are represented by shaded symbols (Kemp et al, 1995).

pore waters), which undergo periodic depletion of supplies. Only the rooted vascular plants have direct access to both nutrient sources.

The output of this model is the simulation of aquatic ecosystem behavior under different physical and chemical environments: an open embayment characterized by rapid exchange with external estuarine waters, a protected cove with more restricted tidal flushin, experimentally fertilized ponds with limited exchange and no tidal mixing, etc.

Knowledge-based large scale aquatic ecosystem restoration model

The purpose of this biogeochemical model is an estimation and choice of the optimum decision from the number of possibilities. The choice of the measures is a very difficult problem, as it is necessary to take into account the influence of the various factors, fuzzy in their interrelations. Fuzziness is displayed as some inherent ecosystem property. Therefore, it is possible to apply the theory of fuzzy sets to the description of natural object properties, and for operations with them using the rules of fuzzy mathematics. Fuzzy description of elements of any system and logic conclusions is entered through the concept of the function (Frolova, 1999). The methodology of using fuzzy sets defines the following sequence of the decision of a task: definition of initial sets of measures on protection of an environment; construction of the membership function of measures in relation to various meanings of parameters of its condition; construction of the membership function of parameters; measurements of the environmental state; definition of operations upon the membership functions; definition of a defuzzification rule.

The suggested fuzzy method for calculation of critical meanings of parameters of loading allows managers to choose the optimum decision and may be used in the fields of soil remediation, water reservoirs restoration or air purification.

The fuzzy model is realized as the module of a logic conclusion in an expert system. The typical structure of an expert system includes the database, mechanisms of logic outputs, knowledge base - user interface, mechanism of expert estimates, knowledge database support and explanation mechanism for management and policy makers.

The expert system can work in two modes: direct and reverse:

The direct mode defines acceptance of a decision on the air protection method depending on the environment parameters.

The reverse mode of an expert system calculates critical meanings of parameters of loading.

The model has been tested at the Kaban Lake, Tatarstan, Russia and a ranking of technologies for the rehabilitation of this lake has been suggested for its management and investment strategy.

For application of this model for water body restoration, the following steps should be taken: create the knowledge base for the natural object; collect the data of analysis in this domain; input the data to an expert system; evaluate the data obtained (limit of measurement error is 20–40%); make calculations of the technology ranks by an expert system; analyze the last result using private appraisals if necessary; use the recommendations of the expert system for natural object restoration and check their usage.

Practically, an expert system can be used as an analyzing system. Users may use the expert system as the recommendation system. This model is suggested to ensure optimization of the restoration technology.

Application of GIS techniques for calculation and mapping of critical loads

Critical loads of pollutants at an ecosystem can be calculated on the basis of the Steady-State Mass Balance (SSMB) biogeochemical model (see Chapter 10). All equations of this model include a quantitative estimation of the greatest possible number of parameters describing pollutant circulation in ecosystems. However, direct use of this method is possible only for the areas in which it is possible directly to measure with necessary accuracy and spatial resolution all parameters, which are included in the equations. It is correct for the West European countries, which have rather small areas and a well advanced network of stations making an estimation of pollution of an environment by selection of tests with necessary accuracy and resolution. But there are regions for which it is impossible to determine with sufficient accuracy for accounting the inputting parameters of the SSMB equations. One such region is Russia. The absence of a regular network of stations on ecological monitoring is characteristic for vast Russian areas. Therefore it is necessary to apply the methods of the definition of inputting parameters of the SSMB model equations through indirect parameters.

One such method suggested by Bashkin *et al.* (1995), allows the researchers to determine critical loads through the internal ecosystem characteristics and their derivative parameters. The method is based on the suggestion that it is enough to calculate critical loads for small well investigated ecosystems (model systems) and then to distribute results of accounts to all mapping areas. All inputting parameters become dependent on internal ecosystem properties (soil type, soil texture, vegetation type, average annual temperature, river-network density etc.).

For realization of calculations, it is necessary to use various sources of a digital cartographic material of different manufacturers and different quality. Just this fact creates certain difficulties, which is necessary to take into account with GIS design. Accordingly the special GIS technology has been developed to run the calculations (Tankanag, 1999).

Using this geoinformation system, mapping and quantitative estimation of critical loads of sulfur and nitrogen for a European part of Russia was carried out (see Chapter 10). As properties and thematic databases determining a situation of ecosystems in space of attributes, the following properties were taken: *Soil types*—FAO–UNESCO SOIL MAP OF THE WORLD, Lat/Long $2' \times 2'$); Temperature and Runoff—NASA, Lat/Long $1° \times 1°$; Land use—RIVM, EMEP $50 \times 50 \, km^2$; *Land cover and Digital Elevation Model*—US Geological Survey, $1 \, km \times 1 \, km$; Deposition—Meteorological Synthesizing Center-East and Meteorological Synthesizing Center-West, EMEP $50 \times 50 \, km^2$. As a cartographic basis the data World Data Bank-2,4 (US Department of State, CIA, 1989-1990) were used.

FURTHER READING

Vladimir N. Bashkin and Heinz-Detlef Gregor (Eds.) 1999. Critical Loads Calculation for Air Pollutants on East European Ecosystems. POLTEX, Moscow—UBA Publishing House, Berlin, 132 pp.

Tom Fenchel T., Gary M. King and Tomas H. Blackburn, 1998. Bacterial Biogeochemistry: the Ecophysiology of Mineral Cycling. Second Edition. Academic Press, 21–29, 284–292

Edward B. Basterter, Agren G.I., and Shaver G.R., 1997. Responses of N-limited ecosystems to increased CO_2: a balanced-nutrition, coupled-element-cycles model. Ecological Applications, 7(2), 444–460.

B.C. Patten and Jorgensen S.E., 1995. Complex Ecology: the Part-Whole Relation in Ecosystems. Prentice Hall PTR, Endlewood Cliffs, New Jersey, 705 pp.

S.E. Jorgensen, Halling-Sorensen and S.N. Nielsen (Eds.) 1995. Handbook of Environmental and Ecological Modeling. Lewis Publishers, Boca Raton, 672 pp.

William H. Schlezinger, 1991. Biogeochemistry: an Analysis of Global Change, Academic Press, 142–194.

QUESTIONS AND PROBLEMS

1. Discuss the physiological processes in living organisms that affect the interactions of biogeochemical cycles in various compartments of the biosphere.

2. Define the "limiting element" in biogeochemical cycling and productivity of ecosystems and individual organisms. Why is phosphorus the most limiting chemical species?

3. Describe the inorganic process with intimate interactions of biogeochemical cycles. Highlight the role of C, N, P, and S in these processes.

4. Estimate the role of carbon-to-nitrogen ratios in enhancement of greenhouse effects. How will the increase of nitrogen airborne deposition affect the greenhouse phenomenon?

5. Explain the relatively similar values of ratios between carbon, nitrogen and phosphorus in Forest ecosystems of various climatic zones.

6. What is the Redfield ratio? Discuss the ratios between main nutrients in aquatic ecosystems and present relevant examples.

7. Discuss the interactions of biogeochemical cycles in microbial mats in aquatic ecosystems. Discuss the peculiarities of individual cycles of C, S, and N, and their interactions in microbial mats.

8. What are the limitation factors of nitrogen content in coastal estuaries?

9. Discuss the stoichiometric aspects of nutrient recycling in terrestrial ecosystems using carbon and nitrogen cycles as the example.

10. Describe the peculiarities of interplay between main nutrients in aquatic biogeochemical food webs. Discuss the applicability of Redfield ratios for grazing communities.

11. Discuss the thermodynamic aspects of biogeochemical cycling. Present examples of thermodynamic calculations in bacterial energetics.

12. Describe the general principles of biogeochemical models. Compare these models with complex ecosystem models.

13. Discuss the areas of application for the MEL model. What type of suggestions can you make up related to carbon sequestration in biomass when the level of CO_2 will be increased?

14. Explain the role of increased nitrogen deposition in sequestration of carbon. Use the description of the MEL model for the discussion.

15. Describe the algorithm and conceptual scheme of the biogeochemical model for aquatic ecosystems. What are the areas of applications for such a model?

16. Characterize the expert-modeling models. Present an example of such a model for lake water rehabilitation.

17. Discuss the application of GIS technology for biogeochemical models. Present the example of applying such a model with the GIS technique.

CHAPTER 6

REGIONAL BIOGEOCHEMISTRY

We have considered the general peculiarities of biogeochemical cycles of various macro- and trace elements in terrestrial and aquatic ecosystems of the World (Chapters 3 & 4) as well as the interactions of these cycles (Chapter 5). However, these general characteristics can provide us only with an integrated pattern of the biogeochemical structure of the biosphere, which is differentiated by many features, especially regarding quantitative parameterization of various ecosystem types. In accordance with climate, geology, soil, vegetation, hydrology, and relief, we can subdivide the global ecosystem into different ecoregions and ecozones (see, for instance, R. Bailey, 1998). Even a priori, based on knowledge of the organization of the World's ecosystem, we can suggest the existence of many peculiarities of biogeochemical cycling of various elements in natural terrestrial and aquatic ecosystems. The known regularities of ecosystem behavior can present us with information on biogeochemical cycling in various regional ecosystems.

1. BIOGEOCHEMISTRY OF ARCTIC ECOSYSTEMS

1.1. Geographical peculiarities

In the Northern Hemisphere the area of arctic and tundra landscapes with plant species' ecosystems is 3,756,000 km^2. In the Southern Hemisphere similar landscapes are completely absent. Most of these landscapes occur in Russia, Fennoscandia, Greenland, Alaska, and Canada.

The climate conditions of Arctic and Tundra ecosystems are the main factor influencing many peculiarities of biogeochemical cycling. Because of the severity of the climate the vegetation season is very short. During the arctic summer the temporary melted soil layer is less than 40–45 cm and the deeper layers of ground are permanently frozen. These permanently frozen grounds are called permafrost. The existence of permafrost mainly determines the qualitative and quantitative parameterization of biogeochemical cycles of all elements. We can say that the biological and biogeochemical cycles are restricted both temporally and spatially in Arctic ecosystems.

The major restricting factor is the ocean. Both continental coastal areas and areas of islands are exposed to cold oceanic currents. The Arctic oceanic basin is separated from the warm influence of currents from the Atlantic and Pacific Oceans owing to the existence of both narrow channels like the Bering Strait and submarine ranges.

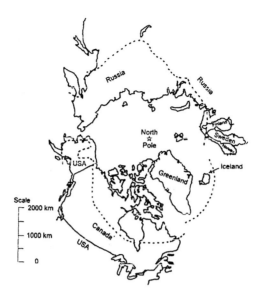

Figure 1. Polar and Tundra ecosystem area in the Northern Hemisphere.

The average precipitation is from 100–200 mm (North American areas) to 400 mm (Spitzbergen Island) and the average temperature of January is between −30° and −38 °C (Figure 1).

The low precipitation and freezing water stage during 10–11 months per annum have led to the development of arid polar and tundra landscapes. The characteristic features of these landscapes are the alkaline soil reaction (pH 7.5–8.0) and even the occurrence of modern carbonate formations.

1.2. Landscape and vegetation impacts

In accordance with the local maximum of precipitation and the relative low winter temperatures, the most favorable climate conditions for biogeochemical processes are in the western part of Spitzbergen Island. Three types of landscapes with corresponding ecosystems are widespread (Dobrovolsky, 1994).

On the wide shore terraces of fjords and on the slopes of hills and low mountains, the Arctic Tundra ecosystems occur. The mosses and lichens are predominant with the twigs of willow (*Salix polaris*), varieties of rockfoils (*Saxifraga oppositifoila, S. polaris, S. caespitosa*, etc), dryad (*Dryas octopetala*), specimens of arctic poppies, buttercups, cinquefoils, various tufted rushes (*Juncus*) and grasses. In some areas the vegetation forms a continuous covering and in others it is confined to depressions enclosing cryogenic polygons. The plant mat covers the soil surface. Most soils are Brown Arctic Tundra soils having only A and C genetic horizons.

Table 1. Chemical composition of different plant species in Spitzbergen island ecosystems (after Dobrovolsky, 1994).

Plants	Ca	K	Na	Fe	Mn	Zn	Cu	Pb	Ni
	Content, ppm by dry plant weight								
Lichen	1170	2000	633	137	6.2	10.0	2.5	7.8	<1.5
Moss	758	2170	867	1240	13.9	8.3	5.2	5.8	1.7
Rockfoils	1460	10000	1833	1751	44.3	50.8	8.8	<1.5	2.6
Arctic willow	1375	6670	658	401	87.8	176.2	8.0	3.7	3.7
Cotton grass	683	20000	442	Nil	Nil	90.0	7.3	2.5	2.5
Rush	400	667	2000	1380	286.0	63.5	5.8	1.5	4.3
Alpine sorrel	1550	8330	2000	3480	172.0	24.5	8.9	1.8	16.7
Heather	9580	22500	10500	1659	106.2	35.6	10.5	1.5	3.7
Laminaria	6000	4330	2033	203	15.0	16.6	3.7	1.5	1.5

The vegetation becomes sparse at the high plateau over 400–500 m above sea level (a.s.l.). The surface coverage is mainly less than 10%. The short grown mosses are predominant. They occupy the depressions with shallow soil accumulation. Lichens grow on mosses and large rock fragments. Only separate specimens of rushes and rockfoil occur. The soils are of the Arctic coarse skeleton type.

The rank *Hyphnum* and *Sphagnum* mosses are mainly represented on the flat variously waterlogged bottoms of glacial valleys. Cassiopes (*Cassiope tetragona*), tufted grasses and rushes grow in the relatively dry sites of these valleys. The given conditions favor peat formation; however, the permafrost layer restricted this process and the peat layer is mainly less than 40 cm. The small lakes place in the wide valley and they are bordered by sedges (*Carex nordina, C. rupestis*), cottongrass (*Eriophorum*) and nappy plant species.

The ash of peat forming plant species contains a predominant amount of silicon. This element is particularly abundant in the *Sphagnum*, where its content achieves 36% by ash weight. Iron and aluminum are the next abundant. The first is accumulated during the peat formation process. The accumulation of calcium and potash is more pronounced than sodium, and the sulfur content is also remarkable. A large amount of mechanically admixed mineral particles (40% to 80% by ash weight) is found in mosses. This is due to the deposition of fine dispersed mineral material from snowmelting waters and atmosphere dust deposition (Table 1).

Table 2. The trace element composition of the Spitzbergen snowmelting water, ppb (Evseev, 1988).

Trace metal	Fe	Mn	Zn	Cu	Pb	Ni	Co
Content	27.5	0.80	31.1	1.7	0.9	0.3	0.3

1.3. Chemical composition of plants

Let us consider the influence of various factors on the chemical composition of plant species in the arctic islands. It seems that the most influential factor is the distance from the ocean shore. For example, in arctic willow growing a few meters from the tide line, the content of Zn, Cu, Pb, and Ni was higher than that of the same plant species growing about 1 km from the coast line and sheltered from the sea by a morain hill. The coastal plants contain also more sea salt cations like Na, Ca, K, and Mg.

The enrichment effect of ocean is mainly related to the chemical composition of aerosols, which determine the chemical composition of snow. For the northern areas of the Eurasian continent and the western areas of Spitzbergen Island we can estimate the average values of sea salt deposition in snow from 3000 to 5000 kg/km^2. The predominant chemical species in the snow water are chlorides (anions) and sodium and calcium (cations). The content of trace elements is negligible. Their origin is connected with long range trans-boundary air pollution from industrial centers of North America, Russia and Europe. This was shown for the Greenland glaciers, where the statistically significant growth of zinc and lead in recent probes in comparison with ancient ice cores has been attributed to environmental pollution (see Chapter 8, Section 2).

The role of air aerosols in the biogeochemical cycle of various nutrients in the Arctic ecosystems has been studied in Spitzbergen Island. The supply of oceanic aerosols is very important in these conditions since the interaction between plant roots and soil or mineral substrates is depressed during a large part of the year. According to the monitoring data the following results are typical for the Spitzbergen snow melting water (Table 2).

For comparison, the mobile forms of trace metals were extracted from local geological rocks, as water-soluble and 1.0 N HCl-soluble forms. The results are shown in Table 3.

We can see that the content of trace metals in water extraction is very low. This means that the direct involvement of these metals in biogeochemical cycles is very restricted. The significant increase of metal contents in acid-soluble form was shown only for Fe, Mn and, partly, for Zn. These data testify to the importance of atmospheric deposition for the Arctic ecosystems as a source of nutrients.

The supply of sea salts and trace metals via precipitation appears to contribute to the elevated content of water soluble forms of alkaline and earth-alkaline elements and trace metals in the uppermost soil layer.

Table 3. Content of mobile forms of trace elements in rocks of Spitzbergen Island, number of rocks = 10 (after Dobrovolsky, 1994).

Statistics	Fe	Mn	Zn	Cu	Pb	Ni	Co
			Trace metal content, ppb				
			Extractant—water				
M	5.71	0.54	0.53	0.11	0.05	0.07	0.03
σ	4.64	0.28	0.21	0.08	0.02	0.08	0.03
V, %	81	52	40	73	40	107	100
			Extractant—1.0 N HCl				
M	1266.6	408.8	7.41	4.64	4.23	0.83	1.04
σ	949.3	148.0	2.40	2.46	2.23	1.07	0.45
V, %	75	36	32	53	53	129	43

1.4. Biogeochemistry of soils

A high amount of various nutrients and trace metals is retained in peat and dead plant residues and thus temporarily eliminated from the biogeochemical cycles. The period of this elimination depends on the solubility of these metals. It has been shown (Dobrovolsky, 1994) that the soluble forms of such metals as iron and zinc accounted for about 70% and 50% of the total contents of these metals in solution, correspondingly, in the upper peat layer with living plants. In the underlying peat layer, the percentage of soluble forms tended to decrease. The similar tendency was recorded for soluble forms of carbon: on leaching from upper to lower peat layer, the concentration of soluble form decreases twice as much in the terrace and still greater, in waterlogged depression.

Electrodialysis of the soluble forms of iron has revealed the predominance of electroneutral forms. A similar distribution has been shown for carbon. The hypothesis that the organic iron-containing complexes are responsible for water-soluble forms of iron in polar peat ecosystems seems logical. Amongst the soluble zinc forms, the percentage of electroneutral forms is somewhat lower that that of charged forms, with the anions present in a larger amount in the upper peat layer.

However, only the smallest part of soluble metals is involved in the biological cycle. Most of these are either lost to water runoff, or retained in the peat organic matter. The latter is the source of gradual remobilization but the whole mineralization may last up to 50 years or even more. The total accumulated retained

Table 4. Airborne input of various trace metals in the
Spitzbergen island ecosystems, mg/year per 100 mm
of precipitation (After Dobrovolsky, 1994).

Trace metals	Fe	Mn	Zn	Cu	Ni
Input	27500	800	31100	900	300

Table 5. The biological productivity of the
Polar Tundra Low Terrace ecosystem.

Productivity	Ton/ha
Total mass of living plants	2.9
Mass of dead plant matter	9.6
Annual net primary productivity	0.6

amount of macro- or trace metals in organic matter of peat is tens and hundreds time
higher than the concentration of annually released soluble forms, which are available
for plants.

1.5. Biogeochemical cycles

The different metal uptake by plants is accompanied by different involvement of these
trace metals and macronutrients in the biogeochemical cycles. A comparison of metal
concentrations in plant tissues and metal concentrations in the aqueous extracts from
soil-forming geological rocks shows that iron and manganese are the most actively
absorbed by plants. The plant to soil metal ratio can be an indicator of this absorption.
These values for Fe and Mn are in a range of $n \times 10^2$ to $n \times 10^3$. This ratio is about
$n \times 10^1$ for Zn, Cu, and Ni. It is noteworthy that the high concentrations of iron and
manganese tend to increase in the dead organic matter of peat.

The systematic removal of elements by runoff and the reimmobilization from
solution by organic matter are continuously counterbalanced by the new input of
chemical species, which maintain both biological and biogeochemical cycles. The
main sources of water-soluble elements are oceanic aerosols deposited on the land
surface and the weathering of rocks. The airborne input of trace metals may be ranked
as follows for the Spitzbergen island ecosystems (Table 4).

We can compare these values with those characterizing the fluxes of trace metals
in biogeochemical cycles. The biological productivity of the Polar Tundra ecosystem
grown on the low terrace in the region of Barentsberg, Spitzbergen Island, is shown
in Table 5.

Table 6. Fluxes of trace metals in the Spitzbergen Island ecosystems (After Dobrovolsky, 1994).

Trace metal	Mean content in plant species, ppb by dry weight	Trace metal fluxes, g/ha/year			Airborne input*, g/ha/year
		In living plant organisms	In dead organic matter	In net annual Production	
Fe	2000.0	5800.0	19200.0	1200.0	82.5–110.0
Mn	150.0	435.0	1440.0	90.0	2.4–3.2
Zn	60.0	174.0	576.0	36.0	93.3–124.4
Cu	6.3	18.3	60.5	3.8	5.1–6.8
Ni	4.3	12.5	41.5	2.6	0.9–1.2
Pb	3.7	10.7	35.5	2.2	2.7–3.6
Co	1.0	2.9	9.6	0.6	0.9–12

*Footnote: The airborne input was calculated per 300 and 400 mm per year in accordance with annual precipitation rates in the western Spitzbergen coast and trace metal rates shown in Table 4.

To be noted for comparison, the annual growth increase for arctic willow (*Salex arctica*) in Cornwallis Island in the Canadian Arctic Archipelago, 75° N, is a mere 0.03 ton/ha (Warren, 1957). The correspond trace metal fluxes are shown in Table 6.

We can see that for iron and manganese the annual fluxes of trace metals are an order of magnitude higher than airborne input. For copper this input is sufficient to supply the annual uptake, and for zinc is even in excess. All these trace metals are essential elements and their input with deposition can be considered as positive for the ecosystem's behavior. The excessive deposition input of lead is rather dangerous owing to an unknown physiological and biogeochemical role of this element in plant metabolism. However, significant amounts of lead can be immobilized in dead organic matter and excluded from biological turnover.

The other output from watershed and slope landscapes positions is related to the surface and subsurface runoff of trace metals. The ecosystems of waterlogged glacial valleys, geochemically subordinate to the above mentioned landscape, can receive with surface runoff an additional amount of various chemical species. This results in 3–4-fold increase of plant productivity in comparison with elevated landscapes and in corresponding increase of all biogeochemical fluxes of elements, which are shown in Table 6. For instance, the accumulation of trace metals in dead peat organic matter of a waterlogged valley was assessed as follows: Fe, $n \times 10^1$ kg/ha, Mn, 1–2 kg/ha, Zn, 0.1–0.3 kg/ha, Cu, Pb, Ni, $n \times 10^{-2}$ kg/ha.

2. BIOGEOCHEMISTRY OF TUNDRA ECOSYSTEMS

The Tundra ecosystems and corresponding tundra landscapes occupy the northern-most strip of the continental area of Eurasia and North America bathed by the seas of the Arctic basin. The climate conditions of the tundra zone provide for a higher productivity of ecosystems and higher activity of biogeochemical cycles of various elements as compared with the Arctic ecosystems. The mosses, lichens, and herbaceous plant species are predominant in the northern part of the Tundra ecosystems and shrubs are prevalent in the southern part.

The edaphic microflora is diversified, and the microbial community is more numerous than that of the arctic soils. The bacterial population varies from 0.5×10^6 to 3.5×10^6 specimens per gram in topsoil horizon.

2.1. Plant uptake of trace metals

The ash contents of the total trace elements and nitrogen are similar in Tundra ecosystem biomass. The highest concentrations, $> 0.1\%$ by dry ash weight, are typical for Ca, K, Mg, P, and Si. We can note the increase of iron, aluminum and silicon contents in the underground parts of any plants.

The uptake of trace metals depends on both the plant species and metal. Such elements as titanium, zirconium, yttrium, and gallium are poorly absorbed owing to their minor physiological role in plant metabolism. Rockfoils (Genus *Saxifraga*) and mosses (genus *Bryophyta*) are especially sensitive to alternations of trace metal concentrations. The bryophytes are capable of sustaining higher concentrations of some trace metals as compared to vascular plants (Shacklette, 1962). Some species of mosses can accumulate enormous amounts of trace elements and can serve as indicators of copper metal ore deposits with elevated copper contents.

2.2. Biogeochemistry of tundra soils

The Acidic Brown Tundra soils (Distric Regosol) are formed under the conditions of the free drainage commonly encountered in slopes and the watershed relief positions. The characteristic features of these soils are related to the accumulation of non-decomposed plant residues and the built-up peat layers. Below the peat horizon the soil profile differentiation is indistinct. In the thin indistinct humus horizon underlying the peat layer, the humus content is from 1 to 2.5% with predominance of soluble fulvic acids. This presents the acid reaction of soils, with values of soil pH < 5.0. The acid geochemical conditions facilitate the migration of many trace elements, phosphorus, nitrogen and many earth-alkaline metals. The migration of chemical species is mainly in the form of Me-organic or P-organic complexes.

The deficiency of oxygen is very common in lowland plains with an impeded drainage. This is favorable to the formation of Gley Tundra soils (Gelic Regosol) with a grey gleyic horizon. This horizon includes the gray and rusty spot-like inclusions of precipitated gels of Fe^{+3} oxides.

Table 7. The partition of Tundra ecosystem biomass, ton/ha (after Rodin and Bazilevich, 1976).

Biomass types	Plant living biomass	Dead organic matter	Annual net production	Annual litterfall
Mass	28	83	20.4	20.3

2.3. Productivity of Tundra ecosystems and cycling of elements

The biomass of Tundra ecosystems gradually increases from 4–7 ton/ha for moss-lichen tundra to 28–29 ton/ha by dry weight for low-bush tundra. In the northern tundra, the plant biomass and dead organic matter are eventually shared. Southwards this percentage tends to diminish, and low-bush living biomass is smaller than dead plant remains mass. A typical feature of the Tundra ecosystems plant species is the prevalence of underground matter (roots) up to 70–80% of the total biomass.

The average mass distribution of Tundra ecosystems is as shown in Table 7.

The biogeochemical turnover of nitrogen is about 50 kg/ha per year. The similar value was shown for the turnover of total mineral elements, 47 kg/ha/yr. The relevant values for various trace and macroelements are shown in Table 8.

The flux of chemical elements per unit area in tundra ecosystems is not proportional to the plant uptake. Presumably, some elements, like Zn and Cu, are taken up selectively, whereas other trace elements, like Ti, Zr, V, or Y, are absorbed passively, depending on their content in the environmental media.

3. BIOGEOCHEMISTRY OF THE BOREAL AND SUB-BOREAL FOREST ECOSYSTEMS

The Boreal and Sub-Boreal Forest ecosystems represent the forests of cold and temperate climate. These ecosystems occupy an extended zone in the northern part of the Northern Hemisphere. The total area is 16.8×10^6 km^2, or 11.2% of the whole World's territory.

The general scheme of biological and biogeochemical cycling in Forest ecosystems is shown in Figure 2

3.1. Biogeochemical cycling of elements in Forest ecosystems

The plant species biomass of Boreal and Sub-Boreal Forest ecosystems accumulates a significant part of living matter of the whole planet. This value is about 700×10^6 tons of dry weight. The biomass per unit area of different Forest ecosystems varies from 100 to 300 ton/ha and even 400 ton/ha in the Eastern European Oak Forest ecosystems. The annual net primary productivity, NPP, varies from 4.5 to 9.0 ton/ha (Table 9).

The overall biomass accumulated in Forest ecosystems per unit area is 20 to 50 times higher than the annual productivity. This means that various chemical species

Table 8. Annual fluxes of chemical species in the Low-Bush Moss Tundra ecosystem (after Dobrovolsky, 1994).

Chemical species	Chemical species symbol	Plant uptake fluxes, kg/ha/yr.
Nitrogen	N	50
Iron	Fe	0.188
Manganese	Mn	0.226
Titanium	Ti	0.031
Zinc	Zn	0.028
Copper	Cu	0.0071
Zirconium	Zr	0.0070
Nickel	Ni	0.00188
Chromium	Cr	0.00165
Vanadium	V	0.00141
Lead	Pb	0.00116
Yttrium	Y	0.00070
Cobalt	Co	0.00047
Molybdenum	Mo	0.00043
Tin	Sn	0.00024
Gallium	Ga	0.00005
Cadmium	Cd	0.00003
Average ash content of plant species, %		2.0
Total uptake of ash elements by vegetation, kg/ha/year		47

Table 9. Net primary productivity of Forest ecosystems, ton/ha.

Ecosystems	Coniferous North Taiga Forest	Coniferous and Mixed South Taiga Forest	Broad-leaved Sub-Boreal Forest
NPP	4.5	8.0	9.0

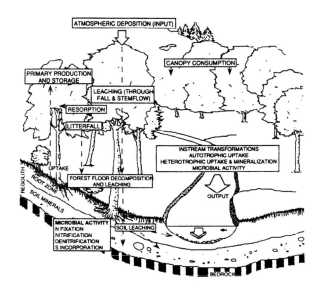

Figure 2. Schematic illustration of the biogeochemical cycling processes in Forest ecosystems (Nihkgard et al, 1994).

are retained during long periods in plant biomass thus being excluded from biological or biogeochemical cycling. The duration of biological cycles may be from 1.5 to > 25 years for Broad-leaved Sub-Boreal Forest and Coniferous North Taiga Forest ecosystems, correspondingly. The slow turnover rates are connected with both a prevalence of aboveground biomass and slow mineralization of plant litterfall on the soil surface.

The microbial activity in forest soils is intense in comparison with Tundra ecosystems. Fungi, bacteria and actinomycetes play a significant role in degradation of carbohydrates of forest litterfall (Box 1).

Box 1. Microbial regulation in Forest ecosystems (after Nihlgard et al, 1994 and Fenchel et al, 1998)

The regulation of biogeochemical cycles by microbial populations is of most direct importance in the cycling of N, S, P, and C. Most of the ecosystem pool of these elements resides as organic forms in forest floor and mineral soil compartments. These organic complexes are subjected to microbial transformations, which regulate nitrate, sulfate and phosphate ions dynamics and uptake. In turn, this influences indirectly the migration of other solutes though maintenance of ionic balances of solutions. For quantification of the role of microbes in forest biogeochemical processes, models like that shown in Figure 3. should be applied.

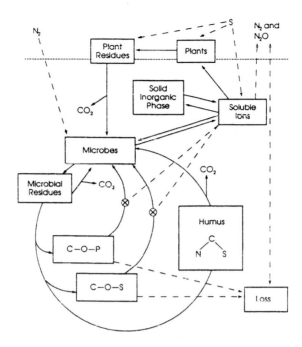

Figure 3. A tentative model illustrating decomposition interactions (Nihkgard et al, 1994).

Nitrogen microbial transformation

Since nitrogen is a nutrient, which limited the productivity of almost all Boreal and Sub-Boreal Forest ecosystems, its microbial transformation is relatively well understood at present. The major N transformations and fluxes are shown in Figure 4.

Processes of dinitrogen fixation, mineralization, immobilization, and nitrification have received the most attention, but there is a paucity of information on denitrification in forest ecosystems. The status and fluxes of nitrogen are strongly regulated by rates of N mineralization and immobilization. The rates of mineralization are greatly enhanced after clearcutting. The influence of clearcutting has been demonstrated in the experiments at Habbard Brook Experimental Forest and in Coweeta (see Likens *et al*, 1977). Over a three-year period after clearcutting a hardwood forest in Habbard Brook, forest floor organic matter decreased by 10.8 ton/ha, soil organic matter declined by 18.9 ton/ha and net N loss from the soil was estimated to be 472 kg/ha with an increased export of inorganic N in the stream waters of 337 kg/ha. Significant alterations of N fluxes have been monitored also at Coweeta. In the first 3 years after clear cut and logging, soil N mineralization increased by 25% and nitrification increased by 200%. However, only a small fraction of this mineralized nitrogen was exported from the ecosystem. The retention was owed partly to rapid revegetation and high rates of nitrogen uptake and partly due to microbial immobilization.

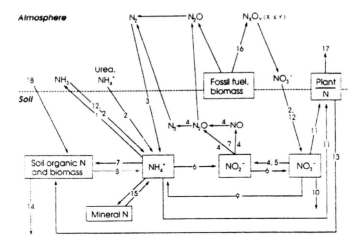

Figure 4. The general nitrogen model for illustrating the microbial role in biogeochemical cycling in Forest ecosystems. Explanations for the fluxes: 1, ammonia volatilization; 2, forest fertilization; 3, N₂-fixation; 4, denitrification; 5, nitrate respiration; 6, nitrification; 7, immobilization; 8, mineralization; 9, assimilatory and dissimilatory nitrate reduction to ammonium; 10, leaching; 11, plant uptake; 12, deposition N input; 13, residue composition, exudation; 14, soil erosion; 15, ammonium fixation and release by clay minerals; 16, biomass combustion; 17, forest harvesting; 18, litterfall (Nihkgard et al, 1994).

When nitrogen input owed to mineralization and atmospheric deposition exceeds the demand of both the vegetation and the microbes in undisturbed maturated forest ecosystems, the phenomenon of nitrogen saturation takes place (see Gunderson and Bashkin, 1994). This phenomenon is accomplished by nitrogen leaching from the forest ecosystems. These nitrogen losses are highly variable, but generally sites in North America and Northern Scandinavia show N loss rates of < 1.4 kg/ha/year, whereas sites in southern Scandinavia and Central Europe exhibit loss rates often > 7 kg/ha/year. Generally, N leaching from undisturbed forest ecosystems starts when the N deposition rates are higher than 10 kg/ha/year.

Fixation of molecular nitrogen, N_2, to ammonia in forest ecosystems can occur on and/or in a variety of forest substrates including plant canopy and stems, epiphitic plants compartments, wood, litter, soil and roots. A recent review of the magnitude of N inputs to forest ecosystems indicates that non-symbiotic fixation ranges from < 1 to 5 kg/ha/year and symbiotic fixation ranges from about 10 to 160 kg/ha/year in early successional ecosystems where N_2-fixing species are present.

Denitrification, a dissimilatory pathway of nitrate reduction (see Chapter 3, Section 3 also) into nitrogen oxides, N_2O, and dinitrogen, N_2, is performed by a wide variety of microorganisms in the forest ecosystems. Measurable rates of N_2O production have been observed in many forest soils. The values from 2.1 to 4.0 kg/ha/yr. are typical for forest soils in various places of Boreal and Sub-Boreal Forest ecosystems.

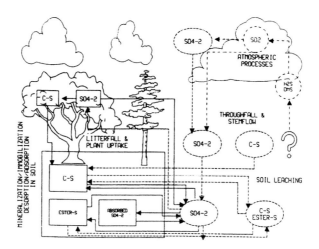

Figure 5. A model illustrating sulfur biogeochemical cycle in forest ecosystems (Nihkgard et al, 1994).

All *in situ* studies (field monitoring) of denitrification in forest soils have shown large spatial and temporal variability in response to varying soils characteristics such as acidity, temperature, moisture, oxygen, ambient nitrate and available carbon.

Sulfur microbial transformation

Microbial transformations are also important in the sulfur cycle in Forest ecosystems. The major sulfur pools and transformation processes are shown in Figure 5.

Similarly to N, most S pools are found in organic form in forest floor and soil humus. However, unlike nitrogen, there are important abiotic processes, especially sulfate sorption processes, which play a critical role in regulating sulfate dynamics in forest ecosystems. An example of this was shown in the Habbard Brook whole-tree harvesting experiment, where the decrease in sulfate output from the watershed was attributed to sulfate adsorption, which was enhanced by soil acidification from nitrification (see above).

Biological pathways of sulfur movement in soils of forest ecosystems are related to microbial transformation of sulfolipids. Back conversion of sulfate-S into organic matter immobilizes the anion and potentially reduces soil cation leaching. Processes of sulfur mineralization and incorporation proceed rapidly in response to several factors, including temperature, moisture, and exogenous sulfate availability in soils and water.

Phosphorus microbial transformation

The microbial regulation of the P cycle in Forest ecosystems is tightly coupled to soil development and the change of phosphorus pools from the predominance of primary

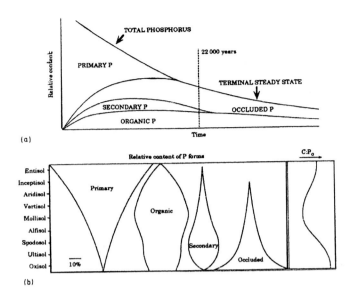

Figure 6. Biogeochemical microbial transformations of phosphorus in forest soils: (a) the long-term relative development of different phosphate fractions in soils of the forest ecosystems; (b) the relative contents of P fractions in different soil types. Also, the idealized carbon to organic P rates (C : P₀) are illustrated (Nihkgard et al, 1994).

inorganic phosphorus (e.g., apatitte) to that of organic-P, secondary-P mineral-P and occluded-P as is illustrated in Figure 6.

The organic phosphorus in forest soils is derived mainly by microbial synthesis and the accumulation of plant and animal residues plays a subordinate role. Much of the organic P occurs in ester linkages (up to 60%) with lesser amounts in other forms. The leaching losses of P are ranged from as little as 7 g/ha/year at Habbard Brook to up to 500 g/ha/year for a glacier outwash in New Zealand. Loss rates generally are greatest in young, base-rich soils and lowest in acidic soils (pH < 5.0) with high content of sesquioxides, which may fix phosphates. Thus, soils at intermediate stages of development have the highest availability of phosphorus, which is partly regulated by microbial mineralization processes.

Carbon microbial transformation

Obviously the fates of N, S and P are tightly coupled not only with each other, but also with C dynamics of soils (see also Chapter 5). For example, it has been suggested that the leaching of dissolved organic species of nitrogen sulfur and phosphorus contributes to the accumulation of these elements in mineral soils. This leaching of organics is an important component of soil formation of Spodosols, which are common especially in Northern Coniferous Forest ecosystems. The ratio between total nitrogen and total carbon, the C : N ratio, is widely applied to predict microbial mineralization-immobilization of nitrogen in soils.

Typically high C : N ratios (> 60) for forest plant organic matter have a major impact on the nitrogen cycle. A number of empirical and theoretical analyses have established a strong linkage between nitrogen mineralization, assimilation and organic matter decomposition. In particular, C : N ratios > 30 decrease mineralization and increase assimilation instead, with the balance between two processes dependent on the nitrogen content in microbial biomass. The latter parameter sets the minimum nitrogen requirement for biosynthesis per unit amount of substrate metabolized. Non-symbiotic nitrogen fixation can ameliorate nitrogen limitation, and to some extent high C : N ratios may be a determinant of soil microbial diversity.

For more details on relationships between N, S, P and C microbial transformation see Fenchel *et al*, 1998.

However, the microbial activity is depressed during long and severe wintertime, and this leads to an accumulation of semi-mineralizable plant residues on the soil surface. With the increasing duration of cold season from south to north, the mass of these half-destroyed remains enlarges from 15 ton/ha of dry organic matter in Broad-leaved Sub-Boreal Forest ecosystems to 80–85 ton/ha in Northern Taiga Forest ecosystems.

In the Northern Forest ecosystems, the relative content of chemical species in dead organic matter of forest litterfall is higher than that in living biomass. In Mixed and Deciduous Forest ecosystems, this is true for the total mass of chemical species, however some elements are more abundant in living biomass. Thus, the general biogeochemical feature of biological turnover in forest ecosystems is the prolonged retention of many chemical species in dead organic matter and exclusion from cycling (Table 10).

The data of Table 11 provide a general characteristic of trace element fluxes in Boreal and Sub-Boreal Forest ecosystems.

Case studies

North American Forest ecosystems
In the USA, two focal points for biogeochemical research have been the forest catchment ecosystems at Hubbard Brook Experimental forest in the White Mountains of New Hampshire and Coweeta Hydrologic Laboratory, located in the southern Appalachians of North Carolina.

The nutrient cycles of the forest catchment ecosystems are to a large extent determined by biota, especially by the primary production of plants and by microbial decomposition. Severe losses from the ecosystem of important nutrients, e.g., Ca, Mg, K and P, are expected to lower the productivity when occurring in the root zone. Most nutrients available for circulation in the temperate forested ecosystems are found in the tree layer or in the accumulated organic mater of soil layer. This is especially true for the most important macronutrients (C, N, P, K, Ca, Mg and S). Nitrogen is

Table 10. Averaged fluxes and pools of biological cycling in Forest ecosystems (after Rodin and Bazilevich, 1965; Dobrovolsky, 1994).

Fluxes and pools	Ecosystems			
	Northern Taiga Spruce Forest	Southern Taiga Spruce Forest	Sub-Boreal Oak Forest	Southern Taiga Sphagnum Swamp
Pools, ton/ha				
Total biomass	100	330	400	37
Nitrogen	0.35	0.72	1.15	0.23
Dead organic matter	30	35	15	100
Annual fluxes, ton/ha/year				
NPP	4.5	8.5	9.0	3.4
Litterfall	3.5	5.5	6.5	2.5
Nitrogen uptake	0.058	0.041	0.095	0.040
Nitrogen return	0.048	0.035	0.057	0.025

almost completely bound to the organic matter and when it is mineralized it is either leached as nitrate or assimilated and immobilized by organisms in the soil. Including the humus horizon, the soil organic matter contains the largest pool of nitrogen in the Boreal Forest ecosystems (Figure 7).

For phosphorus and potassium this pool of organic matter is also of importance, but in the Boreal Forest ecosystems, a relatively higher amount is in the living biomass. The long-term soil development proceeds towards a lower rate of weathering in the root zone and relatively higher amounts in biogeochemical fluxes (Nihlgard *et al*, 1994).

Spruce Forest ecosystem of Northwestern Eurasia
Table 11 presents the averaged data for the whole forest area of Boreal and Sub-Boreal zone. However, there are definite peculiarities of biological and biogeochemical cycles in the individual ecosystems. We will consider the Spruce Forest ecosystem of the Karelia region, Russia. These ecosystems occur in the wide area of the Karelia, south from 63° N.

The dominant species are the spruce (*Picea excelsa*), the birch (*Betula verrucosa, B. pubescens*), the aspen (*Populus tremula*), and the alder (*Alnus incana*). The moss and low bush layer is represented by the blueberry-bush (*Vaccinium myrtiilus*), hypnic

Table 11. Averaged fluxes of trace elements in biological cycling of Boreal and Sub-Boreal Forest ecosystems (after Dobrovolsky, 1994).

Ash/Chemical species	Forest Ecosystems			
	Northern Taiga Coniferous Forest	Sub-Boreal Coniferous and Small-leaved Forest	Sub-Boreal Broad-leaved Forest	Sphagnum Forest Swamp
Average ash content, %	1.7	2.0	3.5	2.5
Total uptake of ash elements, kg/ha/year	76	170	315	85
Trace element turnover, g/ha/year				
Fe	304.0	680.0	1260.0	1500.0
Mn	364.0	816.0	1512.0	75.1
Sr	53.0	119.0	221.0	68.0
Ti	49.4	111.0	205.0	401.2
Zn	45.6	102.0	189.0	62.6
Cu	12.2	27.2	50.4	12.2
Zr	11.4	25.5	47.2	14.3
Ni	3.0	6.8	12.6	13.6
Cr	2.6	5.9	11.0	10.9
V	2.3	5.1	9.4	10.2
Pb	1.9	4.2	7.9	12.0
Co	0.7	1.7	3.1	2.3
Mo	0.7	1.5	2.8	1.0
Sn	0.4	0.8	1.6	—
Ga	0.1	0.2	0.3	0.4
Cd	0.05	0.12	0.22	—

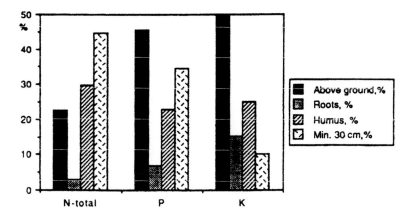

Figure 7. Relative distribution of N, P and K in the Boreal Forest ecosystem. Total amounts for the different fractions are given, expect in the mineral soils down to 30 cm depth, where exchangeable amounts are given for P and K (Nihlgard et al, 1994).

mosses, separate species of cowberry (*Vaccinium uliginosum*) and flowering plants. The biomass of these Spruce Forest ecosystems reaches 10 ton/ha at the age of 100–150 years (Table 12).

We can see from Table 12 that most of the Spruce Forest ecosystem biomass is accumulated in trees, with trunk mass predominating. The values of annual Net Primary Production (NPP) and litterfall production are more connected with needles. In living matter, the mass of moss and bush species makes up to 2–3% of the tree biomass, whereas in dead matter (litterfall), it is up to 10%.

The distribution of trace metals in ash of living matter distinguishes it from that in dead organic matter. About half of the total ash elements contained in the biomass is confined to the deciduous tree organs, whereas for the shedding, most ash elements are found in the needles. About 80% of ash elements are supplied from conifers to the soil in the shed needles, whereas their supply in the deciduous organs of mass and bush species is a mere 10%. The shedding trace metal losses from living plant organs are accompanied by their redistribution into outer bark, where they may be retained for long periods. We can consider the retention of trace metals in bark of the conifers as one of the important sinks in Spruce Forest ecosystems, which exclude these metals from biogeochemical cycles.

The averaged fluxes and sinks of trace metals in biogeochemical turnover in Spruce Forest ecosystems are shown in Table 13.

There are significant differences in plant uptake of trace metals from soils. The coefficient of biogeochemical uptake, C_b, presents the ratio between element content in plant species to its content in soil. Figure 8 shows the averaged values of these co-efficients for most plant species of Spruce Forest ecosystems, Karelia region, Russia.

Table 12. Biomass and total ash mass distribution in Spruce Forest ecosystems of the Karelia region, Russia (After Dobrovolsky, 1994).

Biomass components	Ecosystem parameters		
	Biomass	NPP	Litter production
Woody vegetation, ton/ha	40–100	4.0–9.0	2.0–5.5
Of this:			
Needles, %	10–15	36–38	56–60
Twigs, %	12–17	8–9	8–9
Trunks, %	50–60	41–43	22–25
Roots, %	17–19	12–13	9–10
Moss and low-bush vegetation, ton/ha	0.8–3.5	0.2–0.6	0.2–0.6
Total mass of ash elements (100%)			
Components of woody vegetation			
Needles, %	40–50	79–80	76–81
Twigs, %	13–18	4–5	2–3
Trunks, %	19–33	10–11	4
Roots, %	12–14	4–5	2–3
Moss and low-bush vegetation, %	3–8	5–9	7–10

We can see C_b values for lead, zinc, tin, nickel, and copper are an order of magnitude higher than those for zirconium, titanium, and vanadium. We can observe also that the curves follow a similar pattern independently of the composition of the bedrock, diabasis or gneissic, underlying the forest ecosystems. Simultaneously, various plants absorb the same elements at a different rate. For instance, mosses are better accumulators of poorly absorbed metals, like Ti, Zr, V, than tree and small shrubs. The selective accumulation of metals by plants can be used in prospecting for trace metal ore deposits.

The quantitative characterization of major ash elements involved in the biogeochemical turnover of Spruce Forest ecosystems is illustrated in Table 14.

The results of Table 14 show that for calcium, potassium, and silicon, biogeochemical turnover is within the limits of 10 to 30 kg/ha per year. The turnover for

Table 13. Averaged trace element mass budget for Spruce Forest ecosystems, Karelia, Russia (After Dobrovolsky, 1994).

Ecosystem biomass component and ash content, %	Annual NPP, kg/ha		Annual shedding, kg/ha	
	Dry organic matter	Total sum of ash trace metals	Dry organic matter	Total sum of ash trace metals
Conifer needles, 1.7%	1440–3420	24.5–58.1	1120–3300	19.0–56.1
Conifer bark, 1.3%	200–540	2.6–7.0	100–300	1.3–3.9
Moss and shrub species, 1.7%	200–600	3.4–10.2	200–600	3.4–10.2

magnesium, phosphorus, manganese, sulfur, and aluminum is less than 10 kg/ha per year. These values are about 1 kg/ha per year for iron and sodium.

The trace element concentrations in spruce and blueberry species as the major components of Spruce Forest ecosystems are summarized in Table 15.

The statistical estimation of trace metal concentrations in the Spruce Forest ecosystems of the Boreal zone is the subject of wide variation, with coefficient of variation from 36 to 330%. However, we can note a clear trend in biogeochemical peculiarities of trace metal uptakes by dominant plant species.

First, the concentration of Sr, Ba, and Ti in spruce bark is relatively higher than in needles, while the latter are enriched by Ni and Zn. Second, the concentration of Zn, Ba, Cu, and Cr is higher in blueberry roots, than in the aerial parts. Third, the effect of zonal and local factors is remarkable. For instance, the Sr content in the needles and in the barks of the spruce is appreciably lower than the average content of this trace metal in the shedding of the World's coniferous trees. This should be attributed to the effect of zonal factors influencing the active release of this metal. At the same time, the elevated content of Pb and Zn reflects the specificity of the bedrock since the dispersed sulfide mineralization is a characteristic feature of this Karelia region.

Using the results on biogeochemical uptake and content of trace metals in Spruce Forest ecosystems (Dobrovolsky, 1994), we can calculate the biogeochemical migration rates (Table 16).

We can see that the tree vegetation absorbs annually from soil tens of grams per hectare of Zn and Ba, units of grams of Ni, V, and Co. The absorption of trace metals by low bush species is smaller by an order of magnitude. Simultaneously, similar amounts of metals are released from the living biomass of the Spruce Forest ecosystems.

Of great interest is the calculation of a relative uptake of trace metals by forest species. The C_b values for each metal are similarly independent of the composition of crystalline bedrock and the depth of detrital deposits. For instance, the C_b values

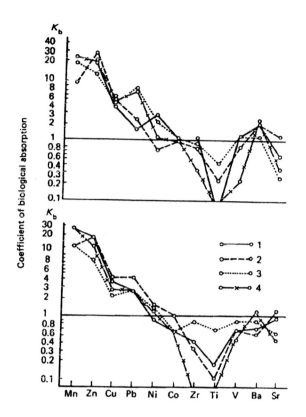

Figure 8. Averaged coefficients of biogeochemical uptake, C_b, of trace metals in Spruce Forest ecosystems of the Karelia: top scheme represents the ecosystem with Podzols on diabase outcrops and bottom scheme represents the ecosystem with Podzols on gneiss outcrops. 1—spruce needles; 2—spruce bark; 3—hypnic mosses; 4—leaves and twigs of blueberry (Dobrovolsky, 1994).

for Zn, Mn, Cu, and Pb are in a range from 2 to 30, they are considered as the elements of intense uptake. Poorly absorbed are Ni and Co, with C_b values about 1. The metals, such as Ti, Zr, and V, are very reluctant to be taken up and their C_b values are less than 1.

Swampy ecosystems of North Eurasia

The remarkable part of the forested area is swampy. In certain regions, like the vast West Siberian plain, swamp and waterlogged ecosystems occupy about 30% of the total area. The biogeochemical cycling in these ecosystems is very complex and specific. The slow rates of biogeochemical turnover, typical for all Boreal Forest ecosystems are more depressed in Swamp ecosystems. For instance, in the *Sphagnum* Swamp ecosystem, the most widespread type of Swamp ecosystems, the annual NPP

Table 14. Averaged major ash element mass budget for Spruce Forest ecosystems, Karelia, Russia (After Dobrovovlsky, 1994).

Major ash element	Major ash element content, kg/ha					
	Ecosystem biomass		Annual NNP		Annual shedding	
	Mean	Limits	Mean	Limits	Mean	Limits
Ca	205	150–260	32	20–45	27	15–40
K	110	50–170	15.5	7–24	13	6–20
Si	52	40–65	14.5	10–19	13.5	9–18
Mg	32	25–40	5.5	4–7	4.5	3–6
P	30	15–45	4	2–6	3	1–5
Mn	20	15–25	3.3	2.2–4.5	2.9	1.9–3.9
S	8	6–10	1.6	1.2–2.1	1.5	1.2–1.9
Al	8.5	5–10	1.5	1–2	1.2	0.7–1.8
Fe	0.7	3.5–8.0	0.8	0.5–1.2	0.8	0.4–1.1
Na	1.5	0.5–2.5	0.3	0.1–0.5	0.2	0.1–0.3
Total	429	300–600	79	47–107	67.6	40–90

is only 10% or less from living plant biomass and less than 1% from the total mass of accumulated dead organic matter of peat. The rate of biogeochemical turnover in such an ecosystem is about 50 years.

There are significant differences between biogeochemical cycling in forest and swampy ecosystems of the Boreal-climate zone. The annual growth (NPP) of moor vegetation is about 3.5 ton/ha, which is twice as small as that in the forest ecosystem. In the bog, the degradation of dead organic matter proceeds at a much smaller rate than in the forest. The mass of peat accumulated within a period of 100 years accounts for some 20% of the organic material formed in bog landscape. This is nearly 10 times higher than a relative mass of dead organic matter, preserved in forests.

In spite of small rates of organic matter degradation, mineralized iron and trace metals are permanently accumulating in stagnant bog waters. This leads to a larger uptake of these metals by swampy vegetation. In an area of 1 ha the annual uptake of Zn, Ba, Pb, and Ti amounts to tens of grams; of Cu, Zr, Cr, Ni and V to units of grams, Co to tenths of gram, by moor plant species.

Table 15. The statistics of trace metal contents in the major plant species of Spruce Forest ecosystems of Boreal zone, Karelia, Russia (After Dobrovolsky, 1994).

Trace metal	Spruce species (*Picea excelsa*)				Blueberry (*Vaccinium myrtillus*)			
	Needles		Barks		Aerial parts		Roots	
	M, ppm	*V*,%	*M*, ppm	*V*,%	*M*, ppm	*V*,%	*M*, ppm	*V*,%
Zn	1250	98	1188	90	751	190	1515	161
Ba	390	65	456	76	465	71	578	22
Ti	102	100	170	330	113	130	86	110
Cu	86	86	88	60	107	90	124	51
Pb	64	60	62	130	84	77	11	85
Sr	43	60	73	88	35	85	33	84
Zr	38	36	33	61	20	27	28	48
Cr	28	51	26	41	28	37	34	56
Ni	23	100	17	120	19	71	25	53
V	16	42	23	55	19	26	14	41
Co	12	80	13	61	13	70	14	70

Note: M is the arithmetic mean, ppm by dry ash weight, V is variation, %.

Broad-leafed deciduous Forest ecosystems of Central Europe

Southward from the northern border of Boreal Forest ecosystems, the natural humidity gradually decreases simultaneously with the shortening of the cold winter season, which inhibits biogeochemical processes. Accordingly, the floral composition and geochemical conditions of ecosystems is altered also. The Coniferous Forest ecosystems grade into Mixed Forest ecosystems and then into Deciduous Forest ecosystems. Herbaceous plant species change the small shrubs and mosses. Biogeochemical processes become more active and annual NPP values are increasing as well. The litterfall amount is increasing but that of dead organic matter decreases. Consequently, changes occur in the chemical composition of plant species and in values of coefficients of biogeochemical uptake, C_b.

The Broad-leaved Forest ecosystems are widespread in regions of the Sub-Boreal zone with a well balanced precipitation:evapotranspiration ratio. The southern periphery of the vast belt of Eurasian boreal and Sub-Boreal Forest ecosystems is represented by Oak Forest ecosystems. These ecosystems exhibit both the largest biomass and

Table 16. Rates of trace metal biogeochemical migration in the Spruce Forest ecosystems, Karelia, Russia.

Ecosystem plant components	Rates of biogeochemical migration of trace metals, g/ha/year						
	Zn	Ba	Sr	Cu	Pb	Ni	Co
Needles							
New grown	30–70	10–50	1–2	2–5	2–4	0.5–1.0	0.3–1.0
Lost in shedding	25–70	8–20	0.8–2.0	2–5	1–3	0.4–1.0	0.2–0.6
Bark							
New grown	4–8	2–4	0.2–0.4	0.3–0.6	0.2–0.4	0.1–0.2	0.03–0.04
Lost in shedding	2–5	0.7–2	0.1–0.3	0.1–0.4	0.1–0.2	0.1–0.3	0.02–0.04
Shrub vegetation							
New grown	3–8	2–5	0.2–0.5	0.3–10	0.3–0.8	0.08–0.2	0.04–0.1
Lost in shedding	3–8	2–5	0.2–0.5	0.3–10	0.3–0.8	0.08–0.2	0.04–0.1
Total plant uptake	37–86	14–59	1.4–2.9	2.6–6.6	2.5–5.2	0.68–1.4	0.37–1.1
Total shedding	30–83	10.7–27	1.1–2.8	2.4–6.4	1.4–4.0	0.51–1.3	0.26–0.74
Accumulation in tree bark	2–3	1.3–2.0	0.1	0.2	0.1–0.2	0.07–0.1	0.01

annual NPP rates in comparison with other forest ecosystems of this zone. However, the dead mass surface organic matter is 2–3 time less than that of coniferous forests.

The content of nitrogen in fallen leaves as a major shedding component is about twice as much as that in the needles of coniferous trees. The total sum of ash elements in the leaves accounts for 3 to 5%, average about 4% on dry weight basis. Accordingly, the concentration of calcium increases from 0.5 to 4.0%, potassium from 0.15 to 2.0% and silicon with a wide variation. The row of nutrient uptake is as the following:

$$C > N > K > Si \cong Mg \cong P \cong S > Al \cong Fe \cong Mn. \qquad (1)$$

The amount of nitrogen in the biomass of the Oak Forest ecosystem reaches 900–1200 kg/ha and the sum of ash elements is about 2000–3000 kg/ha, e.g. greatly in excess of nitrogen. The corresponding values for accumulation of nitrogen in annual growth are 80–100 and for ash elements, 200–250 kg/ha. An essential point is that the green leaves store 70–80% of the mass of uptaken elements, and the fallen leaves,

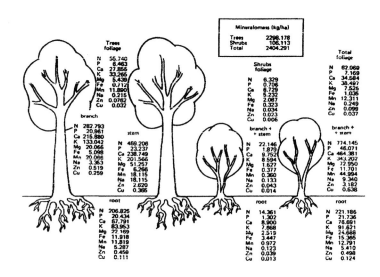

Figure 9. Mass distribution of chemical species in the Oak Forest ecosystem, Central Europe (Jakics, 1985).

80–90% of the mass of elements eliminated from the biological cycle. Nitrogen, as returned in the shedding products, amounts to 40–70 kg/ha, and the ash elements, to 180–200 kg/ha (see Box 2).

Box 2. Biogeochemical fluxes of elements in Oak Forest ecosystem (after Jakucs, 1985)

The distribution and mass budget of chemical species in the Oak Forest ecosystem have been studied in Hungary. The principal timber species is *Quercurs petraea*, accounting for 78% of the wood mass; *Quercus cerris* accounts for 22%. The shrubs are chiefly represented by *Acer campestre* and *Cornus max*, and the grass by the genera *Carex, Dactylis*, and *Poa*.

The mass of trees is 241.03; of shrubs, 6.54; of herbaceous grasses, 0.62, and the mosses, 0.0016 ton/ha. The distribution of chemical species in biomass and annual NPP is depicted in Figure 9 and summarized in Table 17.

Amongst the ash elements the most abundant in biomass is calcium, which is accumulated in leaves, in trunk wood, and in twigs. Potassium is dominant in annual NPP. The masses of trace metals in biogeochemical cycling of this Oak Forest ecosystem are roughly in correspondence to their respective average values for the other Oak Forest ecosystems of the Sub-Boreal climate. An exception to the rule is manganese, with its inordinately large fluxes in pool and annual NPP. Possibly, this phenomenon should be related with local landscape-geochemical peculiarities.

Table 17. The biogeochemical fluxes and pools in the Oak Forest ecosystem of Central Europe.

Element	Annual plant uptake, kg/ha			Pool in ecosystem, kg/ha		
	Trees	Shrubs	Total	Leaves	Stems, twigs	Roots
N	103.39	14.49	117.88	62.07	774.15	221.19
P	11.91	1.58	13.47	7.17	46.07	21.74
Ca	60.87	13.49	74.36	34.58	464.38	76.69
K	79.80	12.28	92.08	38.50	343.20	91.62
Mg	11.89	3.90	15.79	7.53	72.95	24.69
Fe	1.54	1.31	2.85	1.04	11.70	15.37
Mn	19.12	0.86	19.98	12.31	44.99	12.79
Na	1.01	0.14	1.15	0.25	9.34	5.41
Zn	0.28	0.04	0.32	0.10	3.18	0.50
Cu	0.06	0.01	0.07	0.04	0.64	0.12

In may be of interest to compare the fluxes of elements in biogeochemical cycles of the Oak Forest ecosystem with airborne deposition input. The latter were (in kg/ha/year) for N, 17.7; for Ca, 14.7; for Mg, 1.8; for K, 4.2; for Na, 1.4; for P, 1.1; for Fe, 0.07; and for Zn, 0.14. The deposition input of these elements falls into a range of 20% (calcium) to 4.5% (potassium) relative to the respective biogeochemical fluxes (see Table 17). The airborne Fe input accounts for a mere 2.5%. Simultaneously, for some trace metals, like zinc, the deposition input is commensurate with the fluxes of the biogeochemical cycle.

3.2. Biogeochemical fluxes in soils of Boreal Forest ecosystems

The main soils of Forest Ecosystems are Podzols and Podzoluvisols. There are plenty of various soil subtypes, groups and families among these two main soil types. However, all forest soils have a number of common features originating in the similarity of processes occurring therein. The retarded biological cycle provides organic materials for the build up of the covering layer on the soil surface. This layer consists of partly decomposed plant residues and is called forest litter (A_0 horizon). The slow, but permanent, microbial decomposition of these materials leads to the formation of fulvic acids. The predominance of these mobile organic acids in humus structure

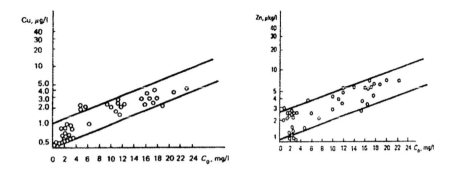

Figure 10. Relationships between organic matter concentration in permafrost soils of East Siberia Taiga Forest ecosystems and concentration of copper (left) and Zinc (right) (Nikitina, 1973).

and the predominance of precipitation over evapotranspiration are favorable to the formation of permeable, readily leached soils.

The second characteristic feature of Forest Ecosystem soils is the accumulation of macro-nutrients in the litter with sharp decreasing to the download horizon. However, the trace elements show the opposite trend and the concentration of micro-nutrients is gradually increasing up to the soil forming rocks.

Two groups of chemical elements can be considered related to their distribution in soil profiles. The first group includes the essential nutrients actively absorbed by vegetation and relatively tightly bound in soil organic matter (Figure 10)

These elements are the main macronutrients, like nitrogen, phosphorus, potassium, and many micronutrients, like copper, zinc, and molybdenum. Their concentration in A_0 horizon is remarkably higher than in soil-forming rocks. For the second group representatives, the concentration in the A_0 horizon, although elevated in comparison with elluvial horizon, fails nevertheless to reach the rocks. The typical elements of the second groups are titanum, zirconium, vanadium, and chromium. The averaged distribution curves are shown in Figure 11.

Various forms of macro- and microelements differ in their ability to migrate and redistribute among the soil profile. The elements contained in clastic minerals are practically immobile. The elements bound to finely dispersed clay minerals are either co-transported with clay particles, or are involved in sorption-desorption processes. Part of the elements are found in concretions and also in very thin coating films of hydrated iron oxides; some elements make a part of specially edaphic organic compounds.

The determination of distribution pattern of various forms of both macroelements and trace metals in soil profile is a very complicated task. We have to know the distribution of organic matter, mineral particle, and microbes, the existence of different barriers, redox conditions etc (Box 3)

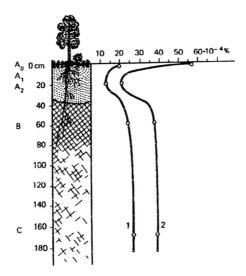

Figure 11. The profile distribution of vanadium (1) and copper (2) in typical Podzol of Boreal Forest ecosystems (ppm, 1 N HCl extraction) (Dobrovolsky, 1994).

Box 3. Distribution of various forms of trace metals in Podzols of Boreal Forest ecosystems (after Dobrovolsky, 1994)

In the Mixed Forest ecosystems a soil fraction less than 1μm contains most of the elements previously confined in the forest litter and gradually involved in the biogeochemical cycle. In this fraction Cu and Mo forms account for 60–70% of the total soil content. The metals, poorly absorbable by plants, for example, Cr and V, occur in finely dispersed soil fraction in smaller amounts, about 20–30%.

Table 18 illustrates the variations of major forms of Cu and Co in typical Forest ecosystem Podzols of Moscow region, Russia.

We can see that the soluble and exchange forms of these metals are present in small amounts accounting merely for a few percent of the total metal content in soil. The content of organometal species is relatively high in the upper profile rich in humic species, whereas it drops sharply in the mineral horizons. Copper is extensively involved in the biogeochemical cycle in the Forest ecosystems and this is less profound for cobalt. It is noteworthy that a large part of metals (in particular, of copper) become bound to iron hydroxides. This is typical for various trace elements, including arsenic, zinc and other elements with variable valence.

The excessive ground humidity is favorable to the formation of gley soils in Boreal forest ecosystems. Clay-podzolic soils with a massive forest litter layer provide the conditions for low water saturation. In high water-saturated conditions, a peat horizon

Table 18. Distribution of copper and cobalt in Podzols of East European Boreal Forest ecosystems.

Soil horizon	Total metal content, ppm	Water-soluble and exchangeable	Metals bound with:		
			Organic matter	Hydrated iron oxides	Mineral soil matter
		% of the total content of metal in soil			
Copper					
A_1	7.4	3.1	32.4	51.3	13.2
A_1/A_2	7.4	3.6	24.3	45.9	26.2
A_2	6.0	3.2	26.7	55.0	15.1
A_2/B	16.8	2.7	3.6	10.5	53.2
B_1	20.6	3.8	4.4	43.2	48.6
B_2	19.4	3.9	4.6	54.1	37.4
BC	19.8	3.6	4.5	47.0	48.5
Cobalt					
A_1	5.5	4.2	12.7	30.9	52.2
A_2	4.5	3.7	13.3	24.4	58.6
B_1	5.8	3.1	3.4	34.5	59.0
C	5.3	4.5	3.8	34.0	58.3

is formed in the top layer of soil forming boggy soils. In the forest litter of Podzols the concentration of fulvic acids is two or three times higher that of humic acids; in contrast, humic acids are predominant in the peat horizon of Boggy soils. This is reflected in a lower pH of aqueous extraction from the forest litter of Podzols and in more active biogeochemical migration of element in the corresponding ecosystems.

The concentration of many nutrients and trace metals in Peat-Boggy soils is often higher than that in Podzols. In boggy soils the relative concentration of elements bound with humic acids is also significantly higher. For instance, in the Spruce Forest ecosystems of Karelia most of the mobile nickel forms are bound to humic acids, and those of copper to the fulvic acids. Consequently copper easily leaches from the soil upper layer with relevant accumulation in the download horizon. The concentrations of the organometal forms of trace elements in the soils of Swamp ecosystems compare with that of Podzols. However, the overall mass of organometal species in boggy soils is appreciably higher.

In the southern direction the Podzols are transferred into Podzoluvisols and Distric Phaerozems. These soils have less acidity and less pronounced migration of various elements.

3.3. Biogeochemical processes in the soil-water system of Boreal and sub-Boreal Forest ecosystems

Almost all biogeochemical processes in Forest ecosystems are connected with aqueous migration in surface and ground water runoff. Low content of dissolved salts and organic species are the most characteristic features of surface runoff waters of Boreal-climate forest ecosystems. This feature is particularly distinct in Taiga Forest ecosystems with permafrost soils where the zone of extensive water exchange in the river valleys is confined to supra-permafrost waters and to ice-thaw waters. The review of leaching processes is shown in Box 4.

Box 4. A brief review of soil leaching processes (after D.W. Johnson (1992))

A concept of anion mobility may be considered a useful paradigm for explaining the net retention and loss of cations from soils. This paradigm relies on the simple fact that total cations must balance total anions in soil solution (or any other solution), and, therefore, total cation leaching can be thought of as a function of total anion leaching. The net production of anions within the soil (e.g., by oxidation or hydrolysis reactions) must result in a net production of cations (normally H^+), whereas the net retention of anions (by either absorption or biological uptake) must result in a net retention of cations.

The major anions in soil solutions are Cl^-, SO_4^{2-}, NO_3^-, HCO_3^-, and organic anions. Chloride may dominate soil solutions only in coastal areas. The four major anions often undergo oxidation-reduction and hydrolysis reactions that cause a net production or consumption of H^+, which in turn strongly affects a net retention or release of base cations from the ecosystem. Carbonic acid, which is formed by the dissolution and hydrolysis of CO_2 in water, is the major natural leaching agent in many temperate ecosystems. Carbonic acid concentrations in soil solution are many times greater than in precipitation or throughfall because of the high levels of CO_2 in the soil atmosphere. On dissociation, carbonic acid forms H^+ and HCO_3^-, H^+ exchanges for a base cation (or causes the dissolution of a mineral), and a bicarbonate salt leaches from the system. Because carbonic acid is a very weak acid, HCO_3^- becomes associated with H^+ to form carbonic acid at low soil pH. Thus, the carbonic acid leaching mechanism is self-limiting, and eventually becomes inoperable during the soil acidification process.

Organic acids are often the major cation leaching agents in extremely acid soils that occupy the boreal regions. These acids are stronger than carbonic acid and can produce low solution pH while providing organic counter-anions for cation leaching. Organic acids are also responsible for the chelation and transport of Fe and Al from surface (E or albic) to subsurface (Bs or spodic) horizons during the podzolization process. In theory, organic acid leaching, like carbonic acid leaching,

Table 19. The concentration of trace metals in atmospheric deposition and ground waters of Siberian Taiga Forest ecosystems, ppb (μg/L) (After Shvartsev, 1978 and Dobrovolsky, 1994).

Trace metal	Ecosystems				
	Tundra Forest of Middle Siberian High Plain			South Taiga Forest of West Siberia	
	Rainwater	Snowmelt water	Ground water	Rainwater	Ground water
Mn	4.0	3.7	12.5	4.2	20.2
Zn	3.0	10.0	8.5	7.2	29.2
Cr	2.2	2.5	1.3	1.1	2.5
Cu	1.0	2.6	3.1	0.2	1.8
Ti	1.0	1.0	1.5	2.4	22.7
V	1.0	0.8	1.0	0.1	0.7
Pb	0.3	1.0	0.4	2.3	0.5
Ni	0.3	0.8	0.8	0.3	1.5

is self-limiting because organic acids are typically weak acids. However, organic acids can provide very low soil and soil solution pH and still remain active in leaching processes.

Sulfate and nitrate are the anions of strong acids, and therefore leaching by sulfuric and nitric acids is not self-limiting because of solution pH alone. The deposition or other inputs of either S or N in excess of biological demands of these elements will ultimately cause an increase in SO_4^{2-} or NO_3^- availability within soil. In case of SO_4^{2-}, inorganic absorption processes may prevent increased leaching, but this is seldom true in the case of NO_3^-. In that soil SO_4^{2-} absorption is strongly pH dependent (increases with decreasing pH); there can be a negative feedback involved in sulfuric soil leaching that is at least partially self-limiting.

The input of chemical species from atmospheric deposition is the main source of feeding the supra-permafrost ground waters. The deposition contributes a significant amount of elements to natural waters of the Boreal ecosystems. The concentration of trace metals in the ground waters of the North Siberia Taiga Forest Cryogenic ecosystems is lower than their respective average values for global riverine waters. In southern Boreal Forest ecosystems, the concentrations of trace metals in ground waters are only several-fold higher than that in precipitation (Table 19).

The higher concentrations of some trace metals, like Zn, Cu, Ni, and Pb, in snowmelt waters in comparison with rain waters is possibly related to the elevated content of solid particles in snow. The deposition fluxes are less important in the biogeochemical mass budget of elements in Southern ecosystems, than in northern Forest ecosystems.

The content of soluble salts increases in surface and ground water runoff of Forest ecosystems from north to the south with parallel decreasing soluble organic acids. To a certain extent, this may be related to the changes of soil-forming geological rocks. The biogeochemical processes in Podzols formed on sandy fluvio-glacial and ancient alluvial deposits are favorable to a decreased solid salt content in waters and simultaneously to the relatively increased content of fulvic acids. Such a geochemical situation is common for the sandy lowlands of the East European plain, for instance, Meshchera and Belarussian Swamp Forest ecosystems. In contrast, the alteration of sandy deposits to clay rocks formed the clay soils and consequently this is accompanied with increase in both pH and solid salt content. This trend is clearly distinct in Table 20.

We can see the rise of trace metal groundwater contents in southward direction with increasing content of total soluble salts. However, this increase is not similar for various metals. For example, Zn is the most abundant metal in the waters of Tundra and North Coniferous Forest ecosystems, whereas in Mixed Forest ecosystems it recedes to the third place after Sr and Mn.

The most important role in the migration of trace metals in the Forest ecosystems is connected with organic complexes and colloidal particles. For instance, the role of ionic forms of iron in water biogeochemical migration is of minor importance as compared with migration in metal-organic complexes accounting for 80% of total content of metal in the soil waters. On average, about 50% of metals are subjected to biogeochemical migration as the metal-organic complexes (see Box 5).

Box 5. The role of metal-organic complexes in biogeochemical migration of trace elements (after Newman and McIntosh, 1991)

Of particular interest to biogeochemical migration of trace metal is the effect of complexation on metal availability in the presence of natural dissolved organic matter. Humic and fulvic acids (symbolized here as Hum) are widely distributed in soil solutions and surface waters, especially in surface waters with visible color. Hum is an important fraction of the dissolved organic matter, even in low color water. Other fractions of natural organic matter, including dissolved nitrogen organic compounds (oligo- and polypeptides) and synthetic organic compounds, such as polyaminocarboxylates (e.g., ethylenetetraacetic [EDTA], nitriloacetic acid [NTA]), may be important complexing agents for trace metal ions in some waters.

Metal ions complexed with natural macromolecular organic matter or strong synthetic chelating agents generally are considered not to be directly available to aquatic organisms, whereas inorganic complexes generally are. Aluminum fluoride complexes are important exceptions among inorganic complexes of interest in freshwaters.

Table 20. The average concentrations of trace metals in ground waters of Siberian Boreal and Sub-Boreal Forest ecosystems, ppb (μg/L) (After Dobrovolsky, 1994).

Trace metal	Forest Ecosystems		
	Tundra	Northern Taiga	Mixed
Zn	23.0	31.8	39.5
Sr	21.3	26.3	163.0
Mn	12.3	17.9	55.6
Ba	10.0	9.1	29.4
Li	3.97	6.09	19.0
Cr	2.52	2.16	4.02
Ti	2.34	4.64	21.9
Ni	1.91	1.63	5.29
Pb	1.88	1.16	2.88
Cu	1.70	2.98	5.11
As	0.73	0.99	4.15
Zr	0.68	1.28	2.27
Mo	0.64	0.92	1.28
V	0.50	0.88	1.45
Ga	0.35	0.49	0.63
Sr	0.22	0.50	0.77
Co	0.40	0.24	0.61
U	0.30	0.34	1.01
Ag	0.21	0.37	0.20
Be	0.02	0.04	0.18

In addition, organometallic forms of several metals (e.g., methylmercury, triorganotin species), which are lipophilic, are bioaccumulated by aquatic organisms much more so than inorganic forms of the metals, and some organometallic complexes are toxic to aquatic biota (see also Chapter 8, Section 2).

For a variety of reasons, it is difficult to measure stability constants of metals with Hum, and the use of stability constants measured under a given set of solution conditions (so called 'conditional constants') for a different set of conditions (e.g., at a different pH or different set of metals and Hum concentrations) must be done

Table 21. The comparative estimation of average content of trace metals in surface waters of Swamp ecosystems, ppb ($\mu g/L$) (After Shvartsev, 1978 and Dobrovolsky, 1994).

Trace metals	Swamp ecosystem relief position	
	Watershed	Subordinate
Mn	41.9	52.5
Ba	11.5	9.23
Zn	5.73	9.86
Cu	0.55	1.20
Ti	2.65	1.10
Pb	0.60	0.89
Ni	0.66	0.72
V	0.10	0.19

cautiously. Significant advances were made during the past decade in ways to model metal-Hum binding, and a sufficient variety of conditional binding constants are now available at least to approximate the metal-binding behavior of natural water and soil solutions containing Hum.

The example of calculated trace metal speciation in Little Rock Lake, Wisconsin, USA is shown below:

Metal	% of free ions		Complexed ions, > 1%
	pH 6.1	pH 4.7	
Al	< 1	18	$ALOH^{2+}$, $Al(OH)_2^+$, $Al(OH)_3^0$, $Al(OH)_4^-$, AlHum, AlF^{2+}, AlF_2^+
Mn	91	95	MnHum
Cd	73	75	CdHum, $CdSO_4^0$
Cu	22	25	CuHum, $Cu(OH)_2^0$
Pb	70	92	PbHum, $PbOH^+$, $PbSO_4^0$
Zn	93	94	ZnHum, $ZnSO_4^0$

We can see that the most chelating properties are shown for Cu, in lesser degree for Cd and Pb, and a minimum for Mn and Zn.

During water migration the trace elements undergo a systematic redistribution with the subsequent elementary landscapes in the catena from the watershed to relief depressions. The lowland swamps are typical subordinate ecosystems in the belt of boreal forests. Migration of metals occurs mainly with soluble organic matter. More than 90% of Fe and a similar percentage of other metals, like Mn, Cu, Zn, Ni, Cr and V, in lowland swamp waters migrate in the form of soluble metal-organic complexes.

Swamp ecosystems are placed both in watershed and subordinate relief positions. An elevated content of metals is monitored in the latter case (Table 21).

The elevated concentrations of trace elements in boggy waters lead to an elevated migration in biogeochemical cycles and finally to an increased accumulation in dead organic matter of peat. The metal content in the low peat lands of the Karelia is higher than that in watershed peat. For instance, for manganese it is higher by 2 times, for cobalt by 3–4 times, for molybdenum and copper by 56 times. In different Forest ecosystems of the World this proportion is similar. However, this is maximally distinguished in the Boreal Forest ecosystems and to a lesser degree in Sub-Boreal ecosystems, where the pH of natural waters tends to increase and the concentration of dissolved organic compounds, to decrease.

4. BIOGEOCHEMISTRY OF STEPPES AND DESERTS

We will consider both Steppe and Desert ecosystems as deficient in atmospheric humidity (evapotranspiration exceeds precipitation). The Sub-Boreal, Semi-arid and Arid Steppe and Desert ecosystems occupy a significant part of the global area. This territory includes Sub-Boreal zones (Steppe, Arid Steppe, and Desert Steppe ecosystems) with total area of $9.23 \times 10^6 \, km^2$, as well as subtropical zones (Shrub Steppe and Desert Steppe ecosystems) with total area of $7.04 \times 10^6 \, km^2$. These areas do not include Subtropical and Tropical Sandy Desert ecosystems ($5.77 \times 10^6 \, km^2$) and Stony Desert ecosystems ($8.96 \times 10^6 \, km^2$). Thus the extra-tropical arid area takes about 20% of the World's terrestrial ecosystems. Most of this area goes to the inter-continental regions of Eurasia and, partly, of North and South America. The biogeochemistry of semi-arid and arid ecosystems shows distinctive parameters, which allow us to consider the quantitative features of element turnover and dynamics in natural fluxes.

4.1. Biogeochemical cycle of nutrients in arid ecosystems

Grasses, small shrubs and shrubs, whose number of xerophytic and ephemeral forms increases with aridity, are the chief representatives of Arid ecosystems. At the north and south peripheries of the arid zone, herbaceous ecosystems are predominant. In regions with a well balanced atmospheric humidity prior to human activities, of wide occurrence were Meadow Steppe ecosystems intermixed with Broad-leaved Forests. With an increase in climate continental properties, the Meadow Steppe ecosystems of Eurasia grade into Forb-Fescue-Stipa and Fescue-Stipa Steppe ecosystems. In turn, the natural vegetation has been destroyed by humans and replaced by pastures and

Table 22. The annual biogeochemical fluxes and pools in Steppe and Desert ecosystems (After Rodin and Bazilevich, 1972).

Ecosystem	Pool on biomass, kg/ha		Annual uptake, kg/ha		Annual litter, kg/ha	
	Nitrogen	Ash species	Nitrogen	Ash species	Nitrogen	Ash species
Meadow Steppe	274	909	161	521	161	521
Dry Steppe	103	242	45	116	45	116
Half-Shrub Desert	61	124	18	18	18	41

crops. Arid Steppe ecosystems are characterized by an annual precipitation of about 300 mm or less; these are Absinthum-Fescue-Stipa Steppe and Fescue-Stipa Steppe ecosystems. With increasing aridity, the small shrubs, mostly those of *Absinthum* species, become relatively abundant. The halophytic species is also of common occurrence in Semi-Arid and Desert ecosystems.

The biomass of arid ecosystems is significantly less than that of Forest ecosystems and changes from 10 to 25 ton/ha, by dry weight, in Steppe, from 4.0 to 4.5 ton/ha in Desert and from 2 to 3 ton/ha in Extra-Desert Ecosystems of Central Asia. The overall biomass of Arid Steppe and Desert ecosystems is an order of magnitude less than that of Forest ecosystems (Rodin and Bazilevich, 1972).

The ash content of Arid Steppe and Desert ecosystem vegetation is about 2 times higher than that of forest species. Accordingly, the biogeochemical fluxes of elements are similar to those in the forest ecosystems, in spite of the smaller biomass (see above). The compartments of biogeochemical turnover in Steppe and Desert ecosystems are shown in Table 22.

The characteristic biogeochemical feature inherent in all Steppe and Desert ecosystems is the most intensive cycling of different chemical species in comparison with forest ecosystems. For a Steppe ecosystem the biogeochemical cycle is 2–3 years and this means that the complete renewal of all ecosystems biomass takes place over this period. Remember that in Forest ecosystems the biogeochemical cycling is about 3–> 25 years and even about 50 years in Forest Swamp ecosystems. The turnover is the highest in Ephemeral Desert and gradually decreases to the north.

The annual biogeochemical turnover of trace ash elements is calculated in Table 23.

Table 23 shows the average values for Arid ecosystems. However, the local geochemical conditions can alter this tendency significantly. A relevant example can be shown for the Colorado Plateau, USA, where the fluxes of selenium are so high that the forage grasses are toxic for cattle. In the areas of volcanic eruptions, arid conditions are favorable to the extensive accumulation of fluorine. In these cases, the annual fluxes of elements are much higher than the figures shown in Table 23.

Table 23. Annual turnover of ash trace metals in Arid ecosystems, kg/ha (After Dobrovoslky, 1994).

Ash trace metals	Ecosystems	
	Fescue-Stipa and Absinthum-Fescue-Stipa Steppe	Absinthum-Saxaul Desert
Average ash content, %	3.5–4.0	4.0
Total uptake of ash elements, kg/ha/year	100–300	40
	Trace element turnover, g/ha/year	
Fe	400–1200	160
Mn	410–1230	164
Sr	65–195	28
Ti	60–180	26
Ba	50–150	18
Zn	5–0.150	24
Cu	16–48	6.4
Zr	15–45	6.0
Ni	4–12	1.6
Cr	3.5–10.5	1.4
V	3–9	1.2
Pb	2.5–7.5	1.0
Co	1–3	0.4
Mo	1–3	0.4
Sr	0.5–1.5	0.2
Ga	0.1–0.3	0.04
Cd	0.07–2.1	0.03

Table 24. The content of trace metals in the aerial parts of plant species of South Ural Steppe ecosystems, ppm, by dry weight (after Skarlygina-Ufimtseva et al, 1976 and Dobrovoslky, 1994).

Trace metals	Plant species				
	Stipa rubens	Festuca sulcata	Poa	Artemisia marshaliana	Veronica incana
Mn	1650.0	450.0	225.0	975.0	650.0
Zn	750.0	278.0	150.0	373.0	550.0
Ti	250.0	934.0	265.0	242.0	900.0
Ba	215.0	210.0	65.0	47.0	35.8
Sr	200.0	131.4	142.1	406.7	253.8
Pb	110.0	94.3	41.8	35.0	112.0
Cu	35.0	26.4	27.9	174.8	39.8
V	20.0	20.9	15.5	19.6	5.8
Ni	8.0	14.0	13.5	9.6	16.7
Ag	0.6	0.3	0.4	0.4	0.2

Case studies

Dry steppe ecosystems of South Ural, Eurasia

The various steppe plant species indicate the individual biogeochemical peculiarities. For example, we can discuss the results from South Ural's region, Russia. Table 24 shows the concentrations of trace elements in typical plant species of Steppe ecosystems.

The samples were taken from the site covered with thin rubble stone of Pleistocene deposits. The annual precipitation was 380 mm, and the annual evapotranspiration was twice as much. Despite the identical growth conditions the accumulation of trace metals depended on plant species. In feathergrass (*Stipa rubens*), the highest concentrations of Mn and Pb were monitored; in the sheep's fescue (*Festuca sulcata*) of Ti and Zn, in wormwood (*Artemisia marshaliana*), of Cu; in veronica (*Veronica incana*) of Ni. Of course, these data might be changed under different geochemical or climate conditions. However, the biogeochemical peculiarities of element specification will have to be taken into account.

Meadow Steppe ecosystems of the East European plain

For these ecosystems we consider the biogeochemical peculiarities of trace metal accumulation in the biomass forming whole plant groups, rather than genera. Such

Table 25. The average accumulation of trace elements in the main botanical groups of Meadow Steppe ecosystems of East European Plain, ppm, by dry weight (After Dobrovolsky, 1994).

Trace metals	Legumes	Grasses	Forage crops
Sr	1616.7	458.8	971.4
Ba	341.7	255.0	729.9
Mn	550.0	1092.4	1093.5
Ti	203.6	627.8	490.6
Ni	20.5	35.0	26.1
Pb	14.9	19.2	20.9
Zr	10.0	13.6	20.0
V	9.1	50.6	31.4

groups are grasses, legumes, and forage grasses. These groups differ in accumulation of trace metals. For instance, the accumulation of Ti, Cu, V, and Ni is characteristic for grasses, Pb and Ba, for forage grasses, and Sr, for legumes (Table 25).

The content of many elements in the roots and in the aerial parts of herbaceous plant species is different. In the root mass of grasses the content of trace metals is higher than that in aerial organs. This is closely correlated to the coefficients of biogeochemical uptake of these metals (Figure 12).

However, in the halo of dispersion of ore deposits many metals (Cu, Mo, Ag, Pg) frequently occur at higher concentrations in the aerial parts (Kovalevsky, 1984).

With aridity increasing, various plant species of forage crops become gradually less numerous and finally disappear. In Dry Steppe ecosystems xerophylic half-shrubs and salt-tolerant plants replace the grasses. However, the ash content is higher in these species (see above). This is attributed not only to a higher concentration of major ash elements in the plant tissue, but also to the occurrence of finely dispersed dust adhered to the plants' exterior (Table 26).

Dry desert ecosystems of Central Eurasia
In Desert ecosystems similar to Steppe ecosystems the plants distinctly exhibit their biogeochemical specificity. We can consider the distribution of trace metals in Dry Desert ecosystems of the Ustyurt Plateau, Kazakhstan, with predominance of wormwood (*Artemisia terrae albae*) and saxaul (*Anabasis salsa*). In rubble stone territories, of common occurrence is the dense shrubbery of *Sasola anbuscula*. Most elements found in the wormwood occur in their highest concentrations. In the roots of the

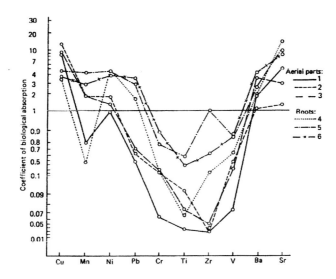

Figure 12. Coefficients of biogeochemical uptake of trace metals by typical plant species of Meadow Steppe Ecosystems of East European Plain. Aerial parts: 1—legumes; 2—grasses; 3—forage crops; roots: 4—legumes, 5—grasses, 6—forage crops (Dobrovolsky, 1994).

Table 26. The content of ash elements in aerial and root parts of plant species from various Arid ecosystems, %.

Ecosystems	Aerial parts	Root parts
Meadow Steppe	3.0	4.5
Dry Steppe	5.5	6.0
Semi-Dry Desert	7.0	10.1
Dry Desert	12.1	16.5

wormwood and saxaul, higher contents of Mn, Cu, Mo, and Sr have been monitored, whereas the aerial parts contain more Ti, V, and Zr. We can see that the root elements are most biologically active and those in aerial parts, more inert. Possibly their presence was related to the dust deposition on the plant exterior (see above).

Despite the quantitative variability of salts and silicate dust particles in the plants of Arid ecosystems, we can easily discern a trend towards the selective uptake of trace elements. The calculation of coefficient of biogeochemical uptake (C_b) shows that the elements contained in the plant species of both Steppe and Desert ecosystems are in equal measure susceptible to the influence of environmental factors. The most

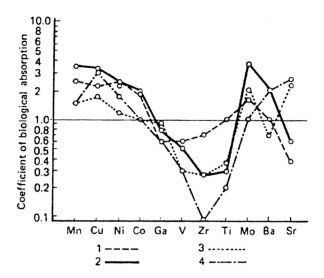

Figure 13. Coefficients of biogeochemical uptake of trace metals by plant species of the Ustyurt Plateau Dry Desert ecosystems. 1—wormwood (Artemisia terrae albae), aerial parts; 2—roots; 3—saxaul (Anabasis salsa), aerial parts; and 4—roots (Dobrovolsky, 1994).

extensively absorbed are Sr, Cu, Mo, and Zn. Their values of C_b are more than unit. The group of other elements, like Ti, Zr, and V, are poorly taken up, with their values of C_b often dropping below 0.1 (see Figure 13 and 14).

The general trend towards increase of ash elements in the plants of grading the steppe ecosystems to Dry and Extra-Dry Desert ecosystems does not seem to affect the C_b values appreciably (see Box 6).

Box 6. Biogeochemical processes in Central Asian Extra-Arid desert ecosystems (after Dobrovolsky, 1994)

The biogeochemical processes that occur under the least favorable conditions for life of Extra-Arid Desert ecosystems are of considerable interest. Such extreme environments extend over a vast territory in the middle of the Eurasian continent. The Gobi is one of the most severe deserts in the World. The rainfall in the western part of the Gobi desert is commonly from 20 to 50 mm, whereas the evapotranspiration is about 1250 mm per annum. The surface of gentle piedmont slopes and intermountain valleys has the aspect of a compact rocky crust, the so-called desert armor. This armor, composed of the pebbles of metamorphic and volcanic rocks with the lustrous black glaze of 'desert varnish', defies even the timid suggestion of an eventual existence of life in this forlorn expanse. Periodically, at an interval of about 10 years, the atmosphere over the Gobi desert becomes invaded with a moist air mass, which discharges profuse rains. The runoff streams erode numerous shallow depressions dissecting the

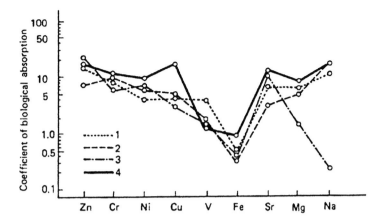

Figure 14. Coefficients of biogeochemical uptake of trace metals by cenospecific plant species of Gobi Extra-Dry Desert ecosystems, Central Asia. 1—Haloxylon ammodendron; 2—Iljina regeli; 3—Ephedra Przewalskii; 4—Anabasis brevifolia (Dobrovolsky, 1994).

surface of the rocky hammada into separate extended stretches. Extra-Dry Shrub and Under-Shrub ecosystems are in a large part of the Gobi desert.

The predominant plant species are haloxylon (*Haloxylon ammodenndron*), aphedra (*Ephedra Przewaskii*), and other shrub species, like *Zygophyllum xanthoxylon* and *Reamuria soongoriea*. The under-shrub species include *Anabasis brevifolia* and *Sympegma regelii*. At the periphery of the Extra-Dry Shrub and Under-Shrub ecosystems, grasses (*Stipa glareosa*) and onions (*Allium mongolicum*) are encountered. In the southernmost regions the extra-arid landscapes of rocky hammana are either entirely devoid of vegetation, or provide a scant residence of rare specimens of *Ilinia regelii*, on average 1.3 specimens per $100\,\text{m}^2$. The annual production of the aboveground biomass in the Iljina ecosystem is 2.2 kg/ha by dry matter and in the Haloxylon-Iljina ecosystem is 2.5 kg/ha. The net annual production of Xerophytic Shrub ecosystems is about 8 kg/ha. The content of chemical species is about 100–1000 ppm for Ca, Mg, Na and Fe; $n \times 10$ ppm for Mn, Zn, Sr, Cr; 1–10 ppm for Cu, Ni and V, and < 1 ppm for Pb and Co.

The annual biogeochemical fluxes of various elements are shown in Table 27.

In plain autonomous ecosystems the fluxes of sodium are less than 40 g/ha/year and those of Mg are less than 10 g/ha/year. For iron these values are close to 1 g/ha/year, and for all trace metals, are between 0.01 and 0.04 g/ha/year. In the geochemically subordinate landscapes (*Naloxylon ammodendron* and *Ephedra przewalskii* ecosystems) which receive additional moisture and chemical elements, the biogeochemical fluxes are 360–912 g/ha/year for Mg and Na, and from 0.44 to 6.65 g/ha/year for trace metals. In the periphery of the Gobi desert, *Anabasis brevifloria* and *Graminaceae* Dry Desert ecosystems show an overall increase of biogeochemical fluxes. The turnover for some elements (Mg, V, Cr) rises but slightly in comparison to their turnover in

Table 27. Annual biogeochemical fluxes in Gobi Extra-Desert ecosystems, g/ha.

Elements	Ecosystems			
	Iljina regelii (desert plains)	*Naloxylon ammodendron* (desert depressions)	*Haloxylon ammodenron* and *Ephedra przewalskii* (desert depressions)	*Anabasis brevifloria* and *Graminaceae* (Dry Desert ecosystem)
Na	39.6	281.5	912.0	2718.0
Mg	7.7	86.5	360.0	603.0
Fe	1.1	22.0	71.3	264.4
Sr	0.08	1.07	3.47	17.39
Mn	0.07	2.05	6.65	12.28
Zn	0.04	0.91	2.94	6.58
Cu	0.01	0.16	0.50	2.98
Ni	0.02	0.14	0.44	1.70
V	0.02	0.35	1.14	0.97
Cr	0.03	0.36	1.16	2.54

Extra-Dry ecosystems, whereas the turnover for other elements (Sr, Zn, Cu) increases several times.

4.2. Soil biogeochemistry in Arid ecosystems

High biotic activity is characteristic for soils of Meadow Steppe ecosystems with relatively high precipitation. An enormous number of invertebrates promptly disintegrate and digest the plant residues and mix them with the mineral soil matter. The presence of the predominant part of plant biomass as the underground material facilitates greatly this process.

The microbial population in soils of Steppe ecosystems is different from that in forest soils. Fungi, which play a decisive role in destruction of plant remains in Forest ecosystems, are changed by bacteria. The microbial and biochemical transformation of organic matter in the steppe soils leads to the predominant formation of low soluble and low mobile humic acids. The accumulation of humic acids in upper soil layer is increasing also due to formation of mineral-organic complexes. Furthermore, the migration of many chemical species is also decreasing due to impeded water regime

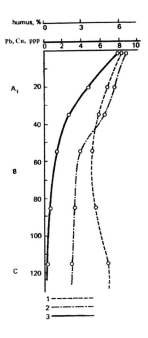

Figure 15. Download distribution of mobile forms (1 N HCl) of zinc (1), copper (2) and humus (3) in Chernozem profile (Dobrovolsky, 1994).

and soil saturation with Ca ions. This provides for a tight coagulation of films of humic acids on the surface areas of mineral particles.

These properties of soils in Steppe ecosystems are favorable to the formation of uppermost humus barrier, where the accumulation of almost all the chemical species occur. The concentration of chemical elements is slightly decreasing downward in soil profile, in parallel with decreasing soil humus content (Figure 15).

The significant part of trace elements in the soils of Steppe ecosystems are bound with highly dispersed mineral-organic particles, to a lesser degree, with only organic matter. We can see that the water soluble and exchangeable forms are less than 1% of the total content. Specific forms of trace elements are bound with carbonate and gypsum in B and C horizons (Table 28).

Role of humidity in soil formation in Steppe and Desert ecosystems

The water deficiency in Arid ecosystems is the main restricting factor for biogeo-chemical processes. We know that many links of the biogeochemical food web are connected in Steppe soils with invertebrates. Their population varies very much in Steppe ecosystems depending on the moisture conditions (Table 29). For instance, the wet biomass of soil invertebrates in the Meadow Steppe and Forest Steppe ecosystems exceeds that for the Extra-Dry Rocky Desert ecosystems by 150–300 times.

Table 28. Distribution of Co in Calcaric Chernozem and Chestnut soil of Meadow Steppe ecosystem in the south part of East European Plain.

Soil horizon	Total content, ppm		Fraction of total Co content bound with, %					
			Humus		Clay matter		Carbonate	
	Chernozem	Chestnut soil	Chernozem	Chestnut soil	Chernozem	Chestnut soil	Chernozem	Chestnut soil
A	9.2	11.3	30.4	25.7	43.5	49.5	—	—
B	9.4	10.4	22.2	14.4	48.0	43.3	—	11.5
C	8.5	9.4	4.7	4.7	61.2	60.7	10.9	6.4

Table 29. The influence of water deficiency on invertebrate biomass and humus content in Steppe ecosystems.

Ecosystems	Invertebrate biomass, kg/ha	Humus content, %
Forest Steppe	700	4–6
Meadow Steppe	750	6–8
Semi-Desert	6	2–4
Extra-Dry Rocky Desert	2–4	< 1

The humus content in Steppe ecosystem soils reflects the total biomass production and humidity.

The content of trace elements in Steppe soils is tightly connected with their contents in geological rocks. In soils of Desert ecosystems water soluble forms play the most important role. We can see an analogy between the increasing content of elements in soil dead organic matter as a function of decreasing water excess in Forest ecosystems and the increasing content of water-soluble species of chemical elements in the soils of Dry Steppe and Desert ecosystems as a function of enhanced aridity. The accumulation of water-soluble species occurs in the upper horizon for almost all elements, with exception of strontium. The main factor responsible for the accumulation of water-soluble forms is connected with evapotranspiration.

The existence of evapotranspiration barrier in the upper soil horizon of Dry and Extra-Dry Desert ecosystems favors the accumulation of alkalinity and alkaline reaction of soil solution. In turn this accelerates the mineralization of organic matter and mobilization of finely dispersed mineral and organic suspensions. This fact provides a plausible explanation of the occurrence of some trace metals, like Zr, Ti, Ga, Yt and their congeneric elements in the aqueous extracts from soil samples of Dry Desert ecosystems.

Table 30. Comparative assessment of trace metal contents in surface waters of various ecosystems from the East European Plain and average values for the similar global corresponding ecosystems, ppb.

Trace metals	Ecosystems					
	Forest		Forest-Steppe		Steppe	
	East European Plain	Global average	East European Plain	Global average	East European Plain	Global average
Mn	158.0	16.0	94.0	9.4	126.0	13.0
Ti	40.0	13.3	94.0	31.3	106.0	35.3
Sr	36.0	0.5	68.0	0.9	128.0	1.6
Zn	11.0	0.6	28.0	1.4	36.0	1.8
Cr	5.4	5.4	8.0	8.0	9.0	9.0
Pb	4.8	4.8	3.0	3.0	3.0	3.0
Sn	3.5	7.0	4.0	8.0	1.5	3.0
Zr	4.2	1.5	9.0	3.5	2.0	0.8
V	2.7	3.0	6.0	6.7	4.0	4.4
Cu	2.4	0.3	12.0	1.7	10.0	1.4

The extraction by 1 N NCl yields 5–10% of total trace metal content. In case of Fe and Mn, these values are even higher. The maximum contents of mobile fractions of trace elements are monitored in the upper horizon. Remember the role of evapotranspiration barrier in biogeochemical migration of elements in Dry Desert ecosystems.

4.3. Role of biogeochemical processes in aqueous migration of elements in Steppe ecosystems

It is well known that the total content of water-soluble solids in natural waters (TSS) is increasing with an increasing aridity. The concentration of some trace elements correlates significantly with the total content of soluble solids (see Chapter 4, Section 4, for boron, for example). Comparison of the trace element contents in surface waters in the area of East European Plain has shown certain relationships between surface water chemistry and ecosystem types. These data are summarized in Table 30.

We can see that for zinc the average content in surface waters of various ecosystems is increasing in a row: Forest < Forest Steppe < Steppe. Similar peculiarities are shown for strontium. However, for some metals this tendency is not confirmed or we can even see the opposite direction, for instance for manganese, whose content is higher in waters of Forest ecosystems.

For understanding these tendencies, we will consider the values of the biogeo-chemical coefficient of aqueous migration. This coefficient C_w is the ratio between the content of anelement in the sum of water-soluble salts and in the geological rocks. The values of C_w for certain chemical species are smaller in Arid ecosystems than those in Forest ecosystems. We can suggest two explanations. First, soils of Forest ecosystems are enriched in water-soluble metal-organic complexes (see above). Second, most chemical species are trapped in the transpiration barrier of upper soil layers of Arid ecosystems.

The smaller C_w values are also connected with water deficiency in Steppe and Desert ecosystems. However, the concentration of various chemical species in rain-water of background regions is higher that that in the Forest ecosystem belt. The major reason is the wind deflation of the soil's surface owing to lack of tree species and only a lean protective layer of grasses and half-shrubs. A large mass of soil particles becomes entrained into the air migration. The most characteristic example is connected with the 'Yellow sand' phenomenon (see Box 7).

Box 7. "Yellow sand" formation and transport in Asia (after Jie Xuan, 1999)

The vast area of Arid and semi-Arid ecosystems of Central and East Asia is a subject of wind erosion. The major natural sources of dust emission are Gobi desert (Xinjiang Province, Northwestern China), Karakum and Kyzylkum deserts (Central Asian areas of Kazakhstan, Uzbekistan and Turkmenistan). The Gobi desert is the second largest desert in the World. The climate is extremely dry and windy. The strongest winds occur during the winter and spring period. These factors influence the deflation of soil surface layers, which are not protected by sparse vegetation, especially during the springtime, after thawing of the frozen upper soil horizon.

The annual dust emission rates for the whole China area are shown in Figure 16.

We can see that due to joint effects of aridity and soil texture, the dust emission rates increase from east to west by as much as 5 orders. The maximum emission rate is 1.5 ton/ha/year. The total dust emission amount of the Gobi desert is estimated as 25×10^6 tons per year and that in spring is 15×10^6 tons per year. The seasonal dust emission amounts in summer, autumn and winter are 1.4×10^6, 5.7×10^6 and 2.9×10^6 tons, correspondingly.

The low rates of aqueous migration of many chemical species in Arid ecosystems and the accumulation of their water-soluble and dispersed forms in the uppermost soil layers play an important role in the geochemistry of aerosol formation and rainwater chemistry. In turn, biogeochemical processes in the soil-plant-air system determine the water chemistry (Table 31).

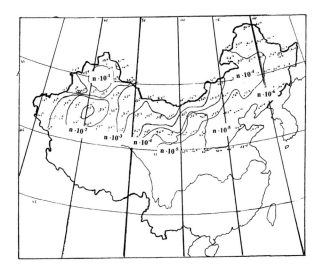

Figure 16. Annual dust emission rates in China (ton/ha/year).

Table 31. Rainwater total salt content and salt deposition rates over various natural ecosystems in Eurasia.

Ecosystems	Total rainwater salt content, mg/L	Total salt deposition rate, kg/ha/year
Forest	17–20	70–100
Steppe	45–50	170–180
Desert	> 150	210–240

Despite the smaller rainfall over the arid areas in comparison to the forest belt, the Arid ecosystems receive more salts from the atmosphere than the areas of excess humidity.

5. BIOGEOCHEMISTRY OF TROPICAL ECOSYSTEMS

The tropical ecosystems occur between 30° N and 30° S. This belt receives about 60% of solar radiation inputting on the Earth's surface. The total area of tropical ecosystems is about 40×10^6 km^2, with exception of the High Mountain and Extra-Dry Sandy Deserts with strongly depressed life processes.

The rates of biogeochemical processes in tropical ecosystems, especially in Tropical Rain Forest ecosystems, are the highest in comparison to other considered ecosystems. This is connected not only with the modern biospheric processes, but, to a great degree, with the history of geological and biological development in these areas.

Table 32. Proportion of African equatorial areas with different precipitation rates.

Annual precipitation, mm	Per cent of equatorial belt
> 1800	22
1000–1800	48
600–1000	12
200–600	16
< 200	2

5.1. Biogeochemical cycles of chemical species in Tropical ecosystems

The different ratios between precipitation and evapotranspiration, duration of dry and wet seasons, relief positions and human activities create a great variability of Tropical ecosystems, varying from African, Australian and American Extra-Dry Deserts to Tropical Rain Forest ecosystems. Due to a prolonged dry season Drought-Deciduous High Grass Tropical Degraded Forest ecosystems are typical in the areas where the annual evaporation exceeds the precipitation. Woody Savanna ecosystems represent clusters of thinly growing tress alternating with the open space of herbaceous vegetation. With increasing aridity, Dry Woody Shrub and Semi Desert Shrub ecosystems become prevalent, where the trees are replaced by thornbrush and tall grasses grade down to low-growing species with shallow soil coverage.

The proportion of areas with different precipitation rates varies from continent to continent. For instance, different Arid ecosystems, from Dry Savanna to Extra-Dry Desert, are predominant in India and Australia. To a lesser degree these ecosystems occur in Central and South America (see Chapter 7 'Biogeochemical Mapping'). In an equatorial belt of Africa, the distribution of areas with different precipitation is shown in Table 32.

We can see that the Tropical Rain Green Forest ecosystems occupy about 1/5 of the African equatorial belt, whereas about 1/2 of this area is Woody and Tall Grass Savanna ecosystems. The rest of the area is occupied by various Dry Steppe and Dry and even Extra-Dry Desert ecosystems, like Sahara, with annual rainfall less than 200 mm.

Biogeochemical cycling in Tropical Rain Forest ecosystems

All types of Tropical Rain Forest ecosystems occupy $20.45 \times 10^6 \, km^2$, or 13.3% of the total global land area. These ecosystems constitute the most powerful plant formation. The abundance of solar energy and water provides for the largest biomass growth, up to 1700 ton/ha. The only restriction factor is the availability of sunlight for every plant species. To maximize the using of sun energy several stories of trees are in Tropical Rain Forest ecosystems, from the upper story of 30–40 m height down to

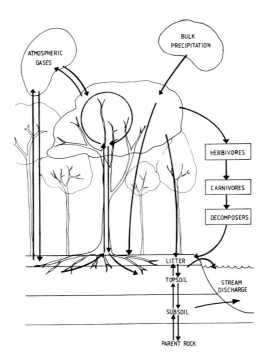

Figure 17. Biogeochemical cycle in Tropical rain Forest ecosystems.

2–5 m trees of height well adapted to stray light. A large part of died-off and fallen leaves from taller trees is entrapped for assimilation by numerous epiphytes. This results in fast re-circulation of chemical elements. The average annual Net Primary Production of these ecosystems is 25 ton/ha.

The main specificity of biogeochemical cycling in Tropical Rain Forest ecosystems is related to its almost closed character. This means that almost the total number of nutrients is re-circulating in biogeochemical cycles (Figure 17).

This type of closed biogeochemical cycling is very sensitive to uncontrolled intervention into ecosystems. For instance, clearcutting leads to the entire destruction of the whole ecosystem with its multi-annual history. In other words, the deforestation will leave behind a barren soil with completely destroyed biogeochemical turnover.

For instance, clearing tropical forests in the Amazon Basin for pasture alters rates of soil nitrogen cycling (Table 33).

The pattern of NH_4^+ and NO_3^- concentrations and net mineralization and net nitrification rates in soils before and after clearing and burning tropical forest indicate:

(1) forest inorganic N pools are either dominated by NO_3^- or contain NH_4^+ and NO_3^- in roughly equal proportions;

Table 33. Inorganic N concentrations, mineralization and nitrification rates and turnover rates of NH_4^+ and NO_3^- in a chronosequence and before and after clearing at Nova Vida, Brazil (After Neill et al, 1999).

Forest cutting	NH_4^+, ppm	NO_3^-, Ppm	Net mineralization, ppm/day	Net nitrification, ppm/day	Turnover, day	
					NH_4^+	NO_3^-
Before cutting	8.8	13.7	2.72	2.60	1.8	4.9
1.5 months after	6.8	12.6	3.65	2.05	1.5	5.0
6 months after	2.1	10.0	10.0	2.38	1.1	5.5
8 month after	1.5	11.5	11.5	2.30	0.8	4.9

(2) net mineralization and net nitrification rates tend to decline after forest clearing and burning;

(3) Slash burning in Amazonian Tropical Rain Forest ecosystems is accompanied by a relatively thorough consumption of leaves and other sources of high quality organic material and a large input to soils of low quality, high C-to-N coarse woody debris. This input was probably responsible for the N immobilization recorded by the net mineralization measurements 1.5 months after the burn. No increases have been observed in net mineralization and nitrification rates after the burn, perhaps because microbial communities were diminished by burning and took time to become reestablished. This absence of any increase in mineralization and nitrification rates suggest that the high ammonium and nitrate concentrations, shown in Table 33, were associated with the elimination of plant uptake, rather than accelerated N cycling.

From these results we can conclude that nitrogen is relatively available in soils of Tropical Rain Forest ecosystems and that forest soils mineralize and nitrify large amounts of nitrogen. P. Vitousek and R. Sanford have shown similar results earlier in 1986 studying nitrogen cycling in moist tropical forests.

The total biomass and its annual distribution for these ecosystems are shown in Table 34.

As a rule the concentration of chemical species in the trunk and twig wood of tropical trees is a few times lower than that in the leaves (Table 35).

The average sum of total ash elements in the biomass of Tropical Rain Forest ecosystems is about 8000 kg/ha. The annual ash element turnover is shown in Table 36.

Geochemical conditions of tropical soils favor the biogeochemical migration of iron and manganese. The turnover of other metals is from 1.1 g/ha/yr. for Cd to 1050.0 g/ha/yr. for Sr.

Table 34. Plant biomass parameters in Tropical Rain Forest ecosystems, ton/ha.

Biomass parameters	Total Biomass	Annual NPP	Annual Litterfall
Total mass	520	32.5	25.0
Nitrogen	2.94	0.43	0.26
Ash elements	8.14	1.60	1.28

Table 35. Concentration of chemical species in tropical tree compartments (After Dobrovolsky, 1994).

Sample	Elements, % by the dry weight									Pure ash	Mineral dust admixture
	N	Si	Al	Fe	S	Na	Ca	Mg	P		
Trunk	0.5	0.05	0.01	0.02	0.03	0.01	0.29	0.02	0.02	0.79	0.40
Twig	0.6	0.07	0.03	0.04	0.04	0.03	0.31	0.04	0.03	0.85	0.45
Leaves	2.0	1.06	1.87	1.48	0.04	0.22	0.45	0.27	0.06	9.87	11.3

Biogeochemical cycling in Seasonal Deciduous Tropical Forest and Woody Savanna ecosystems

The Seasonal Deciduous Tropical Forest and various Savanna ecosystems occupy 14.3×10^6 km^2. The biogeochemical cycling in Seasonal Deciduous Tropical Forest and various Savanna ecosystems is similar to that in the Boreal and Sub-Boreal Deciduous Forest ecosystems. The clear distinction is relates to the reasons of periodical inhibition of biogeochemical activity. In the temporal climate it is connected with the winter temperature drop and in tropical areas it relates to the dry season with significant moisture deficit.

Table 37 compares the contents of trace elements in the ash of various grass and tree species from the Savanna ecosystems of East Africa. We can see that nickel, barium, and strontium accumulate in the tree organs (twigs), whereas the accumulation of other metals is pronounced in grasses.

The aerial parts of grasses in Savanna ecosystems exhibit a high ash content from 6 to 10%. This is partly due to the presence of minute particles of mineral dust, which are discernible under a microscope or, occasionally, even with the naked eye. The mineral dust accounts for 2–3% of the weight of dry mass of grass aerial parts. We can consider that this dust is responsible for the elevated concentrations of some elements, like Ga, which has a low C_b value. This element is contained in wind-blown finely dispersed clay particles. Nevertheless, even with allowance made for the silicate

Table 36. Biogeochemical turnover of trace metals in Tropical Rain Forest ecosystems (After Dobrovolsky, 1994).

Trace metals	Chemical symbol	Annual flux, kg/ha/year
Average ash content, %		4.6
Overall turnover of ash elements		1500
Iron	Fe	6.0
Manganese	Mn	6.15
Strontium	Sr	1.05
Titanium	Ti	0.97
Zinc	Zn	0.90
Barium	Ba	0.67
Copper	Cu	0.24
Zirconium	Zr	0.22
Nickel	Ni	0.06
Chromium	Cr	0.052
Vanadium	V	0.045
Lead	Pb	0.037
Cobalt	Co	0.015
Molybdenum	Mn	0.015
Thin	Sn	0.0075
Gallium	Ga	0.0015
Cadmium	Cd	0.0011

dust content, the total sum of ash elements in grasses of savanna ecosystems is twice as much as that of the grasses from Alpine Meadow ecosystems.

Strontium, barium, manganese, copper, molybdenum, and nickel are the elements of strong accumulation in plant species of African Savanna ecosystems, in spite of different content in soils and soil-forming rocks. The C_b values are > 1. The other elements, like beryllium, zirconium, titanium and vanadium, are less taken up by plants and their C_b values are less than 0.5.

Table 37. Biogeochemical fluxes of trace metals in various plant species of Savanna ecosystems of East Africa (After Dobrovolsky, 1994).

Trace metals	Content, ppm		Coefficient of biogeochemical uptake, C_b		Averaged annual flux, kg/ha
	Grasses	Trees	Grasses	Trees	
Ti	1140	230	0.1	0.03	0.35
Mn	1880	943	1.9	0.9	2.05
V	59	45	0.3	0.2	0.015
Cr	28	12	0.2	0.08	0.017
Ni	39	144	0.6	2.0	0.02
Co	20	12	0.6	0.4	0.005
Cu	85	39	1.5	0.7	0.08
Pb	34	21	1.5	0.9	0.012
Zn	118	79	1.2	0.8	0.30
Mo	57	6	7.1	0.8	0.005
Zr	165	92	0.5	0.3	0.075
Ga	36	4	1.6	0.2	0.008
Sr	450	3340	3.5	25.7	0.35
Ba	440	630	3.0	4.3	0.22

Biogeochemical cycling in Dry Desert Tropical ecosystems

Dry Desert Tropical ecosystems occupy $4.5 \times 10^6 \, km^2$, or 3.0% of total land area of the Earth. These ecosystems have dry period during 7–10 months a year. Not only trees, but also numerous grasses can not grow in such severe conditions. The vegetation is generally represented by thorny drought-deciduous shrubs. These shrubs stay leafless for a larger part of the year thus reducing transpiration. The example of these ecosystems is Tar desert, West India. It is a lowland alluvial plain formed by the Indus. The rainfall for this area is 200–600 mm/year. The scattered tree species (*Acacia, Propopis spicigera, Salvadora persica*), shrubs, and grasses (*Gramineae*) represent the vegetation of this Dry Desert ecosystem. The sandy deposits are devoid of trees, which confers the image of desert activity. The desertification of the territory is a result of human activity connected with overgrazing during a long term period. It is known that in 326 B.C., when the troops of Alexander the Great came to the Indus, sal forests (*Schorea rubista Gaerth. f.*) were widespread in the valley. At present these Forest ecosystems do not exist.

Table 38. Biomass and net primary productivity of the Tar Dry Desert ecosystem, India.

Biomass components	Biomass		NPP	
	Ton/ha	%	Ton/ha	%
Green parts of plants	2.90	11	2.90	42
Perennial aerial parts of plants	10.60	47	0.40	2
Roots	11.30	42	3.53	56
Total biomass	26.80	100	6.80	100

Table 38 characterizes the plant biomass of the Tar Dry Desert ecosystems.

We can see that trees are major contributors to the plant biomass of this ecosystem. They account for 60% of the root and 98% of the aboveground biomass. The monitoring results showed also that the grasses provide a larger part of the annual net primary production. Grasses are responsible for 76% of green organs and 83% of root biomass of NPP.

The biogeochemical fluxes of various chemical species are shown in Table 39.

We can see that the green parts of this Dry Desert ecosystem accumulate more than half of both the total studied nutrients and ash elements. These values for the root biomass are 30 and 38% from the total and ash elements in NPP. No more than 5% is supplied to the trunk and twigs. In the green tree organs, nitrogen, calcium, potassium, silicon and sulfur are the most extensively accumulated. In the roots, the highest relative accumulation is that of manganese and silicon, which are distributed roughly in equal parts in the annual growth of green organs and roots.

5.2. Biogeochemical peculiarities of tropical soils

The high rate of edaphic biological processes is the characteristic property of any tropical ecosystem. In the maximum degree this is related to the Tropical Rain Forest ecosystems. For instance, in the African Rain Forest ecosystems the soil surface receives annually from 1200 to 1500 ton/ha of various plant residues. Edaphic invertebrates and microbes transform this large mass very rapidly. A continuous forest litter is practically nonexistent in the Tropical Rain Forest ecosystems and a thin layer of dead leaves alternates with patches of bare ground. All elements that mineralized from litterfall, are taken up by the complex root system of a multi-storied forest to re-input to the biogeochemical cycling.

Most soil-forming rocks of various Tropical ecosystems are the products of ancient weathering. These rocks contain a very limited number of nutrients available for plant uptake. Mineralized dead plant organic matter is the main source of essential macro- and microelements. The microbial transformation of plant residues of tropical plant

Table 39. Biogeochemical fluxes of chemical elements in Tar Dry Desert ecosystem (After Rodin et al, 1977).

Element	Pool in biomass, kg/ha	Content of chemical species in main parts of annual NPP					
		Green parts		Roots		Overall production	
		kg/ha	%	kg/ha	%	kg/ha	%
N	179.34	42.86	59	26.14	36	72.08	100
Ca	256.25	34.22	57	19.28	32	59.78	100
K	111.21	31.57	61	18.68	36	52.04	100
Si	53.81	22.14	52	20.21	8	42.48	100
Mg	48.69	6.80	51	5.40	41	13.30	100
P	12.19	3.57	56	2.68	42	6.42	100
S	17.60	8.86	72	3.18	26	12.39	100
Al	19.37	2.46	52	1.80	38	4.69	100
Fe	11.37	1.64	53	1.21	39	3.11	100
Mn	2.84	1.01	48	1.07	51	2.09	100
Na	9.52	2.79	58	1.95	40	4.82	100
Cl	14.96	5.16	54	4.27	45	9.51	100
Total ash elements	571.81	120.22	57	59.79	38	210.63	100
Total	751.15	163.08	58	85.93	30	282.71	100

species leads to the dominant formation of soluble fulvic acids. The content of humic acid is 5–7 times less than the former. The typical pH values for soil developed on the leached products of quartz-containing crystalline rock weathering, is about 5. The upper layer of these soils is intensively leached. A different situation is observed when the Tropical Rain forest ecosystems confined to a volcanic region and soil formation is in the young products of volcanic rocks weathering, which are reached in calcium, magnesium, potassium and other alkalies. In this case, the most humic acids become neutralized and they condense into larger, less soluble chemical species. This results in the humus accumulation, up to 6–8%. The pH of soil solution is about 6. However, fulvic acids are prevalent over humic acids in any soils of Tropical Rain Forest ecosystems. Soils derived on the younger rocks are more acid and depleted in both nutrients and exchange cations that are involved in biogeochemical cycling (Table 40).

Table 40. Chemical composition of soils in Australian Tropical Rain Forest ecosystems (after Congdon and Lamb, 1990).

Parameters	Parent materials of soils		
	Basalt Pin Gin low hill	Granite Tyson alluvial	Metamorphic Calmara hill
pH	5.7	4.5	4.9
Organic carbon, %	6.4	2.8	1.7
Organic nitrogen, %	0.42	0.18	0.11
Extractable phosphorus, ppm	10	14	7
Total exchange capacity, meq/100 g, inc.	6.6	3.2	2.5
Ca^{2+}	3.1	0.56	0.20
Mg^{2+}	2.2	0.49	0.34
K^+	0.52	0.13	0.06

The Seasonal Tropical Forest and Woody Savanna ecosystems are common in tropical regions with a short dry period. The characteristic features of soils from these ecosystems are the neutral reaction of soil solution and periodic leaching during wet season. The herbaceous species favor the formation of both sward and humus horizons.

Different conditions are typical for the Dry Tropical Wood, Dry Savanna and Dry Woody Shrub ecosystems in the areas with precipitation rates of 400–600 mm and a prolonged dry season. The microbial activity is suppressed during a dry season too. The soils of these ecosystems have no even periodic leaching, the formation of transpiration biogeochemical barrier in upper soil layer favor the alkaline reaction and accumulation of soluble salts.

The accumulation of trace metals in tropical soils depends on the geological rocks. These soils have been developed mostly on re-deposited products of weathering that suffered a small displacement. Furthermore, most tropical areas occupy the fragments of the ancient super-continent Gondwana, whose surface during the last 0.5 billion years has not been covered by oceanic waters. The resultant effect of these soil-forming conditions was connected with the pronounced influence of geochemical composition of geological rocks on the biogeochemical cycling in all

Table 41. Content of trace metals in soils of two Dry Savanna ecosystems from East Africa, ppm (After Dobrovolsky, 1994).

Trace metals	Soil-forming geological rocks			
	Precambian crystalline rocks (Uzanda)		Cenozoic volcanic rocks (Tanzania)	
	Clark content	Content in humus horizon	Clark content	Content in humus horizon
Ti	1.8	5820	4.5	14900
Mn	2.2	1520	3.1	2140
V	2.0	153	3.6	271
Cr	6.9	234	4.7	160
Ni	2.9	75	3.6	93
Co	6.6	48	9.6	70
Cu	4.7	105	3.3	72
Pb	3.2	52	2.2	35
Zn	2.5	125	3.7	190
Mo	3.8	5	6.2	8
Be	2.4	6	6.4	16
Sc	2.2	24	1.5	16
Y	1.6	58	1.8	64
La	1.8	85	2.1	95
Nb	2.9	59	11.5	224
Zr	1.3	215	3.9	670
Ga	0.8	16	1.1	20
Cr	0.6	129	2.2	510
Ba	0.4	274	0.9	590

Tropical ecosystems and finally on the biogeochemical mapping of the whole tropical area (see Chapter 7).

The example of these relationships with the ancient geological rocks is shown in Table 41.

Both of the soils are enriched in various elements, like zirconium, titanium, beryllium, niobium, and strontium, due to their enlarged content in the alkali basalts and phonolites of the East-African Rift. We can see that soils of Tanzanian Dry

Table 42. Biogeochemical mass balance for the Tropical Flooded Savanna ecosystems, kg/ha/year (After Vegas-Vilarrubia, 1994).

Element	Mass balance items		
	Input	Output	Net
Na	13.92	18.29	−4.37
K	2.62	12.89	−10.67
Ca	7.96	11.26	−3.30
Mg	0.68	7.34	−6.66
P	0.19	0.10	+0.09
Zn	0.60	0.21	+0.29
Cu	0.24	0.06	+0.18

Savanna ecosystems contain niobium 11 times, beryllium and molybdenum 6 times, and titanium and zirconium 4 times as high as compared to the respective crustal concentrations (clarks) of these metals. In the Ugandan soils, the chromium content is 7 times higher than clark value and that of copper, 5 times. Such a large difference in soil enrichment of trace metals is related to their different input into biogeochemical cycles of corresponding ecosystems.

Biogeochemical fluxes of elements in soil-water system

Soil links of biogeochemical processes are ultimately connected with the aqueous links of various biogeochemical cycles. The results of small catchment monitoring are very helpful for understanding the relationships between terrestrial and aqueous links of biogeochemical cycles of various chemical species. Such an experiment has been carried out in Venezuelan Flooded Savanna ecosystems. The site is located between the rivers Arauca and Apure ($7°8'$ N and $68°45'$ W) and it presents a flooded area, where Savanna ecosystems are developed under alluvial sedimentation processes. Soils in the study area are naturally fertile. Dominant species are *Leersia hexandra* and *Himenachne amplexicaulis*, species with relatively high aboveground net primary production, 5.5–9.1 ton/ha per year. The duration of the wet season is 6 months.

The biogeochemical mass budget of various macro- and microelements for this catchment is shown in Table 42.

The input of elements was accounted only as a result of atmospheric deposition. Assuming that most soils are poorly drained after reaching field water holding capacity (FWHC), the percolation of water through a soil profile is minimal. Consequently, losses of elements from the watershed by deep seepage are negligible. Nutrient budgets are therefore calculated as the difference between input with deposition and output in surface runoff. We can see that negative values of budget were calculated for sodium, potassium, calcium and magnesium, whereas these values were positive for phosphorus, zinc, and copper.

The possible explanation of these results is related to the construction of dikes in this area a few years prior to the experiment. This changed the biogeochemical cycles of many nutrients in natural ecosystems too. Furthermore, the input of nutrients with flooded waters was not taken into account.

As a result of microbial formation of metal-organic complexes with fulvic acids in soils of Tropical Rain Forest ecosystems, the surface and sub-surface runoff waters are enriched in some trace metals like manganese and copper. The similar tendency has been shown for boron, strontium and fluorine.

Colloidal suspension is the dominant form of riverine migration in tropical ecosystems. These suspensions are mainly composted of products originated from soil biogeochemical metabolism. However, most of these products never reach the river channels and deposit in the subordinated landscapes of relief depressions. During a wet season, tropical grey and black compacted soils with seasonal waterlogged horizon are formed in these landscapes. Seasonal Bog ecosystems provide conditions for the accumulation of many chemical compounds, which have leached from the surrounding elevated ecosystems. For this reason biogeochemical provinces with excessive content of trace elements in food webs occur in these savanna regions.

5.3. *Biogeochemistry of Mangrove ecosystems*

The Mangrove Forest ecosystems are typical for the tropical coastline. These ecosystems occupy the narrow coastal strips periodically flooded during the diurnal or the big syzygial tides. We can say that Mangrove ecosystems are transitional from terrestrial to subaquatic marine ecosystems. Depending on the temperature as a limiting factor, these ecosystems spread from 32° N up to 44° S. This is shown in Figure 18.

The biodiversity of Mangrove ecosystems is the most profound in the islands and coastline of the Indo-West-Pacific region, where the occurrence of 44 varieties has been reported. In the Atlantic Ocean coast Mangrove ecosystems are especially widespread in the Caribbean region. The chemical composition of plant species and soils of Mangrove ecosystems have been recently discussed (Dobrovolsky, 1994).

The Mangrove Forest ecosystems occur on the surface of compact cavernous reef limestone or on carbonate sands, aleuritic and clayey silts in lagoons and shallow bays. In some places small spots of Mangrove ecosystems are placed on coastal quartz sands derived from weathering materials of crystalline rocks. The botanic genus of *Rhizophora* (Red Mangroves), *Avicelinnia* (Black Mangroves) and *Laguncularia* (White Mangroves) are predominant in the floral composition of the Mangrove Forest ecosystems. The shrubs of genus *Pemphis* are common residents in the Mangrove ecosystems in the atolls of the Indian Ocean with compact reef limestone ledges at the coastal strip.

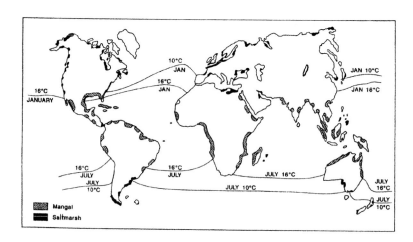

Figure 18. Map showing global pattern of mangroves and saltmarshes in relation to average temperature (Chapman, 1977).

The biogeochemical fluxes of various macro- and microelements are different from those calculated for Tropical Rain Forest ecosystems. The chemical composition of leaves of tree species in Mangrove Forest ecosystems is connected with higher content of Mg, Cl and S − SO_4^{2-} and lesser content of K and Si as compared to the leaves of trees from Tropical Rain Forest ecosystems. The content of Al is 3–4 times higher than that of Si and this can be related to the values of hydrogenic accumulation of these elements in soils (Figure 19).

The total ash content accounts for 11–23% from the dry weight of plant biomass. We can remember for comparison, that these values for the terrestrial forest ecosystems on the similar limestone soils are 5–6% only. These differences can be attributed to the adaptation of Mangrove ecosystems to saline marine waters.

The common biogeochemical feature of the Mangrove ecosystems is connected with small fluxes of trace metals (Table 43).

The Mangrove ecosystems are the example of one of the most productive ecosystems of the World. The biomass pool is in excess of 100 ton/ha of dry matter and annual NPP varies from 10 to 30 ton/ha, including leaf litter production of 8–15 ton/ha. Using average data on annual net primary production and content of various elements, we can estimate the values of biogeochemical cycles (Table 44).

The comparison of biogeochemical fluxes in the Mangrove and Tropical Rain Forest ecosystems shows that the total mass of ash elements per unit area is similar. However, the proportion of various elements is markedly different. The Mangrove plant uptake of Fe and Mn is less and that of Sr is higher than the uptake of these elements in Tropical Rain Forest ecosystems.

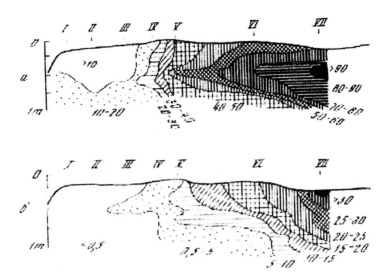

Figure 19. Hydrogenic accumulation of aluminum and silicon oxides in soil solutions of the Mangrove ecosystems of West Africa: a—content of aluminum oxides, mg/L, b—content of aluminum oxides, mg/L (Kovda, 1984).

Table 43. Content of trace metals in the plant biomass of the Mangrove ecosystems of the Indian Ocean islands, ppm on dry ash weight (After Dobrovoslsky, 1994).

Ecosystems	Island	Plant species	Plant organs	Trace metals				
				Mn	Zn	Cu	Pb	Ni
Mangrove Forest of coral islands	North Poivre	*Rhizophora mucronata*	Leaves, twigs	36.8	28.4	18.4	0.1	5.0
	South Poivre		Leaves, twigs	25.0	50.0	35.0	8.8	8.1
	North Farqu-har		Leaves	13.8	42.0	15.8	6.9	6.0
	South Farqu-har		Twigs	32.0	60.0	36.0	6.4	2.5
Mangrove Forest of silicate islands	Silhouette	*Rhizophora mucronata*	Leaves, twigs	71.4	45.3	10.7	0.7	3.8
	Silhouette	*Bruguiera gymnorizha*	Leaves, twigs	306.2	70.0	20.6	0.9	0.9

Table 44. Average annual biogeochemical fluxes of chemical species in Mangrove ecosystems.

Elements	Average ash content, ppm	Annual fluxes, g/ha/year
Sr	1920.0	3670
Fe	113.0	215
Zn	45.0	85
Mn	27.0	51
Cu	26.0	49
Pb	7.4	14
Ni	5.4	10
Total sum of ash elements	—	4094

In Mangrove ecosystems the nutrients accumulate in leaves and accordingly fall with shed or died-off-leaves to the soil surface. The further biogeochemical transformation of these chemical species depends on the surface properties of soils and underlying layers of rocks. When Mangrove ecosystems develop on compact reef limestone the litterfall layer is not produced. In that case, if the underlying layer is carbonate or carbonate-argillaceous silt, soils with weakly developed humus horizon are formed. Sulfate-reducing bacteria play an important role in the microbial processes occurring in soils of the Mangrove ecosystems. Under the conditions of a shallow seawater level (10–15 cm) and minimal tidal interference, a lean peat horizon is formed at the soil surface. The typical example is the Mangrove ecosystems in the lowland littoral plain in the Batabano bay in southwestern Cuba (Dobrovolsky, 1994).

The composition of organic matter in soils of the Mangrove ecosystems is related to the prevalence of the small-size vegetation detritus resistant to decomposition. The humus content is usually about 1% or so, with increase up to 5% in the peat horizon. Independent of the overall content of organic matter in the upper horizon, the easily soluble fulvate compounds are predominant over humates. In soils of Mangrove ecosystems on limestone, the fulvic acids formed undergo a rapid neutralization by carbonates. This may be inferred from the rise of pH values (from 7.2–7.7 to 8 or higher) of aqueous extract of the humus horizon at a depth of 10–20 cm. Dispersed organic detritus can sorb many trace metals and during detritus decomposition these metals become bound to water-soluble fulvis acids. This facilitates the leaching to 15–20 cm depth in soil profile. Further migration is impeded due to the neutralization of fulvic acids yielding water-insoluble calcium fulvates.

Table 45. The content of water-soluble species of iron, aluminum and silicon in soils of West African Mangrove ecosystems, ppm (After Kovda, 1985).

Soil layer, cm	Fe	Al	Si
0–13	756.0	83.2	54.9
14–31	889.0	112.5	27.5
> 32	132.3	117.4	9.2

The characteristic property of soils from Mangrove ecosystems is related to the accumulation of mobile water-soluble forms of iron, aluminum and silicon. The downward increase in soil profile was shown for iron and aluminum and an opposite trend for silicon (Table 45).

The Mangrove ecosystems perform a role of biogeochemical barrier, which decreases significantly the runoff of chemical species from the coast to the ocean waters. This is correlated with the major biogeochemical parameters of these ecosystems such as high productivity and high values of annual biogeochemical fluxes.

Being on the coastal line, the Mangrove ecosystems are subjected to the influence of tidal activity, which triggers periodic changes in biogeochemical fluxes of various elements (see Box 8). The possible climate changes and relevant sea levels rise may affect greatly the distribution of Mangrove ecosystems.

Box 8. Potential change of sulfur biogeochemical cycle in Thai mangrove ecosystems due to sea level rise resulting from climate change scenarios (after Rummasak, 2002)

Increasing concentration of GHG in the atmosphere will lead to climate change and the most probable scenarios are related to sea level rise. According to these scenarios the mangrove ecosystems of South East Asia and Thailand coast, in particular, will change many features, especially those connected with biogeochemical cycle of sulfur.

Mangrove ecosystems are very complex and highly productive, however the quantitative parameterization of various links of their biogeochemical structure is uncertain both in local and regional scale. Several biogeochemical processes in mangrove will be affected by sea level change. Furthermore, at present owing to lack of information, the impact upon biogeochemical reaction of mangrove ecosystems as a result of relative sea level rise is difficult to predict, and the requirement to study such a process quantitatively is of great scientific and political interest. Hence it is important to understand the potential impact of changing climate on mangroves biogeochemical cycling of the most important elements like sulfur, in order to conserve and manage these valuable resources for sustainable management in the future.

For quantitative parameterization of various links of the sulfur biogeochemical cycle, the natural mangrove forest was selected at the mouth of Klong Ngao River in the area of Ranong Biosphere Reserve on the Andaman Sea coast of Thailand. Water and salt (sulfur) biogeochemical fluxes are monitored in the Klong Ngao estuary assuming a two-dimensional flow pattern. These monitoring data are applied to carry out the budget of conservative and non-conservative material. The prognostic model predicting the potential change of biogeochemical cycle of sulfur in this mangrove ecosystems due to sea level rise resulting from climate change scenarios will be developed based on the results of studied budget. The estimation of fluxes in dry and wet season can present the input data for such a model.

Two box models are used for the parameterization of water as conservative material and salts, including sulfur, as non-conservative material. For example, the first estimates were made up for the wet season. For the coastal water body budget, the following fluxes were monitored and calculated: precipitation, evapotranspiration, runoff, groundwater, and residual flow. These budget estimates give the fresh water residence time as 6.2 days with residual flow equal -822.34×10^3 m^{3}d^{-1}. For Klong Ngao salt budget during wet season, similar fluxes were measured. Salt input into the system through mixing salt flux was equal $+14.843 \times 10^3$ kgd^{-1} with similar residual salt flux and the average salinity at high water spring as 18.81 ppt and at low water spring as 12.79 ppt. During the dry season fresh water flow and residence time decreased, whereas salt input increased.

FURTHER READING

Vsevolod V. Dobrovolsky (1994). Biogeochemistry of the World's Land. Mir Publishers, Moscow/CRC Press, Boca Raton-Ann Arbor-Tokyo-London, 221–324.

Robert A. Congdon and David Lamb (1990). Essential nutrient cycles. In: L. J. Webb and J. Kikkawa, Eds. Australian Tropical Rainforests. CSIRO, 105–113.

QUESTIONS AND PROBLEMS

1. Discuss the peculiarities of biogeochemical cycling in Polar ecosystems and especially pay attention to the factors restricting the processes of biological and biogeochemical turnover.

2. Characterize the role of relief in ecosystem formation and behavior in the Arctic Ocean islands. Give characteristic examples.

3. Describe the biogeochemical cycles of heavy metals in Polar ecosystems. Underline the role of peat as a biogeochemical barrier in migration fluxes of chemical species.

4. Compare the biological and biogeochemical characteristics of Tundra and Polar ecosystems. Explain the role of climate and soil chemical composition in the formation of biogeochemical turnover in Tundra ecosystems.

5. Discuss the net primary productivity of Forest ecosystems in Boreal and Sub-Boreal zones. Explain the similarities and differences in distribution of NPP values between various compartments of biomass in Coniferous, Deciduous and Mixed Forest ecosystems.

6. Describe the role of microbial population in biogeochemical cycles of N, P, S, and C in Forest ecosystems. Why is this role so important in soils of Forest ecosystems?

7. Characterize the role of seasonal freezing in biogeochemical cycling rates in various Forest ecosystems. Stress the attention on the ratio between the annual input of organic matter with litterfall and biomass mineralization.

8. Determine the biogeochemical uptake coefficient and discuss the C_b values for various Forest ecosystems.

9. Describe the results of biogeochemical studies in North America using the Hubbard Brook Experimental forest in the White Mountains of New Hampshire and Coweeta Hydrologic Laboratory, located in the southern Appalachians of North Carolina as the examples.

10. Discuss the biogeochemistry of Coniferous Forest ecosystems of Northwestern Eurasia. Place your attention on the role of geological rocks in biogeochemical fluxes of trace elements.

11. Using data from Tables 10 and 11, present a comparative analysis of the biogeochemical fluxes of macro- and trace elements in various Forest ecosystems: Coniferous, Deciduous, Mixed, and Swamp.

12. Describe the main peculiarities of biogeochemical fluxes and pools of various essential elements in Oak Broad-leaved Forest ecosystems.

13. What are the main soils of Forest ecosystems? Indicate their characteristic features and describe the general profile of these soils.

14. Discuss the role of different soil fractions in biogeochemical migration of trace metals. Give relevant examples from different ecosystems.

15. Describe the common chemical composition of surface waters in Boreal forest ecosystems. What types of biogeochemical processes are of the most importance in the hydrochemical composition?

16. Express the significance of organic and inorganic complexes in the aqueous migration of trace metals in the soil-water system of Forest ecosystems.

17. What are the general features of Steppe and Desert ecosystems? Emphasise the role of humidity and aridity in the formation and proceeding biogeochemical cycles of various elements.

18. Discuss the role of various plant species in migration and accumulation of trace metals in Steppe and Desert ecosystems. What parts of herbaceous species are the main accumulators of different elements?

19. Compare the rates of biogeochemical cycles in Forest and Steppe ecosystems. Explain the characteristic differences.

20. Describe the peculiarities of biogeochemical migration in Dry Steppe ecosystems of various regions of the World. Present the similarities and differences in biogeochemical processes of these ecosystems.

21. Discuss the role of invertebrates in biogeochemistry of Steppe ecosystems. Indicate the role of humidity in biomass and number of invertebrates.

22. Compare the biogeochemical cycling in Meadow Steppe and Dry Steppe ecosystems. Note the characteristic differences and explain the reasons.

23. Describe the biogeochemical processes in Extra-Dry ecosystems of central Asian deserts. Compare the role of biotic and abiotic processes in the migration of chemical elements in these ecosystems.

24. Describe typical Tropical Rain Forest ecosystems. Explain why these ecosystems are the most productive ones in our planet.

25. Compare the fluxes and pools of essential chemical elements in Boreal Forest and Tropical Rain Forest ecosystems. Explain the reasons for the intensive rates of biogeochemical turnover in tropical forests.

26. Discuss the soil compartment of Tropical Rain Forest ecosystems. Note the common principles of biogeochemical migration of elements in these soils.

27. Compare the Steppe and Savanna ecosystems. Describe the biogeochemical cycles of essential elements. Express the role of vegetation in biogeochemical processes of both ecosystem types.

28. Characterize the biogeochemical cycling in Seasonal Deciduous Tropical Forest and Woody Savanna ecosystems of Africa. Place attention on the differences in biogeochemical processes during wet and dry seasons.

29. Note the specific features of Mangrove ecosystems. Discuss the role of sulfur in the biogeochemical processes.

30. Discuss the migration of silicon, iron, and aluminum in Mangrove ecosystems. Present the characteristic biogeochemical mechanisms for the speciation of these elements in soil and waters.

CHAPTER 7

BIOGEOCHEMICAL MAPPING

This chapter is devoted to a description of modern approaches to the biogeo-chemical mapping of terrestrial ecosystems on local, regional and continental scales. Biogeochemical structure of natural ecosystems represents the co-evolution of the geosphere and biosphere (see Chapter 2 also). During this co-evolution each elementary geochemical unit (the smallest unit of the Earth's surface organization) was linked to the relevant ecosystems with regular biogeochemical food webs. In turn, these food webs have been adopted to the specific parameters of migration and accumulation of different chemical species in the biosphere. Biogeochemical mapping of the terrestrial ecosystems shows the complex of geological and biological conditions of modern biosphere. During the past 40–50 years the scientific basis of biogeochemical mapping has related to the development of many basic and applied disciplines, like chemistry, geology, biology, geography, biochemistry, geochemistry, ecology, etc. It was also closely connected with the systems of various organisms and understanding the close relationships between these organisms and the environment.

1. CHARACTERIZATION OF SOIL-BIOGEOCHEMICAL CONDITIONS IN THE WORLD'S TERRESTRIAL ECOSYSTEMS

The biogeochemical cycling in different ecosystems is to a large extent determined by biota, especially by the primary production of plants and by microbial decom-position. At present we recognize the development of the intensive biogeochemical investigations of a large number of ecosystems in North America, Europe, Asia and South America.

The biogeochemical cycling picture is designed to summarize the circulation features in various components of ecosystems such as soil, surface and ground water, bottom sediments, biota and atmosphere (Figure 1).

Ecosystem and soil regionalization can be a basis for biogeochemical mapping (Fortescue, 1980; Glazovskaya, 1984, 1990; Ermakov, 1993). The combination of this mapping with the quantitative assessments of biological, geochemical and hy-drochemical turnover gives an opportunity to calculate the rates of biogeochemi-cal cycling and coefficients of biogeochemical uptake, C_b, for different ecosystems (see Chapter 6). In addition to the coefficient of biogeochemical uptake, we can also apply the active temperature, C_t, and relative biogeochemical, C_{br}, coefficients.

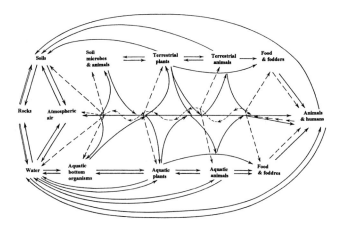

Figure 1. The general scheme of biogeochemical food webs in the terrestrial ecosystems.

The application of these coefficients for the characterization of soil-biogeochemical conditions in various ecosystems is based on the following hypothesis:

a) For northern areas the real duration of any processes (biochemical, microbiological, geochemical, biogeochemical) must be taken into account because they are depressed annually for 6–10 months and the influence of acid forming compounds, as well as any other pollutant, occurs during summer. A process duration term has been derived as the active temperature coefficient, C_t, which is the duration of active temperatures $> 5\,°C$ relative to the total sum.

b) The relative biogeochemical, C_{br}, coefficient is the multiplication of the first two coefficients. This may characterize the influence of temperature on the rates of biogeochemical cycling in various ecosystems. The relative biogeochemical, C_{br}, coefficient is applied as a correction to C_b values

Table 1 shows combinations of soil-biogeochemical and temperature conditions in various geographical regions of the World. The given combination of factors is represented by ecosystem types, FAO main soil types, biogeochemical, C_b, active temperature, C_t, and relative biogeochemical, C_{br}, coefficients. The subdivision of ecosystem types is based on various parameters, including vegetation and soil types, and main climate characteristics like temperature and ratio of precipitation-to-evapotranspiration

The values of C_b for each geographical region were ranged to determine the type of biogeochemical cycling and these ranks are shown in Table 2. Five types of biogeochemical cycling are identified: very intensive, intensive, moderate, depressive and very depressive.

The corresponding values of active temperature coefficients ranged in accordance with the main climatic belts are shown in Table 3.

Table 1. The values of biogeochemical cycling (C_b), active temperature (C_t) and relative biogeochemival (C_{br}) coefficients in various soil-ecosystem geographical regions of the World (Bashkin and Kozlov, 1999).

Ecosystems	Main FAO Soil types	Geographical region	Index Fig. 2	No	C_b	C_t	C_{br}
Arctic Deserts and Primitive Tundra	Litosols, Regosols	North American	1_1	1	10.0	0.06	0.6
		Eurasian	1_2	2	10.0	0.06	0.6
Tundra	Cryic Gleysols, Histosols, Humic Podzols	North American	2_1	3	18.0	0.15	2.7
		Eurasian	2_2	4	18.0	0.15	2.7
Boreal Taiga Forest	Podzols, Podsoluvi-sols, Spodi-Distric, Cambisols, Albi-Gleyic Luvisols, Gelic and Distric Histosols, Rendzinas and Gelic Rendzinas, Andosols, Gleysols	Alaskan-Cordillera	3_1	8	10.0	0.28	4.2
		Laurentian	3_2	3	8.5	0.35	3.0
		North Atlantic	3_3	8	5.5	0.55	3.0
		North European	3_4	3	8.0	0.45	3.6
		European-West-Siberian	3_5	4	8.5	0.35	3.0
		North Siberian	3_6	5	9.5	0.25	2.4
		Central Siberian	3_7	6	9.3	0.30	2.8
		East Siberian	3_8	5	7.5	0.20	1.5
		Kamchatka-Aleutian	3_9	7	5.0	0.25	1.7
Taiga Meadow-Steppe	Planosols	Central Yakutian	$4a_1$	9	10.0	0.35	3.5
		Central Canadian	$4a_2$	9	10.0	0.32	3.2

Table 1. The values of biogeochemical cycling (C_b), active temperature (C_t) and relative biogeochemival (C_{br}) coefficients in various soil-ecosystem geographical regions of the World (Bashkin and Kozlov, 1999) (continued).

Ecosystems	Main FAO Soil types	Geographical region	Index Fig. 2	No	C_b	C_t	C_{br}
Subboreal Forest	Podzols, Dystric and Eutric Cambisols, Umbric Leptosols, Podsoluvi-sols	Eastern North American	$5a_1$	10	2.0	0.65	1.3
		West European	$5a_2$	10	1.5	0.83	1.2
		East European	$5a_3$	11	2.0	0.60	1.2
		Hercian Alpine	$5a_4$	14	2.5	0.72	1.8
		East Asian	$5b_1$	12	2.6	0.67	1.7
		East Chinese	$5b_2$	13	1.5	0.81	1.2
		Coastal Pacific	$5b_3$	12	1.6	0.67	1.1
		New Zealand	$5c_1$	12	1.2	0.78	0.9
		South Chilean	$5c_2$	12	1.2	0.75	0.9
Forest Meadow Steppe	Luvic	Carpathian-North Caucasian	$6a_1$	15	1.2	0.75	0.9
Steppe	Fhaeozems, Cambisols, Chernozems	Central Russian	$6a_2$	15	1.4	0.65	0.7
		West Siberian	$6a_3$	16	1.5	0.47	0.8
		South Siberian	$6a_4$	18	2.0	0.42	1.0
		Amur-Manchurian	$6a_5$	17	1.5	0.65	0.8
		Central Cordillera	$6b_1$	18	1.1	0.70	0.8
		South Canadian	$6b_2$	16	1.3	0.60	0.8
		Central Plain	$6b_3$	17	1.1	0.71	0.8
South American Meadow Steppe	Chernozems, Vertisols	Eastern Pampa	$7a_1$	17	0.8	0.95	0.8
Steppe	Chernozems, Kastanozems Solonetzes	European Kazakhstan	$8a_1$	19	0.7	0.57	0.4
		Mongolo-Chinese	$8a_2$	19	0.8	0.61	0.5
		Central Plains	$8b_1$	20	0.7	0.70	0.5

Table 1. The values of biogeochemical cycling (C_b), active temperature (C_t) and relative biogeochemival (C_{br}) coefficients in various soil-ecosystem geographical regions of the World (Bashkin and Kozlov, 1999) (continued).

Ecosystems	Main FAO Soil types	Geographical region	Index Fig. 2	No	C_b	C_t	C_{br}
Mountain Depression Forest, Bush and Steppe	Vertisols, Eutric Combisols	Mediterranean	$9a_1$	22	0.9	0.87	0.8
		North African	$9a_2$	22	0.8	0.95	0.8
		Texas	$9b_1$	21	0.8	0.95	0.8
		South East African	$10a_1$	22	0.9	0.95	0.9
Desert-Steppe and Desert	Xerosols, Regosols, Arenosols, Yermosols, Solonetzes, Solonchaks	Central Asian	$11a_1$	24	0.4	0.70	0.3
		Pamiro-Tibetan	$11a_2$	26	0.6	0.62	0.4
		Middle Asian	$11a_3$	23	0.5	0.77	0.4
		Preasiatic	$11a_4$	25	0.3	0.89	0.3
		Hindukush-Alai	$11a_5$	22	0.4	0.86	0.3
		Tien Shan	$11a_6$	18	0.6	0.60	0.4
		West American	$11b_1$	24	0.4	0.70	0.3
		Mexican-Californian	$11b_2$	25	0.3	0.95	0.3
		Saharan	$11c_1$	27	0.2	1.00	0.2
		Arabian	$11c_2$	27	0.2	1.00	0.2
		South African	$11d_1$	27	0.3	0.80	0.3
		Andean	$11e_1$	26	0.4	1.00	0.4
		Patogonian	$11e_2$	23	0.5	0.71	0.4
		Central Australian	$11f_1$	27	0.2	1.00	0.2
Savanna, Tropical Forest	Livi-Plinticic Ferrasols, Luvisols, Vertisols, Subtropical Rendzinas, Ferralitic Cambisols, Nitosols, Ferralitic Arenosols, Subtropical Solonchaks	South Asian	$12a_1$	30	0.3	1.00	0.3
		Caribbean	$12b_1$	32	0.3	1.00	0.3
		Brazilian	$12b_2$	31	0.2	1.00	0.2
		East Brazilian	$12b_3$	32	0.3	1.00	0.3
		Cis-Andean	$12b_4$	29	0.3	0.98	0.3
		Somalian-Yemenian	$12b_5$	30	0.3	1.00	0.3
		Sudan-Guinean	$12b_6$	33	0.4	1.00	0.4

Table 1. The values of biogeochemical cycling (C_b), active temperature (C_t) and relative biogeochemival (C_{br}) coefficients in various soil-ecosystem geographical regions of the World (Bashkin and Kozlov, 1999) (continued).

Ecosystems	Main FAO Soil types	Geographical region	Index Fig. 2	No	C_b	C_t	C_{br}
		East African	$12b_7$	32	0.3	1.00	0.3
		Angolo-Zimbabwean	$12b_8$	30	0.3	1.00	0.3
		North Australian	$12c_1$	31	0.2	1.00	0.2
		East Australian	$12c_2$	30	0.3	1.00	0.3
		South Australian	$12c_3$	29	0.4	1.00	0.4
		West Australian	$12c_4$	30	0.3	1.00	0.3
Subtropical and Tropical Wet Forest	Ferrasols, Eutric Subtropical Histosols, Gleyic Subtropical Podzols, Plinthic Gleysols, Nitosols	South East Asian	$13a_1$	34	0.2	1.00	0.2
		Himalayan	$13a_2$	34	0.4	0.80	0.3
		Andean-Equatorial	$13b_1$	37	0.3	1.00	0.3
		Central American	$13b_2$	31	0.1	1.00	0.1
		Malaysian-New Guinean	$13b_3$	38	0.1	1.00	0.1
		Brazilian-Atlantic	$13c_1$	34	0.15	0.98	0.1
		Amazonian	$13d_1$	36	0.1	1.00	0.1
		Congo-Guinean	$13d_2$	36	0.1	1.00	0.1
		South East Latin American	$13e_1$	38	0.1	1.00	0.1
		East Australian	$13e_1$	34	0.1	1.00	0.1

Using the above-mentioned approaches, we may describe the main global ecosystems in various continents. This description is as follows. Many details of biogeochemical cycling of various elements are shown also in Chapter 6 "Regional Biogeochemistry".

1.1. Eurasia

Because of the huge size of Eurasia, all types of ecosystems and climatic belts are presented at the particular area, from Arctic Desert up to Tropical Rain Forest ecosystems (Table 1, Figure 2).

Table 2. The ranges attached to biogeochemical cycling data to assess the migration capacity of soil-ecosystem types.

Ranks	Biogeochemical cycling	Biogeochemical cycling coefficient, C_b
1	very intensive	< 0.5
2	Intensive	0.5–1.4
3	Moderate	1.5–3.0
4	Depressive	3.1–10.0
5	very depressive	> 10.0

Table 3. The ranges attached to temperature regime data to assess the duration of active biogeochemical reactions.

Ranks	Temperature regime	Active temperature coefficient, C_t
1	arctic	< 0.25
2	boreal	0.26–0.50
3	sub-boreal	0.51–0.80
4	mediterranean	0.81–0.99
5	subtropical and tropical	1.00

Arctic Deserts and Primitive Tundra ecosystems

These ecosystems occur in the Eurasian soil-geographical region. This represents the most northern part of the Asian Arctic and includes the northern big island of the North Earth islands and small neighboring islands. The biogeochemical cycling in these ecosystems are characterized by an arctic hydrothermal regime, connected with very severe temperature, low precipitation (50–150 mm annually) and primitive bush-like, algae and lichen vegetation.

The biogeochemical cycling can be ranged as very depressive that relates to 10 and more years of plant residue mineralization and the temperature regime—as the arctic one with an annual relative part of active temperatures $> 5^0C$ as 0.05–0.10. The predominant soil types are Litosols and Regosols, in hollows, Histosols.

Tundra ecosystems

The Tundra ecosystems are represented by the Eurasian geographical region with primitive humid Podzols, Cryic Gleysols, Histosols, Litosols and Regosols.

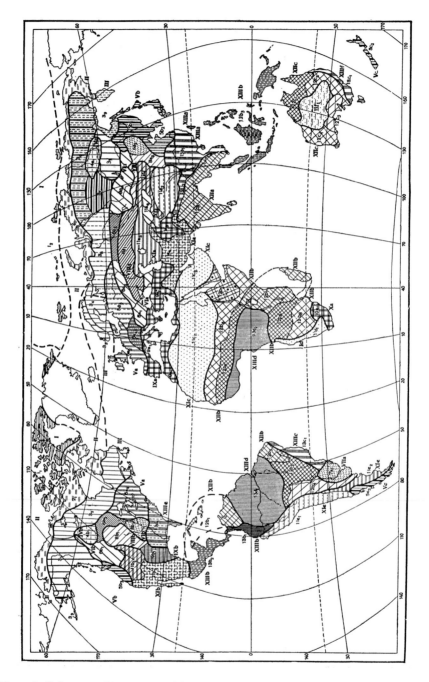

Figure 2. Soil-geographic mapping of the global terrestrial ecosystems (Glazovskaya, 1990).

The biogeochemical processes in these ecosystems are characterized by small heat quantity, a short but very intensive period of active temperatures (the mean value of C_t is equal to 0.15), wide distribution of permafrost, low precipitation, low biological and microbiological activity and low rate of chemical weathering. The mean C_b values are equal to 18 (15–50), which corresponds to a very depressive type of biogeochemical cycling. However, the long winter period enhances the accumulation of various pollutants in snow cover with a sharp increasing of their flash-like influence on different components of ecosystems during short summer period.

Boreal Taiga Forest ecosystems

European-West Siberian, North Siberian, Central Siberian, East Siberian and Kamchatka-Aleutian geographical regions represent these ecosystems in Eurasia. The predominant ecosystem types are Dense Needle Evergreen Forest, Open Needle Evergreen Forest, Dense Mixed Evergreen Forest and Open Mixed Evergreen Forest. In spite of the differences in species composition, the Coniferous Forest ecosystems are characterized by a low organic matter turnover (depressive type of biogeochemical cycling), the C_b values vary from 5.0 (Kamchatkan-Aleutean geographical region) to 9.3–9.5 in Central and North Siberian regions. The predominant soils are Podzols, Podzoluvisols, Histosols, which have low pH, low base saturation and low cation exchange capacity. The mean values of C_t are from 0.25 to 0.35. Under these cold climate conditions the additional stress of acidic deposition to the exposure of plants may tend to make the vegetation in the taiga forest ecosystems more sensitive to the various anthropogenic impacts such as acid deposition (see Chapter 10 for details). The relative biogeochemical coefficients, C_{br}, vary from 1.5 to 3.0 (Table 1) being increased in more continental Central and West Siberian geographical regions.

Taiga Meadow Steppe ecosystems

In Asia these ecosystems are represented by the Central Yakutian geographical region with Planosols. The biogeochemical features of these soils are connected with their occurrence in the inner part of Northern Asia where the climate is the most severe and dry and soils are developed under insufficient atmospheric precipitation; the ratio of precipitation to potential evapotranspiration (P:PE) is equal to 0.45–1.00 and drops during summer period down to 0.20–0.45. In the permanent permafrost, an abundance of carbonate in soil forming rocks has led to the formation of Planosol–Solonchak–Solod–Solonetz complexes. Under the long severe winter period and hot dry short summer, the biogeochemical turnover is depressed in these ecosystems: mean C_b is equal to 10, mean C_{br} is 3.5 (Table 1, Figure 2).

Sub-Boreal Forest ecosystems

These ecosystems are in a monsoon climate. The Dense Deciduous Forest, Dense Deciduous Broad Leaf Forest, Open Deciduous Broad Leaf Woodland are the predominant ecosystem types.

Atlantic-European and West European geographical regions

Between 45 and 55° N the region covers the lowlying plains of the eastern Baltic, great Polish and North German lowlands, Southern Sweden, Jutland, and the British Islands. In the northern part of the region, lying within the territories of Quaternary glaciations, the soil cover changes from southwest to northeast because of differences in the forms of relief and the composition of loose deposits. Soil catena on slopes of sandy hills with shallow occurrence of subsoil water typically includes humus-iron Podzols, humus Podzols, sod-gley soil and peat-gley soil (Podzoluvisols). The hilly moraine landscape typical of eastern parts of the lowland was formed with fluvio-glacial deposits and local soils are more fertile. In the large part of the territory moraine plains have a thin (30–50 cm) cover of silty loess-like loam or sandy loam on the surface, of lighter texture than the underlying moraine loam and clay. Such two-component deposits occur widely in the Great Polish Plain. With them are associated soils very similar to the Podzoluvisols of the southern part of the forest zone of the East European Plain, which were also formed on two-component deposits.

The areas of slightly acidic Cambisols are associated with the occurrence of calcareous moraine or calcareous loess-like loams. Although carbonates are leached from the soil profile their occurrence at shallow depth, and the return of bases including calcium with litterfall from broad-leaved species of local ecosystems, maintain the high degree of base cation saturation (70–80%) and slightly acidic reaction of these soils. With increasing age of the surface (older glacial plains) or on rocks poorer in bases, these soils are replaced by Luvisols.

The ecosystems of this region are characterized by a moderate rate of organic matter turnover with mean values of C_b equal to 2.0; C_t is 0.67 and C_{br} is 1.3. Such moderate rates are favorable to soil acidification under acid rains, which have been abundant here during the 20th century.

East European geographical region

This region covers the zones of Podzoluvisols of the East European (Russian) Plain. In runs southwest to northeast. In a large part of the region, the climate is moderately continental, with an annual rainfall of 500–600 mm, and a summer maximum of rainfall 1.5 times or double the winter rainfall. The average annual moisture coefficient is 1.0–1.3. The temperature of the hottest month is almost the same, 17.5 to 19.5 °C. The temperature and length of winter is less constant. The leached regime of soils is preserved throughout the region. A somewhat shorter warm season than in the Atlantic and West European regions, annual freezing of soils, and no period of winter soil formation result in somewhat better conservation of humus, containing a small quantity (1.5–5.0% of total humus) of calcium humates. This layer plays the role of the biogeochemical barrier in these ecosystems. The biogeochemical cycles of elements are moderate.

Hercynian-Alpine geographical region

Mountain zonal macrostructures complexed by lithogenic macrostructures are characteristic of this region. Hercynian Mountain country consists of low and medium-high hills with an altitude of from 400–600 to 1,000 m, much smoothed by erosion. The climate of valleys and watersheds is very variable. In watersheds it is cold and humid and there are many forests and bogs. In valleys the climate is drier and more continental, and steppes are often found there.

Products of weathering of different Paleozoic and mainly crystalline rocks, very diverse in lithological composition, serve as parent rocks for the soils of this region, giving a variable soil cover. Foothills lie in the zone of Broad-Leafed Oak, Beech and Oak-Hornbem-Beech Forest ecosystems, on Cambisols and Luvisols. Above 400–500 m the nature of vegetation changes to Coniferous Forest ecosystems on mountain Cambisols with acid soil reaction and even Podzoluvisols, often interspersed with skeletal or light phases. Soils of basic volcanic rocks (diabases, tuffs) represent a special variety of soils in Central German Mountain country. Usually these soils have a very dark color and are rather strongly humified. They are characterized by high fertility and moderate degree of biogeochemical cycling (see Table 2).

The fringes of the Alps are much more humid than the central part. The climate of the northern slopes is vastly different from that of the southern slopes. Consequently the boundaries of soil and vegetation zones in the Alps shift like the snowline: in interior parts of the mountains the snowline lies 300–500 m above that in the outer zones.

If we take any part of the Alps the successive change in soil belt will be as follows: the foothills lie in the zone of Broad-Leafed Forest ecosystems with C_b about 2.0 to an altitude of 600–700 m, in the south up to 800–900 m. Above them is the zone of Coniferous Forest ecosystems on mountain Podzols with moderate-to-depressed biogeochemical cycles of elements. The altitude limits are 1,400 in the north and 2,300 m in the south. Above these ecosystems is the zone of Sub-Alpine and Alpine Meadow ecosystems with depressed type of biogeochemical turnover.

East Asian geographical region

East Asian geographical region occupies the continental part of Far East (Russia, China, North and South Korea) and island parts (Russia, Japan). It is characterized by different subtypes of Cambisols (Spodi-Distric, Spodi-Distric Cryic, Humid, Orti-Distric, Distric) and Podzols, especially in Hokkaido island, in Manchurian and Sikhote-Alin mountains in Dark Needle Forest ecosystems.

In the plains, Cambisols place the most drainage areas. These soils occupy the hilly plains and low mountain belts up to the 500–700 m elevation, where they coincide with the Broadleaf and Coniferous-Broadleaf Forest ecosystem types. In the most continental parts the oak forests are dominant. For instance, at the slopes of the Sikhote-Alin range Cedar-Broadleaf Forest and in Korean peninsula, the Oak-Maple Forest ecosystems are predominant. In Japan Beech Forest ecosystems are the most abundant. Heavy precipitation rates during wet season (up to 1000–1200 mm with P:PE equal to 1) favor the increasing base saturation in the Luvic Cambisols. These ecosystems are characterized by a moderate rate of organic matter turnover with mean values of C_b equal to 2.5; C_t is 0.67 and C_{br} is 1.7. Such moderate rates are favorable to soil acidification with deposition input of sulfur and nitrogen acid forming compounds (NIES, 1996, Bashkin and Park, 1998). This process can be especially enhanced in ecosystems with predominant Vitric Andosols where porosity favors rapid chemical and biogeochemical weathering with allophane-kaolinite formation processes. The abundance of free iron and aluminum oxides under acid soil reaction reinforced by acidic deposition leads to a release of Al^{3+} ions and toxic influence on the fine roots of trees (Izuta and Totsuka, 1996).

East Chinese geographical region

East Chinese geographical region occupies the alluvial plains of Central China river watersheds with Eutric Cambisols in the highest elevation relief positions in the northern part and with Ferralitic Cambisols in the southern part. Neutral reaction and very active weathering characterize these natural soils. The biogeochemical cycling rank is moderate, the mean values are of C_b 1.5; C_t is 0.81 and C_{br} is 1.2 which favors to the active biogeochemical migration of various elements. However, the most natural ecosystems have been transformed into agroecosystems with artificial regulation of soil reaction using fertilizers and liming.

Forest Meadow Steppe ecosystems

The biogeochemical cycling of elements in these ecosystems can be characterized as moderate in depressions and as semi-intensive in high mountain forest ecosystems with Cambisols; the average C_b is equal to 2 and C_t is equal to 0.42.

Steppe ecosystems

The main characteristic features of these ecosystems are related to the continental climate and low precipitation, precipitation:potential (and actual) evapotranspiration (P:PE). P:PE ranges between 0.6–0.3. In accordance with the given climatic conditions, the soils of steppe ecosystems (Chernozems, Kastanozems, Solonetzes) are characterized by the presence of a few buffer layers, such as humus, carbonate, and gypsum that makes them insensitive to actual and potential loads of pollutants.

Desert-Steppe and Desert ecosystems

The main soil types of these ecosystems are characterized by high buffering capacity, high pH values, low ratio of P:PE. Thus, in spite of rapid rates of organic matter turnover and nutrient cycling, C_b values lie in the limits of 0.3–0.6, these soils and corresponding ecosystems are insensitive to actual and potential loads of acidity and other pollutants.

Xerofitic Savanna and Tropical Monsoon Forest ecosystems

The main soil types (Table 1) of the Indostan peninsula have a high buffering capacity, high base saturation and low P:PE ratio. Thus in spite of very intensive organic matter (OM) cycling (C_b equal to 0.3; C_t equal to 1.00) most of the soil/ecosystem combinations are sustainable to anthropogenic impacts. These biogeochemical features are complicated in Sri Lanka and in the plain and low plain areas of Mekong and

Menam river basins. These sub-regions are characterized by a monsoon climate with wet summers (1200–1300 mm) and dry winter periods. In the eastern part of the Mekong-Manam geographical sub-region the ecosystems with Nitosols and Rhodic Ferrasols are characterized with very intensive biological turnover and a very high buffering capacity. In contrast, Luvi-Plinthic and Xantic Ferrasols, subtropical Albi-Gleyic Luvisols with plinthite are widespread in the depositional low plains of river deltas. The combination of these moisture conditions with very intensive OM cycling (C_b equal to 0.3) leads to the formation of ecosystems, which biogeochemical cycles are very sensitive to actual and potential acidic deposition potentiating the release of free Al^{3+} in the soil-water system.

Subtropical and Tropical Wet Forest ecosystems

The main characteristic features of these ecosystems are the very old soil parent materials, which are transformed by very intensive geochemical weathering leading to the destruction of all primary minerals except quartz and the accumulation of new formed minerals such as kaolinite, hematite, gibbsite, hydrogillite. The predominant soils are Ferrasols characterized by very low buffering capacity, abundance of free Al^{3+} and Fe^{3+}, acid reaction with soil depth, very intensive biogeochemical cycling of all elements and especially nutrients such as N, P, K, S, Ca, and Mg. The combination of these features with a monsoon and equatorial climate leads undoubtedly to a shift in the original equilibrium towards acidification under the increasing input of acid forming sulfur and nitrogen compounds.

South East Asian geographical region

The region occurs in the northern part of Tropical Rain Green Forest ecosystem zone and is predominantly characterized by Acric Ferrasols. Biogeochemical cycling is very intensive (mean C_b is equal to 0.2) but there are definite differences between hilly plains and low mountains up to 400–500 m a.s.l. (above sea level) and middle elevation mountains (up to 1000 m a.s.l.) where the humus biogeochemical barrier is present in the profiles of Podzolized Ferrasols.

Himalayan geographical region

This region is situated in the eastern part of Tibet and the Chino-Tibetan Mountains. Biogeochemical cycling is very complex with vertical geochemical catenas of different soils, such as evergreen broadleaf forests on mountainous Acric Ferrasols (1400–2000 m a.s.l.), evergreen and deciduous forests on transitive Ferrasol-Cambisol soils (2000–2700 m a.s.l.), mixed coniferous/deciduous forest on Cambisols (2500–2800 m a.s.l.), dark coniferous forest on Histosols, Gleysols and Cambisols (2700–3000 m a.s.l.) and mountainous meadows on mountainous Phaeozems (3000–3200 m a.s.l.).

The soil-ecosystem sequences are connected to climate differences and accompanied by rates of organic matter cycling progressing from very rapid in the lowest parts ($C_b < 0.2$) to moderate in the highest ecosystems ($C_b > 0.5$). The mean values of the biogeochemical cycling coefficients are equal to 0.4 and the active temperature coefficient to 0.8.

Malaysian geographical region

This region is situated in the Malaysian peninsula, and the islands of Indonesia and New Guinea. The predominant ecosystems are Wet Equatorial Tropical Forest with Ferrasols. In accordance with the very intensive biogeochemical cycling ($C_b < 0.1$) and natural acid features of the soils, all components of ecosystems are very sensitive to actual and potential acidic deposition (Bashkin and Park, 1998).

1.2. North America

North America may be divided into four broad thermal belts: Arctic, Sub-Arctic, Temperate, Sub-tropical and Tropical. In the north, in the Arctic and Sub-Arctic belts, the annual radiation balance is 0–10 kcal/cm^2, in the temperate belt it increases to 40–50 kcal/cm^2, and in the Sub-tropical and Tropical belts, to 60–70 kcal/cm^2 per year. The distribution of rainfall and zones of wetness over a large part of the plains of the continent conforms not to latitudinal but to meridional zonality. In the temperate belt, where westerly winds prevail, the maximum rainfall—1,000–2,000 mm or more—is received by the western slopes of the Cordilleras facing to the Pacific Ocean, most of it falling in autumn and winter. East of the mountain barrier, in the central basin and Great Plains, the rainfall decreases to 300–500 mm. On approaching the Atlantic coast, in the region of the Great Lakes and the Appalachians, the rainfall again increases to 1,000–1,200 mm.

In the Sub-tropical and Tropical belts, the maximum rainfall, associated with the trade winds, is received by the southeastern Atlantic part of the continent (1,000–1,200 mm or more). To the west, the rainfall decreases sharply: the Mexican plateau, the central basin and the Californian peninsula, influenced by the cold current, received from 300 to 100 mm or less annual rainfall.

The combination of latitudinal thermal belts and longitudinal zones of wetting in the plains creates a diversity of moisture-temperature conditions and consequently formation of different ecosystems with varying biogeochemical cycling.

Tundra ecosystems

North American tundra geographical region

This part of Tundra ecosystems covers the northern coasts of the continent and the southern part of the North American archipelago. In the north it is bounded by the Arctic desert and in the south by the Boreal Taiga Forest ecosystems.

As in Eurasia, the southern boundary of the Tundra ecosystems in North America deviates considerably from strictly latitudinal direction. In Labrador, cooled by the cold current, and on the shores of cold Hudson Bay, the Tundra ecosystems penetrate south to 54° N. To the west of Hudson Bay, with the increasing continentality of the climate, the boundary between Tundra and Boreal Taiga Forest ecosystems is displaced north. On the meredian of the Great Lakes it lies somewhat north of the Arctic Circle; near the estuary of the Mackenzie River it passes through 68.5° N, which is its northernmost location. The northern half of Alaska falls in the Tundra ecosystems; along the shores of the Chukchi Sea, the boundary of the Tundra ecosystems against shifts south.

The biogeochemical features of Tundra ecosystems are similar to those described for the Eurasian continent (see Table 1).

Boreal Taiga Forest ecosystems

The Boreal Taiga Forest ecosystems running in the latitudinal direction across North America extends from the Atlantic coast to the Pacific coast. Newfoundland and the Labrador Peninsula fall in these ecosystems in the east. Here the southern boundary lies at 48–50° N because of the cooling influence of the cold Labrador Current and Hudson Bay in the eastern North America. In Central Canada in the basin of Mackenzie River and Great Slave Lake the southern boundary of these ecosystems shifts more than 10° north to the latitude of about 61° N. In the western part, which includes a number of high mountain massifs and inter-mountain plateaus, the southern boundary again shifts south to 50° N near the Pacific Ocean. Thus in North America, as in Eurasia, the Boreal Taiga Forest ecosystem occupies the northernmost position in the interior of the continent, where the continentality of the climate gives a warmer summer than at the same latitude in maritime regions.

According to the type of biogeochemical turnover, the North American part of Boreal Taiga Forest ecosystems is subdivided into two large geographical regions: the Laurentian and the Alaskan-Cordillera.

Laurentian geographical region

This region is located in the Canadian Shield, composed of massive Precambrian crystalline rocks, mainly granite-gneisses and granites. In the syncline of Hudson Bay, on the south shore, and on the lowlands of the River Mackenzie, there occur sedimentary deposits of Silurian and Devonian dolomites, limestones and sandstones.

The northern part of this region is covered with Pre-Tundra Spruce, Larch-Spruce and, at places, Larch Thin Forest and Black Coniferous North Taiga Forest ecosystems, in which *Picea mariana* and *Picea canadensis* predominate. *Abies balsamea*, *Larix laricina* are mixed with these, and birch and aspen are also found. This part lies in the permafrost area and permafrost features determine the development of biogeochemical process: C_b is equal to 8.5, C_t, to 0.35 and C_{br}, to 3.0.

The southern part of the Laurentian region is covered by Podzols and Spodi-Distric Cambisols with Black Coniferous Forest ecosystems of *Picea mariana, P. glauca*, and *Abies balsamea*. Larch and birch are mixed with these species. Pine Forest ecosystems of *Pinus banksiana*, bog mosses and lowland bogs occur in this part of the region also. The biogeochemical cycling of the elements is faster than in the northern part, but anyhow it is the depressed type of turnover (see Table 1).

Alaskan-Cordillera geographical region

This region covers the ridges of the Northern Cordilleras and the plateau dividing them. It includes the southern slopes of the Brooks Range, the Yukon plateau and its eastern extension, the Fraser plateau, the Mackenzie Mountains and the Alaska Range. The boundary between Forest and Tundra ecosystems in Alaska fairly closely follows the July 13 °C isotherm. In coastal southwestern Alaska the boundary between Mountain Forest and Mountain Tundra ecosystems is lower than in the dry, more continental interior where the summer is warmer.

Thick Forest ecosystems are found only along the valleys of large rivers, occupying well-drained places of terraces, slopes and low hills, lying at an altitude of 700–750 m. White spruce (*Picea canadensis*), balsam poplar, aspen, and in some places white birch, are trees common to these Forest ecosystems; black spruce, fir and larch are found more rarely. The latter is attracted to swampy areas. On flat, ill-drained surfaces, forest ecosystems are replaced by vast areas of bog mosses and, in relief depressions, sedge-cotton grass bog.

The soil cover of the forest and thin-forest zone of Alaska consists of various sub-types of Gleysols, which are generally similar to analogous soils of North Siberia. Kaolinite and motmorillonite and also chloritized monmorillonite are present in the most intensely gleyed horizons. Lepidocrocite (monohydrate of iron oxide) and a large quantity of amorphous hydrated oxides of silicon and aluminum are found in almost all soils. The ratio of silicon to aluminum in the clay fraction is low, in certain places less than 1.0.

Kaolinite is present in large quantities in soils of higher elements of relief. In the less acidic soils of lower slopes, where solutions containing silica, magnesium and iron can enter from the slopes, the quantity of montmorillonite, especially chloritized montmorillonite and chlorite, increases. The biogeochemical cycle is generally depressed (see Table 1).

Taiga Meadow Steppe ecosystems

These ecosystems are represented by Central Canadian Mountain Meadow Steppe geographical region. This region is located in the northwestern part of the Great Plains of Central Canada, in the basins of the Mackenzie and Peace rivers, in Alberta and northern Saskatchewan. Annual rainfall here does not exceed 300–400 mm. Mean January temperatures are −27 °C, and mean July temperatures, about 15 °C.

The soil-forming rocks are mainly calcareous moraines, often covered with calcareous loess-like loams. The intense weathering of primary minerals and synthesis of secondary minerals have taken place. Solodized soils and solods occur in flat swamped lacustrine plains. They are present in large amounts in the Peace River basin.

This region may be considered as an analog of the cold intra-continental Central Yakutian region of frozen-taiga solodized soils. But in the Canadian interior the climatic conditions are less severe than in Yakutia and there is no permanently frozen horizon in these soils. These insignificant differences are related to the biogeochemical turnover features as well: C_b is equal to 10.0, C_t to 0.35 and C_{br} to 3.5.

Sub-Boreal Forest ecosystems

In North America there are two separate parts of Sub-Boreal Forest ecosystem, belonging to the following geographical regions: the Atlantic North American region and the North Pacific region. Both regions lie between 50° N and 38° N. In the interior they adjoin the Boreal Taiga Forest ecosystems and intra-continental, Forest Meadow Steppe ecosystems of this continent.

Atlantic North American geographical region

Despite the wide occurrence of calcareous glacial, aqueous and deluvial deposits, most soils of the region do not contain carbonates in the soil profiles. Rendzinas formed on moraines rich in limestones or directly on the eluvium of limestones, found in the lacustrine northern part of the region, are an exception. The deep migration of carbonates is associated with intense leaching of the soils and the absence of any period of drying or freezing.

The northern part of the region is occupied by the South Canadian Birch-Beech-Maple Forest ecosystems, which has been largely felled. Large South Taiga Pine Forest ecosystems mixed with Broad-Leafed Forest ecosystems are preserved to a great extent. Cambisols are the main soil type that occurs most widely under Broad-Leafed Forest ecosystems in the plains as well as in the low hill belt.

Biogeochemical cycle is moderate, the leaching processes are predominate and acidity due to acid deposition of nitrogen and sulfur species is of great environmental concern in this region.

Pacific coast geographical region

This region consists of the west coast of North America between 45° N and 55° N. It includes the island chain—the Alexander Archipelago, the Queen Charlotte Islands and Vancouver Island. Mountains rise here to a height of about 2,000 m.

The region lies in the path of westerly cyclones. High mountain ranges obstruct the path of moisture-laden air masses and average rainfall is 2,500 mm a year.

Piedmont plains and mountain slopes up to a height of 1,200–1,500 m are covered with thick forest and grass cover. Different Cambisols with acid reaction are found here.

Biogeochemical turnover is characterized as follows: C_b is equal to 1.6, C_t to 0.67 and C_{br} to 1.1, i.e., this cycling is rather intensive.

North American Forest-Meadow Steppe ecosystems

These ecosystems fall entirely within the continent. It is a crescent opening to the south, with the northern fringe at 50–58° N; the southern tip of the long western arm of the crescent is at 38° N. The other boundary of the Forest-Meadow Steppe ecosystems coincides with the boundaries of the Boreal Taiga and Sub-Boreal Forest ecosystems. The inner edge of the crescent borders the closed Steppe ecosystems of the continent.

Central Cordillera geographical region

High mountain ranges constitute this region and this determines the development of soil-ecosystem complex. Distric Cambisols with acid reaction occur on well-drained mountain slopes. Eutric Cambisols, neutral or slightly alkaline, with a carbonate horizon in the lower part of the profile, occur on less moistened slopes under Dark Coniferous ecosystems. In Eurasia analogs of such soils are found in Tien Shan.

Due to high precipitation rate and high summer temperatures, the biogeochemical cycling is intensive.

South Canadian geographical region

This is a zone of poorly drained alluvial-lacustrine plains, where features of continental salt accumulation are clearly expressed. Broad-Leafed Aspen Forest and Meadow Steppe ecosystems on Phaeozems and Chernozems with association with meadow-bog soils and Solods are characteristic of this region.

The biogeochemical cycles of elements are intensive (see Tables 1 and 2).

Great Plain geographical region

This region forms the southern part of the Forest-Meadow Steppe ecosystems and falls entirely in the Sub-Boreal thermal belt (see Figure 2). Planosols with complexes of Solods and Solonchaks are common here. The natural vegetation under which these soils develop is Tall-Grass Prairie ecosystems, consisting of perennial grasses, with deep root system, mainly from the genus *Andropogon (Andropogon furcatus, A. scoparius, A. gerardi)*, with considerable participation of luxuriant mixed grass, forming a thick closed cover, at places as tall as a man.

Humus content is about 3–5% and in Wisconsin and Iowa it comes up to 10% and this layer plays an important role in biogeochemical migration of various elements. Biogeochemical cycle is intensive.

North American Steppe ecosystems

The vast spaces of the Great Plains in the interior of Southern Canada and the USA, protected to the west by the Cordillera, constitute extensive Steppe ecosystems. The aridity of climate increases from north to south and from west to east. Vegetation and soils coincides with P:PE ratio. The outer, somewhat better moistened, areas represent the Fescue-Couch Grass ecosystems and Feather Grass Steppes on Chernozems. The more westerly dry part constitutes the Short-Grass Steppe ecosystems of the plateau of the Great Plains of tirfaceous grass and dense-tirfaceous Steppe ecosystems on Chestnut soils (Kastanozems).

Biogeochemical turnover is intensive (see Table 1).

West American Desert-Xerophytic Forest ecosystems

These ecosystems cover territories with varied relief. The larger, western part is represented by high uplands and the ranges of the Central and Southern Cordillera.

In the north this is the Great Basin, with the Coast range and the Rocky Mountains bounding it; in the south the Colorado plateau, and still farther south the Mexican Plateau with the western and eastern Sierra Madre bounding it.

Differences in altitude and degree of isolation from moisture-laden air masses, together with diversity of geological rocks have led to the diversity of soils and ecosystems.

West American geographical region

This region is located in the northern part of the area of West American Desert-Xerophytic Forest ecosystems, on the borderline between temperate and subtropical belt (see Figure 2). The lower slopes in this region are covered with Big Tree Forest ecosystems (*Sesquoia semperviverens*) on acidic Cambisols; above, in drier regions, are found Pine Forest (*Pinus ponderosa*) ecosystems and Stiff-Leafed Oak Forest and Shrubs (formation of chaparral) ecosystems on Chromic Cambisols.

In the Rocky Mountains we can see internal-drainage basins, i.e., closed landscape-geochemical areas, in which not only products of physical weathering (proluvial, old-lacustrine and alluvial deposits) but also readily-soluble salts are found.

A large number of salt lakes and areas of Solonchaks deserts surrounding them are characteristic to the central basin of this region. The background plains are formed of Wormwood Desert ecosystems on Eutric Arenosols—analogs of the soils of Semi-Desert ecosystems of Kazakhstan and northern Mongolia from Central Eurasia.

Biogeochemical cycle is intensive, but soil and ecosystems are sustainable to different types of pollutants, like acid rain, due to high buffering cation exchange capacity of soil.

Texas geographical region

In Texas geographical region the main part of these ecosystems is situated in sub-tropical climatic belt with development of Chromic Cambisols, Calcari-Chromic Cambisols and Kastanozems (Table 1, Figure 2). In spite of intensive type of biogeo-chemical cycling (mean C_b is 0.8; C_t is 0.95; C_{br} is 0.8), the high buffering capacity of the above mentioned soils combined with low P:PE ratio influences insensitivity of the given ecosystems to actual and potential pollution loads, including acidic deposition. The exceptions are high mountains with Coniferous Forest ecosystems.

Mexican-Californian geographical region

In Mexican-Californian geographical region, we can see the similar pattern as in the previous region: very intensive type of biogeochemical cycling (mean C_b is 0.3; C_t is 0.95; C_{br} is 0.3) is combined with low P:PE ratio. This leads to the formation of alkaline biogeochemical barriers in soil profiles of dominant Regosols and Xerosols.

1.3. Latin and South America

At the territory of Latin and South America the ecosystems of sub-boreal, subtropical and tropical climatic belts are presented.

Xerophytic Savanna-Forest ecosystems

These ecosystems are widespread in Caribbean, Central Brazilian, East Brazilian and Fore-Andean geographical regions.

Caribbean geographical region

Caribbean geographical region is placed in the northern subtropical part of South America and in the Antilian Islands. Their occurrence in subtropical and tropical belts influences the seasonal rhythm of precipitation and corresponding changes of biogeochemical cycling rates from intensive rank during dry period ($C_b > 0.5$) into very intensive rank ($C_b < 0.3$) during wet season. The predominant soil types are sensitive to acidic deposition Ferrasols and Arenosols, relatively insensitive Nitosols and Rhodic Ferrasols and insensitive humus Rendzinas and Vertisols, for example, in the Pinos island and the Oriente province of Cuba.

Central Brazilian geographical region

Central Brazilian geographical region occupies the inner part of Brazilian hilly land. The dominant ecosystem types are Tropical Savanna, Woodland and Shrub (Campos serrados) and Open Savanna (Campos limpos). In river valleys, there are scattered plots of tropical forests. The soil types are represented by Regosols, Ferralitic Arenosols and Luvi-Plinthic Ferrasols. Due to tropical climatic conditions complicated by seasonal precipitation during winter, the biogeochemical cycling is very intensive and does not depend on soil type ($C_b < 0.2$; C_t is 1.00; $C_{br} < 0.2$).

Eastern Brazilian geographical region

Eastern Brazilian geographical region is placed in a vast depression of San-Francisco river basin, being protected from an influence of wet equatorial deposition. The climate is dry subtropical. The main soil types are represented by strong leached Ferrasols with very low fertility and Arenosols of light mechanical composition. The biogeochemical cycling is characterized as potentially very intensive but due to very low P:PE ratio most ecosystems have the depressed migration of many elements and their accumulation in upper soil horizons.

Fore-Andean geographical region

Fore-Andean geographical region occupies the inner dry plains and low hills of the eastern slopes of the Andes. In spite of very intensive type of biogeochemical cycling connected with very high temperature regime and moderate deposition, the high buffering ability of Calcic Cambisols and low P:PE ratio restrict the migration of chemical species in these ecosystems.

Tropical Rain Forest ecosystems

These are presented in four geographical regions of Latin and South America.

Central American geographical region

Central American geographical region being situated between Atlantic and Pacific oceans is characterized by very complicated bioclimatic, geological and biogeochemical conditions. First, this is related to the relief differences in the eastern and western parts of the given region and these cause the alteration of soils and vegetation species. Correspondingly, the biogeochemical migration and sensitivity to pollution impact reflects mainly the predominant soil type: the least migration is in Calcaric Ferrasols and Chromic Cambisols whereas the most pronounced biogeochemical migration in Tropical Rain Forest ecosystems coincides with Xantic Ferrasols. The biogeochemical migration of elements in the latter ecosystems is very much enforced by very intensive biogeochemical cycling of elements ($C_b \leq 0.1$; $C_t = 1.00$; $C_{br} \leq 0.1$).

Andean equatorial geographical region

Andean equatorial geographical region is placed in the west-eastern part of South America with wet tropical forests on predominant mountainous Andosols, Cambisols and various Ferrasols. The complex soil-ecosystem characteristics of the given mountainous region cause the different values of biogeochemical cycling coefficient varying from 0.1 to 0.8 in dependence upon the elevation and predominant soil type.

Amazonian geographical region

Amazonian geographical region is the largest one in South America with predominant Acric Ferrasols.

Very low buffering capacity, low pH and low humus content characterize these soils. The rates of biogeochemical cycling can be ranked as 'very intensive' with values of $C_b < 0.1$. These combinations of soil features and peculiarities of biogeochemical cycling make the given region very sensitive to input of various pollutants, including acid forming compounds.

Brazilian Atlantic geographical region

Brazilian Atlantic geographical region is placed on the Brazilian hilly plain and its eastern slope. The southern part of this region is characterized by subtropical climate, the active temperature coefficient is a little bit less than 1.0 and biogeochemical cycling coefficient is a little bit higher than in typical Tropical Wet Forest ecosystems (C_b is 0.15). In the northern part the very low fertile Chromic Luvisols are predominant in Tropical Forest ecosystems and in the southern part the coniferous forests grow on Nitosols.

Meadow Steppe ecosystems

The East Pampa geographical region is located in subtropical belt. The natural vegetation is represented by tall grass meadow steppe with Haplic Phaeozems and Vertisols having neutral reaction and high base saturation. These ecosystems are characterized by intensive type of biogeochemical cycling. The combination of these feature allows the conclusion on insensitivity of these natural ecosystems and their anthropogenic modifications to acidic deposition and other pollutants.

Desert-Steppe ecosystems

The ecosystems develop under dry and hot climate on various Xerosols, Regosols, Solonchaks and Calcic Cambisols in the southern subtropical part of the Andean and Patogonian geographical regions. In accordance with chemical properties of these soils, very low P:PE ratio and in spite of very intensive type of biogeochemical cycling these ecosystems are restricted in migration of chemical elements.

Sub-Boreal Forest ecosystems

These ecosystems are in South Chilean geographical region with predominant acid Andosols and intensive rank of biogeochemical cycling: the mean value of C_b is 1.2, the mean value of C_t is 0.75, and the mean value of C_{br} is 0.9. These ecosystems (Deciduous Broadleaf Forest and Steppe) are sensitive and very sensitive to input of pollutants, especially acid deposition.

1.4. Africa

The main characteristic feature of soil-ecosystem distribution in Africa is connected with symmetric tropical and subtropical climatic belts in the North and South Hemi-spheres.

Tropical Rain Forest ecosystems

These ecosystems are represented in Africa by the Congo-Guinean geographical region. The biggest part of this region is covered by Tropical Rain Green Forests with predominant Ferrasols. The biogeochemical cycling rate is very intensive ($C_b <$ 0.1) that in combination with acid reaction of soils and low buffering capacity allows us to consider these ecosystems as very sensitive to both actual and potential pollutant impacts.

Woody Savanna-Dry Savanna ecosystems

These are vastly widespread in the African continent being totally about 35% of the area. The dominant soil types are represented by Ferrasols.

Sudan-Guinean geographical region

The Sudan-Guinean geographical region is located in the North Hemisphere. In accor-dance with annual P:PE ratio changing in the limits 0.6–1.0, low buffering capacity of Ferrasols, low pH values and very intensive type of biogeochemical cycling, the sustainability of these ecosystems to pollutant input is moderate.

Somali-Yemeni geographical region

Somali-Yemeni geographical region is characterized by such soil-ecosystem features that allows the consideration of the whole region as sustainable to anthropogenic inputs like acidic deposition: the predominant soils are Vertisols, Rhodic Ferrasols and Nitosols combined with low P:PE ratio. The very intensive type of biogeochemical cycling does not change this insensitivity.

East African geographical region

East African geographical region is placed in high hilly plains and low mountains of the East equatorial Africa with very complex climate, relief and soil forming rocks. The predominant soils are Ferrasols with low humus content, low buffering capacity and weak acid reaction. In depressions Vertisols are widespread being in soil eatenas with Ferrasols on more elevated part of slopes. The type of biogeochemical cycling is very intensive and being combined with above mentioned soil features, it leads to moderate sustainability of biogeochemical structure to anthropogenic pollution.

The ecosystems of Madagascar Island are represented by wet tropical forests located on the eastern slope with very acid low fertility soils and by savannas in the western slopes with Cambisols. So the biogeochemical migration of elements in western ecosystems is restricted, whereas that in the eastern tropical forest is reinforced by very intensive type of biogeochemical cycling (C_b is 0.1–0.2).

Angolo-Zimbabwean geographical region

Angolo-Zimbabwean geographical region is located in South Africa with predominant Rhodic Ferrasols and Nitosols characterized by acid pH, low buffering ability, low humus content and very intensive type of biogeochemical turnover (C_b is 0.2) under annual precipitation 1200–1300 mm in its northern part. So, this combination of soil-ecosystem features leads to actual and potential damage of acidification or any other airborne anthropogenic loading. This damage is less profound in the southern part of the given region where biogeochemical processes are restricted by relatively less P:PE ratio and salinization of Cambisols and Vertisols.

South an North East African geographical regions

South East African geographical regions are located partly in subtropical belts. The predominant ecosystems are Rain Green Tropical Forest with Ferrasols and subtropical Steppe with Rhodic Ferrasols and Vertisols. The biogeochemical cycling rate is intensive (C_b is 0.9).

Desert ecosystems

These ecosystems are in South-African and Saharan geographical regions. In spite of very intensive type of biogeochemical cycling (C_b values are equal to 0.2–0.3), very low P:PE ratio and salt accumulation in upper soil layers present a barrier in migration of chemical species.

Mediterranean Xerophytic Forest-Shrub ecosystems

These are situated in North-African geographical region (Figure 2, Table 1) with predominant Rhodic and Haplic Nitosols, Calcic Cambisols and subtropical Rendzinas. The given soils are predominantly insensitive to acidic deposition but very intensive type of biogeochemical cycling can change this insensitivity under the combination of definite conditions: high elevation, mixed broad leaf forests and mountainous meadows, annual precipitation up to 400 mm, Cambisols, etc.

1.5. Australia

The ecosystem and soil-biogeochemical mapping of the Australian continent is primarily determined by its position in the systems of latitudinal climatic belts of the Earth. The Tropic of Capricorn practically cuts the continent in two parts, so much of the territory is located in the region of the tropical pressure maximum. This determines the predominance of landscapes of tropical sandy and rocky deserts and semi-deserts, occupied by Sclerophilic Brushwood Savanna and Desert Brushwood ecosystems, in combination with Halophitic Brushwood ecosystems on soils with varying degree of leaching, calcareousness and salination.

Only the northernmost part of Australia (Cape York, Kimberley and the Arnhem Land peninsula) falls in the belt of equatorial monsoons. It is characterized by Tropical Rain Forest, Afforested Savanna and Light Forest ecosystems on Ferrasols and Nitosols, at places, lateritized soils. The southern fringe of the continent, its southwestern and southeastern parts, and also the island of Tasmania fall in the subtropical zone in the belt of westerly cyclonic currents of air masses. They experience winter in the Southern Hemisphere. A dry summer and winter rains impart the characteristics of a 'Mediterranean' subtropical climate and determine the appearance of Dry Forest and Brush ecosystems on different Cambisols.

Savanna-Xerophytic Shrub ecosystems

These ecosystems occupy a large part of Australia. This area is concentric in shape and is bounded to the north, east and southwest by the most arid regions of Central Australia. Paleoclimatogenic and paleohydrogenic macrostructures, which substantially obscure the bioclimatic modern ecosystem mapping, are characteristic of this area. Four geographical regions are distinguished in these ecosystems: North Australian, East Australian, South Australian and West Australian.

North Australian geographical region

This region influences the equatorial monsoon and almost the entire annual rainfall comes in four or five summer months. The Xerophitic Eucalyptus Forest and Light Forest are common for this area and they are replaced in less-drained positions by herbaceous associations. The typical soils are Ferrasols and Nitosols. The biogeochemical cycling is very intensive (see Table 2).

East Australian geographical region

This region stretches in the meridional direction along the western slopes of the Great Divide Range, covering the adjoining piedmont plains. The predominant ecosystems are represented by Wet Afforested Savannas, which alternate with Open Grass associations and 'Brigalow' associations. Soil cover is complex, from Chernozem-like soils, Solonetses and Solods up to Vertisols. The biogeochemical cycling is very intensive: C_b values are equal to 0.2–0.3.

South Australian geographical region

This region covers the aggradational plains of the lower course of the Murray River, the internal-drainage parts of the basins of Lakes Eyre and Torrens, the Nullabor plain and Eyre peninsula. The Mediterranean type of climate, maritime location of the territory, the presence of a series of accumulative maritime and river terraces and old-aggradational and lacustrine plains, the prolonged eolian transport of salts from the ocean and the earlier superaqual regime in aggradational plains have determined the intensive biogeochemical cycling in local Xerophitic Savanna-Shrubs ecosystems.

West Australian geographical region

The region lies in the west Australian peneplain, composed of Precambrian granite-gneisses and diorites. The highest parts of the peneplain are much dissected and devoid of soil cover. Although the middle level of the peneplain was subject to dissection, a ferralitic kaolinitic weathering crust, often topped with lateritic armor, has been preserved on the surface of the vestigial mountains. The surface of the peneplain is dotted with numerous flat closed basins, occupied by periodically dry salt lakes surrounding by Solonchaks. In the dry season crystals of gypsum and other salts, carried by wind, form peculiar salt dunes among the shores of the lakes. The salts are also transported by wind to neighboring denudation surfaces. Biogeochemical cycles of elements are very intensive (C_b is 0.2).

Central Australian Desert ecosystems

The most distinct features of the soil cover of the Semi-Desert and Desert ecosystems of tropical Australia stem from the very wide occurrence of highly laterized old kaolinitic weathering crust, which is silicified in some places. In denudation plains with the weathering crust lying *in situ* or thinly covered by local products of erosion in the upper part, the soils have a bright red color, are non-calcareous and give an acidic or neutral reaction throughout the profile.

Montmorillonitic compact soils, generally calcareous, and in some places solonchakic too, occur on the accumulative surfaces of the old peneplain, in old-lacustrine and aggradational plains, particularly developed in the basins of the Diamantine River and Coopers Creek. Large areas of central Australia are occupied by Sandy Desert ecosystems. The biogeochemical cycling in these ecosystems are very similar to the cycling in the Desert ecosystems of central Eurasia (see above).

We can see interactions of various soil-biogeochemical conditions with relevant ecosystems in Figure 3.

More details in the description of soil geographical regions of the World can be found in Glazovskaya (1982), see Further Reading.

2. BIOGEOCHEMICAL CLASSIFICATION AND SIMULATION OF BIOSPHERE ORGANIZATION

2.1. *Biogeochemical classification of the biosphere*

Biogeochemical mapping is the scientific method for the understanding of the biosphere structure. This method is based on the co-evolution of geological and biological parameters of the biosphere and deals with the quantification of the interrelations between biota and the environment in consequent biogeochemical food webs. The fundamentals of biogeochemical mapping have been intensively developed during the twentieth century in Russia and other countries. Biogeochemical mapping joins the definitions of soil zones (Dokuchaev, 1948) and provinces (Prasolov, 1938), geochemical provinces (Fersman, 1931), biogeochemical provinces (Vinogradov, 1938), climatic zones (Berg, 1938), geochemical landscapes (Polynov, 1946), biogeocenoses (Sukachev, 1964), ecosystems (Odum, 1975), and geochemical ecology (Kovalsky, 1974).

Using these approaches, the biosphere mapping is connected with such units as region, sub-region and biogeochemical province.

Regions of the biosphere are the first order units for mapping of the global biosphere. These regions have the geographic features of soil-climate zones and their combinations. However, the predominant parameters are connected with general qualitative and quantitative conditions of biogeochemical food webs (biogeochemical cycles) and prevalent biological reactions of organisms to the natural chemical composition of the environment. These reactions include the alteration of chemical composition of organisms, element and energy exchange, physiological sensitivity to deficient or excessive content of different chemical species, endemic disease, etc. Every region of biosphere is subdivided into sub-regions of biosphere.

Sub-regions of the biosphere are the second order mapping units. There are two main groups of sub-regions:

1. Sub-regions of the biosphere with the typical combinations of characteristic features of regions, such as the limit concentrations of essential elements, their biochemical and physiological relationships and possible biological reactions;

2. Sub-regions of the biosphere, which feature do not correspond to the general characteristics of regions. These sub-regions occupy the areas over the ore deposits, in depressions with no runoff, in zones of active volcanic activity, and the territories of anthropogenic pollution.

Biogeochemical provinces are the third order units of the biosphere. These are the different areas with the permanent characteristic biological reaction of organisms on

ZONE	COLD DESERT	POLAR DESERT	TUNDRA	BOREAL FOREST	TEMPERATE FOREST CONIFEROUS	TEMPERATE FOREST DECIDUOUS
Soil Order	Entisols	Entisols	Inceptisols	Spodosols	Inceptisols	Alfisols
Climate — Moisture	ultra xeric ultra	ultra xeric	aridic	udic	udic	
Climate — Temperature	pergelic	pergelic	cryic	frigid	mesic	
Specific impact of climate and surface conditions on the horizons — O, A,E, Bh			under-saturated non-mobile organic acids	very under-saturated mobile organic acids	under-saturated non-mobile organic acids	
B		Fe-hydroxide neoformation	smectite, vermiculite formation	non-crystalline aluminosilicates and Fe-hydroxide neoformation	smectite, vermiculite formation by transformation of preexisting phyllosilicates	
C	CaCO$_3$ precipitation	CaCO$_3$ precipitation	Fe-hydroxide neoformation	Fe-hydroxide neoformation	Fe-hydroxide neoformation	
R	nil	nil	nil	nil	nil	
Speed of pedogenesis	extremely slow	very slow	slow	moderate	moderate	

COMPARTMENT

organic and mineral
Horizons: O, A, E,Bh

mineral, upper part
Horizon: B

mineral, middle part
Horizon: C1

mineral, lower part
Horizon: C2

Horizon: C3

(a)

GRASSLAND	DESERT	SAVANNA TREELESS	SAVANNA ARBOREAL	TROPICAL RAINFOREST
Mollisols	Aridisols	Vertisols	Ultisols	Oxisols
ustic	aridic	ustic	udic	perudic
mesic	thermic	isothermic		isohyperthermic
saturated non-mobile organic acids		saturated non-mobile organic acids	under-saturated organic acids	very under-saturated mobile organic acids
Fe-hydroxide neoformation CaCO$_3$ precipitation	Fe-hydroxide neoformation CaCO$_3$ precipitation	smectite neoformation	kaolinite neoformation	Fe- and Al-hydroxide neoformation
CaCO$_3$ precipitation	CaCO$_3$ precipitation	CaCO$_3$ precipitation	Fe-hydroxide neoformation	kaolinite neoformation Fe-hydroxide neoformation
nil	nil	nil		
moderate	slow	fast		very fast

Note: not drawn to scale

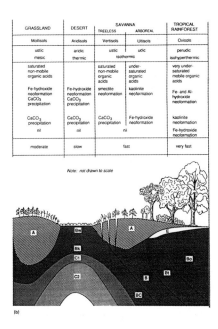

(b)

Figure 3. Soil and soil-forming processes: a global view (after Ugolini and Spaltenstein, 1992).

the excessive or deficient content of essential nutrients in the biogeochemical food webs. The typical forms of these reactions are endemic diseases. There are two types of biogeochemical provinces:

1. The first type represents the natural biogeochemical provinces developed in the areas where the contents of various chemical species in food webs differentiate from those indexes that are characteristic for the surrounding sub-region of biosphere. The development of these natural biogeochemical provinces has been during a long-term co-evolution of geosphere and biosphere.

2. The second type represents the anthropogenic biogeochemical provinces where the excessive content of many chemical species occurred due to industrial or agricultural human activities, like ore explorations, intensive emission of pollutants, application of fertilizers, improper irrigation, etc.

The first step in subdividing the biogeochemical provinces is the existence of correlation between the frequency of the endemic diseases and excessive or deficient content of essential elements in food webs. However, we should also understand the biochemical and physiological mechanisms that drive the given correlation. Four additional notions should be applied:

a) *physiological adaptation curve*;

b) *lower limit concentration;*

c) *optimum concentration;* and

d) *upper limit concentration.*

Physiological adaptation curve is the relationship between an activity of physiological and biochemical functions of organisms and the content of essential elements in biogeochemical food webs. The lower limit concentrations are those that induce the abnormal development of organisms due to deficient content of essential chemical species, and upper limit concentrations are related to those that cause the abnormal development of organisms owed to excessive content of elements. Thus, the limit concentrations show the interval of optimum physiological and biochemical development of organisms in connection with the geochemical conditions (Figure 4).

The physiological and biochemical studies are a necessary step in biogeochemical mapping. Together with geochemical monitoring of essential chemical elements in various environmental media, this can highlight the quantitative aspects of biogeochemical cycling in different biogeochemical provinces. For carrying out biogeochemical mapping we should analyze the content of essential macro-nutrients and trace elements in soil, surface and ground waters, river and lake sediments, vegetation, fodder crops, animals, food, drinking water and human and animal excretions. Using GIS technologies these data are mapped with overlapping various layers and make

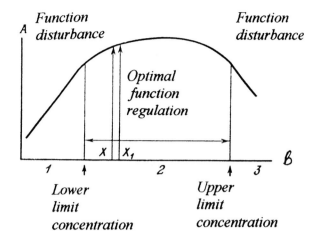

Figure 4. Dependence of the biochemical and physiological processes in the organisms from the content of essential chemical elements in the biogeochemical food webs as a physiological adaptation curve. 1—lower (deficient) content; 2—optimum content, 3—excessive (upper) content.

the final map of biogeochemical provinces, where the endemic diseases correlate to the deficient or excessive content of essential elements in food webs.

2.2. Methodology of biogeochemical cycling simulation for biosphere mapping

The simulation of biogeochemical cycles of chemical species is the basis of the modeling biogeochemical structure of the Earth's ecosystems and biogeochemical mapping. The general scheme of algorithm for simulation of biogeochemical cycles of various chemical species is shown in Figure 5.

We will consider this scheme in details. Each system will be described as a combination of biogeochemical food webs and relationships between them.

System 1: soil-forming rock (I); waters (II); atmosphere (III); soil (IV). This system would not been active without the living matter.

System 5: soil-forming rock (I); soil, soil waters and air (IV); soil microbes (bacteria, fungi, actinomicetes, algae) (V); atmospheric air (III, 25). The activity of this system depends on the activity of living soil biota (V). We can refer to Vernadsky (1932) here: "There is no other relation with the environment, i.e., abiotic bodies, except the biogenic migration of atoms, in the living bodies of our planet". During the consideration of the system organization of the biogenic cycle of a chemical species, the relationship between various links (I, II, V) and the subsequent mechanisms of causal dependence are estimated. Most attention should be paid to the biogeochemistry of soil complex compounds, which include the trace met-

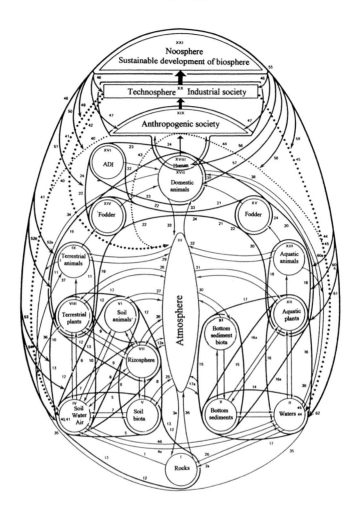

Figure 5. General model of biogeochemical cycles in the Earth's ecosystems. The left part is biogeochemical cycling in terrestrial ecosystems, the right part is aquatic ecosystems and the central part is connected with the atmosphere. The fine solid lines show the biogeochemical food webs (the Latin number I–XXI) and directed and reverse relationships between these webs; the thick solid lines show the primary systems of biogenic cycling organization, usually joining two links of biogeochemical food web, for instance, 7, 11, 18, etc, and secondary more complicated complexes of primary systems, for instance, counters 12, 13, 19, 17, 20, etc.; fine dotted lines show the stage of initial environmental pollution, for instance, soils, 40, waters, 44, air, 43, due to anthropogenic activities; the thick dotted lines show the distribution of technogenic and agricultural row materials, goods and wastes in biosphere, for instance, in soils, 41, in air, 42, in waters, 45, leading to the formation of technogenic biogeochemical provinces; the different arrows show the social stages of human activity, from human being up to the noosphere (After Kovalsky, 1981).

als. The organic substances exuded to the environment by the living organisms are of the most importance. The chemical substances from decomposed died matter play the minor role in biogeochemical migration of chemical species. The vital synthesis and excretion of metabolites, bioligands, is the main process of including chemical species from geological rocks into biogeochemical cycle. When the trace elements are inputting into a cell in ionic form, the formation of metal-organic compounds inside the cell is the first step in the biogeochemical cycle. Ferments, metal-ferment complexes, vitamins, and hormones stimulate the cell biochemical processes. After extraction of metabolites into soil, the formation of soil metal-organic complexes proceeds. These complexes are subjected to further biogeochemical migration.

System 7: soil-soil waters, air (IV); atmosphere air (III, 26); roots-rizosphere microbes (VII); microbiological reactions-metabolisms (VII). The root exudes and microbes of rizosphere provide the organic compounds for the extra-cellular synthesis of metal-organic compounds. Plants can selectively uptake these compounds making the specificity of biogenic migration. This specificity has been forming during plant evolution in definite biogeochemical soil conditions.

System 7, 9, 10: roots-rizosphere (VII); plants (VIII); their biological reactions—metabolism (VIII); soil-soil solution, air (IV); aerosols–atmospheric air (26, 28). In this system, the influence of metal-organic complexes on the plant development and their metabolism is considered. Under deficient or excessive contents of some chemical species, the metabolism may be destroyed (see Figure 4).

System 6: soil-soil solution, air (IV); atmospheric air (III, 27); soil animals (VI); biological reactions of organisms, metabolism, exudes, including microbial exudes (VI); into soils (VI → IV); into waters (II, 4b); into air as aerosols (III, 27). This system is very important for biogeochemical mapping but until now it has not been understood quantitatively.

System 12, soil cycle: soil-forming geological rocks (I); soil (dynamic microbial pattern) (IV); soil solution, air (IV); atmospheric air (III) (aerosols—3a, 3b, 12a, 25, 26, 27); soil organisms, their reactions, and metabolism (V,VI,VII). We should consider the content of essential trace elements in the atmospheric aerosols, both gaseous and particulate forms. These aerosols originate both from natural processes, like soil and rock deflation, sea salt formation, forest burning, volcanic eruption and from human activities, like biomass combustion, industrial and transport emissions. The processes are complicated because of the existence of metal absorption from air and desorption (re-emission) from plant leaves. The first process was studied in more detail. But the second process has not been understood quantitatively and even qualitatively at present. The experimental data in vitro with plant leaves showed the emission of radioisotopes of zinc, mercury, copper, manganese and some other metals. The rates of re-emission are very small, however the fluxes may be significant due to much greater size of leaf surface areas in comparison with soil surface area. For instance, the leaf area of alfalfa exceeds the soil surface 85 times, and that for tree leaves is greater $n \times 10\text{–}10^2$ times. Furthermore, animals and human beings can also absorb trace metals from air as well as exhale them.

System 10: soil (IV); plants (VIII); their biological reactions, endemic diseases (VIII); atmospheric air, aerosols (III, 28). During consideration of System 7-9-10, we have discussed the influence of the lower and upper limits of concentrations on the plant metabolisms, including the endemic disease. The study of link (VIII) should start with the correct selection of characteristic plant species. The following steps should include the different research levels, from floristic description up to biochemical metabolism.

System 13: soil-plant cycle: soil-forming geological rocks (I); soil (IV); soil living matter (community of soil organisms) (V,VI,VII); aerosols, atmosphere air (12a, III); plants (VIII); their biological reactions, endemic diseases (VIII). In the complex system 13, the inner relationships and biochemical and biogeochemical mechanisms are shown for natural and agroecosystems. The system 11 and link IX show the ways for interrelation of system 13 with terrestrial animals.

System 11: terrestrial plant (VIII); wild terrestrial animal (IX); aerosols, atmosphere air (28,29); biological reactions (VIII, IX). System 7-9-10 considers the biological reactions of terrestrial plants on deficient or excessive content of essential elements. System 11 includes the new link of biogeochemical migration, terrestrial animal (IX). The terrestrial plants play the most important role in this biogeochemical food web linking plant chemical composition with the physiological functions and adaptation of herbivorous animals. The links between herbivorous and carnivorous animals should be also set in the given systems 11. The inner relations between content of elements in fodder crops and their bioconcentration in herbivorous animals are connected with the formation of digestible species in the intestine-stomach tract, the penetration through the tissue membranes (suction) with further deposit and participation in metabolism as metal-ferment complexes. The accumulated amount will finely depend on the processes of subsequent extraction from the organisms through kidney (urea), liver (bile), and intestine walls (excrements). These processes depend on both the limit concentrations of elements in animal organism and cellular and tissue metabolic reactions. The development of pathological alterations and endemic diseases are related to the combination of metabolism reaction and element exchange. We should again refer to Figure 4 for the explanation of how to determine the relationships between environmental concentrations and regulatory processes in animal organisms. Between lower and upper limits of concentrations, the adaptation is normal, however the hardness of adaptation is increasing with an approximation to both limit values. Some organisms of population may already show disturbance of metabolism and development of endemic diseases, but the alterations of the whole population will be statistically significant only when the concentrations of chemical species achieve the limits. Under optimal concentrations, there is no requirement in improving the element intake.

System 19: soil (IV); terrestrial plants (VIII); terrestrial animal (IX); forage with accounting for technological pre-treatments (XIV). This system shows the dependence of essential element contents from environmental conditions.

System 21: composition and quantity of crops and forage: food and crops of terrestrial origin (XIV); food and crops of aquatic origin (XV). In many countries, the daily intake standards have been set for humans and animals (see Radojevic and Bashkin, 1999).

Sub-systems 21[1]: foodstuffs of terrestrial origin (XIV) + foodstuffs of aquatic origin (XV); drinking water (39); balanced essential trace element daily intake for domestic animals (XVI).

System 22[2]: foodstuffs of terrestrial origin (XIV) + foodstuffs of aquatic origin (XV); drinking water (39); balanced essential trace element daily intake for humans (XVI).

System 23: balanced intake of various essential elements (XVI); atmosphere air (33); domestic animals—their productivity and biological reactions, endemic diseases (XVII); human, biological reactions (XVIII). The recommendations for balanced essential trace element daily intake for humans are under development in various countries.

System 24[1]: feeding of domestic animals, forage (XIV, XV); balanced essential trace element daily intake (XVI); domestic animals (XVII). The additions of requirement trace elements should be applied for forage in various biogeochemical provinces.

System 24[2]: human nutrition, foodstuffs (XIV); balanced essential trace element daily intake for humans (XVI); human health (XVIII). The research should be carried out on the endemic diseases induced by deficient or excessive content in the biogeochemical food webs of different essential elements, like Ni, Cu, Se, I, F, Mo, Sr, Zn, etc.

System 14: geological rocks (1,2a,2b); waters (II); bottom sediments (X). The chemical composition and formation of natural waters and bottom sediments depend tightly on the geochemical composition of rocks.

System 15: bottom sediments (X); sediment organisms and their biological reactions (XI). The invertebrates of bottom sediment are important in biogeochemical migration of many chemical species in aquatic ecosystems.

System 17: bottom sediments (X); sediment organisms and their biological reactions (XI); waters (II); aquatic plants and their biological reactions (XII); atmosphere air (17a, 30, 31). The chemical interactions between aquatic and gaseous phases play an extremely important role in the composition of both water and air. These interactions determine the development of aquatic ecosystems. The example of oxygen content in the water is the most characteristic one.

System 18: aquatic plants and their biological reactions, endemic diseases (XII); aquatic animals, including bentos, plankton, bottom sediment invertebrates, fishes, amphibians, mammals, vertebrates, their biological reactions and endemic diseases (VIII). Bioconcentration is the most typical and important consequence of biogeochemical migration of many chemical species in aquatic ecosystems.

System 20: aquatic plants—bentos, plankton, coastal aquatic plants (XII); aquatic animals including bottom sediment invertebrates, fishes, amphibians, mammals, vertebrates, their biological reactions and endemic diseases (VIII); aerosols, atmospheric air (31, 32)—foodstuffs, forages (XV). Human poisoning through consumption of fish and other aquatic foodstuffs with excessive bioaccumulation of pollutants is the most typical example of biogeochemical migration and its consequences.

System XVIII, XIX; human being (XVIII); human society (XIX): development of agriculture, industry and transport (XIX); accumulation of wastes in soil (40), air (43) and natural waters (44). Increasing accumulation of pollutants in the environment. We have to remember here that from the biogeochemical point of view, pollution is the destruction of natural biogeochemical cycles of different elements. For more details see Chapter 8 'Environmental Biogeochemistry'.

System XX, modern industrialized 'throwing out' society: intensive industrial and agricultural development, demographic flush—pollutant inputs into soil (41), atmosphere (42), natural waters (45) up to exceeding the upper limit concentrations. Development of human and ecosystem endemic diseases in local, regional and global scale. Deforestation, desertification, ozone depletion, biodiversity changes, water resources deterioration, air pollution are only a few examples of the destruction of biogeochemical cycles in the biosphere. These consequences were predicted by Vladimir Vernadsky at the beginning of the 1940s. He suggested a new structure of biosphere and technosphere organization, the noosphere.

System XXI: noosphere—organization of meaningful utilization of the biosphere on the basis of the clear understanding of biogeochemical cycling and management of biogeochemical structure. The Kingdom of Intellect: re-structuring, conservation and optimization of all terrestrial ecosystems using the natural structure of biogeochemical turnover. We can cite as an example the re-cycling of wastes in technological processes and biogeochemical cycles (46, 48, 49,

50, 52a, 52b, 53, 54, 55, 56, 57, 58, 59, 60a, 61, 62), development of regional and global international conventions, like Montreal Convention on Ozone Layer Conservation, Geneva Convention on Long-Range Transboundary Air Pollution, etc., forwarding the juridical regulation of industrial, agricultural and transport pollution (47), protection of soil and atmosphere (42) as well as natural waters (45) from anthropogenic emissions (41).

Field monitoring and experimental simulation allow the researcher to study the variability of different links of biogeochemical food webs and to carry out the biogeochemical mapping of the biosphere in accordance with above-mentioned classification: regions of biosphere, sub-regions of biosphere and biogeochemical provinces.

3. BIOGEOCHEMICAL MAPPING ON CONTINENTAL, REGIONAL AND LOCAL SCALES

In this section we will present a few examples of different scale biogeochemical mapping on the Eurasian continent. This continent has been studied extensively by various Russian and Chinese scientists during the 20th century. We should remember the names of Russian biogeochemists V. Vernadsky, A. Vinogradov, V. Kovaslky, V. Kovda, V. Ermakov, M. Glazovskaya and many others as well as Chinese biogeochemists J. Luo, J. Li, R. Shandxue, J. Hao etc. The most extensive mapping has been carried out in the Laboratory of Biogeochemistry, which was founded by V. Vernadsky in 1932 and during the 1950s-1980s was led by Prof. V. Kovalsky.

3.1. Methods of biogeochemical mapping

Biogeochemical mapping is based on the quantitative characterization of all possible links of biogeochemical food webs, including the chemical composition of soil-forming geological rocks, soils, surface and ground waters, plant species, animals, and physiological excreta of humans, like excretions, urea, and hairs. These food webs include also fodder and foodstuffs. The biochemical products of metabolism of living organisms, activity of ferments and accumulation of chemical elements in various organs should be studied too.

The subsequent paths of biogeochemical migration of elements on local, regional, continental, and global scales can be figured in a series of maps with quantitative information on content of chemical elements in rocks, soils, natural waters, plants, forage crops, foodstuffs, in plant and animal organisms. The distribution of biological reactions of people to the environmental conditions should be also shown. The geological, soil, climate, hydrological, and geobotanic maps can be considered as the basics for the complex biogeochemical mapping of the different areas. The resultant maps are the biogeochemical maps at various scales. The application of statistical information on land use, crop and animal productivity, population density, average regional chemical composition of foodstuffs and fodder crops, and medical statistics on endemic diseases, will be very helpful.

These mean that biogeochemical mapping requires a complex team of various researchers in fields of biogeochemistry, geography, soil science, agrochemistry, biochemistry, hydrochemistry, geobotany, zoology, human and veterinary medicine, GIS technology, etc.

According to the intended purpose, biogeochemical maps can be drawn for different areas, from a few km^2 (for instance, 20–30 km^2 for the mapping of Mo biogeochemical province in mountain valley in Armenia) up to many thousands of km^2, like boron biogeochemical region in Kazakhstan. The biogeochemical maps can be widespread up to the level of continent or the whole global area. The scale of these maps can vary from 1:50,000–1:200,000 for large scale mapping of biogeochemical provinces, to 1:1,000,000 for the mapping of sub-regions of biosphere, and up to 1: 10,000,000–1:15,000,000 for the continental and global scale.

The large scale mapping of biogeochemical provinces and sub-regions is quite expensive and to reduce the work expenses, the key sites and routes should be selected on a basis of careful estimation of available information on soil, geological, geobotanic, hydrological, etc., mapping. The remote sensing approaches are of usefulness for many regions of the World.

The correct selection of chemical elements is very important for the successful biogeochemical mapping. The first priority is the mapping of sub-regions and biogeochemical provinces with excessive or deficient content of the chemical species, which are known as physiological and biochemical essential elements. These elements are N, P, Ca, Mg, Fe, Cu, Co, Zn, Mo, Mn, Sr, I, F, Se, B, and Li. In different biogeochemical provinces, the role of chemical elements will vary. The leading elements should be selected according to the endemic diseases and the full scale monitoring of these elements should be carried out. The other chemical species can be studied in laboratory conditions using simulation approaches. The main attention should be given to the analyses of biochemical mechanisms driven by this element and its active forms.

Using GIS technology we can compare the different layers of information with spatial distribution of endemic diseases. For subdividing sub-regions of the biosphere into biogeochemical provinces we must study the permanent biological reactions, endemic diseases and the areas with different chemical composition of plant and animal species. We can also foresee the potential areas of technogenic biogeochemical provinces due to chemical pollution of various regions. The comparison of geochemical background with chemical composition of organisms can give a sufficient basis for mapping of biogeochemical provinces.

Biogeochemical provinces with pronounced excessive or deficient content of chemical species are strongly correlated with the similar variations of biochemical processes in living organisms. This gives rise to various alterations in adaptation of morphological, physiological and biochemical processes. Finally this will lead to the formation of new biological species.

The problems related to the biogeochemical mapping are complicated by the technogenic transformation of natural ecosystems and relevant primary biogeochemical provinces and their transition to secondary biogeochemical provinces. During biogeochemical mapping we must analyze carefully the sources of chemical elements to differentiate the natural factors from the anthropogenic ones. For more detail see Chapter 8 'Environmental Biogeochemistry'.

The application of above-mentioned approaches to biogeochemical mapping will be highlighted below on the examples from North Eurasia.

3.2. Regional biogeochemical mapping of North Eurasia

The first order units of biogeochemical mapping are the regions of biosphere. In Northern Eurasia, the regions of biosphere represent the differences of chemical macro- and trace elements in soils, plant species and corresponding endemic diseases. These regions are structural parts of biosphere with high level of ecosystem organization and similar typical peculiarities of ecosystem development. Furthermore, the regions of biosphere are subdivided into sub-regions and biogeochemical provinces (Table 4).

On the basis of data presented in Table 4, a biogeochemical map of Northern Eurasia (the former USSR area) has been carried out (Figure 6).

3.3. Biogeochemical mapping of the South Ural region, Russia

This territory occupies the watershed between rivers Ural and Sakmara, with total area of 14890 km^2. This region is a weakly hilly plain, 200–500 m above sea level. The soil-forming geological rocks are metamorphic, basic and ancient volcanic deposits. The main soils are Chernozems and Kastanozems. In this area, there are two large copper ore deposits with admixtures of other non-ferrous metals, like Zn, Co, Ni, etc. During 1970–1980, the various links of biogeochemical food webs have been monitored to carry out biogeochemical mapping of this region. The results are shown in Tables 5–7.

The results of biogeochemical monitoring are correlated with endemic diseases and morphological alterations in both biogeochemical provinces. In the Baimak Cu-Zn biogeochemical province, the chlorosis, necrosis, alterations of organs, reduction of flowers, sterility and infertility are shown for *Salvia stepposa, Verbascum phoniceum, Astragalus macropus, Galium verum* and *Phomis tuberosa*. In the Uldybaev-Chalil Ni-Cu-Co biogeochemical province, the chlorosis, necrosis, growth depression, alterations of organs, reduction of flowers, sterility and infertility are shown for the chlorosis, necrosis, alterations of organs, reduction of flowers, sterility and infertility are shown for *Salvia stepposa, Verbascum phoniceum, Astragalus macropus, Galium verum* and *Phomis tuberosa, Salvia stepposa, Verbascum phoniceum, Perethrum multifoliatum, Potentilla humiphusa,* and *Phomis tuberosa*. The biological reactions of animals are related to endemic copper toxicology in the Baimak biogeochemical province and the eye diseases, like disturbance of cornea, and atypical skin diseases in the Uldybaev-Chalil Ni-Cu-Co biogeochemical province. The endemic copper anemia is monitored in humans only in Cu-Zn enrichment in Baimak biogeochemical province.

Table 4. Description of regions of biosphere, sub-regions of biosphere and biogeochemical provinces in the area of the Northern Eurasia.

Chemical elements	Distribution of sub-regions and biogeo-chemical provinces	Content of elements in biogeochemical food webs	Biological reactions of organisms and endemic diseases
		Taiga Forest Region of Biosphere	
Co deficit	Everywhere	Low content of Co in Podsoluvisols, Podzols, Arenosols and Histosols. The average Co content in plant species is ≤ 5 ppb.	The decrease of Co content in tissues; decrease of vitamin B_{12} in liver (tr.-130 ppm), in tissue (tr.-0.05 ppm), in milk (tr.-3 ppm). Synthesis of vitamin B_{12} and protein is weakened. Cobalt-deficiency and B_{12} vitamin-deficiency. The number of animal diseases is decreasing in raw: sheep $\rightarrow$ cattle $\rightarrow$ pigs and horses. Low meat and wool productivity and reproduction.
Cu deficit	Everywhere, but especially in Histosols	Low content of Cu in Podsoluvisols, Podzols, Arenosols and Histosols. The 30% of forage samples contents Cu≤ 3 ppm.	The 3-fold reduction of Cu content in blood, 30–40-fold, in liver; $n \times 10$-fold increase of Fe in liver. The synthesis of oxidation ferments is depressed. The anemia of sheep and cattle was shown.
Cu + Co deficit	Especially in Swamp ecosystems	Low content of Cu and Co in Podsoluvisols, Podzols, Arenosols and Histosols. Declining contents of Cu and Co in forage species (Cu from 3 to 0.7 ppm, Co ≤ 5 ppb)	Depressed synthesis of B_{12} vitamin and oxidation ferments. Cobalt-deficiency and B_{12} vitamin-deficiency complicated by Cu deficiency. The prevalent diseases of sheep and cattle.
I deficit	Everywhere	75% of Podsoluvisols, Podzols, Arenosols and Histosols contain I $<$ 1 ppm, 40% of natural waters contains I from 3 till 0.06 ppb. Low content of I in food and forage stuffs; 75% of forage crops contain I $<$ 80 ppb.	Disturbance of I exchange and synthesis of I-containing amino acids and tiroxine by thyroid gland, decreasing protein synthesis. Endemic increase of thyroid gland, endemic goiter. All domestic animals

Table 4. Description of regions of biosphere, sub-regions of biosphere and biogeochemical provinces in the area of the Northern Eurasia (continued).

Chemical elements	Distribution of sub-regions and biogeo-chemical provinces	Content of elements in biogeochemical food webs	Biological reactions of organisms and endemic diseases
Co + I deficit	In the Upper Volga regions	Co + I deficit in Podsoluvisols and Arenosols. The reducing content of both I and Co in foodstuffs and forage.	The disturbance of I exchange and tiroxine synthesis is decreasing by Co deficit. Endemic increase of thyroid gland and endemic goiter is often monitored in sheep and humans.
I deficit, Mn excess	In the Middle Volga regions	Decreased I and increased Mn content in Podsoluvisols and Arenosols.	Disturbance of I exchange due to its deficit is enhanced by Mn excess. Endemic increase of thyroid gland and endemic goiter.
Ca deficit, Sr excess	South of East Siberia and the Tuva region, mainly in river valleys	Deficit of Ca, P, I, Cu, Co, excess of Sr and Ba, reducing Ca:Sr ratio in Podsoluvisols, Arenosols and Histosols. In forage, Ca content is decreased and that of Sr is increased, reducing Ca:Sr ratio.	Disturbance of Ca, P, and S exchanges in cartilage tissues; disturbed growth and formation of bones (midget growth). Reducing Ca:Sr ratio in bones. Urov's diseases are often monitored in humans and domestic animals; wild animals suffer in young age.
		Forest Steppe and Steppe region of biosphere	
Content of chemical elements and their ratios are close to optimum	Phaerozems, Chernozems and Kastanozems. I deficiency is common in river valleys.	Content of many nutrients is optimal in soils and forage crops; in some places, the I deficiency of P, K, Mn, and I occurs.	Endemic increase of thyroid gland and endemic goiter take place in Phaerozems and Floodplain soils.
		Dry Steppe, Semi-Desert and Desert region of biosphere	
Cu deficit, excess of Mo and SO_4^{2-}	Pre-Caucasian plain, Caspian low plain, West Siberian Steppe ecosystems	Food web disturbances were shown in various ecosystems with Meadow-Steppe, Eustric Chernozems, Solonchaks, Arenosols	The reducing Cu content in the central nervous systems, depressed function of oxidation ferments and activation of catalase, demielinization of the central nervous systems, disturbance of motion, convulsions. Endemic ataxia. Lamb disease is predominant.

Table 4. Description of regions of biosphere, sub-regions of biosphere and biogeochemical provinces in the area of the Northern Eurasia (continued).

Chemical elements	Distribution of sub-regions and biogeochemical provinces	Content of elements in biogeochemical food webs	Biological reactions of organisms and endemic diseases
B excess	Aral-Caspian low plain, Kazakhstan	Brunozems, Solonetses, and Solonchaks are enriched in B, up to 280 ppm. The increased content of B in forage species, up to 0,15% by dry weight	Accumulation of B in animal organisms leads to the disturbance of B excretion function of liver, reducing activity of amilase and, partly, of proteinase of the intestine tract in human and sheep. Endemic boron enterites sometimes accomplished by pneumonia. Human, sheep and camel morbidity
$NaNO_3$ excess	Central Asia deserts	Excess of nitrates in forages	Endemic methemoglobinemia
		Mountain regions of biosphere	
I, Co, Cu deficit	Various mountain regions: Carpathian, Caucasian, Crimea, Tien-Shan etc.	Mountain soils	Endemic increase of thyroid gland and endemic goiter, Cobalt-deficiency and B_{12} vitamin-deficiency.
	Azonal sub-regions and biogeochemical provinces, which features differ from the typical features of regions of biosphere		
Co excess	North Azerbaijan	Co enrichment of Kastanozems and Brunozems, and forage pasture species.	Excessive synthesis of B_{12} vitamin
Cu excess	South Ural and Bashkortostan	Cu enrichment of Chernozems, Kastanozems of Steppe ecosystems and Podsoluvisols of Forest ecosystems. High Cu content in food and forage stuffs.	Excessive accumulation of Cu in all organs. Progressive exhaustion. Endemic anemia and hepatitis. Sheep diseases. Human endemic anemia and hepatitis
Ni excess	South Ural and North Kazakhstan	Kastanozems, Solonetses with Ni-enriched soil-forming rocks. 20-fold increase of Ni content in forage pasture species.	Increasing content of Ni in all tissues, especially in epidermal tissues. Excessive accumulation in eye cornea, up to 0.4 ppm. Skin illnesses, Cattle osteodistrophia, lamb and calf diseases

Table 4. Description of regions of biosphere, sub-regions of biosphere and biogeochemical provinces in the area of the Northern Eurasia (continued).

Chemical elements	Distribution of sub-regions and biogeo-chemical provinces	Content of elements in bio-geochemical food webs	Biological reactions of organisms and endemic diseases
Mo excess, Cu deficit or optimum	Armenia	Increasing Mo:Cu ratio in Mountain Kastanozems and Forest Brunozems. High content of Mo (9 ppm) and low content of Cu (1 ppm) in forage species, high Mo:Cu ratio.	Increasing Mo content in tissues, increasing synthesis of xantinoxidase; 2−4-fold level of uric acid. Endemic disturbance of purine exchange in sheep and cattle. Endemic molybdenum gout in humans.
Pb excess	Armenia	25-fold increasing Pb content in Mountain Kastanozems and Forest Brunozems (50-1700 ppm). 7-fold increase of Pb content in plant species (0.5-11.6 ppm). 2-10-fold increase of Pb in food-stuffs.	Daily human food intake of Pb is 0.7–1.0 mg/day. Pb accumu-lation leads to endemic diseases of central nervous system
F excess	Baltic Sea States, Belarus, Moldova, Central Yakutia, Kazakhstan	Excessive content of F in natural waters, > 1.0–1.5 ppm. Low content of F in soil and plants.	Tooth enamel dystrophy. Fluo-rosis and spotted teeth of human and animals.
F deficit	Biogeochemical provinces in different regions of biosphere	Content of F in natural waters < 0.5-0.7 ppm.	Reducing content of F in tooth enamel. Endemic tooth carious in humans and animals
Mn deficit	Biogeochemical provinces in different regions of biosphere	Lowering content of Mn in soils and plant species	Reducing Mn content in bones. Decreasing activity of phos-phatase, phosphorilase, and isocitric dehydrogenase.
Se deficit	Baltic Sea States, Northwestern Russia, middle Volga regions, south of East Siberia	Low Se content in forage plants, 0.01–0.1 ppm	Depressed glutationperoxidase activity. White-colored muscles.
Se excess	Tuva region	Increasing Se content in sandy Dystrict Kastanozems, up to 2–4 ppm. Increasing Se content in plants, up to 13 ppm.	Deformation of hoofs, wool cover losses, hypochromic anemia. Selenium toxicity in sheep and cattle.

Table 4. Description of regions of biosphere, sub-regions of biosphere and biogeochemical provinces in the area of the Northern Eurasia (continued).

Chemical elements	Distribution of sub-regions and biogeo-chemical provinces	Content of elements in biogeochemical food webs	Biological reactions of organisms and endemic diseases
U excess	Issyk-Kul valley, Kirgizia	Increasing U content in soils, plants, food and fodder stuffs.	Morphological alterations in plant species, which accumulate this element.
Zn deficit	Foothills of Turkmenistan and Zerafshan ranges, Uzbekistan and Talikistan	1.5–2 times reducing Zn content in all Serozem sub-types. The plant content is < 7.5 ppm	The reducing Zn content in blood (up to 1.8 ppm) and wool of sheep. Lowering activity of Zn-containing ferments. Endemic parakeratosis
Li excess	Middle and low flow of Zerafshan, Uzbekistan	High content of Li in Serozems and Brunozems. 2.5–3.0-fold increase of Li in plant species	Morphological alterations of plant species
Mn excess	Georgia	Excessive content of Mn in all biogeo-chemical food webs	Plant endemic diseases
Ni, Mg, Sr excess. Co, Mn deficit	South Ural	Unbalanced ratio of essential elements in all biogeochemical food webs	Endemic osteodystrophy in humans and animals
Cu excess	Deserts, Uzbekistan	Serozems	Disturbance of Cu exchange, endemic intero-hemoglobinuria

These results are shown in Figure 7.

Thus, the monitoring of biogeochemical food webs of copper, zinc, cobalt and nickel in biogeochemical provinces, enriched by these elements, showed that, in comparison with the control Steppe Chernozem biogeochemical sub-region of biosphere, the South Ural sub-region of biosphere and corresponding biogeochemical provinces are the areas with straightly altered biochemical and physiological activities of plants, domestic animals and humans. These have led to the endemic diseases, morphological alterations and adaptations.

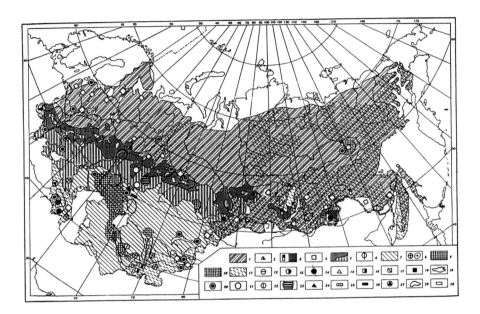

Figure 6. Biogeochemical mapping of North Eurasia. 1–11. Zonal biogeochemical regions, sub-regions and biogeochemical provinces. 1–4—Taiga Forest region of biosphere. Biogeochemical provinces: 1—Co deficit, Cu *deficit, Co* + Cu *deficit, Ca* + P *deficit; 2—I* + Co *deficit; 3—Ca deficit and* Sr *excess; 4—Se deficit. 5–6— Forest Steppe and Steppe region of biosphere. Biogeochemical provinces: 5—I deficit in floodplain soils; 6—disturbed ratio of* Ca:P. *7–10—Dry Steppe, Semi-Desert and Desert region of biosphere. Biogeochemical provinces: 7—Cu deficit; 8—Cu deficit, Mo and sulfate excess; 9—B excess; 10—Co and* Cu *deficit, Mo and B excess. 11—Mountain region of Biosphere. 12–29. Azonal biogeochemical provinces 12—Co excess; 13—I and Mn deficit; 14—Pb excess; 15—Mo excess; 16—Ca and* Sr *excess; 17—Se excess; 18—unbalanced* Cu:Mo:Pb *ratios; 19—U excess; 20—F excess; 21—Cu excess; 22 —disturbed* Cu *exchange; 23—Ni, Mg,* Sr *excess and* Co, Mn *deficit; 24—Ni excess; 25—Li excess; 26—Cr excess; 27—Mn excess; 28—F deficit; 29—Zn deficit.*

Table 5. Biogeochemical cycles of copper, zinc, cobalt and nickel in South Ural sub-region (After Kovalsky, 1981).

Links of biogeochemical food webs	Units	Chemical element	Sub-region	Biogeochemical provinces		Control region
				Baimak	Uldybaev-Chalil	
Rocks	ppm	Cu	20–2000	100.0	33.1	35
		Zn	70–4300	130.0	75.2	70.0
		Co	10–180	45.2	80.2	10.4
		Ni	55–1800	160.0	1250.0	55.4
Soils	ppm	Cu	76.3	98.0	37.1	3.7
		Zn	155.2	201.1	94.3	40.2
		Co	30.6	27.5	51.4	18.3
		Ni	235.3	185.4	663.0	57.3
Waters	ppb	Cu	7.1	14.0	5.6	4.7
		Zn	22.3	31.7	20.3	10.0
		Co	1.8	1.9	1.8	1.1
		Ni	17.4	20.9	16.7	5.1
Air	ppb	Cu	0.72	—	—	0.17
		Zn	2.80	—	—	10.92
Accumulation by microbial biomass	Kg/ha in 20 cm layer	Cu	—	0.161	0.067	0.031
		Zn	—	0.120	0.059	0.034
		Co	—	0.021	0.061	—
		Ni	—	0.097	0.109	—
Terrestrial plants	ppm	Cu	25.3	30.0	19.0	5.3
		Zn	46.5	60.1	32.9	21.1
		Co	1.02	0.65	1.38	0.68
		Ni	14.4	6.5	21.3	6.2
Bottom sediment	ppm	Cu	45.4	98.1	34.5	15.0
		Zn	248.0	185.2	295.1	37.9
		Co	10.3	6.9	11.6	8.1
		Ni	124.0	60.3	124.4	31.1

Table 5. Biogeochemical cycles of copper, zinc, cobalt and nickel in South Ural sub-region (After Kovalsky, 1981) (continued).

Links of biogeochemical food webs	Units	Chemical element	Sub-region	Biogeochemical provinces		Control region
				Baimak	Uldybaev-Chalil	
Aquatic plants	ppm	Cu	18.3	17.3	21.1	15.9
		Zn	70.2	126.9	73.3	24.5
		Co	8.5	6.0	10.5	0.3
		Ni	19.2	16.1	26.4	2.9
Plankton	ppm	Cu	—	—	9.5	—
		Zn	—	—	78.5	—
		Co	—	—	5.0	—
		Ni	—	—	23.0	—
Fish	ppm	Cu	4.1	3.5	5.0	—
		Zn	71.3	69.5	73.1	—
		Co	0.56	0.3	73.1	—
		Ni	1.9	1.5	2.2	—
Bentos	ppm	Cu	23.6	29.6	19.6	—
		Zn	70.0	101.0	50.4	—
		Co	3.2	2.5	3.6	—
		Ni	5.9	3.6	7.1	—

Table 6. Content of metals in forage crops from Uldybaev-Chalil Ni-Cu-Co biogeochemical provinces of South Ural sub-region of biosphere, ppm by dry weight.

Plant species	Cu	Zn	Co	Ni
Mix-grasses	25.1	62.2	2.8	25.0
Legumes	8.2	27.3	1.2	15.2
Herbaceous	22.3	38.4	0.8	18.2
Silage	15.2	62.3	0.9	8.0
Dry mixture	8.3	117.0	0.7	6.5

Table 7. Content of copper in foodstuffs from Baimak Cu-Zn biogeochemical provinces of South Ural sub-region of biosphere, ppm by dry weight.

Foodstuffs	Bairak biogeochemical province	Control region
Wheat	20.16 ± 0.91	3.73 ± 0.36
Potato	7.15 ± 1.26	2.05 ± 0.85
Meat	3.45 ± 0.38	0.43 ± 0.04
Milk	1.33 ± 0.39	0.30 ± 0.01

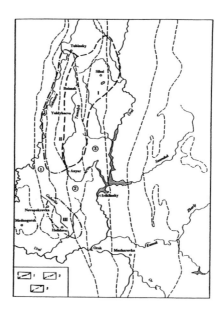

Figure 7. Biogeochemical mapping of South Ural sub-region of biosphere, Russia. 1—the Baimak Cu-Zn biogeochemical province (I); 2—the Uldybaev Ni-Cu-Co biogeochemical province (II), the Chalil Ni-Cu-Co biogeochemical province (III).

FURTHER READING

Viktor V. Kovalsky, 1974. Geochemical Ecology. Nauka Publishing House, Moscow, 325 pp.

Maria A. Glazovskaya, 1984. Soils of the World. American Publishing Co., New Delhi, 401pp.

Vladimir N. Bashkin, Elena V. Evstafieva, Valery V. Snakin et al., 1993. Biogeochemical Fundamentals of Ecological Standardization. Nauka Publishing House, Moscow, Chapter 1, 1–24.

Miroslav Radojevic and Vladimir N. Bashkin, 1999. Practical Environmental Analysis. Royal Society of Chemistry, UK, Chapter 1.1 and Appendices IV.

QUESTIONS AND PROBLEMS

1. Discuss the applicability of soil and ecosystems mapping for characterizing global and regional biogeochemical fluxes.

2. Present the definitions of biogeochemical uptake coefficient, active temperature coefficient and relative biogeochemical cycling coefficient, and give examples of these coefficients for various ecosystems.

3. Characterize the biogeochemical cycling in Europe. Analyze the alteration of biogeochemical parameters from north to south and from west to east.

4. Estimate the role of climate in the transformation of biogeochemical features of ecosystems in Asia. Select the ecosystem and consider the relevant alterations.

5. Discuss the similarities and differences in biogeochemical fluxes of boreal Forest ecosystems in Asia and Europe. Explain the possible reasons.

6. What are the characteristic features of biogeochemical fluxes in Mountain Asian ecosystems? Describe the role of relief in manifestation of biogeochemical cycles.

7. Discuss the geographic alterations of biogeochemical cycling in the Tropical Rain Forest ecosystems of Latin and South America. Describe the role of precipitation and soil types.

8. Despite some similarities, the biogeochemical cycles of various elements are different in Eurasia and North America. Explain these differences.

9. Discuss the peculiarities of biogeochemical cycling in Arid and Extra-Arid ecosystems of North America.

10. Indicate the mechanisms of biogeochemical cycling in the Savanna ecosystems of Africa. Empasise the role of wet and dry seasons in migration of various chemical species.

11. Describe the role of endemic biological species in the formation of biogeochemical cycles in Australian ecosystems. Compare these ecosystems with similar one in other continents.

12. Discuss the basic principles of biogeochemical mapping and define the classification units for this mapping.

13. Draw your attention to the lower, optimum and excess levels of elements in the biogeochemical food web. Present the theoretical curve and discuss the limit values.

14. Describe the general model of biosphere and the role of various links in the biogeochemical food webs.

15. What are the principal routes of biogeochemical migration of various species in terrestrial ecosystems? Present the main systems and explain their inter-relationships.

16. Discuss the main links of the biogeochemical food web in aquatic ecosystems. Characterize the interplay between terrestrial and aquatic links of biogeochemical cycles. Give examples.

17. Provide examples of biogeochemical mapping on a continental scale. What are the principal approaches to mapping the continents?

18. Describe the methods of biogeochemical mapping. What information is required for multi-scale biogeochemical mapping.

19. Present the description of main biogeochemical units in North Eurasia. Compare zonal and azonal biogeochemical units and explain the differences between them.

20. Describe the alteration of biogeochemical food webs in areas over the geological ores of trace metals. Consider the typical food web of a selected trace metal.

CHAPTER 8

ENVIRONMENTAL BIOGEOCHEMISTRY

During the last century, enforced anthropogenic activity led to an increasing pollutant loading at all biosphere compartments and relevant terrestrial and aquatic ecosystems. The biogeochemical structure of biosphere was altered to a most significant extent. We can consider the biogeochemical structure as the most flexible indicator of pollution loading. This flexibility, nevertheless, assumes an existence of a definite homeostatic interval, within which the input of polluting elements will be in the limits of natural deviations of various links of biogeochemical food webs (see Chapter 7, Section 2). However, considering the modern state-of-the-art of biogeochemical cycling, we can conclude that in the most natural biogeochemical sub-regions and provinces, the pollutant loading has led to the formation of technobiogeochemical and agrogeochemical structural units.

In modern literature we can find different definitions of environmental pollution, most of which speculate on the increasing rates of pollutant inputs and accumulation in the various media, like soil, waters, air, fodder and foodstuffs. However, varying geochemical backgrounds make the quantitative approach to the pollution problems quite controversial. We should take into account the biogeochemical cycling in various terrestrial and aquatic ecosystems with parameterization of biogeochemical pools and fluxes in local, regional and global scale. Thus, from the biogeochemical point of view, *environmental pollution is the reversible and/or irreversible alteration of biogeochemical structure of ecosystems.* This alteration leads to many ecological problems and the major one of them is human and ecosystem health.

Accordingly, we can determine environmental biogeochemistry as a biogeochemistry of the modern biosphere. Environmental biogeochemistry is a rapid developing branch of the classic biogeochemistry dealing with the quantitative assessment of transformation of natural biogeochemical cycles of various elements under the loading of pollutants. There are plenty of international and national scientific meetings devoted to these problems, like biannual international conferences on the environmental biogeochemistry of macro- and trace elements. Many questions arising under discussion of global problems of greenhouse effects, ozone depletion, acidification, transboundary air and water pollution, etc. are closely related to environmental biogeochemistry.

In this chapter we will consider the environmental biogeochemical cycles of three elements, which natural biogeochemical turnovers have been altered in the most significant degree. They are nitrogen, mercury, and lead.

1. ENVIRONMENTAL BIOGEOCHEMISTRY OF NITROGEN

Human activity has greatly altered the nitrogen cycle on land, in aquatic systems, and in the atmosphere. Currently, global fixation of atmospheric N_2 for fertilizer, in combustion of fossil fuels, and by leguminous plants exceeds that by all natural sources, and changes in land use cause large additional amounts of nitrogen to be released from long-term reservoirs to both vegetation and soil (see Chapter 3, Section 3) These disturbances have been linked to a number of environmental concerns, including excessive accumulation of nitrates in ground waters, inland and coastal water eutrophication, acidification of freshwater lakes and rivers, climate changes, saturation of forest ecosystems and decline of forest biodiversity and productivity. However, some of these environmental problems can be considered as positive. For instance, accelerated nitrogen cycling has also been suggested to increase the sequestration of carbon in forest ecosystems, slowing the rise of atmospheric carbon dioxide.

The foundations of environmental biogeochemistry were based in part on studies of small-scale watersheds (less than 500 ha) in Europe and North America. Recently, much more attention has been given to evaluating material fluxes through the landscape at larger scale. However, we are only beginning to understand the rules that govern regional exchanges of nitrogen between atmospheric, terrestrial and aquatic ecosystems. These exchanges are especially critical to the functioning of freshwater and coastal ecosystems, as even relatively small changes in the processing and retention of nitrogen applied to the terrestrial ecosystems could have a large impact on the downstream aquatic ecosystems. In fact, several studies have noted much greater fluxes of nitrate in rivers over the past few decades. Nitrate concentrations in the Mississippi River have more than doubled since 1965 and that in many European rivers (Danube, Rein, Volga, Seine, *et al*) has probably increased 5- to 10-fold during the 20th century. The similar trends are now shown for main East Asian rivers (Yellow, Yangtze, Mekong, Chao-Phraya, Han, *et al*).

The pollution of atmosphere and transboundary transport of nitrogen (and sulfur) species is of great concern for the whole North Hemisphere. This leads to the increased atmospheric deposition of nitrogen.

Increased use of nitrogen fertilizers, cultivation of nitrogen-fixing crops and depositions are accompanied with enforced denitrification and production of N_2O. The latter is known as a species that accelerates ozone destruction in the stratosphere.

These are the milestones of problems related to the environmental biogeochemistry of nitrogen. We will consider the environmental biogeochemistry of nitrogen in various scales, from local and regional up to the global. The examples will be given for the most altered regions, such as North Atlantic region and East Asian region.

1.1. Environmental biogeochemistry of nitrogen in the North Atlantic region

The North Atlantic region can serve as an example for several reasons. First, this is the most studied area of the World and we therefore have the best chance to show reasonably accurate estimates of anthropogenic changes in natural biogeochemical pools and fluxes of nitrogen. Second, this region has been and continues to be significantly

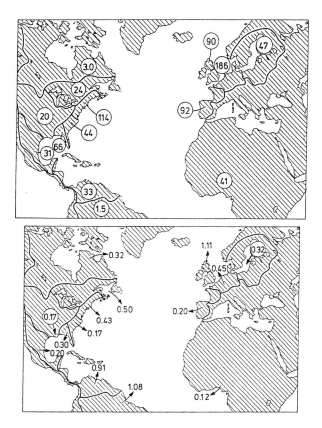

Figure 1. The schematic map of the North Atlantic Ocean. The top is population density of each of the 14 regions in individuals per km^2 and the bottom is water discharge per area in m^3/m^2 per year (Howarth, 1996).

affected by human activity. Third, the region is comprised of a great diversity of environments, thereby allowing comparisons of many of the World's ecosystems within its boundaries (Figure 1).

The watershed of the North Atlantic Ocean was divided into 14 regions. Seven of these regions are in North America, two are in Central and South America, four are in Europe, and one in Africa. Regions are selected to coincide with discrete portions of the coastal ocean (Table 1).

Overall, these regions make up 18% of the surface area of all land on the Earth and have 15% of the global population. Population density in the regions averages 29 individuals per km^2. Freshwater discharge from the North Atlantic watershed contributes 30 to 33% of the World's discharge into the oceans, estimated as 42 to 46 $\times$ 10^{15} L per year. Half of the North Atlantic water discharge comes from the Amazon.

Table 1. Summary of regional data for the North Atlantic Ocean watershed (Howarth et al, 1996).

Regions	Regional data		
	Drainage area, 10^6 km^2	Water discharge, 10^{12} L per year	Population, individuals
Baltic Sea	1.50	475	70
North Sea	0.84	380	156
NW European coast	0.34	378	30
SW European coast	0.55	110	50
Western Europe — total	3.23	1.340	306
North Canadian rivers	3.98	1.260	12
St. Lawrence basin	1.60	801	38
NE coast of US	0.48	208	55
SE coast of US	0.35	59	16
Eastern Gulf of Mexico	0.35	106	23
Mississippi basin	3.23	546	64
Western Gulf of Mexico	1.42	285	44
North America — total	11.4	3.260	252
Caribbean Islands and Central America (incl. Orinoco)	2.28	2.070	75
Amazon & Tocantins basin	6.49	7.000	10
Central & South America — total	8.77	9.070	85
NW Africa	3.53	416	146
Total — all basins	26.9	14.100	790

The North Atlantic regions vary greatly in discharge and population density. The Amazon makes up 24% of the total drainage area feeding into the North Atlantic and 50% of the freshwater discharge, but has only 1% of the population. At the other extreme, the drainages flowing into the North Sea comprise only 3% of the total

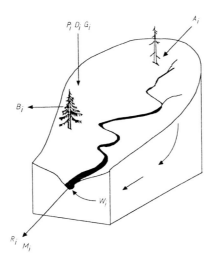

Figure 2. Conceptual scheme of a small catchment ecosystem. Fluxes of element i between a hydrological basin and its surroundings: Wi, total weathering of rocks; Pi, wet atmosphere deposition; Di, dry deposition; Ai, possible anthropogenic inputs (e.g. fertilization); Ri, surface and subsurface runoff of soluble substances; Mi, possible water erosion of solid substances; Bi, biomass export (harvesting) (Moldan and Cherny, 1994).

drainage area and contribute 3% of the freshwater water discharge but have 20% of population. The regions vary in population density from 1.5 person per km^2 in the Amazon to 186 person per km^2 in the North Sea region. Freshwater discharge per area of watershed varies from 0.12 m^3/m^2 per annum in northwestern Africa to 1.1 m^3/m^2 per year for the northwestern European coast (see Figure 1).

Small catchment studies on environmental nitrogen biogeochemistry

The small catchment is a drainage basin or watershed with surface area usually less than 500 ha (Figure 2).

It has an easily recognizable natural topographic boundary, which is defined as the watershed divide. Most often the catchment is situated in a comparatively undisturbed landscape covered by natural or semi-natural Forest ecosystems. These studies have also been conducted within other environments, such as High Mountain, Degraded Forest, Meadow and Agro-ecosystems and even urbanized or semi-urbanized areas.

Forest ecosystems

Biogeochemical cycles of individual elements, including nitrogen, have been intensively studied at the Hubbard Brook experimental Forest (HBEF) in New Hampshire, USA, since 1963, pioneering the use of the small catchment concept for understanding biogeochemical processes in the ecosystems (see Likens *et al*, 1977). The numerous examples of North American and European catchment research on the environmental biogeochemistry of nitrogen have been reviewed in SCOPE project (Moldan and Cherny, 1994).

The atmospheric N input in Europe and North America has increased dramatically during the few last decades due to emission of NO_x from combustion processes and of NH_3 from agricultural activities. The N deposition to Forest ecosystems generally exceeded in the 1980s 20 kg/ha/yr. in the Central part of Europe and even reached 100 kg/ha/yr. in some impact areas (Gundersen & Bashkin, 1994). Forest ecosystems may accumulate to considerable amount of N in biomass and soil organic matter, but there is an increasing concept that forest ecosystems may be overloaded with N from atmospheric deposition that leads to increasing leaching of nitrates and biodiversity changing and so called 'nitrogen saturation' can occur. The nitrogen saturation can be defined as a situation in which the supply of inorganic nitrogen exceeds the nutritional demand of biota and is operationally measured as increasing leaching on N below the rooting zone (Aber *et al*, 1989). P. Gundersen and V. Bashkin (1994) indicated that the values of nitrogen accumulation may be in the limits of 1–142 kg/ha/yr. However, the most statistically significant values for the forest ecosystems of North America and Europe were between 6–24 kg/ha/yr.

At present great attention is being given to an assessment of wetlands as intermediate ecosystems between terrestrial and aquatic ones. Relevant case studies in Europe have shown that the N removal in wetlands depends mainly on denitrification. In addition to denitrification, sedimentation can be quantitatively important during water flooding periods. Large-scale establishment in the Agricultural ecosystems in southern Sweden of the great number of wetlands within a catchment may reduce the nitrogen transport by up to about 15%. In most cases, the result will probably be less. Furthermore, in addition to researches of nitrogen removal in wetlands, suitable for use in farmland areas, a study of the retention capacity of forest wetlands showed that N losses from forest (1–5 kg/ha/yr.) is considerably lower than from arable lands (Jansson *et al*, 1994). But, in case of southern Sweden, where the forest ecosystems make up a major part of the catchments draining to the Baltic Sea coast, nitrogen derived from forest areas constitutes as much as 50% of the total N input. The authors concluded that the N removal capacity of forest wetlands is more or less unknown but could be considerably higher under the increasing input values of nitrogen.

The values of retention of atmospheric deposition N in different ecosystems obtained experimentally can achieve significant figures, up to 30–40 kg/ha/yr. in Agricultural and managed Forest ecosystems. However, both computed and experimental reasonable values were in limits of 5–25 kg/ha/yr.

The leaching of nitrogen from Forest ecosystems due to disruption of the natural biogeochemical cycle is shown in Table 2.

Since nitrogen leaching occurs when the fluxes exceed plant uptake, it is obvious that a decrease in plant uptake induced by harvest or forest decline may as well cause leaching of nitrate. Nitrogen leaching from harvested or windfelled stands is related to the abrupt disturbance of the ecosystems and not directly to the status of the N cycle. It seems evident, though, that the N losses by leaching (and denitrification)

Table 2. Nitrogen leaching from disturbed Forest ecosystems (small catchment studies).

Disruption of biogeochemical cycle	Cause of disruption	N leaching, kg/ha/year	Nitrate concentration in surface waters, mg/L N − NO_3^-
Disturbance and removal of canopy cover	Harvest clear-cut	97–142	24–116
	Insect pests	0.5	50
Forestry manipulation	Liming	45	24
	Fertilization	14–45	12
	Reforestation	8	5–20
Forest decline	Reduced growth rate	6–24	5–12
Nitrogen saturation	N_2-fixing in Alder Forest ecosystems	51	22
	High atmospheric deposition	10–87	18–25

from harvested areas will increase with the degree of nitrogen saturation of the Forest ecosystems before the harvesting. The potential losses in the first years after a disturbance may approach the mineral N biogeochemical flux density of the soil. The correlation was monitored between nitrate leaching and the amount of nitrogen in annual turnover.

Many causes and mechanisms may trigger a disruption of natural biogeochemical cycle of nitrogen and the extent and duration may vary over wide ranges. To make safe conclusions on real changes in the nitrogen cycle and their causes or disruption mechanisms, long-term monitoring (decades) is required. Short-term monitoring may lead to misleading conclusions. For instance, the first portions of the 25-years monitoring records from one of the watersheds at the HBEF project, USA, could be interpreted as evidence of N saturation of Forest ecosystems from increased atmospheric deposition, since the nitrogen output in stream water increased from 1.5 kg/ha/year in the mid-1960s to 5–6 kg/ha/year of N in the mid-1970s approaching the input flux. But looking at the full record, the output decreased again and was constantly low during the 1980s, in spite of absence of the remarkable declining trends in N deposition.

There are some indicators of a qualitative importance for nitrogen saturation and leaching (Table 3).

The physical factors (precipitation, soil texture and depth) regulate the flow of water as the transport media for nitrate and the contact time of the root system with inorganic nitrogen species. It is clear that the atmospheric loading is important.

Table 3. Possible factors influencing the risk of N saturation and nitrate leaching after disturbance of natural biogeochemical cycle of nitrogen in Forest ecosystems.

Factor	Increasing risk	Decreasing risk
Precipitation surplus	High	Low
Soil texture	Coarse-sandy	Fine
Soil depth	Shallow	Thick
Atmospheric N loading	High	Low
Vegetation	Coniferous	Deciduous
Rooting depth	Shallow	Deep
Internal N flux	High	Low
Nitrifying capacity of soils	High	Low
Other nutrient fluxes	Low	Big

Similarly, the forest type is important, as the deposition to the Coniferous ecosystems generally is higher than to deciduous ones. Further, the rooting system of coniferous trees is shallower than for deciduous trees, which may affect the ability to retain nitrogen. Not all soil types are able to nitrify. Non- or poorly nitrifying soils seem to be characterized by high $C : N$ ratio and very low content of other nutrients, like P, K, Ca, Mg and micro-nutrients. The available pool of essential nutrients may also limit the primary productivity and thereby the capacity to store and circulate nitrogen. Water or light may also become limiting. Regression analysis of these factors comparing available plot and catchment studies may reveal the relative importance of different factors in N saturation and leaching.

Agricultural ecosystems

The input of increasing amounts of technogenic nitrogen into small agricultural catchments affects all components of the biogeochemical nitrogen cycle. This has transformed the natural biogeochemical cycle of nitrogen into agrogeochemical with open nutrient cycle characterized by (i) high nitrogen export in crops; (ii) appearance of nitrate in groundwater and (iii) increased denitrification.

Small catchments with similar soil and climate conditions and similar anthropogenic loading are most suitable for investigation of biogeochemical cycles in agroecosystems. The small catchment concept is widely used in European studies to evaluate the effects of different soil types, crops and management strategies on

Table 4. Biogeochemical nitrogen mass balance of a small agricultural catchment (46 ha), Chechejovka river, Czech Republic, kg N/ha/yr.

Process	Monitoring years		
	1st	2d	3d
Input			
Fertilizer	253.5	244.5	200.0
N-fixation	No	No	110.0
N deposition	15.0	16.5	17.0
Output			
Crop harvest	131.9	122.7	183.2
Surface runoff	2.7	14.4	14.0
Subsurface runoff	No	27.2	36.5
Balance	133.9	96.7	93.3

nitrate leaching. The significance of this approach, which includes quantification of all possible pools, sources and sinks of nitrogen, as well as the rate of circulation between them, is apparent; nutrient budgets are a synthesis of experimental and theoretical knowledge on biogeochemical cycling in agroecosystems under the high fertilizer and deposition rates of nitrogen.

Nitrogen inputs to agricultural and mixed catchments include mineral and organic fertilizers, atmospheric deposition, irrigation water, lateral migration between landscape elements, symbiotic and non-symbiotic nitrogen fixation. Nitrogen export from agroecosystems results from removal of agricultural production (harvest), surface, sub-surface and groundwater runoff, denitrification and volatilization of ammonia (Bashkin, 1987).

Biogeochemical nitrogen mass balance has been calculated in a small catchment study in the Czech republic. The data collected in this monitoring has been under analysis for three years (Table 4).

In the first year, under conditions of low precipitation and low hydrological flow, N accumulation was substantial, about 134 kg/ha/year. Accumulated N decreased up to 97 kg/ha/year with greater runoff outflow during the second year, while in the third year, high water flow leached 50 kg/ha/year, but the agroecosystem still accumulated nitrogen due to a leguminous species in crop rotation.

Nitrogen accumulation due to excessive application of mineral fertilizers has been demonstrated in many small agricultural catchments with varying soils and climates. The remainder of applied N is taken up by plants, leached in the groundwater or evolved into the atmosphere through denitrification. Cropping, cultivation and erosion have been shown to deplete N pools in farmlands of the southern plains in the USA. Some pastures were fertilized for 20–22 years (45 kg/ha/year), but these differed from non-fertilized fields in N content only in the uppermost 5 cm of the soil. Thus, nitrogen was not accumulated but was removed from the agroecosystems by cropping. The N accumulation rate appears to be considerably slower than N depletion rate under past farming practices (Compare with N saturation in Forest ecosystems).

Maximum accumulation of nitrogen seems to occur in subordinate landscapes. A comparison of N accumulation in various parts of an agroecosystem with cereal-potato-vegetable rotation was made on the slope and floodplain soils of the Oka River valley, Russia. Due to heavy application of mineral fertilizers on the upper slopes, nitrogen was leached by lateral runoff into the floodplain soils below and accumulated. The accumulations in the agroecosystems of the upper, middle and lower parts of the slope were 85, 64 and 101 kg N/ha per year. Fertilizer application in the upper part of the slope (165–220 kg N/ha/year) was the main input and denitrification in the lower, saturated area was the main output (71–128 kg N/ha/year). Similar results were obtained for heavily fertilized Grapefruit ecosystems with yellow-red soils and the Grass Savanna ecosystems with the same soils on Pinos Island, Cuba. In humid climates nitrogen leached from eluvial (upper part of slope) and transeluvial (middle part) landscapes and lateral runoff might cause accumulation in the lower saturated zone (Figure 3).

In water-saturated zones, the environmental fate of accumulated nitrogen is connected with leaching to surface and ground waters and denitrification. Drainage and cultivation as well as fertilizers and soil organic matter influence denitrification losses from agroecosystems. Cultivation limited denitrification though soil aeration. In drained land, direct-drilled soil lost 9 kg/ha/year, while plowed soil lost only 3 kg/ha/year. Drainage reduced denitrification losses by 50% in ploughed soils but had no effect on direct drilled treatments. Losses from denitrification amounted to 1–6% of fertilizer N applied in condition of clay soils in England.

A significant share of N is lost as nitrous oxide (N_2O) from coastal plain soils in USA through denitrification because most of them are acidic (see below). Wetting and drying cycles did not appear to influence denitrification rates in soils of agroecosystems, but warm temperatures increased them; highest rates occurred during initial spring thaw. Microbial assay for nitrification and denitrification activity indicated that the main nitrate sources are well-aerated soils and the main nitrate sinks are water-saturated soils (Cooke and Cooper, 1988). The relevant losses were up to 20% of fertilizer rates.

The largest sources of atmospheric NH_3 are from fertilizer application and live-stock washes. Studies of long-term trends imply a 50% increase in ammonia emission in Europe during 30 years between 1950 and 1980. A substantial amount of N is lost as volatile NH_3 from plants after fertilizer application and during the senescence

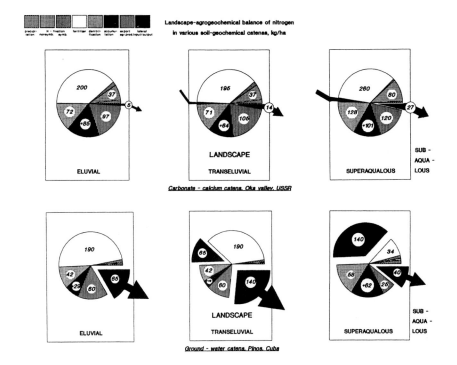

Figure 3. Landscape-biogeochemical balance of nitrogen in various soil-geochemical catenas.

period; about 16 kg N/ha of the applied fertilizer was lost as volatilized ammonia (Patron *et al*, 1988).

Excessive accumulation of nitrogen in small agricultural catchments leads to enhancement of surface and subsurface runoff as well as leaching into groundwater. Nitrogen mobility in catchment plot experiments with ^{15}N-labelled fertilizers followed (Bashkin, 1987). Nitrate was the most prevalent form of migrant, its movement was closely related to the hydrological flow. Application of N fertilizers enhanced the nitrogen mineralization capacity of soils, releasing nitrate from soil organic matter and subjecting it to leaching in much greater extent than the fertilizer nitrogen itself.

Surface runoff in agricultural catchments is greater than in forested areas, with N being exported mainly in soluble form (up to 97%), mostly nitrate. Concentration of N species in surface runoff is variable and discontinuous though more export occurs during high surface flows. It is beneficial to use long-term nitrate leaching monitoring studies with drained agricultural fields under different cropping systems and climate conditions; crop type is the most important factor in nitrate leaching. Grass and clover lays of several years standing and a two-year-old grass lay showed mean nitrate losses 4–6 times smaller compared to losses from cereal production under similar discharge conditions.

Nitrate leaching from the root zone is controlled by an increasing suite of factors, including: climate (precipitation, irrigation, evapotranspiration), soil (topography, texture, nutrient levels, porosity), land use (crop, cultivation and harvest practice) and N fertilization (type, application procedures and times). Nitrate leaching from agricultural land in the Rhine Valley of Germany was estimated to be 46–593 mg $N - NO_3^-/L$ using a soil water simulation model (Simon *et al*, 1988). Nitrate concentration and loading in relation to fertilizer application to sandy soils of the coastal plains in the USA was evaluated; concentrations of 9 ppm under 87 kg N/ha applied and 24 ppm under 226 kg N/ha applied were reported. Fertilizer application rates seem to be the most important single factor in nitrate leaching from agricultural catchments (Table 5).

A variety of factors can influence the rate of leaching of nitrogen from agricultural systems into surface waters, including water runoff, rate of fertilizer application, type and texture of soil, and the type of land use. Table 5 summarizes a number of small watershed or lysimeter studies aimed at quantifying nitrogen losses to the hydroecosystem from argoecosystems with the North Atlantic watershed.

In experimental agricultural watersheds with different soil types and varying rates of fertilizer application, runoff-weighted average concentration of total nitrogen in leachate is plotted against fertilizer inputs (Howarth, 1996). Collectively, these data clearly illustrate the much higher susceptibility to nitrogen leaching of sandy versus loamy or clayey soils, and of arable land compared to grassland. For runoff rates between 0.2 and 0.7 m^3/m^2/year, and for loamy or clayey soils, nitrogen export varies between 3 and 10% of fertilizer input for grassland, and between 10 and 40% for arable land. On sandy soils, corresponding figures are 15 to 50% for grassland and 25 to 80% for arable land. We can note that these percentages can be misleading, because other sources of nitrogen that added fertilizer (such as atmospheric deposition and nitrogen fixation) contribute to the nitrogen biogeochemical cycle.

Mixed catchments

Mixed catchments contain both agricultural and forest ecosystems and occur most frequently in small river basins. Biogeochemical fluxes and pools are altered by application of fertilizers and deposition, but the former are usually dominating. A few examples can be presented from both Europe and North America. In the Gorodnyanka River Basin, Russia, the highest N, P, and K accumulation occurs in catchments containing the highest proportions of agricultural land receiving the heaviest of fertilizers. Excess N can reach 50% of input values (100–200 kg N/ha/year) in many catchments and may be leached into groundwater (Kudeyarov and Bashkin, 1984). Groundwater pollution under mixed farming and woodland areas occurred deeply in coarse sandy soils under farmland and the borders of forests. Lateral transport of excess N can occur though surface runoff between landscape elements. Loading of inorganic nitrogen on farmlands can directly influence N export from wetlands associated with them in the catchment (Richardson, 1987).

Table 5. Nitrate leaching in agricultural catchments in relation to soil, crops and fertilizer rates in temperate climate of Europe and North America.

Soil	Permeability	Crop	N fertilizer rate, kg/ha/yr.	Method	Leachate (kg N/ha/yr.)
Brown/humic/ clayey/loam	Poor	Grasses	0	L	2.2
			100		3.3
			200		6.8
Sand	High	Spring cereals	O	L	67
			60		77
			120		81
			240		88
Loam	Poor	Spring cereals	0	L	36
			100		43
			200		45
			Slurry		48
			FYM		51
Clay	Poor	Corn	100	D	18.1
Loam	Poor	Forest	0	D	1.8
Organic	Poor	Pasture:cereals		D	
		67:33	41		13
		23:69	88		10
		18:73	108		7
		8:61	118		29
Clay/loam	Mean	Cereal/pota-to/ corn	174	D	22
Loam	Poor	Cereals	80		15
		Corn	35		8
Sand	High	Potato	40		7

Note: L-lysiweter, D-draivage.

Regional watershed studies on environmental nitrogen biogeochemistry

For consideration of regional biogeochemical mass balance of nitrogen, the input and output items should be estimated. On the basis of modern state-of-the-art in the environmental biogeochemistry of nitrogen, we will consider the following major anthropogenic fluxes into and out of the regions in the North Atlantic Ocean watershed. The input fluxes are application of nitrogenous fertilizers, fixation by leguminous crops, net export of nitrogen and atmospheric deposition of nitrogen from pollution sources. The output fluxes are river discharge, denitrification, volatilization of ammonia and net import of nitrogen. Sewage and animal wastes are not considered as input to regions as they do not represent newly fixed or imported nitrogen, but rather a redistribution or recycling of nitrogen within a region.

Methods and databases for estimation regional biogeochemical mass balance of nitrogen

The general way for the estimation of fertilizer input is the statistical data on national and regional application of fertilizers under different crops. The national statistic yearbooks and international FAO statistic is of usefulness.

Almost all types of agricultural land—including pastures—exhibit some amount of nitrogen fixation. For regional estimates in North Atlantic Ocean watershed, only five main classes of leguminous crops were taken into account: i) soybeans; ii) other beans (including broad and dry beans); iii) peas and lentils; iv) peanuts, and v) alfalfa, vetch and lupine. Fixation rates may be estimated from reviewed data (Messer *et al*, 1999). Total fixation inputs for each region are derived by multiplying the rate of fixation for each crop type by the crop area within the watershed. Crop areas can be taken from FAO statistics; those for regions within the USA come from annual Bureau of Census data.

Estimates for net import of nitrogen in agricultural products (food and feed) can be taken from FAO statistics on per country imports and exports. The following agricultural categories should be considered: i) cereals, ii) leguminous crops, iii) meat, and iv) milk. Average nitrogen content for each category can be taken from relevant sources, for instance from Blanck (1995), and per country values are then scaled by watershed areas as above. At the present time, these data are more available for North America and Europe, and accordingly only these regions have been used for the estimation of the export-import values.

The calculated values of deposition of atmospheric nitrogen to each region are described in detail for the global scale and for individual domains in various papers (Prospero *et al*, 1996; Dentener *et al*, 2000). The authors presented modeled estimates of wet and dry deposition of oxidized (NO_y) and reduced (NH_x) forms of nitrogen. The resolutions vary from $2.5° \times 3.25°$ LoLa (Latitute-Longitude) for oxidized nitrogen to $10° \times 10°$ LoLa for ammonia species. These deposition estimates are based on a complex model of nitrogen emissions, transformation in atmosphere, transport and deposition.

Sewage and human wastes

During calculation of regional mass balance of nitrogen, the sewage and human/animal wastes are not considered as input to regions as they do not represent newly fixed nitrogen. However, human sewage and wastewaters are obvious sources of nitrogen to rivers. In an analysis of 42 major world rivers, Gote *et al* (1993) concluded that sewage inputs alone are sufficient to account for the increased flux of nitrate observed in rivers whose watersheds have a higher population density. Although they acknowledged that deforestation, atmospheric deposition, and fertilizer application can all contribute significantly to nitrogen export from rivers, the authors stated that "watersheds with moderate to high human population will likely be dominated by sewage" rather than other inputs. On the other hand, agricultural sources of nitrogen and nitrogen deposition from the atmosphere are frequently cited as the major causes of increased nitrogen loading to rivers and estuaries (see for instance, National Research Council, 1993).

For the purposes of regional mass balance estimates, the contribution of sewage and wastewater to the export of nitrogen from each of the regions was calculated from the data on what percentage of the population in each region is sewered. Where possible, data from OECD (1991) were used. Elsewhere, the data from WRI/UNEP (1988) on access to waste treatment facilities in urban areas can be applied. Actual nitrogen fluxes due to sewage wastewater can be calculated from sewered population estimates by assuming a per capita nitrogen load in wastewaters of 3.3 kg N/year per person (Maybeck *et al*, 1989).

The calculations (Howarth, 1996) showed that for the North Atlantic Ocean watershed the percentage of sewage and wastewaters to total nitrogen input varies among the regions, from a high of 34% in the North Sea region to virtually none in the Amazon basin (Table 6).

The percentage contribution of sewage and wastewaters to total nitrogen flux from a region is strongly correlated with population density ($p < 0.05$; compare Table 6 and Figure 1. Nonetheless, this analysis suggests that nitrogen sources other than sewage and wastewater inputs dominate fluxes in all regions, and quite strongly in most regions. Overall, nitrogen from sewage and wastewater inputs is only 11% of the total nitrogen inputs to the North Atlantic Ocean from land (Table 6).

Nitrate fluxes in the major rivers of the World are correlated with human population density (Figure 4).

However, the slope of the relationship for the total nitrogen (TN) fluxes is less than that for nitrate fluxes in major world rivers by almost a factor of 2. This suggest that disturbance associated with human population density preferentially mobilizes nitrate over other nitrogen forms. These correlations between population density and either nitrate or total nitrogen fluxes could reflect correlations between human population density and accelerated nitrogen biogeochemical cycling through fertilizer use, food movements, atmospheric pollution, and land disturbance. This latter explanation is perhaps more likely since non-point sources of nitrogen dominate fluxes of nitrogen in the rivers of the North Atlantic basin.

Table 6. Sewage and wastewater inputs to the North Atlantic and its river basins.

Regions	Sewage N input, 10^6 ton/yr	% of Total N input
Baltic Sea	0.10	14
North Sea	0.42	34
NW European coast	0.073	16
SW European coast	0.047	23
Western Europe — total	0.64	25
North Canadian rivers	0.022	7
St. Lawrence basin	0.072	11
NE coast of US	0.13	26
SE coast of US	0.034	14
Eastern Gulf of Mexico	0.049	21
Mississippi basin	0.16	9
Western Gulf of Mexico	0.082	10
North America — total	0.55	12
Caribbean Islands and Central America (incl. Orinoco)	0.140	13
Amazon & Tocantins basin	0.0003	0.01
Central South America — total	0.14	3
NW Africa	0.096	6
Total — all basins	1.43	11

Human alteration of regional biogeochemical nitrogen cycle

Input items of regional biogeochemical nitrogen cycle

The total net anthropogenic inputs to the watersheds of the North Atlantic basin are about 30×10^6 tons per annum. These data were calculated from input data in Table 7 and areas in Table 1. The given value includes leguminous crop fixation and net fluxes of food and feed for temperate but not tropical regions. This represents an extremely large flux of nitrogen into these watersheds; it is 2.5 fold greater than total river flux to the North Atlantic Ocean (13.1×10^6 ton/year), and roughly one-third of global nitrogen fixation from all natural sources. Fertilizer and deposition dominate total inputs,

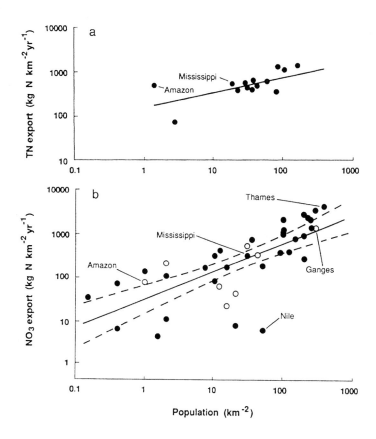

Figure 4. Relationship between log of population density and log of total nitrogen export from regions of the North Atlantic basin (top, a) and of nitrate in the major world rivers (bottom, b). Both relationships are significant, but the relationship for nitrate fluxes in the World's rivers (bottom) is more significant, has lesser scatter and has a steeper slope. For TN fluxes in the North Atlantic basin (top), $\log TN = 2.2 + 0.35 \log(population\ density)$; $r^2 = 0.45$; $p = 0.01$. *For nitrate fluxes in the World's rivers (bottom),* $\log NO_3^- = 1.15 + 0.62 \log$ *(population density)*; $r^2 = 0.53$; $p = 0.00001$ *(Howarth, 1996).*

accounting for 21.8 and 8.2 × 10^6 tons/year, respectively. Fixation by leguminous crops (4.6 × 10^6 tons/year) and net food and feed movement (−3.8 × 10^6 tons/year) nearly balanced at this scale (Table 7).

For atmospheric inputs of fixed nitrogen to each region from human activity, we can consider only deposition of oxidized nitrogen (NO_y, including both wet and dry deposition), and only the increase in this deposition due to human activity. These estimates of human activity can be given by subtracting modeled values of pre-industrial deposition from those for current levels. Estimates for modern and pre-industrial deposition of both NO_y and NH_x are shown in Table 8.

Table 7. Major net anthropogenic nitrogen inputs to the North Atlantic watersheds by region, kg/km^2/year.

Region	Anthpogencmic NO$_y$ deposition	N fertilizer	Leguminous crop fixation	Import (+) or export (−) of food and feed	Total inputs
Baltic Sea	479	1.730	26.7	20.7	2.256
North Sea	1.085	5.960	4.7	−4.2	7.044
NW European coast	1.087	2.870	55.0	−324	3.688
SW European coast	462	3.370	14.7	−64.0	3.783
Western Europe — total	698	3.230	21.9	−36.5	3.913
North Canadian rivers	72	161	32.9	−50.2	216
St. Lawrence basin	612	331	256	−31.3	1.168
NE coast of US	1.204	600	748	998	3.550
SE coast of US	1.023	1.170	369	454	3.016
Eastern Gulf of Mexico	763	1.260	248	576	2.847
Mississippi basin	620	1.840	1.060	−1.300	2.220
Western Gulf of Mexico	318	1.254			1.572
North America — total	431	878	397	−317	1.389
Caribbean Islands and Central America (incl. Orinoco)	140	342			482
Amazon & Tocantins basin	111	63			174
Central & South America — total		118	136		254

Note: Total N inputs for the Western Gulf region and all of Latin America do not include estimates of crop fixation or of net food and feed movement; thus, the continental totals are based on somewhat different components for each continent.

Table 8. Atmospheric deposition of oxidized (NO_y) and reduced (NH_x) forms of nitrogen to the watershed regions. Modern values are total current deposition as reported in Prospero et al (1996), pre-industrial values are modeled estimates (Howarth et al, 1996), and anthropogenic values are the difference between modern and pre-industrial. All values are in $kg/km^2/year$.

Region	Modern NO_y	Pre-industrial NO_y	Anthropogenic NO_y	Modern NH_x	Pre-industrial NH_x	Anthropogenic NH_x
Baltic Sea	490	11.3	479	560	47.5	513
North Sea	1.092	7.3	1.085	742	57.5	684
NW European coast	1.092	4.8	1.087	742	60.2	682
SW European coast	476	14.0	462	322	53.9	268
North Canadian rivers	84.0	11.6	72.4	70.0	28.7	41.3
St. Lawrence basin	630	18.3	612	266	43.3	223
NE coast of US	1.232	28.0	1.204	238	43.7	194
SE coast of US	1.078	54.6	1.023	350	55.5	295
Eastern Gulf of Mexico	812	49.0	763	280	58.5	222
Mississippi basin	658	37.8	620	266	45.8	220
Western Gulf of Mexico	364	46.2	318	224	28.7	195
Caribbean Islands and Central America (incl. Orinoco)	210	70.0	140	182	83.2	98.8
Amazon & Tocantins basin	196	85.4	111	140	86.7	53.3

NH_x deposition is excluded from the estimates of anthropogenic nitrogen input to a region because ammonia and ammonium do not travel far in the atmosphere before being deposited back to the ground, and because the principal sources of high levels of ammonia in the atmosphere are volatilization from fertilizers and animal waste products. Both of these processes are accounted for as inputs to a region in previous estimates of fertilizers, nitrogen fixation by crops, and net movement of nitrogen in feeds. Adding NH_x deposition would be a form of double accounting and would cause an over-estimate of total nitrogen inputs at the scale of regions since most of the NH_x deposition is driven by NH_x volatilization in the same region.

In contrast to NH_x, the principal sources of NO_y are probably from combustion of fossil fuels and are not otherwise accounted for as inputs of nitrogen to a region. Some NO_y may come from soil processes, such as denitrification, but this is small in comparison with the fossil fuel source. That NO_y deposition is largely independent of other source estimates and is an additional input to each region, while NH_x deposition is related to other inputs, can be seen by regressing both NO_y and NH_x deposition against the estimates for fertilizer inputs to a region: NO_y deposition shows no relationship, whereas at the scale of these large regions, NH_x deposition and fertilizer inputs are significantly correlated ($r^2 = 0.53$; $p < 0.05$). It is important to note, however, that while at the scale of an entire region NH_x deposition is not an additional input of nitrogen, volatilization and subsequent deposition of agricultural nitrogen can cause significant redistribution of nitrogen within a region, much of which may fall on forest and other natural ecosystems and be transported into rivers.

Human alteration of the nitrogen biogeochemical cycle is clearly much greater in the temperate portions of the North Atlantic basin, with nearly 90% of total anthropogenic nitrogen inputs occurring in the temperate zone of North American and European watersheds (Table 7). The enormous range among the watersheds in the intensity of human disturbance becomes especially clear when inputs are expressed on a per area unit basis: in the Amazon America and the Central America/Caribbean region, total net anthropogenic inputs average 174 and 482 kg N/km^2/year, while in North America they average 1.389 kg N/km^2/year. From this perspective, the European regions are the most altered: their average total net anthropogenic input of 3.913 kg N/km^2/year is more than twice that of the North American watersheds and an order of magnitude greater than the tropical regions. There is a tremendous range in total inputs among individual watershed regions even within the temperate zone, from a low in Northern Canada of 216 kg N/km^2/year to a high of 7.044 kg N/km^2/year in the North Sea (Table 7).

The relative importance of different anthropogenic nitrogen sources also varies substantially among individual watersheds. Nitrogen fertilizer is the major input to most regions, although inputs through atmospheric deposition of NO_y are greater in the St. Lawrence basin and the northeast and southeast coastal regions of the United States. The net movement of nitrogen in the agricultural food stocks is important only in the northeastern USA, where it is a major input, and the Mississippi basin, where it is a major export. While never the major input, nitrogen fixation by leguminous crops is often a significant input in the North American regions but is generally quite minor in the European regions. That agricultural sources are dominant input to the Mississippi basin is expected since this region drains the heartland of American agriculture. More surprising is the fertilizer dominance in Europe, despite notoriously high rates of atmospheric deposition in some areas; fertilizer inputs dominate in all of the European regions, accounting for 83% of total net anthropogenic inputs to Europe as a whole. Much of the atmospheric deposition in Europe is deposition of NH_x which originates from fertilizer and animal wastes. However, this nitrogen is

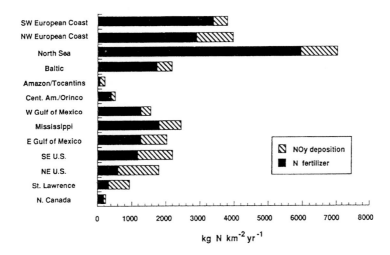

Figure 5. Application of nitrogen in fertilizer and by atmospheric deposition of NO$_y$ *to 13 of the watershed regions. Values are in kg N/km^2/year. Fertilizer inputs dominate net anthropogenic nitrogen inputs in most, but not all, regions of the North Atlantic basin (Howarth, 1996).*

recycling within the region (see above). Per area rates of fertilizer application are far higher in Europe than elsewhere (Figure 5).

Output items of regional biogeochemical nitrogen cycle

The output fluxes of nitrogen biogeochemical cycle in The North Atlantic Ocean watershed are mainly riverine fluxes and denitrification. We will consider these fluxes in connection with input of nitrogen in various regions. Output fluxes of total nitrogen per area from the temperate regions of the North Atlantic basin are strongly correlated with net anthropogenic inputs of nitrogen per area to these regions ($R^2 = 0.73$; $p = 0.002$; Figure 6a).

Note that Figure 6a shows the relationships between linear river fluxes and linear inputs, rather than the weaker log-log relationship between river fluxes of nitrogen and human population density shown in Figure 3. To analyze which nitrogen inputs are most related to riverine fluxes, we can separate the total anthropogenic fluxes into two categories: those related to the combustion of fossil fuels (represented by NO$_y$ deposition) and agricultural inputs (consisting of fertilizer inputs, nitrogen fixation by crops, and the net movement of nitrogen in feeds). The resultant two variable linear models (see Figure 6b) shows that the sum of the agricultural inputs and NO$_y$ deposition are significant predictors of river nitrogen export ($R^2 = 0.89$, $p = 0.0005$).

Given that NO$_y$ deposition is a much smaller input than agricultural sources in most regions, and that nitrogen limitation of production is prevalent in the natural ecosystems throughout most of these regions, we find it especially intriguing that

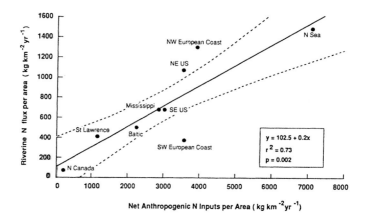

Figure 6a. Plot of river nitrogen export versus net human-derived nitrogen inputs to each of the temperate regions (all European and North American region, except the Western Gulf of Mexico); net inputs are equal to the sum of anthropogenic NO_y deposition, fertilizer inputs, nitrogen fixation by crops, and the net import or export of nitrogen in food and feed. (Howarth, 1996).

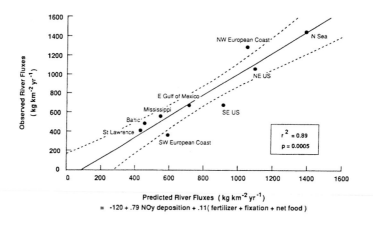

Figure 6b. Modeled versus observed river nitrogen export for same set of regions as in Figure 6a. The model is a multiple linear regression of river nitrogen flux against "agricultural" inputs (fertilizer, plus nitrogen fixation by leguminous crops, plus net import or export in food and feed) and inputs from combustion of fossil fuels (anthropogenic NO_y deposition) (Howarth, 1996).

deposition is so well correlated with river fluxes. This point is further illustrated in Figure 6c and Figure 6d, which compare simple regressions of river nitrogen export against NO_y deposition ($R^2 = 0.79$, $p = 0.0006$) and fertilizer ($R^2 = 0.39$, $p = 0.05$), respectively.

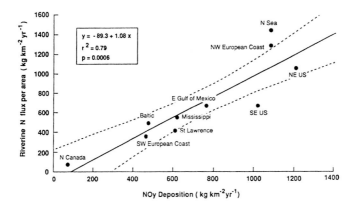

Figure 6c. Simple regression of river nitrogen export per area versus anthropogenic NO$_y$ *deposition for the same regions as in 6a. (Howarth, 1996).*

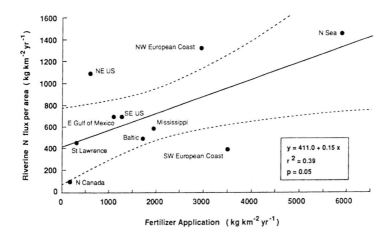

Figure 6d. Same as for 6c, but the regressor is application of nitrogen fertilizer to each region. (Howarth, 1996).

While neither leguminous fixation nor the net movement of nitrogen in feeds was a significant individual variable in any of the analyses, it is clear from the regional data that they may be important in some watersheds, such as the Mississippi (see Table 7).

The strength of these correlations seems somewhat surprising, considering the wide range in magnitude of fluxes and inherent limitations with the quality of data. Given the enormous range in nitrogen inputs being considered, it is even more surprising that the correlations suggest a *linear* relationship between net anthropogenic inputs and riverine fluxes. The capacity for processes within the ecosystems to provide a 'sink' from excess nitrogen before it reaches the coasts seems inherently limited; therefore, one might except to see river nitrogen fluxes increase exponentially with increasing nitrogen inputs, be denitrified or stored in the ecosystems; the total amount of nitrogen removed or stored in the ecosystems is even greater than this since some nitrogen is also input to each region through natural processes. Careful mass balance studies of partially forested watersheds with substantial agricultural activity have reached similar conclusions, finding some 80% of net anthropogenic nitrogen inputs stored within the watershed or denitrified and only 20% exported in rivers (Bashkin, 1987; Jaworski *et al*, 1992; Lowrance and Leonard, 1988).

It is crucial to identify the processes behind the apparent pattern of anthropogenic nitrogen loss upstream from the coastal estuaries, as some pathways may be relatively stable across a large range of inputs whereas others may quickly saturate. Denitrification loss as N_2 and N_2O is the most benign and perhaps the least limited pathway; Neff *et al* (1994) presented evidence of increasing denitrification in an upland terrestrial system with increase in nitrogen deposition. The only limiting factor may be presence of available organic carbon as a source of energy. However, changes in the ecosystems such as removal of wetlands and riparian areas could reduce the total denitrification capacity of a given region.

Average concentration of nitrates in the major European rivers was shown in Figure 7.

Other nitrogen sinks include storage in groundwater and storage in organic matter of terrestrial ecosystems. Both of these might tend to saturate with time and continued high rates of nitrogen input, perhaps eventually resulting in higher rates of riverine nitrogen flux to the North Atlantic (see above in discussing small catchment results).

Storage of nitrogen in groundwater

In assessing disturbance of various links of regional biogeochemical mass balance of nitrogen in the North Atlantic drainage basin, the special attention should be given to the increased accumulation of nitrogen in groundwater. A rapid increase in nitrate concentrations in groundwater has become a major concern in most areas with intensive agriculture in both Europe and North America (Table 9).

The major increase in West Europe occurred in the 1970s; similar patterns were found in East and Central Europe in the 1980s. This also seems to be the case in the eastern part of the United States. The data of Table 10 show a maximum value of nitrate accumulation rate in groundwaters of 36–38 μM/year.

The average nitrate concentration in shallow groundwater under farms in the sandy regions of the Netherlands for 1992–1995 was 144 mg/L (33 mg $N - NO_3^-$/L). Large differences were found in average nitrate concentration between farms and between years (Figure 8).

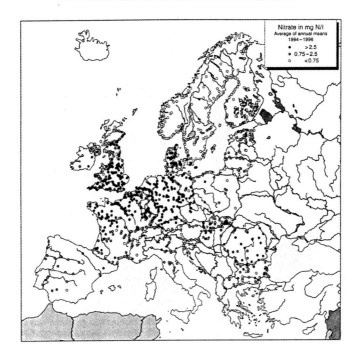

Figure 7. European river nitrate concentrations in 1994–1996 (EEA, 1998).

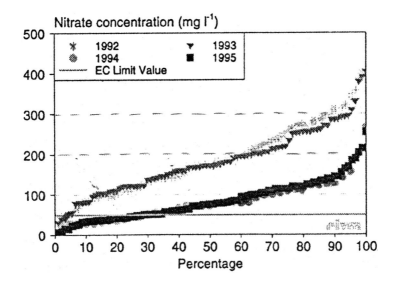

Figure 8. Cumulative frequency diagram of measured farm average nitrate concentrations in shallow groundwater in the snady regions of the Netherlands in the period 1992–1995 (Fraters et al, 1998).

Table 9. Rate of increase of nitrate concentration in ground waters during 1980s in different areas in Europe and North America.

Region	Aquifer characteristic	Nitrate increase (μM/year)	Reference
Europe			
Jutland (DK)	Sand	6.2	Overgaard, 1984
Limberg (NL)	Sand	36.0	Strebel *et al*, 1989
Bayern (Ge)	Sand/gravel	24.0	Resch, 1991
Wallongy(Br)	Sand	26.0	Foundation..., 1992
Wallongy(Br)	Chalk	24.0	Foundation..., 1992
Wallongy(Br)	Limestone	10.0	Foundation..., 1992
Champagne (Fr)	Chalk	34.0	Strebel *et al*, 1989
Central Russia	Loam	38.0	Bashkin, 1987
North America			
Mississippi basin	Sandstone	5.0	Tumer & Rabalais, 1991
South Wisconsin	Sandstone	7.9	Mason *et al*, 1990
Nebraska	Alluvial	36.0	Schepers *et al*, 1983

The extent of groundwater affected by this increase is difficult to evaluate. A crude estimate of groundwater reservoirs by continents and hydrological zones was presented by UNESCO (1978). Taking into account the two uppermost zones, located above sea level and characterized by the most active water exchange, the average groundwater stock is estimated as 47×10^3 m^3/km^2 for Europe and 78×10^3 m^3/km^2 for North America. These values may seem low when compared with local estimates for important aquifers : in Europe, the three most important aquifers are those associated with Permo-Traiassic sandstone, Cretassic chalk, and Tertiary sand and gravel. These have stocks of water, which range between 100 and 650×10^3 m^3/km^2. However, these highly productive aquifers cover less than 20% of the area of the continent.

Combining the estimates for the maximum nitrate concentration increase and the groundwater stock yields, a maximum estimate of the rate of nitrogen storage in groundwater of intensively cultivated areas of the order of 25 kg N/km^2 per annum for Europe and 40 kg N/km^2 per annum for North America. This is only 1% of the net anthropogenic inputs of nitrogen to Europe and 3% for North America. Groundwater storage may, however, be significant in some localized areas where high rates of

Table 10. Estimates of possible leaching of nitrogen to the groundwaters in the Central Russia (Bashkin, 1987).

Year	Small river watersheds											
	Gorodnyanka			Skniga			Itska			Sokhna		
	1	2	3	1	2	3	1	2	3	1	2	1
1969	15	5	33	4	1	25	0.4	0.2	50	3	1.9	63
1977	37	8	22	11	4	36	5	3	60	0.3	0.1	33
1978	21	10	48	12	6	50	6	4	66	0.8	0.4	50
1979	55	21	38	38	10	33	13	6	46	1.2	0.7	58
1984	54	22	40	40	15	50	16	8	50	1.7	0.9	59

Note: 1 accumulation of excessive nitrogen in river drainage basin, kg/ha/year; 2 - leaching of nitrogen to groundwaters, kg/ha/year; 3 - leaching of nitrogen to groundwaters, % of accumulated values.

increase in nitrate concentrations are observed in some large aquifers, as in the region of intensive agriculture in northwestern Europe. The sand aquifer of Central Belgium is an example, with an estimated storage of nitrate of approximately 200 kg/km^2/year. The upper Rhine aquifer is another example.

Considering the part of accumulated nitrogen that can be leached to the groundwater, we can estimate that in some case these values are of 22 to 66% of total nitrogen accumulated in ecosystems (Table 10).

Regional biogeochemical mass balance of nitrogen in the North Atlantic watershed

The overall nitrogen fluxes within and between the terrestrial, groundwater and river systems of the North Atlantic watershed are summarized graphically in Figure 9.

Assuming that terrestrial primary producers take up some 5,000 to 15,000 kg N/km^2/year in the North Atlantic basin as a whole, net external anthropogenic inputs of nitrogen represent as much as 7 to 22% of internal cycling. Fertilizer application dominates these inputs, but anthropogenic atmospheric deposition and nitrogen fixation associated with crop vegetation also contribute significantly. Nitrogen is transferred from terrestrial ecosystems to the hydroecosystems through soil leaching in both agricultural and forest ecosystems and through direct wastewater discharge. Overall for the North Atlantic watershed, sewage and wastewater discharge only represents about 10% of total riverine delivery. Leaching from animal feedlots, which we have not quantified, may also be important. Storage of nitrogen in groundwater, while difficult to quantify, is probably at most a few per cent of the rate of input of anthropogenically derived nitrogen. On the other hand, storage of nitrogen in forests

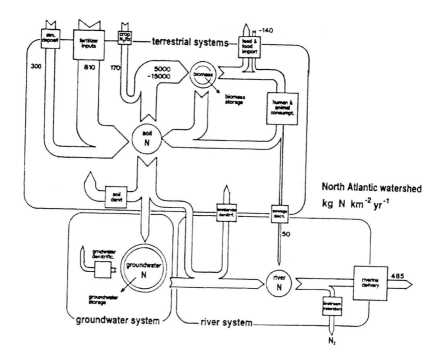

Figure 9. Schematic representation of nitrogen circulation within and between the terrestrial ecosystems, the groundwater systems, and the river system of the North Atlantic basin as a whole. Values in kg N/km²/year. Width of arrows, although not strictly proportional, suggests the relative magnitude of the corresponding fluxes (Howarth et al, 1996).

may be significant, perhaps accounting for up to 25% of net anthropogenic nitrogen inputs to the temperate portions of the North Atlantic basin; the redistribution of ammonia from agricultural to forest ecosystems through atmospheric transport may be important in the ability of forests to store this amount of nitrogen. Nonetheless, much of the net anthropogenic nitrogen input to the North Atlantic basin is not stored in forests or groundwater and does not flow to the ocean in rivers; by contrast with values for the temperate areas, we calculate that on average at least 340 kg/km²/year is probably denitrified or stored in wetlands and aquatic ecosystems, or one third of net anthropogenic input to the temperate region. To the extent we have overestimated storage in forests, denitrification will be even more important. Denitrification in both wetlands and aquatic ecosystems is probably of importance.

1.2. Environmental biogeochemistry of nitrogen in the East Asian region

At present the estimation of regional fluxes of pollutants gives a powerful tool for understanding of man-made changes in different scale, from landscape to a global

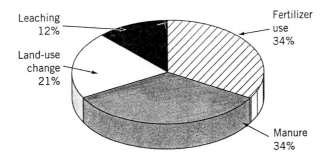

Figure 10. Present-day emissions of N_2O *from agriculture (5.0* $\times$ *10^6 tons per year) in the global scale (Oonk and Kroeze, 1999).*

one. Characteristic examples are the calculation of nitrogen cycling in the North Atlantic Ocean and its watershed (Howarth, 1996), in Baltic (Bashkin *et al*, 1997a) and Mediterranean (Bashkin *et al*, 1997b) drainage basins.

Nitrogen is a key element of many biogeochemical processes and can be both a nutrient limiting the productivity of terrestrial and aquatic ecosystems and a pollutant, which excessive accumulation in biogeochemical food chains leads to many environmental problems. The main anthropogenic sources of nitrogen pollution are related to fertilizer application, waste production and emission of gaseous species.

The modern projections indicate that potentially large increases in emissions may occur during the next 25–50 years namely in East Asia in accordance with planned development patterns. According to present estimates, at the current growth rate of energy consumption, by the year 2020 NH_x, NO_y and SO_2 emissions will surpass the emissions of North America and Europe combined. The primary man-made source of acidifying and greenhouse compounds in East Asia is fossil fuel of low quality, with high content of sulfur (up to 7% in Thai lignite, Chinese brown coal etc.) and heavy oil. The multiple effect of acidification and increased N deposition in East Asia may cause decreases in base cations, leading to nutrient imbalances in forest vegetation. This increases their vulnerability to diseases, attacks of insects and parasites. Nitrogen deposition changes natural vegetation, bacterial and mycorrhizal composition into more nitrophilic communities, with less of diversity of species, especially endemic ones. The NH_x compounds can be transported to a great distance in East Asia. The spatial scale of atmospheric NH_x pollution and its effects depend on its form: gaseous NH_3 is important near sources up to 100 km; aerosol NH_4 is transported over longer distances leading to effects up to 1000 km (Carmichael *et al*, 1997). At present, N_2O warrants also a priority on the policy and research institutes. The increasing N_2O concentration in the atmosphere mainly results from nitrification and denitrification processes due to increasing application of mineral fertilizers in various regions of the Earth and especially in East Asia (Figure 10).

This N species contributes to the greenhouse effect (about 6%) and to the destruction of stratospheric ozone (Erisman *et al*, 1999). There is agreement both nationally and internationally that long-range transboundary air pollution is not limited to the geographical limits of individual East Asian countries. It is known that during the winter the major weather patterns in East Asia facilitate the transboundary transport of air pollutants from west to east, from land to sea and the reverse in summer. Pollutants can thus be transported from country to country in the whole region of East Asia. For instance, it has been calculated that during the 1990s about 50% of oxidized and about 60% of reduced nitrogen deposition over Republic of Korea were imported from abroad. It is therefore impossible for individual countries to solve the problem of air pollution and acid rain alone. There is need for regional intergovernmental cooperation. Currently, regional/sub-regional agreements on the issue of pollutant emission abatement strategy do not exist at all or are in the initial stages (Bashkin and Park, 1998).

The abatement strategy for reduction of N emission and deposition as well as decreasing of undesirable losses from agroecosystems due to excessive application of fertilizers has to be based on the calculations of its regional biogeochemical budgets with quantitative parameterization of different fluxes in terrestrial and aquatic ecosystems.

General description of East Asian domain

The waters of four oceans wash the Asia continent with respect to the oceanic slopes. The East Asian domain is presented by Pacific ocean slope of 11,905,000 km^2 or 27,4% of total area of Asia (Figure 11).

The four main river basins are placed on this area: Amur, Huanghe, Yangtze and Mekong rivers. There are also plenty of middle and small rivers in East Asia. These rivers are drained into East (Japan), Yellow, East China and South East China Seas of the Pacific ocean.

The Yellow Sea is a typical epicontinental sea surrounded by the continent of China and the Korean peninsula and connected with the East China Sea to the south and with Bohai Sea in the north (Figure 12).

The mean water depth is about 44 m and the surface area is 420,000 km^2. The rivers (the Huanghe, the Aprock, the Han, the Keum, the Haihe, the Luanhe) drain freshwater of about >160 km^3 and suspended materials of 1.1×10^9 ton annually into Bohai and Yellow Sea system. Many industrial complexes and large cities are along the coastlines, to which great quantities of pollutants are discharged into rivers or directly into coastal waters. Further, the westerly and northwesterly winds, which prevail in winter and spring, deliver great amounts of mineral dust and anthropogenic material. As this coastal system receives large amounts of terrestrial material produced by natural weathering and human activities, it is a suitable study site for the investigation of the biogeochemical cycle of nitrogen in the regional scale.

Yellow Sea is a semi-enclosed basin. Wide coastal area (< 40 m water depth) are located along shorelines nearby both continents with a channel (> 60 m water depth), which is developed in NW-SE direction. The southward flows in both coastal areas

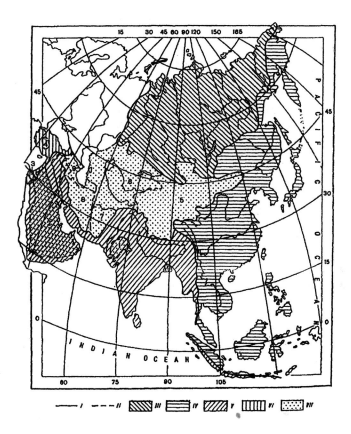

Figure 11. Ocean drainage basins in Asia.

and northward flows of warm water in the channel are the general circulation pattern in winter, but northward flow may disappear in summer, which results in the formation of the Yellow Sea cold water in the channel. The tidal fronts near the boundary of shallow coastal area and channel are developed during summer when thermal stratification is established in the channel and vertically homogenous water mass by strong tidal currents is sustained in shallow coastal areas (Seung and Park, 1990). These fronts may affect the transport of terrestrial materials (including nitrogen) to offshore, and hence the biogeochemical activity.

Billions of tons of terrigenous materials are discharged annually through rivers (Huaghe, Aprock, Han, Keum etc) and tens of millions of tons of mineral dusts ('Yellow sand') during a year are deposited into surface seawaters from the atmosphere. Most particles derived from the Huanghe river are deposited in the Bohai Sea (Cheng and Chao, 1985), and it was suggested that the basin-wide atmosphere dust flux might be comparable to the river input (Gao *et al*, 1992; Zhang *et al*, 1992; Choi *et al*, 1998).

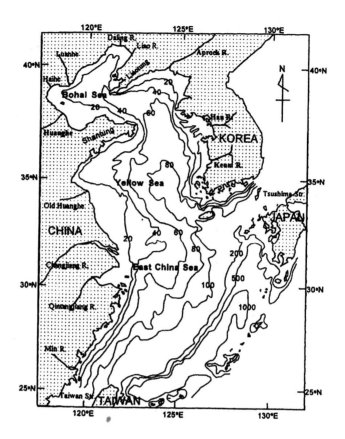

Figure 12. Bathymetry and geographic system including the east China Sea, the Yellow Sea and Bohai Sea system.

General description of natural terrestrial ecosystems in Republic of Korea

Republic of Korea is in the southern part of mountain Korean peninsula between 126° E–130° E and 34° N–38° N. Geographically this is the northern temperate zone of the Eastern Hemisphere. The overall area of Republic of Korea (ROK) comprises 99,022 km².

In ROK the principal native vegetation is Forest ecosystems. Very little of the original, virgin composition remains, mainly in the plateau of high mountain areas. The dominant vegetation consists of forest trees with varying undergrowth of shrubs and small plants.

The forest vegetation in the country is divided according to climate into two kinds, the temperate and subtropical zone forests. The secondary vegetation, which occurs in the extensive areas in the western and southern regions, consists mostly of shrubs, grasses and conifers, associated with deciduous trees. Grassland with shrubs also appears commonly as a kind of climax vegetation in the higher elevation and plateau remnants.

In accordance with land use type, the following natural ecosystems are presented in South Korean peninsula: Coniferous Forest, Deciduous Forest, Mixed Forest, Evergreen Forest, Plain Shrubs and Mountain Shrubs with Grasslands. The agricultural land use includes about 20% of total South Korean area.

Mountainous topography occupies 2/3 of the total South Korean area. The mountain range of Taebaeg on the Gangweon-Do runs southward along the east coast with lateral branches and spurs expending in a south westerly direction. The slopes to the east are steep while those to the west are gentle. The mountain range slopes towards the south, thus making the southern part of the country fairly level. On the contrary, the northern part is mountainous and hilly land.

About 80 per cent of ROK are in regions where the altitude of the summit ranges from 300 to more than 1,000 meters. The land bears a strongly dissected relief reworked by numerous erosion cycles. The low lands include both coastal plains clayey materials and the continental alluvial plains and valley flood plans of the interior.

According to USDA Soil Taxonomy, the soils in Republic of Korea are classified as 6 orders and 14 suborders (Um, 1985). The dominant soils are Inceptisol order with 4 suborders: Andepts, Aquepts, Ochrepts and Umbrepts (5,840,441 ha). The Entisol order consists of 4 suborders (Psamments, Aquents, Fluvents and Orthents) and occupies 2,849,102 ha. The Altisol order is presented by Aqualfs and Udalts suborder on the area of 309,677 ha and Histosol order is divided between Saprist and Hemist suborders with total area of 384 ha.

Data sources

The statistical characteristics of different parts of East Asian domain were extracted from both national and international sources (see, for example, Environmental Statistics Yearbook, 1998. Ministry of Environment, Republic of Korea, or UNESCO, 1978; ESCAP, 1998). Data on content of nitrogen species in river waters were selected both from our own and literature studies.

Application of biogeochemical mass balance approaches to regional studies

Any type of budget for biogeochemical turnover of pollutants depends on the availability of data. As usual the more precise calculations can be made for the small watershed with homogenous (Moldan and Cerny, 1994) or variable (Bashkin *et al*, 1984; Gunderson and Bashkin, 1994) land use. With currently available data, it is unable to fully account for the fate of both natural and anthropogenic nitrogen added to the East Asian domain as a whole and in its individual parts like North-East Asia or Yellow Sea basin. Relatively more accurate estimates might be carried out for the South Korean agroecosystems and for the total area of Republic of Korea (Bashkin *et al*, 2002).

The SCOPE project constructed mass balance for reactive nitrogen under anthropogenic influence at the regional scale (Howarth, 1996). For inputs, for the terrestrial

Table 11. Links of nitrogen biogeochemical cycle accounting for N mass balance calculations in East Asia (+ is considered; − is not considered).

Links	ROK, Agroecosystem	ROK, country scale	Yellow Sea basin	East Asian domain
Deposition	+	+	+	+
Fertilizers	+	+	−	+
N fixation	+	+	+	+
Import	−	+	−	−
Riverine N	−	−	+	−
Crops	+	−	−	−
Denitrification	+	+	+	+
Volatilization	+	+	+	+
Runoff	+	+	−	+
Sedimentation	−	−	+	−
Sea exchange	−	−	+	−

ecosystems the SCOPE N analysis considered application of N fertilizers, N fixation by agricultural crops, if any, NO_y depositions, and import or export of N in food and animal feedstocks. Output items considered crop uptake, denitrification and volatilization and river discharge. For marine ecosystems, this analysis includes also seawater exchange.

Depending upon the region, the following items were taken into account for various types of regional biogeochemical mass balance of nitrogen (Table 11).

Quantitative parameterization of fluxes for South Korean peninsula

The input links of regional N biogeochemical mass balance in South Korean agro ecosystems are assumed to include the deposition, mineral and organic fertilizers, and biological fixation (Table 11). Output items considered crop uptake, river discharge, denitrification and volatilization. All calculations were conducted for 1994–1997.

N *deposition*

The nitrogen depositions were calculated by two models, HEMISPHERE (Sofiev, 1998) and MOGUNTIA (Dentener & Crutzen, 1994), and showed similar values (10.7–11.0 kg/ha/yr.). Application of mineral fertilizers averaged 226 kg/ha/yr in South Korean agriculture being the maximum input source in mass balance of nitrogen.

Biological N fixation

Assessment of biological fixation (nonsymbiotic only, since the area under symbiotically fixed crops was very small in ROK) was carried out using the following data (Environmental Statistics Yearbook, 1998; Cleveland *et al*, 1999; Zhu *et al*, 1997).

Agricultural land use
Area of rice plantation — 1009560 ha
Rate of fixation — 45 kg/ha/yr
Annual flux — 45430 tons

Area of other crops — 966280 ha
Rate of fixation — 15 kg/ha/yr
Annual flux — 14494 tons

Forest land use
Forested area — 5072600 ha
Rate of fixation — 1 kg/ha/yr
Annual flux — 5072 tons

TOTAL: 64996

Denitrification

Denitrification values were calculated on a basis of the following data (Environmental Statistics Yearbook, 1998; Freney, 1996; Lin *et al*, 1996; Mosier *et al*, 1998; Zhu *et al*, 1997):

Agricultural land use
Losses from fertilizers

Area of rice plantation — 1009560 ha
Rate of denitrification — 32% from fertilizer rate
Annual flux — 73011 tons

Area of upland crops — 966280 ha
Rate of denitrification — 15% from fertilizer rate
Annual flux — 31887 tons

Losses from manure
Rate of denitrification — 13% from manure N rate
Annual flux — 20528 tons

Losses from soils

Rate of denitrification — 3 kg/ha/yr
Annual flux — 5916 tons

Table 12. Annual accumulation of nitrogen in human and animal excreta in Republic of Korea, mean 1994–1997.

Items	Rate, kg/capita/yr.	Population, thousand	Tons/year
HUMAN			
Population*			
Urban	0.44	33745	14848
Rural	0.69	5494	3791
Subtotal			18639
LIVESTOCK			
Cattle	11.35	3267	37085
Horse	9.79	6.7	6552
Pig	3.22	6691	21525
Sheep	0.7	1.6	1
Goat	0.7	653	458
Poultry	1.0	85623	88483
Subtotal			154113
TOTAL			172752

*Only adults.

Agricultural recycled N

Agricultural recycled N was considered for reginal biogeochemical budget in South Korean agroecosystems as organic fertilizer nitrogen. The values of organic fertilizer N were assessed using the statistical data on human and animal/poultry population and rates of N in excreta (Table 12).

Anthropogenic NH_3 emission

These values have been estimated by S-U. Park (1998). The modified European calculation factors (IPCC, 1997) were applied. The average total value was 142,123 ton and NH_3 emission from fertilizers was predominant (35% from the total value).

Table 13. Annual riverine fluxes of nitrogen from the area of Republic of Korea, mean 1994–1997.

Species	Content, mg/L N	Fluxes, tons/year
$N - NO_3^-$	1.470	90552
$N - NO_2^-$	0.045	2784
$N - NH_4^+$	1.540	94864
N-SPM	0.085	5220
TOTAL		193142

N *surface water discharge*

In addition to the input/output items for agroecosystems, we estimated the N fluxes with river runoff for calculating the N budget for the whole South Korean area. The mean annual water discharge was 61.6×10^{12} L. In accordance with statistical data, about half of wastewater was untreated in Republic of Korea in 1994–1997. As a consequence, the content of reduced nitrogen in surface waters was almost the same as the content of oxidized N. Nitrite-N was also monitored in South Korean rivers and its mean content was 0.045 mg/L. The dissolved organic nitrogen (DON) content in most monitored rivers and water reservoirs was negligible (< 1 mg/L) due to both intensive mineralization and algae uptake as well as low content of organic matter in South Korean soils. The fluxes of suspended matter were significant, totaling 1.1×10^9 ton/yr., especially during summer monsoon period, with discharge-weighted mean N content of 0.085%. The total fluxes of DIN and N-SPM were 193142 tons per year in 1994–1997 (Table 13).

Regional mass balance of nitrogen in Republic of Korea

The quantitative parameterization of different input and output links of N biogeochemical cycle in South Korean peninsula allows us to make up the calculation of two mass balance estimates, for agroecosystems ($\sim 2.0 \times 10^6$ ha) and for the whole area of Republic of Korea. The input/output items for different types of mass balance were estimated in accordance with characteristic fluxes shown in Table 11.

Biogeochemical budget of nitrogen in South Korean agroecosystems

On a basis of above-mentioned data the N budget in South Korean agroecosystems is as a follow (Table 14).

Since the South Korean agriculture combines the features of developed countries (great amount of synthetic fertilizers applied) and developing ones (N recycling),

Table 14. Nitrogen budget in agroecosystems of Republic of Korea, 1994–1997.

Items	Ton/year	% from input
INPUT		
Fertilizers	446081	65.0
Manure	157904	23.0
Biological fixation	59924	8.7
Deposition	21692	3.3
Subtotal	685601	100
OUTPUT		
Crop production	259779	37.8
Denitrification	132211	19.3
NH_3 volatilization	142123	20.7
Subtotal	524113	77.8
BUDGET*	+151488	22.2

Note: Accumulated nitrogen is distributed between following items: surface runoff; groundwater leakage, and increasing N content in crops, mainly vegetables.

it was of interest to compare the corresponding values of N mass balance in some European countries and China with local fingers (Table 15).

The general comparison of mass balance values between South Korean agro ecosystems and those for developed European countries testifs to similar processes such as great surplus of nitrogen (typically more than 100 kg/ha/yr.). The similar situation appears in modern China agriculture. However, N crop uptake efficiency in Asian countries is less than in European ones and Republic of Korea is last in the row with only 38 per cent of efficiency. This really supports an idea that an increasing non-sustainability within the agriculture, human nutrition and waste management complex has occurred both in European and Asian countries. It has been leading to a disturbance of N biogeochemical cycle.

In order to estimate the fate of nitrogen accumulated in agroecosystems, we assessed the annual N accumulation in municipal waste for the urban area of the country (Table 16).

Table 15. Nitrogen mass balance (kg/ha/yr.) and N crop uptake efficiency (% from total input) in different developed and developing countries*.

Country	Agricultural area, 10^6 ha	Input	Output	Surplus	Crop uptake efficiency
Denmark	2.9	217	30	187	59
Germany	12	215	51	164	73
UK	18.1	127	17	110	—
Netherlands	2.3	463	96	365	63
Norway	1.0	147	80	67	71
Sweden	3.7	121	21	100	63
S. Korea	2.0	347	51	296	38
China	94.9	294	95	199	51

*Data for European countries from K.Isermann (1991) and for China from G.X.Xing and Z.L. Zhu (this volume).

Table 16. Distribution accumulated N between various waste treatment types in ROK.

Items	Tons/year	% from total
Landfill	207740	75
Incineration	13849	5
Agricultural use	5540	2
Recycling	27699	10
Damping at sea	22159	8

To calculate these values, the following data were used (Environmental Statistics Yearbook, 1998):

Population, thousand — 46164
Annual N accumulation in sewage & wastes — 6 kg/capita
Annual accumulation flux — 276987 tons

One can see that the main amount of N was deposit in landfill with subsequent transformation and leaching to surface and ground water as well as denitrified.

*Table 17. Biogeochemical budget of nitrogen in
South Korea.*

Items	Ton/year	% from input
INPUT		
Deposition	108160	13.3
Fertilizers	446081	54.8
N fixation	64996	8.0
Import		
Foods	184110	22.6
Goods	10377	1.3
Subtotal	813724	100
OUTPUT		
River discharge	189124	23.2
Denitrification	132211	16.2
NH_3 volatilization	142123	17.5
Sea waste damping	22159	2.7
Subtotal	485617	59.7
BUDGET	+328107	+40.3

In its turn, this leads to further pollution of drinking water, eutrophication of surface waters, mainly, water reservoirs, and increasing input of N_2O to atmosphere. For instance, the DIN contents in many South Korean water reservoirs are 4–10 mgN/L in summer season and most of these reservoirs are eutrophied (Environmental Statistics Yearbook, 1998).

Regional N *mass balance in Republic of Korea*

Taking into account the values of various input/output items of biogeochemical N cycle as well as literature data (Environmental Statistics Yearbook, 1998; Park, 1998; Freney, 1996; Mosier *et al*, 1998; Zhu *et al*, 1997), the regional mass budget was calculated for the whole South Korean territory (Table 17).

The dominant input items were related to the application of mineral fertilizers and import of food and goods (about 80 per cent from total input). Deposition

Table 18. Distribution of accumulated nitrogen in Republic of Korea.

ACCUMULATION	Ton/year	% from total
Landfill	207740	63.2
Forest uptake	16455	5.0
Groundwater leakage*	103912	31.8

* as a difference

(about 55% from abroad through transboundary air pollution) and non-symbiotic N fixation were responsible for the other 20% of input. The output was connected with N volatilization via direct NH_3 volatilization and biological denitrification (33.7% from total input) and river discharge (23.2%). The sum of output was about 60% from input.

Thus, this budget was positive (+ 40.3% from total input) and the estimated distribution of accumulated N is shown in Table 18.

As it has been shown already during the analysis of N balance in agroecosystems, the main part of excessive nitrogen is stored at landfills with corresponding prolonged problems of environment pollution. The values of N forest uptake were calculated on a basis of data on net primary productivity and N content in tree stems and branches. Groundwater leakage was assessed as a difference between total N accumulation and sum of annual landfill storage and plant uptake in forest ecosystems. It is to note that generally the decomposition of waste residues in landfills lasts longer than one year and this approach to estimating groundwater leakage can be used.

Nitrogen mass balance in Yellow-Bohai Sea system

In accordance with approaches shown in Table 11, the total riverine N fluxes were assessed for the Yellow-Bohai sea system (UNESCO, 1978; Zhu, 1997; ESCAP, 1998; Choi, 1998; Cha *et al*, 1998). These data are shown in Table 19.

The analysis of the International SCOPE N Project suggested that without the influence of humans on the landscape, the flux of N from land to coastal waters would likely be of the order of 130 $kgN/km^2/yr$ when expressed per area of watershed (Lewis *et al*, 2002). In comparison, actual fluxes from Yellow-Bohai Seas drainage basin (areas of China and Koreas) were some 8-fold larger for the China part of drainage basin and 13-fold larger for the drainages from South Korean area in 1994–1997. Due to huge amounts of SPM transport in Yellow river, the N discharge from China area was mainly presented by solid matter (87%) whereas the opposite picture was for South Korean area: only 6% of total N flux was discharged with SPM and 94% was discharged as DIN.

Table 19. Assessment of total nitrogen riverine fluxes to Yellow-Bohai seas system.

River	Discharge, km^3/yr	N species	N content, mg/L	N fluxes, tons/yr
Han	25.0	NO$_3^-$	1.33	33188
		NH$_4^+$	1.31	32800
		NO$_2^-$	0.037	930
		N-SPM	0.064	1600
		Total		68518
Keum	6.4	NO$_3^-$	1.61	10317
		NH$_4^+$	1.54	9859
		NO$_2^-$	0.045	298
		N-SPM	0.056	1040
		Total		21514
Aprock	33.6	NO$_3^-$	0.20	6920
		NH$_4^+$	0.41	13840
		N-SPM	0.098	3460
		Total		24220
Liaohe	14.8	NO$_3^-$	0.20	2960
		NH$_4^+$	0.39	5920
		N-SPM	0.16	2664
		Total		11544
Haihe	22.8	NO$_3^-$	0.50	11400
		NH$_4^+$	0.65	14820
		N-SPM	0.18	4104
		Total		30324
Yellow	59.2	NO$_3^-$	0.87	51495
		NH$_4^+$	1.39	76947
		N-SPM	18.3	1083177
		Total		1211619

Table 19. Assessment of total nitrogen riverine fluxes to Yellow-Bohai seas system (contiued).

River	Discharge, km^3/yr	N species	N content, mg/L	N fluxes, tons/yr
Yellow & Bohai seas	161.8	DIN		271614
		N-SPM		1096045
		Total		1367959

Joining these data with those existing in literature (Cha *et al*, 1998; Choi, 1998; Park, 1998; Nixon *et al*, 1996), we calculated the N fluxes in the Yellow Sea (Table 20).

It has been shown that the main part of both soluble (53%) and particulate (> 99%) nitrogen was inputting into the Yellow sea from Northern China in spite of only 30% excess of river water discharge from the Chinese area. Totally riverine input from watershed drainage basin was equal to 52% from input values (2,581 kt/yr). Both wet and dry depositions were responsible for 20 and 22 % from input values with relatively small values of N fixation in marine waters. Denitrification is the main output items (37%) with similar values of sedimentation and water exchange with East the China Sea (7 and 8% from total input) and with negligible values of losses as N$_2$O (< 1%).

So, annual accumulation of nitrogen in the Yellow Sea was 1,229 kt/yr. (+47% from input) and the residence time of nitrogen was 1.5 year. This means that the N content in marine water was doubling every 3 years during 1994–1997. It leads to excessive eutrophication and pollution of the Yellow Sea.

Preliminary assessment of N mass balance in East Asian domain

This preliminary estimates of N fluxes in the whole East Asian domain were also conducted. This regional mass balance was presented as a comparison of input (N deposition, agricultural N fixation and fertilizers only) and output values (river discharge, volatilization and denitrification in agroecosystems only). It was based upon three assumptions:

i)   in regional scale anthropogenic N fixation during combustion and the Haber reaction process might be equal to volatile losses (denitrification and NH$_3$ volatilization) of N to atmosphere and crop uptake;

ii) biological N fixation is equal to denitrification of soil N; and

iii) N soil immobilization is equal to N soil mobilization.

These assumption are workable under steady state mass balance calculations in the regional scale like the East Asian domain where the information sources are scare and uncertainty of available data is high.

Table 20. Sub-regional fluxes of nitrogen in Yellow-Bohai seas system.

Items	Tons/year	% from input
INPUT		
Wet deposition	504000	20
Dry deposition	557000	22
N fixation	152880	6
Riverine soluble N	271614	10
Riverine SPM-N	1096045	42
Subtotal	2581539	100
OUTPUT		
Denitrification	946680	37
Losses as N_2O	23520	1
Sedimentation	170000	7
Water exchange	212000	8
Subtotal	1352200	53
ACCUMULATION	+1229339	+47
N pool in marine water	1818000	
Residence time, yr	1.47	

Statistical data for N fertilizer application was produced from national and international yearbooks, N fixation was assessed on a basis of Cleveland *et al* (1999) data. Denitrification and N volatilization were estimated using national data for China (Xing G. X. and Zhu Z. L., this volume), Republic of Korea and Russia.

Since the main uncertainty is related to N deposition input and N riverine output, these data are considered separately.

N *deposition input*

The assessment of N deposition was also done on a basis of two existing model results (HEMISPHERA model — M. Sofiev, 1998; MOGUNTIA model — P. Zimmermann, 1988; F. Dentener & P. Crutzen 1994) (Table 21).

Table 21. Assessment of nitrogen deposition in East Asian domain, 1990's.

Country/Region	Area, km^2	HEMISPHERE		MOGUNTIA	
		Rate, kg/km^2/yr.	Flux, tons/yr.	Rate, kg/km^2/yr	Flux, tons/yr.
Russia*	1244000	140	174000	200	248800
China	9600000	830	8008000	1010	9696000
N. Korea	120410	1490	170000	800	96330
S. Korea	99390	1070	107000	1100	109320
Japan	377800	1140	429000	800	302240
Taiwan	36000	—	—	600	21600
Vietnam	330000	—	—	500	165000
Thailand	511000	—	—	400	204400

*Amur river basin only.

These results are mainly similar (China, South Korea) or differentiate insignificantly (Russia, Japan, North Korea). However, the existing national data for South Korea (Park, 1998, Lee and Young, 1998) and Taiwan (Chen, 1998; Lin, 1998) deviate significantly, sometimes by 2–3-times. Since the national data are not at present available for the whole East Asian domain, only the international data on deposition were applied.

Riverine discharge of N *to Pacific Ocean*

We calculated the river N discharge using national and international data sets (UNESCO, 1978; ESCAP, 1998; Zhu, 1997; Choi, 1998; Shen *et al*, 1998; Cha *et al*, 1998; Satake *et al*, 1998; Neudachin, 1999). The results on N content in various rivers were compared with those published in the beginning of the 1980s (Meybeck, 1982). These results are shown in Table 22.

One can see that during a 20-years period both content of individual N forms and DIN have been increasing, especially ammonium content. This might be related to output of untreated wastewater to rivers. In main degree this emphasis in China rivers and in less degree in Mekong and Chao Phraya. Using these data and water discharge values, the annual fluxes of various N species were calculated for different East Asian rivers (Table 23).

These data were grouped in the scale of various sea basins in Table 24.

Table 22. Comparison of N content in main East Asian rivers in 1980 and 1996, mgN/L.

River	Cited from M. Meybeck, 1982		Present content	
	$N - NH_4^+$	$N - NO_3^-$	$N - NH_4^+$	$N - NO_3^-$
Yellow	0.01	0.24	1.39	0.87
Yangtze	0.04	1.94	1.30	0.87
Pearl	< 0.01	0.59	1.12	0.75
Mekong	—	0.24	0.07	0.16
Chao Phraya	—	0.63	0.90	0.60

The annual total flux of dissolved and suspended N was equal to 5,576,714 tons in 1994–1997 and DIN was responsible for 65% of N discharge with river water. The amount of discharged nitrogen was mainly related to water discharge values.

Regional N mass balance

Comparison of the input and output data allows us to make a preliminary assessment of N regional budget in East Asia (Table 25).

This regional budget was positive but the values of interregional transformation like crop uptake (N in crop harvest — 22,510 k tons/yr) were not accounted.

Thus, the following links of nitrogen biogeochemical cycle were accounted for mass balance calculations in Northern-East Asia and the whole East Asian domain: input — deposition, fertilizers, biological N fixation, import of food and products, riverine fluxes and output — crop uptake, denitrification, volatilization, runoff, sedimentation and sea water exchange. All calculations were conducted for 1994–1997 and the mean values were used.

The values of biogeochemical chains were assembled for the N mass balance calculations in both agroecosystems and the whole area of Korean peninsula (in the borders of Republic of Korea). N budget in South Korean agroecosystems was positive (+151 kt/yr. or 22% from input) with prevalence of fertilizer N (65%) in input and more similar distribution of output values (crop production — 38, denitrification — 19 and ammonium volatilization — 21% from input).

Relatively more nitrogen was accumulated in the whole South Korean area (+328 kt/yr or 40% from total input) due to increasing fertilization, deposition and import rates which were not equilibrated by output due to river discharge, gaseous losses and sea water damping. This gave a rise in N storage in landfill and groundwater as well as gradually increasing riverine N discharge to the Yellow Sea.

Table 23. Assessment of total nitrogen riverine fluxes to Pacific Ocean from East Asian domain.

River	Discharge, km^3/yr	N species	N content, mg/L	N fluxes, tons/yr
Amur (East sea)	355	NO_3^-	0.43	152650
		NH_4^+	0.03	10650
		NO_2^-	0.46	163300
		N-SPM	0.001	355
		Total		163655
Han	25.0	NO_3^-	1.33	33188
		NH_4^+	1.31	32800
		NO_3^-	0.037	930
		N-SPM	0.064	1600
		Total		68518
Keum	6.4	NO_3^-	1.61	10317
		NH_4^+	1.54	9859
		NO_2^-	0.045	298
		N-SPM	0.056	1040
		Total		21514
Aprock	33.6	NO_3^-	0.20	6920
		NH_4^+	0.41	13840
		N-SPM	0.098	3460
		Total		24220
Liaohe	14.8	NO_3^-	0.20	2960
		NH_4^+	0.39	5920
		N-SPM	0.16	2664
		Total		11544
Haihe	22.8	NO_3^-	0.50	11400
		NH_4^+	0.65	14820
		N-SPM	0.18	4104
		Total		30324

Table 23. Assessment of total nitrogen riverine fluxes to Pacific Ocean from East Asian domain (continued).

River	Discharge, km³/yr	N species	N content, mg/L	N fluxes, tons/yr
Yellow	59.2	NO_3^-	0.87	51495
		NH_4^+	1.39	76947
		N-SPM	18.3	1083177
		Total		1211619
Yellow & Bohai seas	161.8	DIN		271614
		N-SPM		1096045
		Total		1367959
Yangtze (East China sea)	924	NO_3^-	0.87	803880
		NH_4^+	1.30	1201200
		DIN	2.17	2005080
		N-SPM	0.526	486000
		Total		2491080
Pearl	326	NO_3^-	0.75	244500
		NH_4^+	1.12	365120
		DIN	1.87	609620
		N-SPM	0.300	97800
		Total		707420
Langgang	42	NO_3^-	0.70	29400
		NH_4^+	1.05	44100
		DIN	1.75	73500
		N-SPM	0.200	8400
		Total		81900
Mekong	510	NO_3^-	0.16	81600
		NH_4^+	0.07	39270
		DIN	0.23	122400
		N-SPM	0.50	255000
		Total		377400

Table 23. Assessment of total nitrogen riverine fluxes to Pacific Ocean from East Asian domain (continued).

River	Discharge, km^3/yr	N species	N content, mg/L	N fluxes, tons/yr
Small rivers	210	NO$_3^-$	0.06	12600
		NH$_4^+$	0.21	44100
		DIN	0.27	56700
		N-SPM	?	?
		Total		> 100000
South China sea	1088	DIN		1249820
		N-SPM		461200
		Total		1711020

The main part of both soluble (53%) and particulate (> 99%) nitrogen was inputting into the Yellow sea from Northern China in spite of only 30% excess of river water discharge from the Chinese area. Totally riverine input from watershed drainage basin was equal to 52% from input values (2,581 kt/yr.). Both wet and dry depositions were responsible for 20 and 22% from input values with relatively small values of N fixation in marine waters.

Denitrification is the main output items (37%) with similar values of sedimentation and water exchange with the East China Sea (7 and 8% from total input) and with negligible values of losses as N$_2$O (< 1%). So, annual accumulation of nitrogen in the Yellow sea was 1,229 kt/yr. (+47% from input) and the residence time of nitrogen was 1.5 year doubling the N content in marine water every 3 years during 1994–1997. It leads to excessive eutrophication and pollution of the Yellow Sea.

The regional nitrogen balance in the whole East Asian domain was positive (+12, 103 kt/yr.) or +36% from total input of 33816 kt per annum during 1994–1997. These values do not include the N uptake in agroecosystems. The latter makes regional balance negative.

2. ENVIRONMENTAL BIOGEOCHEMISTRY OF MERCURY

Mercury occupies a unique and infamous place in environmental biogeochemistry. It was the first chemical species for which a direct connection was proven between relatively low concentrations in a natural water system, bioaccumulation up the biogeochemical food webs, and a serious health impact on the human population at the top of the food chain.

Table 24. Nitrogen riverine fluxes in East Asian domain, 1990's.

Sea	River discharge, km^3/yr	N species	Fluxes, tons/yr
East	335	DIN	163300
		SPM	355
		Total	163655
Yellow & Bohai	162	DIN	271614
		SPM	1083177
		Total	1211619
East China	924	DIN	2005080
		SPM	486000
		Total	2491080
South China	878	DIN	1193120
		SPM	361200
		Total	1554320
The whole East Asian domain	2319	DIN	3633114
		SPM	1943600
		Total	5576714

2.1. Speciation of mercury

The biogeochemistry and bioavailability of mercury is strongly influenced by speciation. Mercury can exhibit a variety of aqueous and particulate species (Figure 13).

The usual form of mercury in aqueous solution is the Hg^{2+} ion. Mercury has two oxidation states, Hg(I) and Hg(II), but the first of these, that contains the unusual ion $^+Hg - Hg^+$, is stable only as insoluble salts such as Hg_2Cl_2. It disproportionates in solution as follow

$$Hg_2{}^{2+}(aq) \longrightarrow Hg^{2+}(aq) + Hg(l). \tag{1}$$

We can see from the reaction that reduction of Hg^{2+} under anaerobic conditions, for example, in bottom sediments, gives the metal in liquid form.

Mercury(II) is a very soft Lewis acid, which forms stable complexes preferentially with soft Lewis bases such as sulfur ligands. You should remember here that the major natural form of mercury is sulfides. Increasing the pH of the aqueous solution due to

Table 25. Regional budget of nitrogen in East Asian domain, 1994–1997.

Items	K tons/year	% from input
INPUT		
Deposition	8776	26
Fertilizers	23094	68
N fixation	1947	6
Subtotal	33817	100
OUTPUT		
Riverine runoff	5577	16
Denitrification	7512	22
Volatilization	8652	26
Subtotal	21714	64
BUDGET	+12103	+36

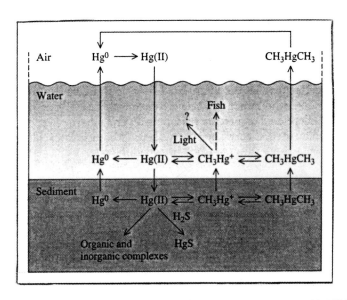

Figure 13. The mercury cycle in air-water system (Winfray and Rudd, 1990).

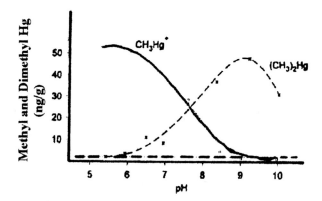

Figure 14. Methylation of Hg^{2+} *in sediments over two weeks (After Bunce, 1994).*

pollution or river water discharge to the marine water leads to precipitation of HgO. We know that HgO has finite solubility in water, and the solution may be described in term of mercury(II) hydroxide as follows.

$$HgO(s) + H_2O \longrightarrow Hg(OH)^0$$

$$Hg(OH)^0 + H^+ \longrightarrow HgOH^+ + H_2O$$

$$HgOH^+ + H^+ \longrightarrow Hg^{2+} + H_2O.$$

The major complicating factor in environmental biogeochemistry of mercury and its speciation is the biological methylation of Hg^{2+} to CH_3Hg^+ and $(CH_3)_2Hg$. This process converts inorganic mercury to organo-mercury, which is both more lipophilic and toxic (see below).

Mercury is methylated in nature by the attack of methylcobalamin (vitamin B_{12}) upon Hg^{2+}. Methycobalamin contains a methyl group bonded to a central cobalt atom, making a methyl group somewhat carbanion-like. Representing methylcobalamin as $L_5Co - CH_3$, the simplified equation is

$$L_5Co - CH_3 + Hg^{2+} \longrightarrow L_5Co^+ + CH_3Hg^+. \tag{2}$$

CH_3Hg^+ occurs mostly as CH_3HgCl. In shellfish, CH_3HgSCH_3 is also found, since mercury has a strong affinity for sulfur (Bunce, 1994). While CH_3Hg^+ derivatives predominate when mercury is methylated in sediments below pH 7, further methylation to $(CH_3)_2Hg$ becomes important as the pH rises (Figure 14).

The carbon-mercury bond is intrinsically weak, about 200 KJ/mol. However, it is almost completely nonpolar. Neither nucleophiles nor electrophiles react readily with such a center, and so organomercurials tend to be kinetically unreactive.

Reduction of Hg^{2+} to Hg^0 and alkylation to form methyl- or dimethylmercury can both be viewed as detoxication process, because all of the products are volatile and can be lost from the aqueous phase (see Figure 13). Organisms can also convert the methylated forms to Hg^0, which is more volatile and less toxic. However, both the methylated and reduced species are more toxic to humans and other mammals than is Hg^{2+}.

The most biogeochemical processes of mercury speciation are driven by microbial activities. Bacteria can facilitate the mobilization of mercury from mineral deposits by oxidizing sulfide and thereby allowing mercury, which has been sequestered in the extremely insoluble solid cinnibar (HgS) to dissolve.

2.2. Anthropogenic mercury loading

Mercury is a relatively rare chemical element. In lithosphere it occurs mainly as sulfides, HgS. Mercury sulfide comes in two forms: cinnibar, which is black, and vermillion. In some places mercury exists in small proportion as the free chemical species.

Mercury refining involves heating the metal sulfide in air in accordance with the following reaction:

$$HgS + O_2 \longrightarrow Hg + SO_2. \tag{3}$$

Gaseous mercury is condensed in a water-cooled condenser and redistilled for sale.

At present industrial mercury uses are connected with electric batteries, electric tungsten bulb, pulp bleaching and agrochemical production.

Mercury batteries are used widely in everyday life, in applications such as camera and hearing aids. About 30% of U.S. production of mercury are used in this way, the reason being the constancy of the voltage in the mercury battery, almost to the point of complete discharge.

The electrical uses of mercury include its application as a seat to exclude air when tungsten light bulb filaments are manufactured. Fluorescent light tubes and mercury arc lamps used for street lightning and as germicidal lamps, also contain mercury.

Mercury is consumed in the manufacture of organomercurials, which are used in agriculture as fungicides, e.g., for seed dressing.

2.3. Biological effects of mercury

There is no known biochemical reaction in organisms that applies Hg as an essential element. Mercury is the only metal which is a liquid at ordinary temperatures. The boiling point of this metal is 357 °C. This temperature is relatively low for metals and its vapor pressure is significant even at room temperature. The threshold limit value (TLV) of elemental mercury is 0.05 mg/m^3, a value that is less than the equilibrium vapor pressure at ambient temperatures. However, in the mercury miners in Sicily, where the mercury occurs in shales, the miners are exposed to elemental mercury vapor, which content in the air may reach toxic levels of about 5 mg/m^3. Another source of exposure in mines is the mercury-containing dust.

We can see accordingly, that in Sicily mines the TLV was exceeded by 10 fold. The TVL of organic mercury is set at 0.01 mg/m^3, in recognition of their greater toxicity.

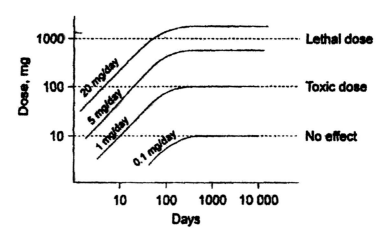

Figure 15. Accumulation curves for different levels of mercury in daily intake (After Bunce, 1994).

Mercury-containing mineral, vermillion, has for centuries been used as a pigment for oil based paint. Mercury poisoning among artists has occurred as a result of licking the brush to get a fine point.

Organic derivatives of mercury are more hazardous than the simple inorganic salts because they are lipid soluble and hence bioconcentrate. These species are able to cross the blood-brain barrier, thereby causing the neurological symptoms associated with mercury intoxication.

The greater toxicity of the lipophilic and bioconcentratable forms of mercury is shown in the following values of LD_{50}, lethal doze for 50% of exposed population, for birds: $HgCl_2$, 5000 mg/kg: C_6H_5HgOAc (seed dressing), 1000 mg/kg; C_2H_5HgCl, 20 mg/kg. Alkylmercury compounds, especially short chain alkylmercury derivatives, are able to cross the blood-brain barrier, and this explain why mercury poisoning is accompanied with mental disturbance.

Elemental mercury is mainly hazardous as vapor. There is less danger of absorbing the metal for the digestive tract. Like the alkylmercurials, elemental Hg affects the central nervous system, accompanied with such symptoms as tremors, irritability, and sleeplessness. Kidney damage is also reported due to influence of inorganic mercury salts due to a complexation of mercury by the protein metallothionein, which accumulates in the renal tubules.

Consumption of mercury in the diet leads to the accumulation of this metal in the body (Figure 15).

Mercury acts as a cumulative poison because the rate of clearance of mercury from the body is low. The single intake at a specified level may cause no ill effects, however the same concentration of metal in the meal of a steady diet can lead to sickness or even death.

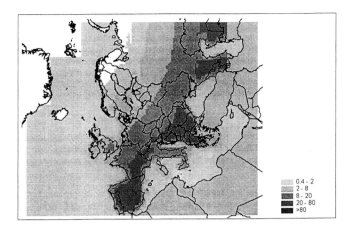

Figure 16. Spatial distribution of natural mercury emissions in Europe, g/km^2 per year (Ryaboshapko et al, 1999).

2.4. Environmental biogeochemical cycling of mercury

Unlike most heavy metals, the natural and anthropogenic cycles of mercury are dominated by atmospheric transport. We can show this on a few characteristic points. First, metallic mercury has the highest vapor pressure of any heavy metals, and it is released in biogeochemically significant amounts by volcanic eruptions, volatilization of methyl- and alkylmercurials from land and ocean surfaces. Volatilization processes are also of great importance for industrial emissions such as smelting of minerals and burning of fossil fuels. As an example, Figure 16 and Figure 17 show the natural and anthropogenic emission of mercury in Europe.

The highest density of mercury emission is observed in industrial regions of Europe such as northwestern Germany, southern Poland and the eastern Ukraine. The spatial distribution of anthropogenic mercury emissions reflects mainly the level of coal consumption in different European regions. Total anthropogenic emission in Europe in 1996 was estimated as much as 325 ton/year. The diversity of sources of mercury emissions to the atmosphere and their sporadic character makes it difficult to estimate their intensities. The difference in modern estimates often exceeds an order of magnitude. However, the averaged values for European domain was 145 tons/year from the land surface and 26 tons/year from the sea surface, totally 171 tons/year. Thus, anthropogenic emission exceeded the natural one by 1.9 times at the end of the twentieth century.

The calculation of atmospheric budgets allows us to obtain a general idea on the effect of anthropogenic activity on the environmental biogeochemical cycle of mercury. For mercury, it is interesting to consider budgets of its individual species because depending on speciation, the environmental effects are appreciably different. If elemental mercury, Hg^0, is sufficient inertial for biota, oxidized inorganic forms,

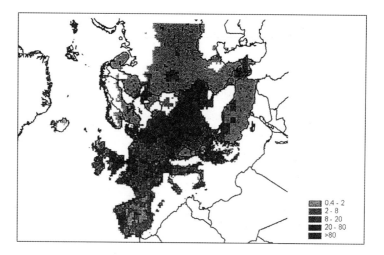

	0.4 - 2
	2 - 8
	8 - 20
	20 - 80
	>80

Figure 17. Field of anthropogenic emission of mercury in 1996 with spatial resolution $50 \times 50\,km$, g/km^2 per year (Ryaboshapko et al, 1999).

Hg^{2+}_{gas} and Hg^{2+}_{part} and particularly organic forms, dimethylmercury, DMM, easily interact with living matter. These calculations include the air reservoir over European domain, $37.5 \times 10^6\,km^2$ and 4 km height. (Table 26).

The consideration of these budget components shows the predominant emission of the gaseous elemental mercury to the atmosphere from both natural and anthropogenic sources. The dry deposition is represented also by elemental mercury, however the wet deposition is mainly connected with gaseous and particulate Hg^{2+}.

The deposition pattern for total mercury (dry and wet deposition) in European region in 1996 is shown in Figure 18.

Mercury deposition field is formed by wet and dry deposition of three different mercury forms (DMM is scavenged only by photochemical decomposition). A pronounced gradient of depositions is observed from Central Europe to the north. Maximum values of depositions reach $500\,g/km^2/year$. With the distance from pollution sources, the deposition intensity is abruptly decreased and in the periphery of domain it is $5–20\,g/km^2/year$. The lowest values are shown for the northern periphery of Europe.

Catchments study

The small catchment study of biogeochemical mass balance of mercury has been carried out in southern Sweden in the early 1990s. The fluxes of methyl $Hg(Hg_m)$ and total $Hg(Hg_t)$ were monitored (Figure 19).

Much of the Hg_t pool was found in the upper part of the soil, which is rich in organic matter. This pattern is likely due to an elevated atmospheric deposition of Hg_t over the extended period and immobilization of mercury by organic functional

Table 26. Mercury atmospheric budget in Europe, 1996, tons/year.

Budget items	Mercury species				
	Hg^0	Hg^{2+}_{gas}	Hg^{2+}_{part}	DMM	$\sum Hg$
Total emission	418	95	57	27	597
Natural emission	197	0	0	22	219
Direct anthropogenic emission	174	95	57	0	326
Re-emission	45	0	0	5	50
Advective transport in the region	8559	0	60	0	8619
Vertical influx from the free troposphere	9712	0	65	0	9777
Total deposition	148	88	262	0	498
Dry deposition	129	29	17	0	175
Wet deposition	18	60	245	0	323
Advective transport from the region	9101	4	177	1	9283
Vertical outflow to the free troposphere	9082	3	157	0	9242

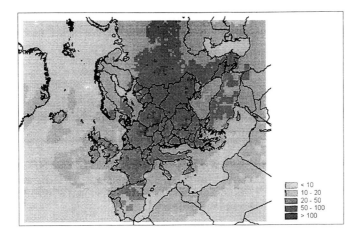

Figure 18. Field of total (wet and dry) mercury deposition in Europe in 1996, g/km^2/year (Ryaboshapko, 1999).

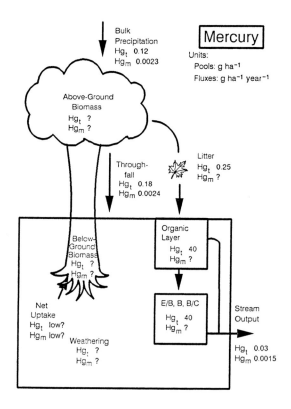

Figure 19. Biogeochemical mass budget of mercury in the experimental forest catchment in South Sweden. The fluxes of methylmercury (Hg_m) and total mercury (Hg_t) are shown in g/ha/year (Driscoll et al, 1994).

groups and accumulation of organic matter as part of the soil-forming process. The retention of mercury in the more humus layer was almost complete due to the very strong association between Hg_t and humic substances.

The runoff export of mercury was about 0.03 g/ha of Hg_t per year. This value lies within the range of 0.008 to 0.059 g Hg/ha/year, derived from a number of catchment studies in Sweden. The output of methylmercury from catchment area was 0.0015 g/ha/year of Hg. This is substantially lower than the input to the catchment. There appears to be an ongoing net accumulation of Hg_m in the terrestrial ecosystems, similar to the pattern previously shown for the total mercury.

There exists evidence of a coupling between the total and methyl-mercury concentrations in surface runoff water. The concentration of methylmercury in surface runoff is of special interest, since this pathway is a major component of the total mercury loading from drainage terrestrial Forest ecosystems to the aquatic ecosystem of lakes. Moreover, methylmercury is the form of Hg that is enriched in the aquatic biogeochemical food web and subsequently transferred to the human population through fish consumption. A close correlation was found between the water color (i.e., dissolved humic substances) and the concentration of methylmercury and total mercury. This supports the assumption that the transport of dissolved organic matter from the soil with drainage water is regulating the flux of both Hg_m and Hg_t from the terrestrial to the aquatic ecosystem.

Driscoll et al (1994) have studied the mercury species relationships among water, sediments, and fish (yellow perch) in a series of Adirondack lakes in New York state, USA. In most lakes, approximately 10% of the total mercury loading was in the form of $C_2H_5Hg^+$. Mercury concentrations increased as pH fell, but the best correlation was found between [dissolved Al] and [dissolved Hg] suggesting that the same factors are responsible for mobilizing both these metals. Methylmercury concentrations correlated strongly with the dissolved organic carbon content in the water. Fish muscle tissue was analyzed for mercury and showed an increase with age. However, the study was unable to resolve the question of whether the principal source of mercury to these lakes was atmospheric deposition or dissolution from bedrock due to acid rains.

2.5. Global mass balance of mercury

As we can see above, the anthropogenic sources are predominant in mercury loading in comparison with natural emission for European domain as one of the most polluted regions of the World. The comparison of pre-industrial and modern biogeochemical fluxes and pools has been made up in the late 1970s. These biogeochemical cycles are shown in Figures 20 and 21, correspondingly.

Mercury has been used by man about 2000 years. The comparison of pre-industrial (natural) and anthropogenically modified biogeochemical cycle of mercury allows us to make some general remarks. First, the fluxes between atmosphere and both terrestrial and oceanic ecosystems are much greater than between terrestrial and oceanic ecosystems. Second, the human activity changed significantly the pre-industrial balance fluxes between land and atmosphere. Modern fluxes from terrestrial ecosystems to the atmosphere are about 40% higher than pre-industrial ones. In the most extent, the riverine transport has been changed, up to 4 fold. The results of Figures 20 and 21 show that the average residence times for mercury in the atmosphere, terrestrial soils, oceans, and oceanic sediments are approximately 11 days, 1000 years, 3200 years and 2.5×10^8 years, respectively.

The up-to-date results can be compared with those of the late of 1970s. The review of natural sources of mercury emission has been carried out in some researches. Table 27 presents values of natural emissions obtained by J. Nriagu in 1989.

Data of other authors differ appreciably from the estimates shown in Table 27. For example, Thornton et al (1995) believed that integral value of natural emissions to the atmosphere reach 180×10^2 tons/year, whereas those of Rasmussen (1994) were 350×10^2 tons/year. According to data of Geological Survey of Canada (GSC, 1995) the natural inflow of mercury to the atmosphere can exceed even 1000×10^2 tons/year.

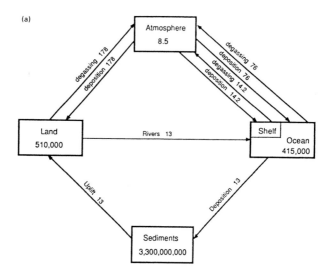

Figure 20. The pre-industrial global biogeochemical cycle of mercury. Units are 10^2 tons (pools) and 10^2 tons/year (fluxes). (Adapted from National Academy of Sciences, 1978).

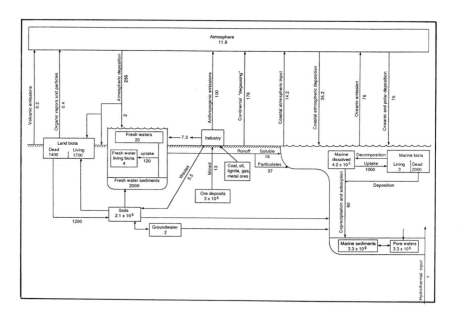

Figure 21. The global biogeochemical cycle of mercury at the present environment. Units are 10^2 tons (pools) and 10^2 tons/year (fluxes). (Adapted from National Academy of Sciences, 1978).

Table 27. Mean median values and possible scattering range (in brackets) of mercury emissions to the atmosphere from different natural sources (After Nriagu, 1989).

Natural source	Mercury emissions, 10^2 tons/year
Wind-borne soil particles	5(0–10)
Seasalt spray	2(0–4)
Volcanoes	1(0.3–20)
Wild forest fires	2(0–5)
Biogenic particulates	3(0–4)
Continental volatilies	61(2–120)
Marine volatilies	77(4–150)
Total natural emission	25(0.1–49)

Table 28. The most important anthropogenic emission sources of mercury to the atmosphere on the global level in 1983 (After Nriagu and Pacina, 1988).

Source type	Mercury emission flux, 10^2 ton/year
Fossil fuel combustion	7–35.0
Wood burning	0.6–3.0
Metallurgical processes	0.5–2.0
Waste incineration	2.0–22.0
Total — median value	36.0
— scattering range	9.0–62.0

A similar comparison can be conducted with results of environmental biogeochemical cycle of mercury. The most important anthropogenic emission sources of mercury are listed in Table 28.

We can see a significant range of scattering in Table 28. Some author presented a significantly narrower range of global emission estimates for mercury, from 36 to 45×10^2 tons/year.

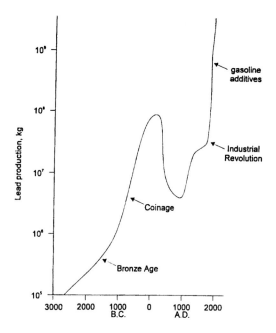

Figure 22. Historical production and consumption of lead (Bunce, 1994).

These modern data allow us to make some correction in the environmental bio-geochemical cycle of mercury, however these can not change the general conclusion on the dramatic alteration of natural biogeochemical cycle of this element under anthropogenic activity.

3. ENVIRONMENTAL BIOGEOCHEMISTRY OF LEAD

Human activity has changed the intensity of natural biogeochemical fluxes of lead during industrial development. However, the history of lead use is the longest among all metals. The period of relatively intensive production and application of lead is about 5000 years. Lead has been used as a metal at least since the times of the Egyptians and Babylonians. The Romans employed lead extensively for conveying water, and the elaborate water distribution systems and bending of the soft metal lead. Through the Middle ages and beyond, the malleability of lead encouraged its use as a roofing material for the most important constructions, like the great cathedrals in Europe. The modern production of lead is $n \times 10^6$ tons annually (Figure 22).

The noticeable changes of lead content in the environment are dated as long as about 2700–3000 years. These results come from glacier and peat monitoring using deep drilling.

3.1. Speciation of lead

The common compounds of lead derive from the +2 oxidation state. As a number of the periodic group IV-A, lead also forms tetravalent compounds, which are covalent. The most important are the tetraalkylleads, which are used as gasoline additives. The Pb − C bond is very non-polar, and the organolead species tend to be kinetically inert, like organomercurials.

The speciation of lead(II) in aqueous solution involves several polymeric hydroxo-complexes. Below pH 5.5, Pb^{2+}(aq) predominates. However, with increasing pH, $Pb_4(OH)_8^{4+}$, and $Pb_3(OH)_4^{2+}$ form consequently with following deposition of $Pb(OH)_2(s)$.

3.2. Anthropogenic lead loading

Lead occurs in nature as the sulfide, galena, PbS. Lead is more electropositive than mercury, and roasting the sulfide in air forms lead oxide.

$$2PbS + 3O_2 \longrightarrow 2PbO + 2SO_2. \tag{4}$$

The oxide is then reduced to the metal with coke. The impure metal is refined by electrolysis.

Major anthropogenic sources of lead include the use of Pb as a petrol additive, Pb mining and smelting, printing, Pb paint flakes, sewage sludge and the use of pesticides containing Pb compounds, like lead arsenate.

The famous use of lead is also the familiar lead-acid storage battery. This device is an example of a storage cell, meaning that the battery can be discharged and recharged over a large number of cycles. The lead-acid battery is familiar as a battery in your car.

An important disadvantage of the lead-acid battery is its heavy mass, on account the high Pb density. The second disadvantage is that the used car batteries distribute a lot of lead into the environment; despite recycling, they are the major source of lead in municipal waste. Recently, the recycling of lead-acid batteries has created problems in the local environment around recycling plants. Most of these plants are located in developing countries of Asia and Latin America and they process batteries imported from industrialized nations. Levels of Pb as high as 60,000–70,000 ppm have been measured in soils in the vicinity of Pb-battery recycling plants in the Philippines, Thailand and Indonesia. The relevant health effects have been observed. This appears to be one example where trying to conserve resources and minimize pollution has gone seriously wrong. In California, soil contaminating 1000 ppm of Pb is considered to be hazardous waste and its disposal is strictly regulated.

3.3. Biological effects of lead

The half-life of lead in humans is estimated to be about 6 years for the whole body burden and from 15 to 20 years for skeleton. Thus, an excretion from the skeleton is very slow. Lead, like mercury, is a cumulative poison. The skeletal burdens of lead increase almost linearly with age. This suggests that the Pb steady state is not normally reached. Chelation of Pb^{2+} with ethylendiaminetetraacetic acid (ADTA) has been found beneficial in reducing Pb body burden for clinically affected patients.

Lead, like mercury, causes neurological diseases. The organolead compounds are more toxic than mineral lead salts, since they are non-polar, lipid-soluble, and more readily cross the blood-brain barrier. This disease is related to mental retardation is children, lower performance on I.Q. tests, and hyperactivity. Severe exposure in adults causes irritability, sleeplessness, and irrational behavior. Some have gone as far as to blame anti-social behavior and criminality on sub-clinical Pb poisoning. A correlation between Pb in blood and Pb in air, dust and soils has been observed in many studies. The U.S. Centers for Disease Control has proclaimed a goal of reducing blood lead contents in children below $10\,\mu g/100mL$.

Lead is a well-known poison, but the effects of exposure to lower levels have been contentious. There is growing evidence of sub-clinical Pb poisoning, especially among young children who play in polluted parks, gardens and streets. Contaminated soil or dust particles may be transferred to children's hands and ingested accidentally. Humans are exposed to Pd from various sources and road dust and soils can contribute to the total lead exposure. Approximately one half of lead ingested in food is absorbed.

3.4. Environmental biogeochemistry of lead

The long-term uses of lead explain why this element should be so widely dispersed in the environment. In this relation we should answer the question as to what is the natural background level of lead. At present this is a question of controversy. Lead levels in modern people are frequently 10% of the toxic level. Some analyses of ancient bones and ancient ice cores seems to suggest that this relatively high level is not new and has been existed in the environment. Accordingly, the assumption was carried out that life has evolved in the presence of this toxic element.

However, recent researches have challenged this viewpoint, claiming that these lead analyses in ancient samples are the results of inadvertent contamination of the samples during their collection and analysis. Dr. C. C. Patterson of the California Institute of Technology argues, for example, that the ice cores are contaminated by lead from drilling equipment. His data of chemically careful Pb analysis on Greenland ice cores show the increasing trend of lead pollution (Figure 23).

Similar data reported on the content of lead in the meticulously preserved old skeleton contain 0.01 to 0.001 times as much lead as contemporary skeleton.

A different perspective is provided in the analysis of pre-industrial and contemporary Alaskan Sea otter skeletons. The total concentrations of lead in the two groups of skeletons were similar, but their isotopic compositions were different. The pre-industrial skeletons contained lead with an isotopic ratio corresponding to natural deposits in the region, while the ratio in the contemporary ones was characteristic of industrial lead from elsewhere (Smith *et al*, 1990).

Lead emission and deposition in Europe

The most important source of lead transport is connected with air emission, transportation and deposition. Among the other metals, lead emissions are most uniformly

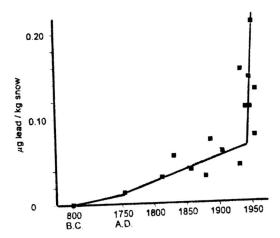

Figure 23. Increase of lead in Greenland snow, 800BC to present (Bunce, 1994).

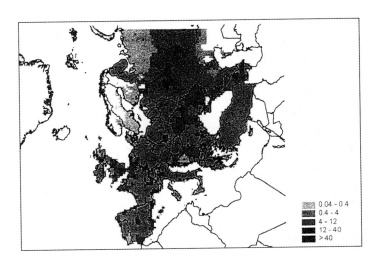

Figure 24. Field of anthropogenic emission of lead in 1996 with spatial resolution of $50 \times 50\,km$ for the European domain, kg/km^2 per annum (Ryaboshapko et al, 1999).

distributed over space. It is associated with an essential input of road transport abundant everywhere.

The spatial pattern of lead emission in Europe is shown in Figure 24. The spatial resolution is $50 \times 50\,km$.

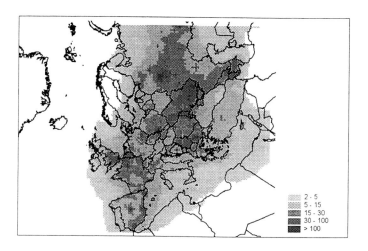

Figure 25. Mean annual air concentrations of lead in surface air in Europe in 1996, ng/m³ (Ryaboshapko et al, 1999).

The most emissions were monitored in UK, northern parts of France, the Netherlands and Germany, southern Ukraine and central Russia. The northeastern part of Russia and northern parts of Scandinavia have had the lowest emissions.

The levels of emission should coincide with levels of air concentrations and depositions of lead. Lead surface air concentrations are shown in Figure 25.

As evident from Figure 25, the most polluted regions are located in southeastern countries of Europe, Ukraine, Romania, Bulgaria and Yugoslavia. Lead concentration in the surface air of this part of Europe reach $100 \, ng/m^3$. It is worth recalling that we are speaking about concentrations on the regional levels beyond cities. It is clear that in cities and in the vicinities of point sources the concentrations can be by orders of magnitude higher. With the increase of distance from major emission sources concentrations decrease rapidly.

Comparatively high pollution levels from 15 to $100 \, ng/m^3$ are characteristic of central Russia, UK, France and Mediterranean coastal area of Spain. The countries of North Europe are the less polluted.

The map of total wet and dry deposition is demonstrated in Figure 26.

Comparing the emission map (Figure 24) and deposition map (Figure 26), we can see that the regions of maximum emission correspond to the regions with maximum depositions. As in case of air concentrations, depositions demonstrate a gradient from Central Europe to the Arctic. In countries of southeastern Europe as well as in Great Britain, France, Spain and Poland, the deposition rates can exceed $10 \, kg/km^2$ per year. In the countries of central Europe, such as Germany, Austria, and Italy, deposition rates range from 0.5 to $10 \, kg/km^2$ per year. In northern countries, the maximum values are not higher then $2.5 \, kg/km^2$ per year.

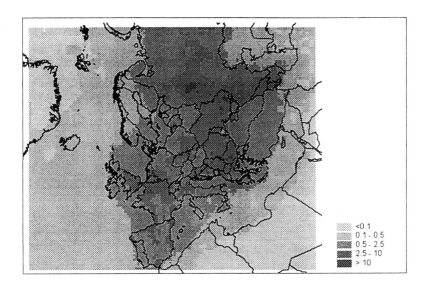

Figure 26. Total (wet and dry) deposition of lead in Europe in 1996, kg/km^2 per year (Ryaboshapko et al, 1999).

The calculation of atmospheric budgets allows us to obtain a general idea on the effect of anthropogenic activity on the environmental biogeochemical cycle of lead in European domain. These calculations include the air reservoir over European domain, 37.5×10^6 km^2 and 4 km height (Table 29).

It follows from Table 29 that the major contribution to the pollution of Europe by lead is caused by direct anthropogenic sources. Natural emissions and the inflow from outside are only 8%. The bulk of lead influx to the European domain fell out within it. The main process responsible for scavenging of this metal from the atmosphere is wet deposition. According to the data of Table 29, wet deposition scavenges about 85% of total removal of lead.

The total export of lead outside of Europe exceeds its advective and vertical influx. Thus Europe is a 'net exporter' of lead to the Earth's atmosphere, 8000 tons annually. However, this value is significantly lower than that of direct anthropogenic emission.

Small catchment study of environmental lead biogeochemistry

The input of airborne lead to the Forest ecosystems has been studied at the Hubbard Brook Experimental Forest in New Hampshire. The small catchment approach has been used to study the lead biogeochemical cycle since 1963 (Lickens *et al*, 1977; Driscoll *et al*, 1994). By monitoring precipitation inputs and stream output from small watersheds that are essentially free of deep seepage, it is possible to construct accurate

Table 29. Atmospheric budget for lead over European domain in 1996.

Budget items	Lead flux, ton/year
Total emission	40525
Natural emission	912
Direct anthropogenic emission	39613
Advective transport in the region	2370
Vertical influx from the free troposphere	40
Total deposition	32606
Dry deposition	4793
Wet deposition	27813
Advective transport from the region	5605
Vertical outflow to the free troposphere	4829

lead mass balance. The detailed study of soil and soil solution chemistry and forest floor and vegetation dynamics supplemented the deposition monitoring.

The biogeochemical mass balance of lead is shown in Figure 27.

The atmospheric deposition of lead was 190 g/ha/year and this value was connected with declining of leaded petrol use in USA from 1975. The mineral soil and forest floor were the major pools of Pb in the ecosystem. Mineral soil pools ($<$ 2 mm size fraction) are generally the largest element pools for the HBEF, however this includes relatively unreactive soil minerals. Deposition and accumulation of Pb in the forest floor have been the focus of a number of investigations. It has been shown that at HBEF, much of lead entering the ecosystem from the atmosphere appears to be retained in the forest floor. Concentrations and fluxes of lead in bulk deposition are much greater than in Oa horizon leachate. Solution concentrations and fluxes of Pb decrease through the soil profile and losses in stream water are low. There was a strong correlation between concentrations of Pb and dissolved organic carbon (DOC) in soil solution and stream water at Hubbard Brook (Driscoll *et al*, 1994).

Pools and uptake of lead in vegetation at Hubbard Brook were insignificant. Lead is not a plant essential nutrient and therefore it is not surprising that uptake was low.

The calculated weathering release of lead at HBEF is negative (-174 g/ha/year). This pattern is likely due to changes in mineral soil lead pools over the study period. There has been marked decline in lead concentration of the forest floor; concurrently with decline in atmospheric inputs during 1975–1990 (Figure 28).

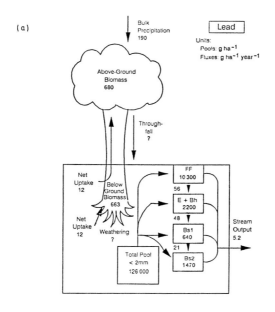

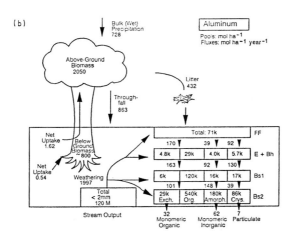

Figure 27. Biogeochemical mass balance of lead in Forest ecosystems of Hubbard Brook Experimental Forest, USA (Driscoll et al, 1994).

These long-term small catchment study results suggest that stream water concentrations are very low and not a water quality concern. In addition, a study of lead in soil solution and stream water following a commercial whole-tree harvest at Hubbard Brook showed that Pb was not released to drainage waters from clearcutting activities (Fuller *et al*, 1988).

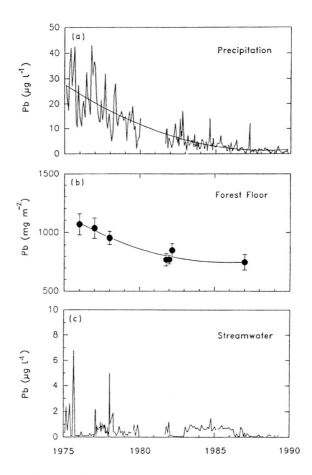

Figure 28. Temporal pattern of the concentration of Pb in the biogeochemical reference watershed at the Hubbard Brook Experimental Forest, NH, USA: (a) bulk precipitation; (b) the forest floor; (c) stream water (After Driscoll et al, 1994).

3.5. Global mass balance of lead

V. Dobrovolsky has conducted a detailed consideration of global lead mass balance in 1994. The pools and fluxes of this heavy metal are shown in Table 30 and 31.

We can compare the natural and anthropogenic emissions of lead in the global cycle. Table 32 shows the natural sources and Table 33 demonstrates the major anthropogenic ones.

Data of other researchers differ appreciably from the data of Table 32. For example, Thornton *et al* (1995) considered the integral natural emission of lead as much as 35×10^3 tons/year. According to data of Geological Survey of Canada (GSC, 1995), the natural inflow of lead to atmosphere can reach 330×10^3 tons/year.

Table 30. Pools of lead in the Earth's bio-sphere (After Dobrovolsky, 1994).

Pools	Lead mass, 10^6 tons
Earth's crust	
Sedimentary shell	32×10^6
Granite layer	131.2×10^6
Terrestrial ecosystems	
Troposphere	0.003
Vegetation	3.1
Soil humus	9.0
Aquatic ecosystems	
Troposphere	0.001
Marine biota	0.004
Soluble marine salts	41.0
Suspensions	0.14
Total	163.200×10^6
Without Earth's crust	53.248

As evident from Table 33, the uncertainty level of the estimates is significant. The mid-1990s estimates are given by Thornton *et al* (1995) and the global anthropogenic emission was calculated as much as 450×10^3 tons per year. The latest estimates were made for various continents by Pacina *et al*, 1998 (Table 34)

We can see that anthropogenic emissions of lead are most important in developed industrial countries because they are mainly associated with fossil fuel combustion for power generation and with transport. The highest emission per individual is in Australia due to non-ferrous industry emissions predominating on this continent. As to the pollution density per area unit, however, Europe is ahead of Australia, Africa and South America by an order of magnitude. The total global emission was estimated as much as 210×10^3 tons per year with averaged emission of about 40 g/capita per year.

Thus, the contemporary anthropogenic emission of lead exceeds the natural emission in global scale in almost 30 times.

Table 31. Biogeochemical fluxes of lead in contemporary Earth's biosphere (After Dobrovolsky, 1994 with authors' additions).

Items	Annual fluxes, 10^3 tons per annum
Terrestrial ecosystems	
Dry deposition	10–75
Wet deposition	110–330
Oceanic ecosystems	
Dry deposition	20–100
Wet deposition	400–2500
Fluxes from terrestrial to oceanic ecosystems	
River discharge	4100
Wind transport	40
Fluxes from terrestrial ecosystems to the atmosphere	
Natural emissions	18–330
Anthropogenic emissions	289–450
Living matter uptake	
Terrestrial ecosystems	210
Oceanic ecosystems	110
Municipal and industrial waste	350
Total	5621–8605

Table 32. Mean median values and possible scattering range (in brackets) of lead emissions to the atmosphere from different natural sources (After Nriagu, 1989).

Natural source	Lead emissions, 10^3 ton/year
Wind-borne soil particles	3.9 (0.3–7.5)
Seasalt spray	1.4 (0.02–2.8)
Volcanoes	3.3 (0.54–6.0)
Wild forest fires	1.9 (0.06–3.8)
Biogenic particulates	1.3 (0.02–2.5)
Continental volatilies	0.21 (0.01–0.38)
Marine volatilies	0.24 (0.02–0.45)
Total natural emission	12.0 (9.7–23)

Table 33. The most important anthropogenic emission sources of lead to the atmosphere on the global level in 1983 (After Nriagu and Pacina, 1988).

Source type	Lead emission flux, 10^6 ton/year
Fossil fuel combustion	2.7–18.4
Wood burning	1.2–3.0
Metallurgical processes	31.1–83.8
Waste incineration	1.6–3.1
Mobile sources	248.0
Other human activities	4.0–19.6
Total — median value	332
— scattering range	288.7–376.0

Table 34. Anthropogenic lead emissions to the atmosphere as of 1989 on different continents (After Pacina et al, 1998)

Continent	Emission, 10^3 tons per year	Relative emission per:	
		Capita, g/yr/capita	Area unit, g/km^2/yr
Africa	17.5	30	600
Asia	74.3	20	1700
Australia	5.4	200	700
Europe	69.6	70	7000
North America	36.8	90	1500
South America	15.1	50	800
World	208.6	38	1390

FURTHER READING

Vladimir N.Bashkin (1987). Nitrogen Agrogeochemistry. ONTI Publishing House, Pushchino, 267 pp.

Samuel S. Butcher, Robert J. Charlson, Gordon H. Orians and Gordon V. Wolfe, Eds. (1992). Global Biogeochemical Cycles. Academic Press, London *et al*, Chapter 15.

Bedrich Moldan and Jiri Cherny (Eds.) (1994) Biogeochemistry of Small Catchments, J. Wiley and Sons, Chapter 13.

Vsevolod V. Dobrovolsky, (1994). Biogeochemistry of the World's Land. Mir Publishers, Moscow and CRC Press, Boca Raton-Ann Arbor-Tokyo-London, 206–211.

Robert W.Howarth (Ed.) (1996). Nitrogen Cycling in the North Atlantic Ocean and its Watersheds, Kluwer Academic Publishers, 304 pp.

Miroslav Radojevic and Vladimir N. Bashkin (1999). Practical Environmental Analysis. Royal Society of Chemistry, UK, Chapter 5.13.

QUESTIONS AND PROBLEMS

1. Present a definition of environmental biogeochemistry and define the process of environmental pollution from a biogeochemical point of view.

2. Explain the choice of nitrogen as a main element for consideration in environmental biogeochemistry. Give relative examples.

3. In this chapter the North Atlantic region has been considered to as an example of the environmental biogeochemistry of nitrogen. Explain this selection.

4. Describe the application of small catchment approach for the nitrogen mass budget research. Give an example from North American and European domains.

5. Discuss the environmental chemistry of nitrogen in small forest catchments. Define the process of nitrogen saturation in Forest ecosystems and give characteristic examples.

6. In the major extent the natural biogeochemical cycle of nitrogen has been changed in agroecosystems. Discuss this problem by taking into account the small catchment approach.

7. Consider the disturbance of biogeochemical cycle of nitrogen and explain why leaching plays the most importance role in environmental biogeochemistry of this element.

8. Discuss the biogeochemical cycle of nitrogen in mixed ecosystems. Stress your attention on wetlands and upland landscapes.

9. Discuss the peculiarities of a regional approach to estimating environmental biogeochemical processes in the North Atlantic region.

10. Explain different approaches to a consideration of the role of various airborne nitrogen species during regional mass balance calculations.

11. Consider the correlation between atmospheric deposition of oxidizing N species in various regions of the North Atlantic watershed and riverine runoff of nitrogen from the drained areas. Explain the reasons and causes.

12. Discuss the accumulation of nitrate nitrogen in ground waters of different regions. Why may these waters be considered as biogeochemical barrier in environmental biogeochemical cycle of nitrogen?

13. Present the example of regional biogeochemical mass balance of nitrogen. Discuss the ratio between input and output items for different regions.

14. Consider the environmental biogeochemistry of nitrogen in East Asia. What are the similarities and differences between North Atlantic and East Asian domains regarding the disturbance of natural N cycling?

15. Describe the regional biogeochemical mass balance of nitrogen on the example of South Korea. Stress your attention on the characteristic features of nitrogen cycle in this country.

16. Discuss the regional mass budget of N in Bohai-Yellow Seas. Consider the role of riverine discharge in nitrogen fluxes and pools of this marine system.

17. Compare the environmental biogeochemistry of nitrogen in the North Atlantic watershed and East Asia. What are the main reasons of N accumulation in both regions?

18. Discuss the behavior of mercury in the modern environment. Why is this metal of special attention for environmental biogeochemistry?

19. Describe the role of speciation in the biogeochemical cycle of mercury. What forms of mercury are the most dangerous and why?

20. Consider the budget of mercury over European domain. Characterize the ratio between different forms of this metal in emission and deposition processes.

21. Compare the pre-industrial and modern cycles of mercury in global scale. What are the consequences of disturbance of the natural biogeochemical cycle of mercury?

22. Discuss geochemical characteristics of lead and the distribution of this element in ancient and modern environments. Present evidence of accelerated migration of lead during the last centuries.

23. Compare the ecotoxicology of mercury and lead and discuss their migration in biogeochemical food webs.

24. Describe the biogeochemical mass balance of lead over Europe. Present the main sources of lead emission in Europe.

25. Discuss the biogeochemical behavior of lead and mercury in small catchments of Europe and North America. Explain the role of complexation in biogeochemical migration of these metals.

26. Consider the global fluxes and pools of lead. What processes are of main importance in natural and anthropogenic biogeochemical cycles of lead?

CHAPTER 9

HUMAN BIOGEOCHEMISTRY

In the previous chapters we have considered the natural biogeochemical cycles of various chemical elements and their alterations under anthropogenic activity. We can conclude that at present the alteration of biogeochemical cycling under the influence of humans is undoubtedly accompanied by changes in food webs. This leads to the reversible and/or irreversible transformation of human health both at the individual and population levels. For an understanding of the interactions between changing natural biogeochemical cycling and human health, the researcher should take into account the numerous data on the migration of essential and non-essential elements in biogeochemical food webs and estimate the correlation with biochemical and physiological indexes of human organisms. This interaction is the subject of human biogeochemistry. *Human biogeochemistry is the rapidly developing branch of modern biogeochemistry dealing with the quantitative assessment of relationships between migration of chemical species in food webs of natural and technogenic biogeochemical provinces and human health.*

We will consider the natural biogeochemical provinces (see Chapter 6), new-formed biogeochemical provinces owed to anthropogenic influence and the migration of elements in food webs for various multi-scale regions of the Worlds.

1. BIOGEOCHEMICAL AND PHYSIOLOGICAL PECULIARITIES OF HUMAN POPULATION HEALTH

The geochemical heterogeneity of the biosphere and co-evolution of biosphere and geosphere have led to the formation of various biogeochemical provinces with different food webs and definite sustainability or sensitivity of living organisms, including human, to physiological disturbance and diseases (see Chapter 7).

V. Vernadsky initiated the study of the interactions between chemical heterogeneity of the biosphere and human health in his Biogeochemical laboratory in 1928. The main tasks of this laboratory were related to the study of living organisms and their origin depending upon geographic, geological and biological characteristics of the permanent environmental media. In 1932 researchers began to monitor the geochemical areas with different diseases caused by the peculiarities of chemical composition of soil, waters, plant and animal species, or in other words, by biogeochemical food webs.

In 1938 Vinogradov published the book 'Biogeochemical provinces and human endemic diseases'. He analyzed the relationships of organisms and populations within

Table 1. The lower, optimum and upper limit concentrations of essential trace nutrients in soils of North Eurasia.

Trace nutrient	Number of samples	Limit concentrations, ppm		
		Deficit/lower	Optimum	Excess/upper
Co	2400	< 7	7–30	> 30
Cu	3194	< 15	15–60	> 60
Mn	1629	< 400	400–3000	> 3000
Zn	1927	< 30	30–70	> 70
Mo	1216	< 1.5	1.5–4	> 4
B	879	< 6	6–30	> 30
Sr	1269	< 600	600–1000	> 1000
J	491	< 5	5–40	> 40

the geochemical structure of the biosphere. This book initiated the biogeochemical mapping of North Eurasia in 1940–1980s headed by Kovalsky (see Chapter 7).

In the USSR Kovalsky showed that productivity in cattle can be correlated to the excess or deficiency of boron, cobalt, copper, molybdenum and selenium in animal feed. Similar studies were carried out in England and Ireland by Webb (Webb, 1964; Webb *et al*, 1966) and in USA, by Ebens (Ebens *et al*, 1973).

As has been reported, the levels of trace metals in drinking water and foodstuffs of local production can affect human health. The Canadian biogeochemist Warren (1961) showed the relationships between thyroid gland malfunction and iodine deficiency in North America. Shacklette (1970) related the level of trace metals in soil and plants to the incidence of cardiovascular diseases in Georgia, USA. The Finnish geochemist Salmi (1963) reported a correlation between the concentration of lead in rocks and the incidence of multiple sclerosis. Dobrovolsky (1967) developed the method of geochemical mapping for hygienic and prophylactic health management.

Biogeochemical mapping brings together the biological reactions of living organisms and their adaptation to environmental conditions with chemical composition of geological rocks, soils, natural waters, feed and food. The most important feature of biogeochemical mapping is the estimation of the upper and lower limits of essential elements in biogeochemical food webs, for instance in soils, waters, crops, feed and food daily intake. The limit concentrations are the values, lower or upper of which the regulatory mechanisms of exchange processes in living organisms (plant, animals and humans) will be disturbed (see also Chapter 7, Figure 4). Tables 1 and 2 show the example of limit concentrations of some essential micronutrients in soil and forage crops.

Table 2. The lower, optimum, and upper limit concentrations of essential trace nutrients in forage crops for domestic animals in North Eurasia.

Trace nutrient	Number of samples	Mean content in pasture crops	Limit concentrations, ppm		
			Lower (deficit)	Optimum for animal organisms	Upper (excess)
Mn	819	73.0	< 20	20–30	> 30
Zn	519	21.0	< 20	20–60	> 60
Cu	937	6.4	< 3	3–12	> 12
Mo	537	1.25	> 0.2	0.2–2.5	> 2.5
Co	859	0.32	< 0.25	0.25–1.0	> 1.0
J	397	0.18	< 0.07	0.07–1.2	> 1.2

Using similar data the biogeochemical mapping of North Eurasia was carried out (see Chapter 7, Figure 6).

According to the specific characteristics of North Eurasian biogeochemical provinces, the alterations of the biogeochemical food webs were studied for I (endemic goiter), Si (the Urov disease), B (endemic enteritis), Se (endemic myopathia), and Mo (endemic gout). On a basis of these results, the recommendations have been produced for correction of daily intake of essential elements.

During last decades the researches in human biogeochemistry were concerned with the cancer development in various regions of the World. It has been shown that most human cancer diseases are related to the environmental conditions such as the content of various macro- and microelements in biogeochemical food webs.

In accordance with existing knowledge, the inorganic and organic species of Ni, Be, Pt, Cd, Pb, Co, Zn, Mn, Fe, Ti and Hg are suspected carcinogens or procarcinogens (see Box 1. for classification of carcinogens).

Box 1. Biochemistry of carcinogens (After Manahan, 2000)

Cancer is a condition characterized by the uncontrolled replication and growth of the body's own cells (somatic cells). Carcinogenic agents may be categorized as follows:

1. Chemical agents, such as many chemical organic and inorganic compounds;

2. Biological agents, such as hepadnaviruses or retroviruses;

3. Ionizing radiation, such as X-rays;

4. Genetic factors, such as selective breeding.

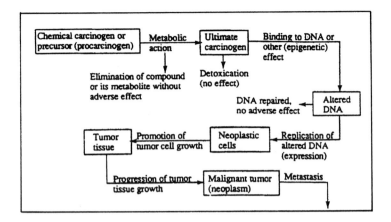

Figure 1. Outline of the process of carcinogenesis (Manahan, 2000).

Clearly, in many cases, cancer is the result of the action of synthetic and naturally occurring chemical species. The role of chemicals in causing cancer is called chemical carcinogenesis. It is often regarded as the single most important facet of toxicology and clearly the one that receives the most publicity.

Large expenditures of time and money on the subject in recent years have yielded a much better understanding of the biochemical bases of chemical carcinogenesis. The overall process for the induction of cancer may be quite complex, involving numerous steps. However, it is generally recognized that there are two major steps in carcinogenesis: an initiation stage followed by a promotional state. These steps are further subdivided as shown in Figure 1.

Initiation of carcinogenesis may occur by reaction of a DNA-reactive species with DNA or by the action of an epigenetic carcinogen that does not react with DNA and is carcinogenic by some other mechanism. Most DNA-reactive species are genotoxic carcinogens because they are also mutagens. These substances react irreversibly with DNA. They are either electrophilic or, more commonly, metabolitically activated to form electrophilic species, as is the case with electophilic $^{+}CH_3$ generated from dimethylnitrosoamine (see Chapter 3, Section 3). Cancer-causing substances that require metabolic activation are called procarcinogens. The metabolic species actually responsible for carcinogenesis is termed an ultimate carcinogen. Some species that are intermediate metabolites between procarcinogenes and ultimate carcinogens are called proximate carcinogens. Carcinogens that do not require biochemical activation are categorized as primary or direct-acting carcinogens. Most substances classified as epigenetic carcinogens are promoters that act after initiation. Manifestations of promotion include increased number of tumor cells and decreased length of time for tumors to develop or in other words, shortened latency period.

Most metal, whose compounds are carcinogenic are from group IV of the Periodic Table of Elements. In biological systems, carcinogenic metals can form stable complexes and biological availability of these complexes determines the carcinogenic potential of various metal compounds.

The carcinogenicity of many metals, like As, Cd, Ni, Be, Cr, Pb, Co, Mn, Fe and Zn, depends on their concentrations in the food webs. These concentrations are the sum of natural content and input owed to pollution. In natural and technogenic biogeochemical provinces, the content of pollutants depends on both geochemical background and anthropogenic inputs. The understanding of relationships between concentrations of carcinogenic metal compounds in food webs and cancer mortality is very important. However, the role of other strong factors like smoking and professional work activity should be also accounted for.

During biogeochemical and physiological studies of cancer diseases, we should take into account also the combined influence of various carcinogens, for instance asbestos, PAH, PAN, metals, agrochemicals, etc.. The combined action may change the risk of cancer due to synergetic, additive or antagonistic effects. For instance, the combinations of benzo(a)pyrene and Ni_3S_2, benzo(a)pyrene and chromates, benzo(a) pyrene and As_2O_3, are synergetic for the development of lung cancer in rats. The carcinogenic actions of nitrosoamines and Ni and Cd compounds are additive.

The antagonistic influence on the cancer development due to organic carcinogens may be induced by chemical species of Se, As, Al, Co, Cu, Zn and Mo. We should note that the antagonistic effects of latter metals depend strongly on their concentrations (see above on limit concentrations). Similar interactions are characteristic for the combinations of metals in biogeochemical food webs of various biogeochemical provinces.

1.1. Cancer diseases in the Carpathian mountain sub-region of the biosphere

Extensive biogeochemical studies of cancer illnesses have been carried out in different biogeochemical provinces of North Eurasia. The influence of various trace elements has been studied during 1980–1990's in the Forest Steppe and Mountain regions of biosphere, Ukraine. Three natural biogeochemical provinces, Carpathian, Pre-Carpathian and Forest Steppe, were monitored to study the migration of trace elements in food webs and human cancer distribution (Table 3).

We can see that the aborigines of Carpathian and Pre-Carpathian biogeochemical provinces are relatively seldom ill with lung and stomach cancer compared to the Forest Steppe province. These differences are related to the chemical composition of soils and ground waters in various provinces. For instance, in the Carpathian biogeochemical province with Cambisols and Podsoluvisols enriched in titanium and depleted in vanadium, strontium and manganese, predominant human diseases are strong leukemia, hemorrhagic vasculitis, Fe-deficit anemia, and thrombosis. Lung and stomach cancer seldom occur.

In the Pre-Carpathian biogeochemical province with prevalent Eutric Podsoluvisols, enriched in lead and barium and depleted in chromium and vanadium, the predominant diseases are myloukemia, chronic lymphatic leukemia, hemorrhagic

Table 3. Lung and stomach cancer distribution and average content of trace elements in soils and drinking waters of various biogeochemical provinces.

Province	Average content of trace elements, ppm								Rank of cancer morbidity
	Ti	V	Cr	Mn	Cu	Sr	Ba	Pb	
Soils, dry weight									
Carpathian	1490	123	61	482	14	36	74	21	Low
Pre-Carpathian	1297	110	62	690	15	49	99	21	Low
Forest-steppe	1216	110	66	955	14	42	116	23	High
Ground waters, ppm per dry salt content									
Carpathian	—	53	16	1714	23	6992	69	5	Low
Pre-Carpathian	—	18	16	1107	27	1422	41	5	Low
Forest-steppe	—	14	16	754	18	2927	12	6	High

vasculitis, hypoanemia with a relatively low number of sharp leukemia, lung and stomach cancer.

In Forest Steppe biogeochemical province with Eutric Phaerozems and Distric Chernozems, enriched in all trace metals, such illnesses as lung and stomach cancer, tumor of cerebrum and spinal cord, and nephritis are predominant, whereas the Addison-Bearmer anemia, progressive myopia and glaucoma are relatively seldom.

1.2. Cancer diseases in Middle Volga silicon sub-region of the biosphere

Similar biogeochemical studies of cancer development have been carried out in the Middle Volga silicon sub-region of the biosphere (Kovalsky and Suslikov, 1980, Ermakov, 1993). The biogeochemical map of this sub-region is shown in Figure 2.

In the Chuvash administrative region of the Middle Volga drainage basin, three sub-regions of biosphere (Pre-Kubnozivilsk, Pre-Sura, and Pre-Volga), and three natural biogeochemical provinces (silicon, nitrate and fluorine) have been mapped. We will consider the biogeochemical food webs and typical endemic diseases in these sub-regions and provinces.

Pre-Kubnozivilsk sub-region of the biosphere

This sub-region is in the central and east part of the Chuvash administrative region. Most of the sub-region is occupied by Steppe ecosystems with some small spots of Broad-Leafed Forest ecosystems. The predominant soils are Phaerozems. The biogeochemical food web of this sub-region is presented in Figure 3.

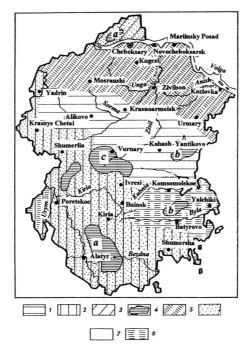

Figure 2. Biogeochemical mapping of the Chuvash administration region, Russia.

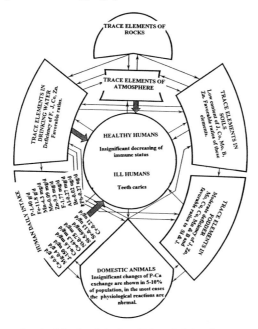

Figure 3. Biogeochemical food web in Pre-Kubnozivilsk sub-region of the biosphere.

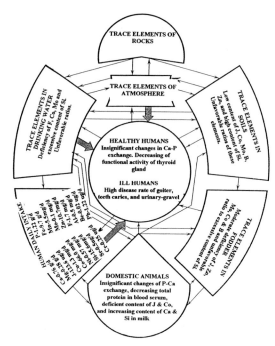

Figure 4. Biogeochemical food web in Pre-Volga sub-region of the biosphere.

We can see from Figure 3 that the moderate deficit of I, Co, Zn, Cu, Mo, B, and Mn with optimal ratios of trace metals to I and Si, is characteristic for all links of the biogeochemical food web. These biogeochemical peculiarities favor the optimal physiological regulation of exchange processes in animal and human organisms. However, a moderate deficit of essential trace nutrients decreases the human immune system. In spite of the latter point, this sub-region can be considered as a natural control biogeochemical area for understanding the endemic diseases of humans and animals.

Pre-Volga sub-region of the biosphere

This sub-region of the biosphere occupies the northern part of the Chuvash administrative region with predominance of Broad-Leafed Forest and Meadow Steppe natural ecosystems and agroecosystems. The prevalent soils are Podsoluvisols. The scheme of the biogeochemical food web is shown in Figure 4.

The deficit of I, Co, Mn, and Zn, moderate excess of Si and disturbed ratios of trace metals to I and Si are characteristic features of the biogeochemical food web of this Pre-Volga sub-region of biosphere. Most natural water sources have a decreased content of flour. These peculiarities of biogeochemical food web favor the occurrence of tooth decay (caries) and endemic goiter.

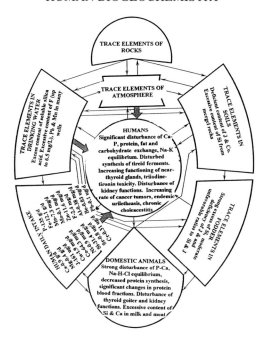

Figure 5. Biogeochemical food web in Pre-Sura sub-region of the biosphere.

Pre-Sura sub-region of the biosphere

This sub-region is in southwestern and south parts of the Chuvash administration region, Russia (see Figure 2). Most of its ecosystems are Coniferous Forests with Arenosols and Podsoluvisols. There is some silicon ore (trepels) in this area and silicon-enriched geological rocks are typical as soil-forming materials. Both soils and river waters are enriched in silicon and this has led to a specific form of biogeochemical migration, which can be called silicon biogeochemical food web (Figure 5).

We can see that in all links of the biogeochemical food web in the Pre-Sura sub-region of the biosphere, the large excess of silicon, moderate deficit of iodine and cobalt, and unfavorable ratios of essential trace elements to silicon are characteristic features. These peculiarities of biogeochemical migration of elements lead to the disturbance of phosphor-calcium and protein exchange in human and animal organisms and to the development of endemic goiter, uric gravel diseases and other endocrine illnesses.

In addition to these three sub-regions, we can see three biogeochemical provinces: a) silicon; b) fluorine, and c) nitrate. The origin and development of these provinces are connected with local geological and agricultural conditions. The silicon and fluorine provinces were developed over dispersed areas of silicon and apatite ores. The nitrate biogeochemical province is related to the intensive application of nitrogen fertilizers in agroecosystems (Figure 2). The technogenic nitrate biogeochemical province (agrogeochemical province) occupies some places inside of the Pre-Sura silicon sub-region

Table 4. Correlation between water hardness and cancer rates in silicon sub-region.

Hardness, meq/L	Per cent from monitored drinking water sources	Rank of cancer illnesses
< 7	27	Low
8–14	62	Medium
15–29	11	High

of the biosphere. The increased rates of methemoglobinemia and stomach cancer have been monitored in the latter province.

Medical statistics indicate that in the Pre-Sura silicon sub-region, the number of malignant tumor formations is three times higher than in the control Pre-Kubnozivilsk sub-region. Significant increase of cancer has been registered in a young women. For instance, in the silicon sub-region, the different types of uretheral cancer were from 0.117 to 2.63 cases per 1000 women, whereas these diseases were not detected at all in the control sub-region. The clinical monitoring of chemical blood composition showed an increasing content of K ion in 88.64%, Ca ion, in 70.05%, and phosphorus, in 57.2% of monitored people above physiological optimal limits. The average content of K^+ in blood of people from the silicon sub-region was 2 fold, Ca^{2+} 1.14 fold, and phosphorus 1.45 fold higher than that of people from the control biogeochemical sub-region.

In this biogeochemical sub-region the rate of carcinogenecisis was also correlated with the drinking water hardness. The water hardness is connected with the content of Ca, Mg and Fe compounds (Table 4).

The minimum cancer number was monitored at content of Ca^{2+} 4.5–4.8 meq/L, Mg^{2+}, 2.5 meq/L, Ca : Mg ratio, 1.65–1.90 and water alkalinity, 6.0 meq/L.

From the physiological point of view the development of excessive cancer rates and various endemic diseases in the silicon biogeochemical sub-region and especially in the silicon biogeochemical province can be explained by transmission from the norm to pathological state through pre-pathology state during long term influence of small intensity factors on human organisms. The human heath risk assessment showed that about 50% of local population are living in the conditions of prelithiasis, or in other words, in conditions of the initiation stage of carcinogenesis.

1.3. Cancer diseases in the boron biogeochemical sub-region of the biosphere

The influence of natural biogeochemical factors and technogenic pollution can be demonstrated on the example of the boron sub-region of the biosphere. This sub-region occupies a vast area in the Caspian low plain, the watershed areas between lower Volga and lower Ural rivers of Russia, the Ustyurt plateau and Syr-Daria and Amu-Daria watershed area in north western Kazakhstan (see Chapter 4, Figure 13)

*Table 5. Alteration of chemical composition of
the Syr-Daria river water used for drinking by
aborigines.*

Chemical species	Content in river water, ppm	
	1956	1985
Nitrate	1.1	66.0
Nitrite	0.0	5.6
Chloride	40.0	480.0
Sulfate	141.0	900.0

*Table 6. Concentration of pesticides in the Syr-Daria river
water used for drinking by aborigines, μg/L.*

Sampling point	DDT	DDE	HCH
Middle flow	0.026–0.095	0.001–0.012	0.068–0.14
Kzyl-Orda city	0.04–0.09	0.009–0.06	0.065–0.14
Kazalinsk city	0.03–0.11	0.00–0.012	0.05–0.09
Delta, Aral Sea	0.068	0.116	0.140

The natural biogeochemical peculiarities (see Chapter 4, Section 4 and Chapter 6, Section 4), which favor the development of endemic diseases were complicated by anthropogenic activity. In the area of the Volga-Ural watershed and the Ustyurt plateau, this activity is connected with oil explorations and oil-chemical industry. The watershed of Amu-Daria and Syr-Daria rivers is the agricultural area with intensive irrigated cotton production with application of heavy rates of pesticides and fertilizers.

The historical epidemiological statistics have been connected with the oil exploration and oil treatment that were initiated more than 100 years ago. For a long time, the cancer diseases (mostly, esophagal cancer) were more prevalent in Russian people, who professionally worked in oil industries than in aboriginal Kazakh people. The disturbance of environmental media and social conditions of the Kazakh population from the beginning of the 1930s have led to a sharp increase of cancer rates among aborigines. The statistics of cancer rates and environmental pollution are shown in Tables 5–8.

Table 7. Dynamics of cancer diseases in the Kzyl-Orda administrative region (Syr-Daria drainage area).

Malignant tumors	1955	1960	1965	1970	1975	1980	1985	1995
			Cancer case per 100,000 individuals					
All tumors	80.1	91.7	186.8	117.6	137.7	140.0	130.2	213.8
Stomach cancer	—	12.4	32.6	23.6	19.7	12.1	19.1	32.7
Esophagal cancer	—	28.1	86.1	51.8	56.3	52.9	49.6	69.5
Lung cancer	—	4.1	9.5	8.1	9.1	7.9	13.0	21.4

Table 8. Growth of cancer diseases in Kazakhstan in 1970–1990, % of 1970 rate.

Administrative region	Lung cancer	Breast cancer	Skin cancer	Intestinal cancer	Limphomas
Guriev	143	257	209	54	26
Kzyl-Orda	100	285	—	275	113
Semipalatinsk	130	91	25	129	96
Chimkent	71	90	13	—	3
Dzhambul	166	96	93	133	285

We will consider the development of cancer in two administration regions: a) the Guriev region with technogenic pollution by oil and oil products and b) the Kzyl-Orda region with agricultural pollution by nitrogen fertilizers and pesticides. Both regions belong to the boron biogeochemical sub-region of the biosphere.

The results of Table 7 and 8 show two peaks of cancer diseases in both regions. The first peak was in the 1950s as a consequence of environmental and social disturbances in the 1930–1940s and the second peak was observed in the late 1980s to the earlier 1990s as a result of the strong environmental pollution and formation of technogenic and agrogenic biogeochemical provinces.

There is a correlation between content of carcinogens in various links of biogeo-chemical food webs and frequency of stomach cancer. The most significant correla-tion was monitored in agrogenic biogeochemical nitrate province of the Kzyl-Orda administrative region. The multiple increase in the application of different agro-chemicals, such as fertilizers, pesticides, and defoliants,induced the development of carcinogenecisis. The rate of cancer increased from 80.1 in 1955 to 213.8 in 1995. For a short time (1982–1987) the lung cancer rates were increased by four times and liver cancer rates, by two times.

We should stress the role of natural biogeochemical and climate conditions in the development of various cancer diseases in the boron sub-region of the biosphere, especially esophageal cancer. The average rates of esophagal cancer in developed countries are about 5 per 100,000 individuals. The traditional reasons, which are highlighted in news media, are smoking and strong alcohol drinking. However, there are some countries, like China, Kazakhstan, Iran, Pakistan, etc., with moderate/low smoking and drinking, where the rate of esophageal cancer exceeds 100 per 100,000 individuals. We can assume that in this Central Asian biogeochemical region with Dry Steppe and Semi-Arid and Arid Desert ecosystems, the environmental conditions favor cancer development.

This is an area of strong continental climate with dry and hot summer and severe winter with strong winds that transport dust at short and long distance, for instance 'yellow sand' phenomenon in northwest China. During air transport these soil particles absorb numerous pollutants-carcinogens, like benzo(a)pyrene and heavy metals (Ni, Cd, Co, Zn, Pb, As) both from industrial emissions into atmosphere and waste landfill sites.

These atmospheric particles inhaled through the respiratory tract may damage human health. Relatively large particles are likely to be retained in the nasal cavity and in the pharynx, whereas very small particles are likely to reach the lungs and be retained by them. The respiratory system possesses mechanisms for the expulsion of inhaled particles. In the ciliated region of the respiratory system, particles are carried as far as the entrance to the esophagus and gastro-intestinal tract by a flow of mucus. Macrophages in the non-ciliated pulmonary regions carry particles to the ciliated region.

The respiratory system may be damaged directly by particulate matter that enters the blood systems or lymph system through the lungs. In addition, the particulate material or soluble components of it, heavy metals, for instance, may be transported to organs. Particles cleared from the respiratory tract are to a large extent swallowed into the gastrointestinal tract.

Nitrate biogeochemical provinces and cancer diseases

Nitrogen species, such as nitrate and nitrite, are known to be confirmed procarcinogens. The excessive input of these nitrogen compounds with food and drinking water in the presence of tertiary amines, for instance, from medicines, can lead to the formation of carcinogenic N-nitrosoamines.

At present the extensive application of nitrogen fertilizers in many agricultural regions of the World has led to the formation of agrogenic nitrogen biogeochemical provinces. The characteristic feature of these provinces is the accumulation of nitrate and other nitrogen species in soils and drinking water. This accumulation accompanies an excessive input of nitrate in food webs (see Box 2).

Box 2. Nitrates in food and drinking water (After Radojevic & Bashkin, 1999, and Bunce, 1994)

The nitrate (NO_3^-) content in crops is one of the most important indicators of farm production quality. Nitrate content in food is strictly regulated because of its toxicity, especially in young children. The actual toxin is not the nitrate ion itself but rather the nitrite ion (NO_2^-), which is formed when nitrate is reduced by intestinal bacteria, notably *Escherichia coli*. In adults, nitrate is highly absorbed in the digestive tract before reduction can take place. In infants, whose stomach are less acidic, *E. coli* can colonize higher up the digestive tract and therefore reduce the NO_3^- to NO_2^- before it is absorbed.

The nitrite ion is toxic because it can combine with hemoglobin with the resulting formation of methemoglobin. The association constant for methemoglobin formation is larger than that for oxyhemoglobin complex formation. Thus, the nitrite ion binds with hemoglobin, depriving the tissues of oxygen. Severe cases of disease called methemoglobinemia can result in mental retardation of the infant and even a death from asphyxiation.

At stomach pH, nitrite is also converted to $H_2NO_2^+$, which is capable of nitrosating secondary amines and secondary amides. The resulting N-nitrosoamines may be carcinogenic. For example, N-nitrosodimethylamine (or dimethylnitrosoamine) is carcinogenic in many experimental animals, although it is not a confirmed human carcinogen. Dimethylnitrosoamine can also contaminate drinking water supplies, both as a result of industrial activity and also because the compounds may be present in the discharge waters of sewage treatment plants, where it is formed by the microbial degradation of proteinaceous materials.

The average human daily intake of nitrate/nitrite is 95 mg/day in adults. Estimates of the relative contributions of nitrate from drinking water and food to the daily intake vary considerably, depending on how they are calculated. Nevertheless, they show that between 50% and 90% of nitrates in human intake may originate from vegetables, conserved meat (sausages, canned meat, smoked meat, etc.) and even milk products. Vegetables tend to concentrate nitrate ion, especially if they are grown using high rates of nitrogen fertilizers. The concentration of nitrate in vegetables can vary extensively. Lettuce, spinach, cabbage, celery, radish and beetroot can contain as much as 3000–4000 mg/kg of fresh weight. These levels could have potential health effects. The problem of nitrate accumulation seems to be especially severe in leafy vegetables grown in greenhouses under winter conditions, owing to intensive application of nitrogen fertilizers and low light intensity, which retard nitrate metabolism in crops.

Another source of nitrate and nitrite accumulation in food is their use as food additives. Nitrate and nitrite salts ($NaNO_2$, $NaNO_3$, KNO_2, KNO_3) are added to meats and other food products as a curing salt, color fixative (preventing the meat turning brown) and as food preservative to prevent the growth of the dangerous bacterium *Clostridium botulinum*, which produces the highly poisonous botulism toxin.

Cured meats, bacon, ham, smoked sausages, beef, canned meat, pork pies, smoked fish, frozen pizza and some cheeses contain nitrate and nitrite additives, typically at levels of 120 mg/kg. Although without them there would certainly be many deaths due to growth of toxic microbes in meat, excessive intakes of these salts may cause gastroenteritis, vomiting, abdominal pain, vertigo, muscular weakness and an irregular pulse. Long-term exposure to small amounts of nitrates and nitrites may cause anemia and kidney disorders. The level of these additives is strictly controlled (for example, 500 mg/kg as $NaNO_3$ in the UK), and the addition of nitrates and nitrites in baby food is now banned in many countries. The acceptable daily intake (ADI) for $NaNO_3$ is 0–5 mg/kg body weight. For KNO_2 and $NaNO_2$ the ADI is 0–0.2 (temporary), while for KNO_3 is not specified. The value of ADI is zero for baby food. The WHO sets the ADI at 220 mg an adult.

Since vegetables are a major source of ingested nitrates, the most rational way of reducing the problem is to grow crops with safe levels of nitrates. Most counties do not have actual standards but some kinds of guideline or criteria value based on the ADI. Criteria values of nitrate content in the same kind of vegetable consumption and in vegetable production practices. For instance, the nitrate level (mg/kg of wet weight) in spinach in different countries is: USA, 3,600; Switzerland, 3,000; the Netherlands, 4,000; Czech Republic, 730; Russia, 2,100.

Nitrate content is also one of indicators of fodder quality. Numerous cases of cattle poisoning by nitrates present in fodder have been reported in various countries, and many heads of cattle were lost from affected herds. Feed beetroot, cabbage, mustard, sunflower, oat, as well as various types of ensilage used as green fodder may contain potentially toxic levels of nitrate. The toxic level of nitrate for animals is 0.7 mg/kg of body weight. Among farm animals, cattle and young pigs are the most sensitive to nitrates, while sheep are more resistant. Also, consumption of nitrates at sub-toxic levels by cattle has been reported as a cause of reduced milk production and weight gain, vitamin A deficiency, abortions, stillbirths, cystic ovaries, etc.. Nitrate-N concentrations of 0.21% in the feed are considered to be toxic to farm animals.

Agriculture is usually the major source of nitrate ion in drinking water. Through manure seepage from feedlots, seepage from the holding tanks used to contain liquid manure from intensive hog production, and excessive use of fertilizers are important. The higher crop yields obtained today compared with 20–40 years ago are largely due to increased use of chemical fertilizers. However, low crop prices combined with high land and machinery cost encourage farmers to cultivate fields up to their margins, thus promoting runoff from fields to water bodies and ground waters. The accumulation of nitrates in ground water has become an issue of great concern in many European countries and many States of USA (see Chapter 8, Section 1).

In 1945, H. H. Comly first estimated the correlation between nitrates in drinking water and the incidence of methemoglobinemia. Research shortly afterwards showed that no cases of methemoglobinemia have been reported in any area of the United States where the water supply contained less than 45 ppm of nitrate ion. This value has become accepted in USA as the upper limit for the nitrate concentration in drinking

Table 9. Perennial concentrations of nitrogen species in food web of nitrate biogeochemical province of the Desert region of the biosphere, Uzbekistan.

Links of food web	Nitrate	Nitrite
Surface water, ppm	2.03–12.10	0.03–0.42
Drinking water, ppm	3.62–150.80	0–0.09
Vegetables, ppm of wet weight	500–8600	12.4–328.3

water. At present the WHO limit is also 45 ppm of nitrate but the value of 22 ppm of nitrate has been set for EC countries.

The main reasons of nitrate accumulation are the increasing application of nitrogen fertilizers and increasing generation of nitrogen-containing municipal wastes.

Let us consider in brief the nitrate biogeochemical provinces, where a significant growth of cancer incidents has been observed. It is known that the absorption of nitrate and nitrite from water in the intestinal tract is two times as much as from food products. This means that the accumulation of nitrate in drinking water sources should be of special concern during consideration of food webs in any relevant biogeochemical province.

As an example, we will consider the formation of nitrate biogeochemical provinces in the Desert region of the biosphere in the Zerafshan river watershed that occupies most of the Samarkand administrative region of Uzbekistan. This is an agricultural area with irrigation production of cotton and some other crops like cereals, vegetables and fruits. The natural biogeochemical provinces are characterized by iodine, copper, and zinc deficiency and lithium excess.

The concentration of nitrogen species in various links of biogeochemical food webs of this province is shown in Table 9.

We can see that the concentrations of both nitrate and nitrite in drinking water and vegetable are excessive. The accumulation of nitrate in food web of this biogeochemical province was accompanied with the high content of residual pesticides in soils of agroecosystems (Table. 10).

The accumulation of pesticides is connected with their long-term application in heavy rates, 15 kg/ha in average during 1970–1980s. The elevated concentrations of PCB are related to irrigation by polluted waters from the Zerafshan river (Galiulin and Bashkin, 1996). The mean content of residual DDT and its metabolites DDE and DDD is higher than the corresponding standard in soils of Cotton and Tobacco agroecosystems. The mean content of total amount of PCB is also higher than the standard in soils of Tobacco agroecosystems.

Table 10. The residual content of persistent orgauochloriue compounds in soils of nitrate biogeochemical province of the Desert region of biosphere, Uzbekistan, ppb.

Xeno-biotic	Cotton		Tobacco		Orchards	
	Mean	Limits	Mean	limits	Mean	Limits
DDT	8.3	tr–47.5	27.2	tr–157.7	24.7	tr–70.9
ΣDDT*	113.4	tr–970.0	282.7	tr–1715.0	19.9	tr–432.5
ΣPCB	28.0	2.3–100.6	121.1	28.4–425.8	24.9	9.2–67.5

Note: ΣDDT − DDT + DDE + DDD; the corresponding standard for DDT content in soil is 100 ppb, and temporary criterion for PCB is 60 ppb.

Table 11. Growth of liver and stomach cancer rates in the Samarkand nitrate biogeochemical province of the Desert region of the biosphere, Uzbekistan, cases per 100,000 individuals.

Period	Liver cancer	Stomach cancer
1950–1955	1.4	12.5
1956–1960	1.5	17.8
1961–1965	2.4	18.9
1966–1970	2.9	23.2
1971–1975	3.9	25.3
1980–1985	4.7	27.8
1985–1990	5.3	32.6
1990–1995	6.0	34.6

Thus, the combination of natural biogeochemical features of food web (deficiency of J, Zn, and Co with excess of Li) with anthropogenic pollution has led to increased cancer rates (Table 11).

2. HUMAN HEALTH INDICES IN TECHNOGENIC AND AGROGENIC BIOGEOCHEMICAL PROVINCES

2.1. *Physiological indices for human biogeochemical studies*

Traditionally, an assessment of population health is based on the medico-demographic statistical rates of mortality, morbidity, and life period expectance. Among these parameters, the morbidity is the most informative one. The known toxicological and hygienic data describe physiological mechanisms of trace metals, pesticides and nitrate action at cellular, organ and systemic levels. Depending on chemical structure of these pollutants, the molecular mechanisms are variable (Blain, 1992, Trachtenberg, 1994). However, the generalization of these data allows us to reach the conclusion that a common link of function failure is hypoxia. This physiological phenomenon is an oxygen starvation of tissues (Evstafieva, 1996). At long-term action of small doses of pollutants, the systems that required more oxygen will start suffering in the first turn. Amongst them the central nervous system, myocardium, and reproductive system should be referred. The metabolic transformation and excretion of chemical substances from an organism is one of the physiological mechanisms neutralizing the toxic action of pollutants, and the liver and kidneys should be also considered as indicator organs and systems.

Recent experimental data show that the chronic influence of pollutants in low doses produces a nonspecific action on the human organism resulting in different forms of pathology of all organs and systems. In turn, for a selection of the indicating forms, we should recognize the nosological profiles (morbidity specification) of population in various technogenic and agrogenic biogeochemical provinces and correlate these profiles with the technogenic loading. The example of such a correlation is shown in Table 12.

To test this hypothesis, different biogeochemical regions of biosphere in the territory of Ukraine were selected:

1) The Western Forest Steppe region of the biosphere with predominant agricultural activities, including the Lvov, Ivano-Frankovsk, Vinnitsa, Ternopol and Khmelnitsky administrative regions;

2) The Eastern Steppe region of the biosphere with strong industrial loading, including the Kharkov, Donetsk, Dnepropetrovsk and Zaporozhye administrative regions;

3) The Crimean Dry Steppe region of the biosphere with mixed industrial, agricultural and recreational activities.

In division into zones, the maps of an ecological situation in Ukraine were applied. These zones are considerably different in biogeochemical food webs and in the manifestation of technogenic or agrogenic loading. The nosological profiles of population living in these regions of the biosphere have been calculated on a basis of the National Report of the Ukrainian Ministry of Statistics on the rate of morbidity in 1994 (Table 12). The standardized values of average morbidity for these regions in 27 groups of nosologies encompassing the main diseases were counted. They were normalized with respect to the mean value overall for Ukraine, using the formula

$$Z_i = (X_i - X_m)/s, \tag{1}$$

where Z_i is the standardized deviation; X_i is the value of morbidity in estimated region; X_m is the average morbidity for the whole Ukrainian area; and s is number of observations.

Table 12. The statistical estimation of the morbidity in different regions of the biosphere, Ukraine, using the deviations from normalized mean values of morbidity (σ-values) (After Evstafieva et al, 1999).

Diseases	Crimean Dry Steppe region	South-eastern Steppe region under strong industrial pollution	Western Forest Steppe region under agricultural influence
Infectious and parasitic	+2.40	−0.23	−0.55
Neoplasm	+1.53	+0.37	−0.83
Endocrine and immune	−0.82	−0.05	+0.49
Thyroid gland	−0.73	−0.91	+1.09
Blood and hemophoietic organs	−0.55	−0.54	+1.41
Nervous system and organs of sense	−0.89	+0.28	+0.10
Cardiovascular systems	−0.78	−0.20	−0.19
Peripheric nervous system	−1.64	−0.49	+0.63
Systems of the blood circulation	+3.43	−0.26	−0.23
Chronic rheumatic heart disease	−0.70	−1,30	+0.68
Hypertension	+0.66	−0.01	−0.43
Ischemia heart disease	−0.36	−0.60	+0.04
Acute infarct	+2,22	+0.76	−0.66
Stenocardia	+1,14	−0.06	−0.01
Cerebral-vascular distresses	+1,34	+0.60	−0.94
Insulite	−0.26	+0.01	−0.36
Respiration systems	−0.25	+0.83	+0.27
Digestion systems	−1,39	−0.62	+0.55
Stomach and duodenal ulcer	−0.75	+0.28	−0.57
Chronic hepatite	−0.89	−0.38	+0.83
Gallstone disease	+0.26	−0.62	+0.50
Urological disease	−0.73	+0.01	+0.19
Chronic pyelonephritis	+1.15	+0.52	−0.21
Skin disease	+0.78	+0.88	−0.61
Bone and muscular system	−1.06	−0.10	−0.05
Arthrosis	−0.27	+0.32	+0.43

The application of this approach can eradicate the difference in the morbidity rates of different nosologies and estimate an extent of a deviation of parameters from an average for the given disease over the country as a whole. The results showed that in selected zones of the country, there are remarkable deviations of morbidity rates from the average values overall for Ukraine. For instance, in the industrial zone a significant increase of the morbidity rate of the respiratory system ($R^2 = 0.83$) was marked, while in other zones the level of the significance for the morbidity of the respiratory system was lower than average in Ukraine. The excess over the average Ukrainian level took place in the cases of dermal diseases (on 0.88s), diseases of an excretory system (on 0.52s), cerebro-vascular distresses (on 0.60s), on 0.76s—acute infarction, on 0.37 s—neoplasms. At the same time, in the agricultural zone the average Ukrainian level of the morbidity rate was exceeded in such forms of nosologies as a total morbidity of digestive system (0.55s), chronic hepatitis (0.83s), cholelithiasis (0.50s), disease of a peripheral nervous system (0.63s), disease of blood and hemophoietic organs (1.41s), thyroid gland (1.09s), chronic rhematoid diseases of heart (0.68s), and deforming arthrosis (0.42s). An extremely unfavorable situation occurred in the Crimea, where the infection and parasitogenic diseases are higher than the average Ukrainian level as much as 2.41 times, the acute infarction takes place more often as much as 2.22 times, tumors on 1.53s, the attacks of a stenocardia on 1.15s, chronic pyelonephritis on 1.15s, dermal diseases on 0.78s, hypertonia on 0.66 are higher 1.53s, and cerebro-vascular distresses as much as 1.34 times higher.

Thus the nosological profiles of the reviewed regions are variable. The distinctive feature of industrial zones is the high morbidity of the respiratory system and dermal diseases, tumors, and also negative influence on excretory and cardiovascular systems, including vascular distresses of the brain. This generally corresponds to known susceptibility of an organism system to an action of atmosphere pollutants of an industrial zone. In agricultural zones the diseases of an alimentary system dominate, that is clear, because the inflow of pollutants with food and water through this system predominates in the rural areas. Besides this, the lesions of a peripheral nervous system take place here, which can be connected with the specific exposing of action of pesticides on an organism in professional contacts, that is in high usage in agriculture. As to the Crimea, the pattern of morbidity is characterized not only by the highest level of series of pathologies, but also the amplitude of these deflections is maximal in the whole Ukraine. It is known the Crimea is characterized by a big diversification of natural biogeochemical conditions, as well as technogenic loading. Probably, in this case we are dealing with a consequence of this diversification.

However, these results are not sufficient to estimate quantitative dependence on a level of the particular pollutants in the environment. The second working hypothesis on the relationship between pollutants loading and morbidity dynamics in the whole area of North Eurasia has been considered. This hypothesis was estimated using a complicated statistical treatment (Box 3). The considered pollutants were fertilizers, heavy metals, pesticides, radionuclides, oil products, and polycyclic aromatic hydrocarbons, PAH.

Box 3. Statistical assessment of morbidity in technogenic and agrogenic biogeochemical provinces of North Eurasia (After Evstafieva et al, 1999)

The correlation between the indicator forms of diseases for both heavy metals and pesticides was estimated using the results on morbidity in Russia for the 10-year's period of observation in 1980–1990s. The estimated pollutants were fertilizers, heavy metals, pesticides, radionuclides, oil products, and polycyclic aromatic hydrocarbons, PAH. The pollutant loading was estimated for different biogeochemical regions of the biosphere in North Eurasia. According to these data, all pollutants were ranked owing to their toxicological effects on human health. In the assessment of land pollution by heavy metals at rural areas, three levels (increased, mean and low) were taken into consideration. The gradations for city soils of four categories were introduced separately: extremely dangerous, dangerous, moderately dangerous, and low dangerous. In pesticides loading, the extent of the danger of contamination of arable lands of four degrees and summary index of pesticides loading on arable lands of three levels of pollution were considered. The data file of morbidity by a bronchitis, nasopharyngities, pneumonia, asthma was analyzed, and also the ulcer of a stomach and duodenum, gastritis, cholecystitis (incidents per 100 000 residents) for the period from 1982 until 1991. To compare the level of morbidity on various regions (administrative and territorial units) a preliminary analysis of the data was made. All the massif of observations (5110 values) was divided into two 5-year's sets (1982–1986 and 1987–1991). Inside these sets, the 10-level ranging of each parameter was conducted. Then the received ranging was converted in 2 ranks: low (−1) and high (+1) levels of morbidity with value of initial ranks with 1 on 5 and with 6 on 10 correspondingly. Thus, each administrative unit was described by nonparametric 2 level parameter of morbidity (−1, +1) on all nosologies for each year with 1982 on 1991. To characterize the territorial zones on definite nosologies, the average parameters for all 10-year's period was used. Received values characterize level of morbidity on studied territories on scale from −1 (low) up to +1 (high) and do not depend on the size of the chosen territories, that is recommended for a similar type of studies. The statistical treating of results was done on the basis of a computer system SAS by a method of correlation analysis with calculation of standard, rank and signed correlation.

Let us consider the results of this biogeochemical study. The total morbidity for gastrointestinal and respiratory systems did not find a reliable connection with a total anthropogenic loading determined as the sum of scores for each pollutant. But the separate nosologies discovered such a connection. For instance, the pneumonia and asthma morbidity was correlated with total soil pollution. The level of a total morbidity of alimentary and respiratory systems was connected with content of petroleum, PAH, and heavy metals in urban soils. The asthma discovered inter-coupling with most considered pollutants. Apparently, large susceptibility of a respiratory system to negative influence of soil pollution can be explained by correlation of the last index with air pollution in large cities. Here the inhalation of chemical pollutants is prevalent. In rural areas, the main route of pollutants is via food chains and the interaction with morbidity of a digestive system was shown with the highest correlation to pesticides content in soil. We can note that about 40% of variations in asthma morbidity is related to soil pollutants ($R^2 = 0.378$; $n = 70$) and this is compounded with the similar regional epidemiological studies (Gryaznova, 1989; Diakovich and Evimova, 1989; Zhakashov, 1990).

Thus the total soil pollution was connected with a respiratory system and a digestive tract. These systems were also both sensitive to such urban pollutants as heavy metals and PAH. For radionuclides the correlation with the given nosologies was not revealed. The asthma morbidity was mostly connected with soil pollution rates. This circumstance, apparently, can be related to nonspecific action of pollutants on a human organism, because the etiology of asthma is connected with human immune defense system and allergy state (Roite, 1991). The last was shown for pesticides (Nikolaev *et al*, 1988) and heavy metals (Drouet, 1990). The sensitized immune system is, apparently, responsible for chronic toxic effects of other pollutants at low doses (Sidorenko *et al.*, 1991; Novak, Magnussen, 1993).

2.2. Case study of interactions between human health indexes and pollution in Crimea Dry Steppe region of the biosphere

As we have seen the morbidity cases of respiratory and alimentary systems highlight the inter-correlation between biogeochemical food web, pollution loading and human health. However, the physiological response of central nervous and cardiovascular systems to heavy metals and pesticides is also known and the relevant analysis of morbidity of cardiovascular system and psychological distresses was carried out in 1991–1997 in the industrial area of Crimea—Armyansk city, Ukraine (Evstafieva *et al*, 1999). The correlation between morbidity of a respiratory system and pollutants emissions was found (Figure 6).

Simultaneously during these years the progressive increase of morbidity of cardiovascular system and psychological distresses was noted (Figure 7).

However, the specific weight of social factors in an etiology of these diseases was predominant during the observation period. Probably, the assessment of the state of these systems as the indicator can be more justified if we utilize preventive methods of an evaluation, namely, to determine functional parameters describing pre-morbidity states and adaptation to the environment. There are functional methods for pre-morbidity diagnostic, which have been tested in monitoring of the selected population groups. However, for wide screening these methods seems very expensive. Accordingly, a monitoring of population health can be based on the clinical examination data, which allowed the researcher to reveal physiological parameters, most reactive to various types of technogenic pollution. These data showed that radionuclide pollution, industrial loading, and pesticides rates are the most significant factors.

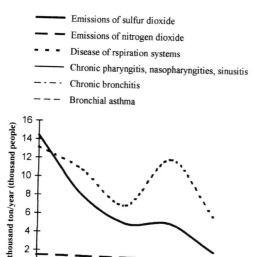

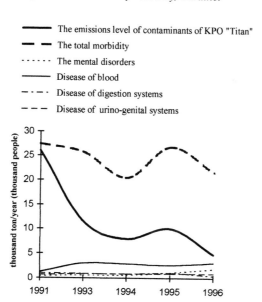

Figure 6. The dynamic patterns of SO$_2$ and NO$_x$ emissions and the rates of respiratory morbidity over a period of 1991–1996 in Armyansk City, Ukraine.

Figure 7. The dynamic patterns of SO$_2$ and NO$_x$ emissions and the rates of total morbidity over a period of 1991–1996 in Armyansk City, Ukraine.

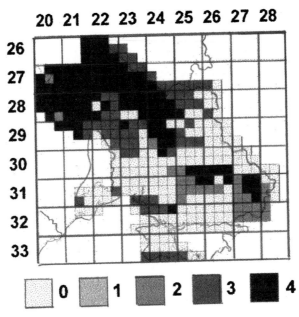

Figure 8. Exceedances of critical loads of sulfur and nitrogen for terrestrial ecosystems of Ukraine in 1992 deposition.

In turn, physiological parameters related to the red blood and oxygen-supplying tissues were the most sensitive. This may be explained by hypoxic action of the most common pollutants (see above).

It is known that the level of pollution may be also estimated using critical loads of pollutants as biogeochemical standards (see Chapter 10 for more detail). For the calculation of critical loads of various pollutants on the human and ecosystem health, the third working hypothesis has been considered. This hypothesis is connected to the assessment of sensitivity of various human physiological parameters to the environmental biogeochemical factors. In this case interaction may be established by statistical exploration of the dependence between loading and various types of morbidity. The critical loads of sulfur and nitrogen at various ecosystems and their exceedances during 1992–1996 were compared with human respiratory system morbidity both in the Crimea and the whole Ukraine. In Ukraine, the respiratory cases were correlated with the exceedances of critical loads (Figures 8 and 9 and Table 12).

Here we should stress also that the higher level of asthma and chronic bronchitis morbidity took place in recreational zones (Yalta, Alushta) of South coast, where the calculated exceedances of critical loads of a two technogeuic sulfur and nitrogen compounds were the greatest among the whole area of the Crimea Dry Steppe biogeochemical region of the biosphere.

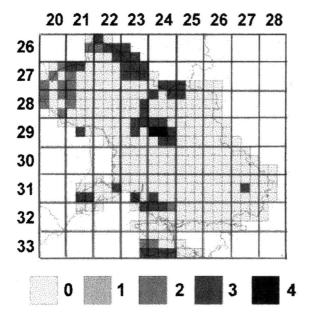

Figure 9. Exceedances of critical loads of sulfur and nitrogen for terrestrial ecosystems of Ukraine in 1996 deposition.

FURTHER READING

Vladimir N. Bashkin, Elena V. Evstafieva, Valery V. Snakin *et al.*, 1993. Biogeochemical Fundamentals of Ecological Standardization. Nauka Publishing House, Moscow, Chapter 2, 70–125.

Viktor V. Kovalsky, 1974. Geochemical Ecology. Nauka Publishing House, Moscow, 325 pp.

Stanly E. Manahan, 1994. Environmental Chemistry. Sixth Edition. Lewis publishers, Boca Raton etc., 649–673.

QUESTIONS AND PROBLEMS

1. Present a definition of human biogeochemistry. Give a historical review of development of research in this field.

2. Discuss the connection between biogeochemical mapping and human biogeochemistry. Stress the attention on the key questions related to organization of the food web.

3. Why is cancer considered as a biogeochemical-binding disease? Give characteristic examples.

4. Describe the chemical and biochemical parameters of carcinogenecesis. Explain the role of primary, confirmed, and suspected carcinogens.

5. Describe the role of the chemical composition of soils and natural waters in the development of cancer diseases. Present examples of these relationships.

6. Characterize the role of silicon as a carcinogen. Present a description of relevant biogeochemical regions and provinces.

7. Nitrogen is known as an essential nutrient and prominent pollutant. Describe the connection between nitrogen species in biogeochemical food webs and cancer development.

8. Describe the synergetic, additive and antagonistic reactions between different carcinogens. Explain these relationships from biochemical, physiological and biogeochemical positions.

9. Explain the unity of biogeochemical food web composition and physiological parameters of living organisms.

10. Discuss the role of industrial and agricultural pollution in development of various diseases. How and to what extent may we use the morbidity and mortality statistical data?

11. Present examples of interactions between anthropogenic pollution and biogeochemical regionalization in development of different illnesses.

12. Present a list of human and animal diseases, the development of which is connected with biogeochemical food webs.

CHAPTER 10

BIOGEOCHEMICAL STANDARDS

During previous chapters we have described many theoretical aspects of modern biogeochemistry. The main attention has been given to the quantitative parameterization of biogeochemical cycles of various chemical species. We have also concluded that from the biogeochemical position, environmental pollution is a process of reversible and/or irreversible disturbances of biogeochemical structure of both terrestrial and aquatic ecosystems. To prevent this disturbance, the anthropogenic loads of pollutants must be decreased significantly. There are different approaches in environmental chemistry and ecotoxicology aiming to set various criteria, threshold levels, and standards to control the pollution of various biosphere compartments and decrease the rate of human and animal diseases. These methods are generally based on experimental modeling with various animals and there are many uncertainties in the implication of the results to the real environmental conditions.

In Chapter 9 we have considered the application of biogeochemical approaches for the assessment of human health. This chapter deals with the application of a biogeochemical concept of critical loads forwarding the impact-oriented reduction of pollutant inputs to the terrestrial and aquatic ecosystems.

1. CRITICAL LOAD CONCEPT FOR IMPACT-ORIENTED EMISSION ABATEMENT STRATEGY OF SULFUR AND NITROGEN ACID-FORMING AND EUTROPHICATION COMPOUNDS

Over the last three decades the debate on air pollution control has assumed both national and international dimensions. With the establishment and quantification of the long-range transport of air pollutants in the mid-1970s, air pollution and acid rain have been seen as an international problem leading to increased research into the impact of air pollutants on the environment.

1.1. Critical load approach: challenges and biogeochemical fundamentals

Political requirements

The first agreement to control transboundary air pollution was connected with the signing of the United Nations Economic Commission for Europe (UN ECE) Convention on Long-Range Transboundary Air Pollution (LRTAP) by 32 European countries, plus Canada and the USA, in 1979. In a protocol added to this convention in 1985,

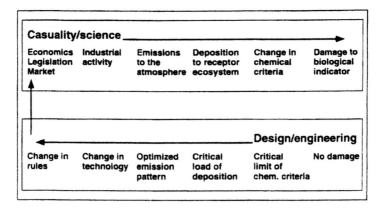

Figure 1. A comparison of the scientific approach to establish causality in the environment.

about 20 countries agreed to reduce their 1980 levels of sulfur emissions by 30% by 1993.

At the same time, initial international actions were taken within both the European Union and UN ECE relating to the drafting of the directives and protocols on the reduction of sulfur dioxide (SO_2) emissions. Considerable progress has been made to reduce SO_2 emissions within the member countries of the European Union through such legislation as the sulfur in gas, oil and large combustion plant directives. All these initiatives, however, were based on a flat rate reductions approach (target loads) and the adoption of best available technology. Over the past few years it has been recognized gradually that such strategies do not necessarily give the best environmental benefits or the lowest control costs.

It should be stressed that in both the sulfur agreement and the subsequent UN ECE protocols to freeze emissions of NO_x (signed in 1988 by 26 countries), the controls agreed were not based on any accurate scientific assessments of reductions required to protect the environment from the harmful influence of these emissions. The general differences in these approaches are shown in Figure 1.

Since 1988 a new concept has been agreed both nationally and internationally to try to improve this situation by optimizing costs and benefits. This is the critical loads/levels approach and it has been adopted as the basis for 1994 UN ECE Second Sulfur Dioxide Protocol, which commits signatories to the Convention on LRTAP to further SO_2 reductions up to and beyond the year 2005. Action on this Protocol will produce significant reductions in the emissions of SO_2 and sulfate deposition over the next decade, till 2005.

Over the same period, emissions of nitrogen oxides (NO_x) were predicted to remain at current levels or to increase, depending specially on the growth of motor vehicle transport. Such predictions were not initially viewed with any sense of environmental concern because of the accepted nutrient properties of deposited nitrogen and its possible fertilizing action on the terrestrial environment. However, as understanding of the impacts of total acidification on the environment has improved, it has become evident that total nitrogen deposition is an important and significant contribution to total acidity. In view of the predicted decline in SO_2 emissions over the 1990's, this contribution of nitrogen deposition to the acidification process is forecast to increase. Furthermore, although nitrogen is an essential nutrient, enhanced deposition can give rise to eutrophication of semi-natural and forest ecosystems and fresh waters, which may lead to changes in species composition of vegetation and fresh water biota (see Chapter 8, Section 1).

Total nitrogen deposition comprises both oxidized (NO_x) and reduced (NH_x) species. The oxidizing species are almost entirely generated by motor vehicles and industry, while the sources of reducing nitrogen are almost entirely agricultural. The deposition of nitrogen in a variety of forms, plus the complex biological interactions and transformations, which nitrogen compounds undergo in the nitrogen biogeochemical cycle, make the prediction and interpretation of the impacts of pollutant inputs of nitrogen considerably more difficult than that of sulfur. In addition, in the presence of volatile organic compounds (VOC) and sunlight, nitrogen oxides play a certain role in the formation of tropospheric ozone.

Definition of critical load

It is well known that biogeochemical cycling is a universal feature of the biosphere, which provides its sustainability against anthropogenic loads, including acid forming compounds. Using biogeochemical principles, the concept of *critical loads* (CL) has been developed in order to calculate the deposition levels at which effects of acidifying air pollutants start to occur. A UN/ECE (United Nations/Economic Committee of Europe) working Group on Sulfur and Nitrogen Oxides has defined the critical load on an ecosystem as: "*A quantitative estimate of an exposure to one or more pollutants below which significant harmful effects on specified sensitive elements of the environment do not occur according to present knowledge*" (Nilsson & Grennfelt, 1988). These critical load values may be also characterized as "*the maximum input of pollutants (sulfur, nitrogen, heavy metals, POPs, etc.), which will not introduce harmful alterations in biogeochemical structure and function of ecosystems in the long-term, i.e. 50–100 years*" (Bashkin, 1998).

The term *critical load* refers only to the deposition of pollutants. Threshold gaseous concentration exposures are termed *critical levels* and are defined as "*concentrations in the atmosphere above which direct adverse effects on receptors such as plants, ecosystems or materials, may occur according to present knowledge*".

Correspondingly, regional assessments of critical loads are of concern for optimizing abatement strategy for emission of both N and S compounds and their transboundary transport (Figure 2).

The critical load concept is intended to achieve maximum economic benefit from the reduction of pollutant emissions since it takes into account the estimates of differing sensitivity of various ecosystems to acid deposition. Thus, this concept is considered to be an alternative to the more expensive BAT (Best Available Technologies) concept (Posch *et al*, 1996). Critical load calculations and mapping allow

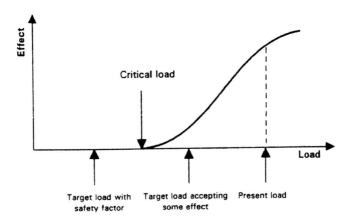

Figure 2. Illustration of critical load and target load concepts.

the creation of ecological-economic optimization models with a corresponding assessment of minimum financial investments for achieving maximum environmental protection.

The critical load concept

In accordance with the above-mentioned definition, a critical load is an indicator for sustainability of an ecosystem, in that it provides a value for the maximum permissible load of a pollutant at which risk of damage to the biogeochemical cycling and structure of ecosystems is reduced. By measuring or estimating certain links of biogeochemical cycles of sulfur, nitrogen, base cations and some other relevant elements, sensitivity both biogeochemical cycling and ecosystem structure as a whole to acidic deposition and/or eutrophication deposition can be calculated, and a 'critical load of acidity', or the level of acidic deposition, which affects the sustainability of biogeochemical cycling in the ecosystem, can be identified, as well as "critical nutrient load", which affects the biodiversity of species within ecosystems. According to the political and economic requirements of the protocol for reduction of N and S emissions and deposition, as well as the parameters of subsequent optimizing models, the definitions of critical loads are given separately for sulfur, nitrogen and for total acidity, which is induced by both sulfur and nitrogen compounds. Hence, critical loads for acidity can be determined as the maximum input of S and N before significant harmful acidifying effects occur. When assessing the individual influences of sulfur and nitrogen, it is necessary to take into account the acidifying effects induced by these elements and the eutrophication effect caused only by nitrogen. In this case, critical load for nitrogen can be determined as the maximum input of nitrogen into the ecosystem, below which neither significant harmful eutrophication effects nor acidifying effects together with sulfur occur during long-term periods (de Vries, 1991).

Both the ratio of base cations to aluminum, and the aluminum concentrations, are used as indicators for steady-state geochemical and biogeochemical processes. By assigning established critical loads to these indicators (for example, the concentrations of aluminum in soil solution should not exceed 0.2 meq/L and the base cations to aluminum ratio should not be less than 1), it is possible to compute the allowable acidification for each ecosystem. An extensive overview of critical values for the ratio of base cations to aluminum for a large variety of plants and trees can be found in H. Sverdrup's papers (see Box 1)

Box 1. Biogeochemical model PROFILE for calculation of critical loads of acidity (After Warfvinge and Sverdrup, 1995)

The biogeochemical model PROFILE has been developed as a tool for calculation of critical loads on the basis of steady-state principles. The steady-state approach implies the following assumptions:

a) the magnitude of capacity factors such as mineral abundance and cation exchange capacity is constant;

b) long-range average values for precipitation, uptake, water requirement and temperature must be used as input;

c) the effect of occasional variations in input variables such as soil carbon dioxide, nitrification rate and soil moisture content can not be addressed;

d) the rate of change in soil chemistry over time can not be taken into account.

The application of these assumptions allows the researchers to use the PROFILE model for calculation of critical loads in Europe (Downing *et al*, 1993; Posch *et al*, 1995; 1997; 1999) and Asia (World Bank, 1994; Shindo *et al*, 1995; Lin, 1998; Hao *et al*, 1998). In spite of visible limitations connected with the numerated assumptions, running the PROFILE model can give comparable results for different ecosystems on regional and continental scales.

Since biogeochemical model PROFILE includes such important characteristics as mineral abundance, another model UPPSALA has been created. This model allows the researcher to calculate the soil mineralogical composition on the basis of total element content. The combination of these models (PROFILE and UPPSALA) gives the possibility to use existing soil and ecosystem data bases for calculating critical loads of acidity in broad-scale regions.

Model characterization

PROFILE is a biogeochemical model developed specially to calculate the influence of acid depositions on soil as a part of the ecosystem. The sets of chemical and biogeochemical reactions implemented in this model are: (1) soil solution equilibrium, (2) mineral weathering, (3) nitrification and (4) nutrient uptake. Other biogeochemical processes affect soil chemistry via boundary conditions. However, there are many important physical soil processes and site conditions such as convective transport of solutes through the soil profile, the almost total absence of radial water flux (down through the soil profile) in mountain soils, the absence of radial runoff from the profile in soils with permafrost etc., which are not implemented in the model and have to be taken into account in other ways.

1) Soil solution equilibrium

Soil solution equilibrium is based on the quantification of acid neutralizing capacity, ANC, which has been defined as:

$$[ANC] = [OH^-] + [HCO_3^-] + 2[CO_3^{2-}] + [R^-] - [H^+]$$
$$- \sum_{m=1}^{3} m[Al(OH)^{m+}_{3-m}], \tag{1}$$

where $[R^-]$—organic acid anions.

With the ambient CO_2 pressure (4×10^{-4} atm) and no dissolved organic carbon (DOC) present, the ANC attains the value 0 at pH values in the range 4.6–5.6 and may thus attain positive or negative values, alkalinity or acidity, correspondingly. With the other [DOC] and P_{CO2} values the ANC-pH dependence is much more complicated.

2) Mineral weathering

Chemical weathering is calculated on a basis of the following equation:

$$R_w = \sum_{i=1}^{horizons} * \sum_{I=1}^{minerals} \times r_i \times A_{exp} \times c_i \times q \times z, \tag{2}$$

where r_i is mineral weathering rate for every i mineral ($Keq/m^2/s$), A_{exp} is exposed mineral surface area (m^2/m^3), c_i is part of i mineral in total mineral mass (%), q is volumetric water content (m^3/m^3), z is soil layer thickness (m).

3) Nitrification

The nitrogen reactions in the PROFILE model are very simple, since only nitrification and uptake are included explicitly.

4) Nutrient uptake

Nutrient uptake includes base cation uptake (BC_u) and nitrogen uptake (N_u). BC uptake assigns annual average uptake of Ca^{2+}, Mg^{2+} and K^+. The data represent an annual net uptake in keq/ha/yr. and storage in stems and branches calculated over rotation. This includes the nutrients in the biomass compartments that are expected to be removed from the site at harvest.

It should be stressed that PROFILE needs the nutrient uptake limited to PROFILE-acceptable layer (0.5–1 m depth) for simplicity, whereas the real nutrient uptake takes place down to the 3-5-7-10 m depth corresponding to tree roots distribution. So, the nutrient uptake in the PROFILE-acceptable layer is always less than the whole nutrient uptake. This might be a source of uncertainty in critical load calculations.

5) Critical leaching of Acid Neutralizing Capacity of soil solution—$ANC_{le(crit)}$

The second most important output parameter in the calculation of the critical acid load by PROFILE is the ANC in water leached from the soil system. This parameter characterizes the difference between basic and acidic compounds (between base cation and strong acidic anion contents or alternatively between OH^-, HCO_3^-, CO_3^{2-}, R^- and $H^+ + Al^{3+}$ contents) in soil solution. Thus, positive ANC is called alkalinity and the negative ANC is called acidity. Critical ANC (μeq/L) was calculated on the basis of critical molar BC/Al ratio equal to 1. High amount of cations may alleviate Al toxicity or low catons may aggravate its toxicity. The molar ratio of Ca/Al or $(Ca + Mg + K)/Al$ is an important index to calculate critical loads. Warfvinge & Sverdrup (1995) proposed a critical point of 1.0 for calculating critical loads as expressed as the molar ratio of $(Ca + Mg + K)/Al$ that was mainly from the data of European species. Kohno *et al* (1998) have shown also the applicability of this ratio for Asian conditions in the experimental studies with Japanese species and soils. Below BC/Al ratio = 1, irreversible changes in ecosystem functioning can happen.

In spite of almost global attraction of the critical load concept, the quantitative assessment of critical load values is connected till now with some uncertainties. The phrase "significant harmful effects" in the definition of critical load is of course susceptible to interpretation, depending on the kind of effects considered and the amount of harm accepted (De Vries & Bakker, 1998). Regarding the effects considered in terrestrial ecosystems, a distinction can be made in effects on:

i.   soil microorganisms and soil fauna responsible for biogeochemical cycling in soil (e.g., decreased biodiversity);

ii. vascular plants including crops in agricultural soils and trees in forest soils (e.g., bioproductivity losses);

iii. terrestrial fauna such as animals and birds (e.g., reproduction decrease);

iv. human beings as final consumers in biogeochemical food webs (e.g., increasing migration of heavy metals due to soil acidification exceeding acceptable human daily intake etc.)

In aquatic ecosystems, it is necessary to consider the whole biogeochemical structure of these communities and a distinction can be made in accounting for the whole diversity of food webs:

Select a receptor		Select a receptor
⇓		⇓
Select the environment quality objectives for corresponding ecosystems		Determine the actual load for the corresponding ecosystem
⇓		⇓
Select a computation method (model)		Select a computation method (model)
⇓		⇓
Collect input data		Collect input data
⇓		⇓
Calculate the critical loads		Calculate the steady-state concentrations
⇓		⇓
Compare with the actual load and calculate the exceedances		Compare with environmental quality objectives (MPC)

Figure 3. Flowchart for calculating critical loads (left) or steady-state concentrations (right) of acid-forming and eutrophication S and N compounds.

i.   aquatic and benthic organisms (decreased productivity and biodiversity);

ii. aquatic plants (e.g., decreased biodiversity, eutrophication);

iii. human beings that consume fish or drinking water (surface water) contaminated with mobile forms of heavy metals due to acidification processes (e.g., poisoning and depth).

1.2. General approaches for calculating critical loads

The possible impact of a certain load on soil and surface water quality can be estimated by determining:

— the difference between actual load and critical load;

— the difference between the steady-state concentration (that will occur, when the actual load is allowed to continue Maximum Permissible Concentration, MPC) and increasing levels of pollutant concentration in soil or surface water under permanent pollutant input.

In the first, critical load, approach, the single quality objective is used to calculate a critical load. The second, steady-state, allows comparison with various quality objectives. Both approaches, which are the reverse applications of the same model (Figure 3), have their advantages and disadvantages.

One can see that both algorithms are similar, but steady-state approaches based on MPC values do not practically take into account either ecosystem characteristics or their geographic situation. Furthermore, there are many known drawbacks of traditional approaches applying MPC (Bashkin *et al*, 1993; Van de Plassche *et al*, 1997). Since the steps in the steady-state approach are similar but in reverse order, they will not be further elaborated and only the various steps of the critical load approach are summarized below.

1. Select a receptor

A receptor is defined as an ecosystem of interest that is potentially polluted by a certain load of acid forming or eutrophication compounds of sulfur and nitrogen. A receptor is thus characterized as a specific combination of land use (e.g., forest type, agricultural crops), climate, biogeochemical regionalization and soil type or as an aquatic ecosystem, such as a lake, a river or a sea, accounting their trophic status and hydrochemistry. Regarding the terrestrial ecosystems, one should consider information (environmental quality criteria, methods and data) for both agricultural soils (grassland, arable land) and non-agricultural (forest, bush) soils, where the atmospheric deposition is the only input to the system. Similar information has to be collected for aquatic ecosystems.

2. Select the environmental quality objectives

Quality objectives should be based on insight in the relation between the chemical status of the soil or the surface water and the response of a biological indicator (an organism or population). According to the definition, the critical load equals the load that does not cause irreversible changes in biogeochemical cycling of elements in ecosystems, thus preventing "significant harmful effects on specific sensitive elements of the environment". Consequently, the selection of quality objectives is a step of major importance in deriving a critical load.

3. Select a computation method (model)

In this context, it is important to make a clear distinction between steady-state and dynamic models. Steady-state models are particularly useful to derive critical loads. These models predict long-range changes in biogeochemical structure of both terrestrial and aquatic ecosystems under the influence of acid deposition such as weathering rates, base cation depletion, nutrient leaching etc. either in soils or surface waters. Dynamic models are particularly useful to predict time periods before these changes will occur. These models are necessary to determine an optimal emission scenario, based on temporal change of pollutant status.

4. Collect input data

This includes soil, vegetation, water (surface and ground), geology, land use *etc* data, influencing acidification and eutrophication processes in the considered ecosystem.

For application on a regional scale it also includes the distribution and area of receptor properties (using available digitized information in geographic information systems, GIS).

5. Calculate the critical load

This step includes the calculation of critical loads of sulfur, nitrogen and the total acidity in a steady-state situation for the receptors of choice or for all receptors in all cells of EMEP or LoLa grid (150×150 km; 50×50 km; 25×25 km; $1 \times 1°$, $10 \times 10'$ etc.) of a region using a GIS (to produce critical load maps).

6. Compare with actual load

The amount by which critical loads are exceeded and the area in which they are exceeded (using a GIS) can be also included in the calculation when the actual loads (for example, atmospheric deposition data in case of forest) are known. Furthermore, these exceedance values are used for an ecological-economic optimization model in running the scenario of emission reduction.

1.3. Environmental risk assessment under critical load calculations

At present it is agreed both nationally and internationally that the process of quantitative prediction of the probability of an adverse response in ecosystem health due to exposure to one or more pollutants is collectively known as Environmental Risk Assessment, ERA (US EPA, 1992).

In accordance with this definition, an environmental risk assessment process is used especially in cases when the probability component appears during the calculation of various parameters due to any of a number reasons: uncertainty of input information; uncertainties in applied algorithm due to lack of knowledge, insufficient knowledge and/or simplification of input information; uncertainties in the defined geographic boundaries of pollutant influence; uncertainties in both computer calculations and management operations based on these calculations.

The principal scheme of ERA (Smith *et al*, 1988) is shown in the Figure 4.

The analysis of this principal scheme of ERA is restricted so that the quantitative risk assessment is possible based only on the whole operational flowchart.

Suggested ERA frameworks for development of acidification oriented projects

ERA in general is a process, as is EIA (Environmental Impact Assessment), and not the occasional report or document that is published at various steps. The framework for the orderly process which has been developed for various environmentally sound projects can be applied also for acidification oriented projects and especially for an evaluation of ecosystem sensitivity to acid deposition and critical load calculations. The close link to management is an essential feature.

Hazard identification is akin to the qualitative prediction of impacts in EIA and is largely accomplished when the EIA is performed independently of, or prior to,

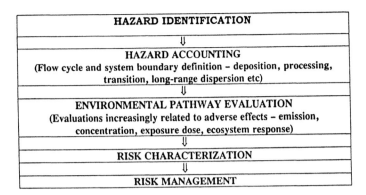

Figure 4. Recommended environmental risk assessment framework.

an ERA. Potentially significant risks are often identified because of experience elsewhere with similar materials, processes, ecosystems, and conditions. This step is immediately useful to management and helps to sharpen the question posed in the further stages, for example, in the stage Term Of Reference, TOR, for the ERA (ADB, 1991).

Hazard accounting considers the total system of which the acidification influence is a part and sets practical boundaries for the assessment. For example, Figure 5 shows the acidification and eutrophication processes identification in both terrestrial and aquatic ecosystems of concern in any of several points. These two steps are the basis for writing TOR for the ERA to be performed.

The environmental pathway evaluation considers various routes by which ecosystems could be exposed to acid deposition (Figure 5).

Associated with this step is a determination of the degree to which measurements of the hazardous acidification and/or eutrophication effects can be directly related to ecosystem/human health.

Risk characterization estimates the frequency and severity of adverse events and presents the results in a form useful to management, for example, in the form of various scenarios for emission abatement strategy on local or regional scale.

Risk management is the selection and implementation of risk-reduction actions. If the recommendations and findings of the ERA leave important questions unanswered, an iteration to hazard accounting can change the boundaries of analysis and refine the assessment. Although risk management is the goal of assessment, it must be integrated to guide the process efficiently.

The purpose of providing risk information to investment responsible managers is to improve decision-making about development of acidification reduction strategy. These decisions are not just a go or no-go statement about a project proposal because the financing agency is usually fully involved in the design of most projects. EIA and ERA advise of unwanted consequences to the environment. If the lender is not comfortable with predictions, there are opportunities to change the project plans such

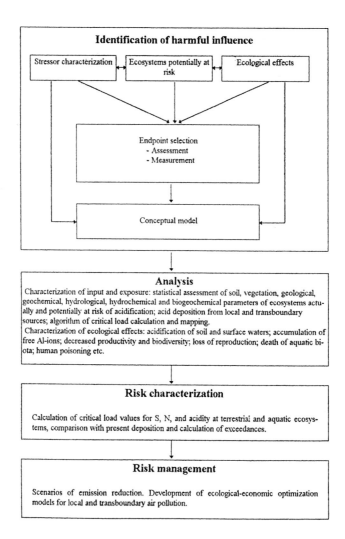

Figure 5. Comparative application of CL and ERA analysis of acidification loading at ecosystem.

as different sites, alternative technologies, risk reduction measures, and emergency responses. The changes may be made a condition for loan approval or a covenant in the loan.

So, regarding the ecosystem acidification effect assessment, risk management is the evaluation of alternative emission reduction measures and implementation of those that appear cost-effective. Management concerns that arise because of substantial uncertainties about major environmental consequences determine the scope

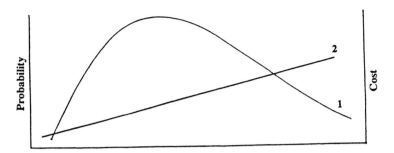

Figure 6. Scheme of comparative analysis of probability distribution function of critical load values of acidity at ecosystems and cost of emission reduction 1- probability distribution function of critical load values 2- cost of emission reduction.

of detailed risk assessment. Projects are undertaken for obvious and direct benefits of economic growth, employment, and exploitation of natural resources in various countries of the World but the ecosystem sensitivity is the main environmental concern of many projects. Achievement of these benefits always entails risk, but the risk must be acceptable to the funding agency and the country. Reduction of risk costs money but so does incurring the unwanted impacts (Figure 6).

Avoiding one risk may create a new risk; net risk is always a consideration. Thus, the risk assessment analyses trade-off in risk, compares risk levels, and evaluates cost-effectiveness of risk reduction alternatives.

The inquiry into the presence of hazards is also part of the preliminary assessment for the EIA. It is by the explicit identification of significant uncertainties that the need to extend an EIA to include the ERA is determined. Of course, if uncertainties can be resolved by readily acquiring more information, then the assessor should proceed to do so.

The ecosystem acidification and critical load calculation processes are only partly scientific exercises being connected closely with economic development of all countries. So, in different projects the hazards of concern include ecosystem damage due to acidification and eutrophication processes (e.g., decreased productivity and biodiversity, soil erosion, drinking water quality, reproduction losses etc), firstly, on a local scale and, secondly, on a regional scale that may lead to transboundary pollution. For more details see Figure 5.

Under critical loads calculations the uncertainties arise from:

– lack of understanding of important cause-effect relationships, lack of scientific theory (e.g., biogeochemical cycling of elements; bioaccumulation of toxic chemicals in food chain, reaction of trees and crops to air pollutants);

– models that do not correspond to reality because they must be simplified and because of lack of understanding (see above);

— weakness of available date due to sampling and/or measurement problems, insufficient time-series of data, lack of replication;

— data gaps such as no measurements on baseline environmental conditions at a study site;

— toxicological data that are extrapolated from high dose experiments to relatively low exposure;

— natural variations in environmental parameters due to weather, climate, stochastic events.

Consequently, risk assessment process is the obligated continuation of the process of quantitative calculation and mapping of critical loads of sulfur, nitrogen and acidity at various natural and agricultural ecosystems. This is connected with numerous uncertainties *a priori* included in the computer algorithm for CL calculations:

— at *the receptor selection* step the uncertainty is related to the determination of the most sensitive receptor, which protection will definitely protect other, less sensitive, ecosystems;

— at *the select environmental quality criteria* step the uncertainty is connected with an assessment of biogeochemical structure of ecosystems and quantitative characterization of biogeochemical cycles of individual elements;

— at *the select computer method (model)* step the uncertainty is related to the applicability of steady-state models to dynamic systems requiring the definite simplification of these systems;

— at *the calculate critical loads* step the uncertainty is usually minimum and related mainly to the possibilities of modern computer tools;

— at *the compare with actual load* step the uncertainty is connected with an assessment of modern deposition and their spatial and temporal conjugation with definite ecosystems at the selected resolution scale.

Comparative analysis of CL and ERA calculations of acidification loading at ecosystems

The existing uncertainty at all steps of the algorithm for critical load calculation and mapping influences the probabilistic character of these values and requires the joining of both approaches. This is illustrated in Figure 5. In the maximum degree the given conjugation is required at *the risk management* step in the ERA flowchart. The probabilistic approach to the critical loads of acid forming compounds allows us to run the set of emission reduction scenarios to minimize the financial investments for ecosystem protection (see Figure 6).

Within the defined areas, critical loads are calculated for all major combinations of tree species and soil types (receptors) in the case of terrestrial ecosystems, or water biota (including fish species) and water types in case of freshwater ecosystems. These combinations include the great variety of different ecosystems, the sensitivity of which to both acidification and eutrophication inputs by atmospheric pollutants differs greatly, determining the necessary reduction needs when CLs are exceeded by modern deposition levels.

This information on ecosystem sensitivity can be compared with a pollutant deposition map, to determine which areas currently receive deposition levels, which exceed the area's CL. The areas of "exceedance" indicate where present levels of pollutant deposition increase the risk of damage to ecosystems.

1.4. Sensitivity of European ecosystems to acid deposition

The calculation and mapping of CLs of acidity, sulfur and nitrogen form a basis for assessing the effects of changes in emission and deposition of S and N compounds. So far, these assessments have focused on the relationships between emission reductions of sulfur and nitrogen and the effects of the resulting deposition levels on terrestrial and aquatic ecosystems.

Critical load calculation models

Critical loads of sulfur and nitrogen, as well as their exceedances are derived with a set of simple steady-state mass balance (SSMB) equations. The first word indicates that the description of the biogeochemical processes involved is simplified, which is a necessity when considering the large scale application (the whole of Europe or even large individual countries like Russia, Poland or Ukraine) and the lack of adequate input data. The second word of the SSMB acronym indicates that only steady-state conditions are taken into account, and this leads to considerable simplification. These models include the following equations.

The maximum critical load of sulfur, CLmaxS

$$CLmaxS = BC_{dep} - Cl_{dep} + BC_w - BC_u - ANC_{le(crit)}, \qquad (3)$$

where BC_{dep} is Base cation deposition, Cl_{dep} is Correction for sea salt deposition, BC_w is Base cation weathering, BC_u is Base cation uptake by plants, $ANC_{le(crit)}$ is Critical leaching of acid-neutralizing capacity of soil.

This equation equals the net input of (sea-salt corrected) base cations minus a critical leaching of acid neutralizing capacity.

The minimum critical load of nitrogen, CLminN

$$CLminN = N_u + N_i, \qquad (4)$$

where N_u is net nitrogen uptake, N_i is nitrogen immobilization in soil organic matter as long as the deposition of both oxidized and reduced nitrogen species, Ndep, stays below the minimum critical load of nitrogen, i.e.

$$Ndep \leq CLminN = N_u + N_i. \tag{5}$$

All deposited N is consumed by sinks of nitrogen (immobilization and uptake), and only in this case CLmaxS is equivalent to a critical load of acidity.

The maximum critical load of nitrogen, CLmaxN

$$CLmaxN = CLminN + CLmaxS/(1 - f_{de}), \tag{6}$$

where f_{de} is the denitrification fraction

The maximum critical loads for nitrogen acidity represents a case of no S deposition. The value of CLmaxN not only takes into account the nitrogen sinks summarized as CLminN, but considers also deposition-dependent denitrification as a denitrification fraction f_{de}. Both sulfur and nitrogen contribute to acidification, but one equivalent of S contributes, in general, more to excess acidity than one equivalent of N, since nitrogen is also an important nutrient, which is deficient in the most natural ecosystems.

Critical load of nutrient nitrogen, CLnutN

$$ClnutN = CLminN + N_{le(acc)}/(1 - f_{de}), \tag{7}$$

where $N_{le(acc)}$ is acceptable leaching of nitrogen from terrestrial ecosystem

Excess nitrogen deposition contributes not only to acidification, but can also lead to the eutrophication of soils and surface waters.

Therefore, no unique acidity critical load can be defined, but the combinations of Ndep and Sdep not causing "harmful effects" lie on the so-called *critical load function of the ecosystem* defined by three critical loads, such as CLmaxS, CLminN, and CLmaxN. In addition, the critical loads of nutrient nitrogen should be also accounted, CLnutrN. An example of such a trapezoid-shaped critical load function is shown in Figure 7

These four CL values have been calculated for all available natural Forest, Steppe and Heath terrestrial ecosystems for the whole European area. In the European integrated assessment modeling efforts, one deposition value for nitrogen and sulfur, respectively, is given for each $150 \times 150 \, km^2$ EMEP grid cell. In a single grid cell, however, many (up to 100,000 in some cases) critical loads for various ecosystems, mostly forest soils, have been calculated. These critical loads are sorted according to magnitude, taking into account the area of the ecosystem they represent, and the so called *cumulative distribution function (CDF)* is constructed (see Posch *et al*, 1999 for the description of this statistical procedure). From this CDF, percentiles are calculated which can be directly compared with deposition values. The application of a 5-percentile value shows the protection of 95% of the ecosystems in a grid cell, 3-percentile, 97%, 1-percentile, 99%, etc.

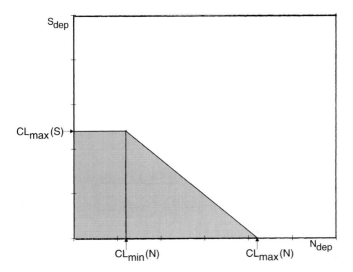

Figure 7. Example of a critical load function for S and N defined by the CLmaxS, CLminN, CLmaxN *and* CLnutN. *Every point of the grey-shaded area below the critical load function represents depositions of* N *and* S, *which do not lead to the exceedance of critical loads (UBA, 1997).*

Critical load exceedance

If only one pollutant contributes to an effect, e.g., nitrogen to eutrophication or sulfur to acidification, a unique critical load (CL) can be calculated and compared with deposition (Dep). The difference is termed *the exceedance of the critical loads*: Ex = Dep − CL.

In the case of two pollutants no unique exceedance exists, as illustrated in Figure 8.

But for a given deposition of N and S an exceedance has been defined as the sum of the N and S deposition reductions required to achieve non-exceedance by taking the shortest path to the critical load function (see Figure 8). Within a grid cell, these exceedances are multiplied by the respective ecosystem area and summed to yield the so called *accumulated exceedance (AE)* for that grid cell. In addition, *the average accumulated exceedance (AAE)* is defined by dividing the AE by the total ecosystem area of the grid cell, which has thus the dimension of a deposition (for detailed explanations see Posch *et al*, 1999).

Maps of critical loads and their exceedance

In this section, we present European maps of critical loads and their exceedance. These values have been used for multi-pollutant, multi-effect Protocol of UNECE Long-Range Transboundary Air Pollution Convention signed in Goteborg in December 1999.

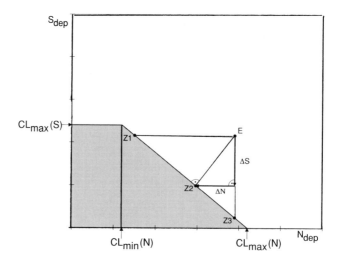

Figure 8. Critical load function for S *and acidifying* N. *It shows that no unique exceedance exists. Let the point* E *denote the current deposition of* N *and* S. *Reducing* Ndep *substantially one reaches point* Z1 *and thus non-exceedance without reducing* Sdep; *but non-exceedance can also be achieved by reducing* Sdep *only (by a smaller amount) until reaching* Z3. *However, an exceedance has been defined as the sum of* Ndep *and* Sdep *reductions (*$\Delta N + \Delta S$*), which are needed to reach the critical load function on the shortest path (point* Z2*). (Posch et al, 1999).*

Figures 9 and 10 are maps of 5th percentiles of the maximum critical loads of sulfur, CLmasS, the minimum critical load of nitrogen, CLminN, the maximum critical load of acidifying nitrogen, CLmaxN, and the critical load of nutrient nitrogen, ClnutN. They show that values of CLmaxS and CLmaxN are lowest in the northwest and highest in the southwest. The low values of CLminN, as compared to ClnutN, in the south (Italy, Hungary, Croatia) indicate low values of nitrogen uptake and immobilization, but relatively high values for N leaching and denitrification.

Figure 11 shows snapshots of the temporal development (1960–2010) of the exceedances of the 5th percentile maximum critical load of sulfur, CLmaxS. The exceedance is calculated due to sulfur deposition alone, implicitly assuming that nitrogen does not contribute to acidification. Although this is probably true at present in many countries, as most of the deposited nitrogen is still immobilized in the soil organic matter or taken by vegetation, the long term sustainable maximum deposition for N not to contribute to acidification is given by CLminN. However, the main purpose of Figure 11 is to illustrate the changes in the acidity critical load exceedances over time. As can be seen from the map, the size of area and magnitude of exceedance peaked around 1980, with a decline afterwards to a situation in 1995, which is better than in 1960.

As mentioned in the previous section, a unique exceedance does not exist when considering both sulfur and nitrogen, but for a given deposition of S and N one

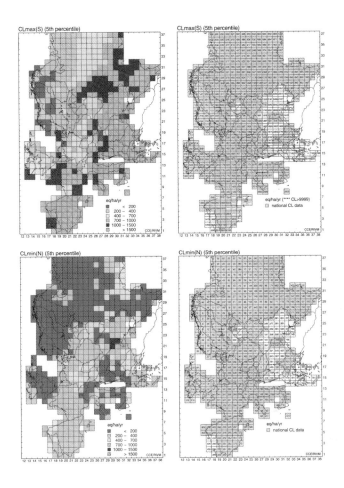

Figure 9. The 5th *percentiles of the maximum critical loads of sulfur,* CLmaxS, *and of the minimum critical load of acidifying nitrogen,* CLminN *(Posch et al, 1999).*

can always determine whether there is non-exceedance or not. The two maps on the top of Figure 12 show the percent of ecosystem area that is protected from acidifying deposition of S and N in 1990 and 2010. In 1990 less than 10% of the ecosystem area is protected in large parts of central and western Europe as well as on the Kola peninsula, Russia. Under the scenario of the 1999 multi-pollutant, multi-effect Protocol of UNECE LRTAP Convention (CDR 2010), the situation improves almost everywhere, but still far from reaching complete protection.

To compare the deposition of S and N with the acidity critical load function, an exceedance quantity has been defined (see previous sections). This average accumulated exceedance (AAE) is the amount of excess acidity averaged over the total ecosystems

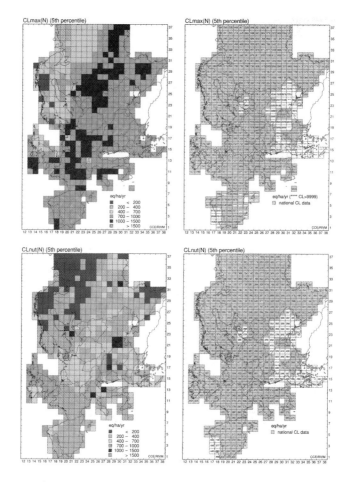

Figure 10. The 5th *percentiles of the maximum critical loads of acidifying nitrogen, CLmaxN, and of the minimum critical load of nutrient nitrogen, CLnutN (Posch et al, 1999).*

area in a grid square. The two maps at the bottom of Figure 12 show the AAE for 1990 and 2010 (CRP scenario). In 1990 the highest acidity excess occurs in central Europe, the pattern roughly matching with the ecosystem protection percentages for the same year. Under the CRP scenario in 2010, excess acidity is reduced nearly everywhere, with a peak remaining in the "Black Triangle" of Germany, Poland and the Czech Republic.

1.5. *Sensitivity of North American ecosystems to acid deposition*

Since the late 1970's, precipitation monitoring programs have been placed in the USA and Canada; eleven Canadian networks (approx. 110 sites) and two large-scale US

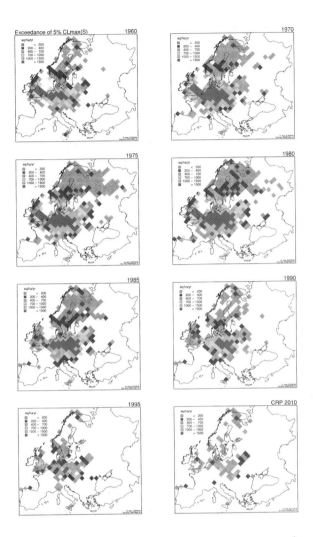

*Figure 11. Temporal development (1960–2020) of the exceedance of the 5*th *percentile maximum critical load of sulfur. White areas indicate non-exceedance or lack of data (e.g. Turkey). Sulfur deposition data were provided by the EMEP/MSC-W (Posch et al, 1999).*

networks (approx. 220 sites) are currently operational. The various networks have now accumulated information for well over 15 years about ion concentrations in precipitation and wet deposition.

Acid rain was recognized as a problem in the North America in the 1950s. Two decades later, scientists noted losses of fish populations in some highly acidified lakes of the East Coast of USA and Canada. The reason was related to acid rain caused by

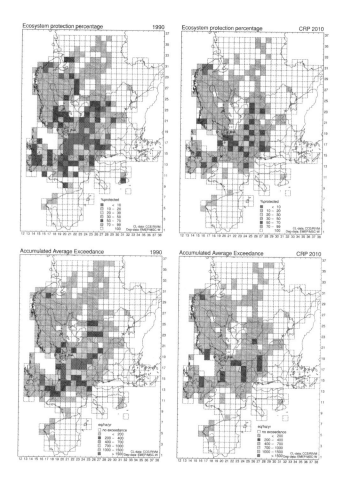

Figure 12. Top: The percentage of ecosystem area protected (i.e. non-exceedance of critical loads) from acidifying deposition of sulfur and nitrogen in 1990 (left) and in the year 2010 according to current emission reduction plans in Europe (right). Bottom: The accumulated average exceedance (AAE) of the acidity critical loads by sulfur and nitrogen deposition in 1990 (left) and 2010 (right). Sulfur deposition data were provided by the EMEP/MSC-W (Posch et al, 1999).

pollutants such as sulfur dioxide and nitrogen oxides, which in the atmosphere are chemically converted to sulfuric and nitric acids. At present, acid rain is the major environmental problem in USA and Canada.

Acidic precipitation has been most recognized as a serous environmental problem in areas of granite rocks, namely Northern and Eastern Canada and the Northeastern United States, where the forests are under assault and the lakes have been becoming

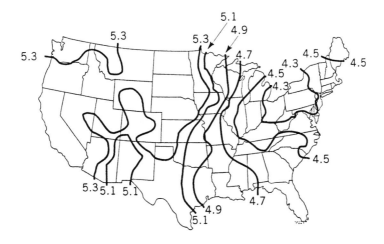

Figure 13. Averaged values of pH *in precipitation in North America in early 1990s (Smith, 1999).*

progressively acidified during the 1980s. The content of base cations and alkalinity in these soils and surface waters is low. Correspondingly, the buffer capacity of the ecosystems to acidity loading is also low. In these poorly buffered lakes a "normal", natural pH would probably be in the range 6.5 to 7.0. In the mid-1990s, many lakes in these areas record pH levels of 5.0 and lower.

The averaged values of pH in precipitation are shown in Figure 13.

Acidifying emissions in Canada and the USA

Sulfur dioxide emissions in both Canada and the USA peaked in the early 1970s and have declined ever since with year to year variability. Actions to reduce acid deposition have been focused mainly on SO_2 emissions because they play generally a much higher role in rainfall acidification than nitrogen oxides. However, this is not the case in some areas of North America, like California, where nitrogen emissions are predominant and consequently contribute the major part in acidity as well.

Since approximately half of the acid precipitation in eastern Canada has come from American sources, the Canada-United States Air Quality Agreement was signed in 1991 to reduce sulfur emissions and also set up a framework for dealing with nitrogen oxides and other pollutants that commonly cross the USA-Canada border. As a result, SO_2 emissions in two countries have declined substantially. Under the current programs, total emissions from the two countries are expected to drop from 28.2 million tons (MT) (Canada 4.6 + USA 23.6) to 18.3 MT (Canada 2.9 + USA 15.4) by the year 2010. In Canada alone, sulfur dioxide emissions have declined considerably over the 1980-1990s and, by 1995, had been reduced to 2.65 MT, lower than the agreed upon limit of 2.9 MT (Ro and Vet, 1999).

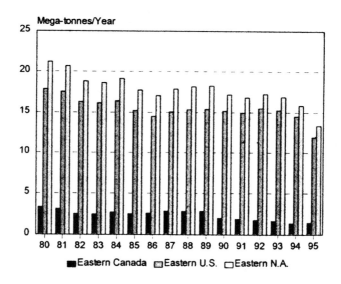

Figure 14. Sulfur dioxide emissions in eastern Canada, eastern USA and total North America (Ro and Vet, 1999).

Since environmental damage due to acid deposition has largely been limited to the eastern parts of Canada (east of Manitoba-Ontario border) and the USA (east of the Mississippi River), most of the emission reductions have occurred in those areas. Figure 14 illustrates the SO_2 emission totals in eastern Canada, eastern USA and total North America.

In contrast to the situation with sulfur emissions, neither Canada nor USA has made significant progress in reducing NO_x emission, the other major acidifying pollutant. Although technological innovations such as catalytic converters have greatly reduced NO_x emissions from individual sources, the gains have been offset by a continuous increase in the number of emission sources, particularly cars and trucks. In 1995, eastern Canadian NO_x emissions stood at 1 MT, while the eastern US sources were responsible for 11 MT. During recent years, these levels have not changed appreciably in either country.

Wet deposition of sulfate in eastern North America

In theory, significant reduction in SO_2 emissions should, over a long term period and large areas, produce detectable reductions in the amount of wet sulfate deposition. The acid rain monitoring data in North America have been gathered by Environment Canada and stored in the National Atmospheric Chemistry (NatChem) Database, details of which can be found at www.airquality.tor.ec.gc.ca/natchem. Analysis of the deposition chemistry data has confirmed that wet sulfate deposition did indeed

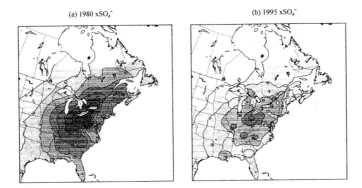

Figure 15. 1980 (a) and 1995 (b) wet sulfate deposition pattern in eastern North America, kg/ha/yr. of $SO_4{}^{2-}$ (Venkatesh et al, 2000).

decline in concert with the decline in SO_2 emissions in both eastern Canada and the eastern USA. This is shown in Figure 15.

We can see from Figure 15 that wet deposition declined markedly. In fact, close inspection reveals the total area that received $\geq$ 20 kg/ha/yr. in 1980 had virtually disappeared in 1995, a total area reduction of 87%.

However, in accordance with insignificant reduction of NO_x emissions, the wet deposition of nitrate has been practically unchanged (Figure 16.)

Ecological impacts of acid deposition in Eastern North America

It is roughly estimated that there are more than 1,200,000 water bodies in eastern North America that are currently affected by acid deposition. A subset of these lakes has been sampled since early 1980s in order to monitor the changes in lake water chemistry induced by the declining sulfur dioxide emissions and wet sulfate deposition loading. Sampling at several hundred of these lakes during 1980–1990s indicated that water quality improvement has been slow and inconsistent. For instance, of 202 lakes in southeastern Canada that were consistently monitored from 1981 to 1994, 56% showed no improvement in acidity, 11% became more acidic and 33% became less acidic (Figure 17).

Most of the lakes showing improvement were located in the area around Sudbury, Ontario, where smelter emissions declined dramatically, from 2,000,000 tons in 1970 to 265,000 tons by 1994 (Bunce, 1994).

There are several possible explanations for the slow and uneven recovery of these lakes (Ro and Vet, 1999):

(1) insufficient time for major recovery to become apparent;

(2) lakes continuing to receive sulfate wet deposition well above critical loads, i.e., reductions in deposition have been insufficient;

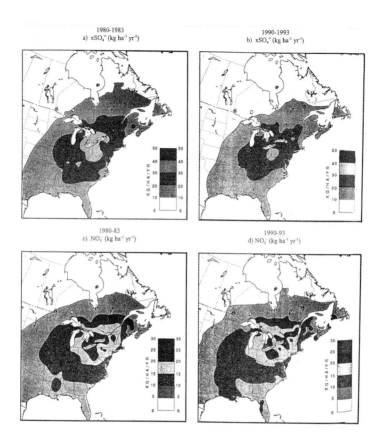

Figure 16. The 4-year average patterns of wet nitrate deposition in eastern North America in 1980–1983 and 1990–1993 (Ro and Vet, 1997).

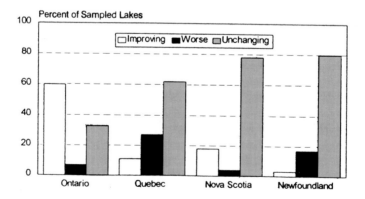

Figure 17. Trends in lake acidity between 1981 and 1994 (Environment Canada, 1997).

(3) lakes experiencing declining base concentrations, which, in turn, have reduced their ability to neutralize acids;

(4) acidification due to nitrate deposition in watersheds that nitrogen saturated (see also Chapter 8, Section 1).

Forest damage caused by acid deposition has been found in many areas of eastern North America. For example, acid fogs have been found to cause damage in Birch Forest ecosystems in the Bay of Fundy area of New Brunswick and in high elevation areas of eastern Canada, where they prevent the germination of pollen in certain birch species, and reduce the frost hardiness of red spruce trees. Red Oak, Red Pine and Sugar Maple Forest ecosystems in acidic soils on Quebec and Ontario have exhibited damage due to acid deposition. Other areas of Ontario have experienced accelerated nutrient loss and declining acid-neutralizing capacity of soils. Reduced growth rates have been found over the past 30 years in Sugar Maple Forest ecosystems covering large areas of northern Ontario and Quebec.

The ability of forests to withstand acid rain depends on the capacity of the soil to neutralize the inputting acidity. This is largely determined by local geology, in much the same way that it affects the acidification of lakes. Acidification is mainly a problem in areas where the underlying rocks provide poor buffering capacity. Rocks such as granite offer little buffering protection. Chalk and limestone neutralize added acid, and so soils, lakes and streams in limestone areas are fairly insensitive to acidic precipitation.

$$2H^+ + CaCO_3 \longrightarrow Ca^{2+} + CO_2 + H_2O. \qquad (8)$$

The mechanisms underlying this reaction is that H^+ ions react with the HCO_3^- ions which are responsible for the alkalinity of the water, and solid $CaCO_3$ from bottom sediments dissolves to restore equilibrium. As a result, the pH in the aqueous phase is not changed significantly by the addition of the acidic rainwater.

Most of eastern North America is underlain by granite geological rocks, whose acid-neutralizing capacity is low and is exhausted rapidly under acidity loading. As a result, areas of forest sensitivity correspond closely to the areas where aquatic ecosystems are also sensitive. In southeastern Canada, these areas are in Central Ontario, southern Quebec and the Atlantic provinces. Estimates have been made that Forest ecosystems in these areas receive about twice the level of acid rain that these ecosystems can tolerate without long-term damage. The damage depends very much on the soil type, geological rocks and type of ecosystems. Documented damage in eastern Canada includes loss of nutrients in forest soils, declining growth rates (NPP), excessive aluminum uptake, cuticle damage, decreased photosynthesis, interference with pollen germination, defoliation, and reduced ability of trees to cope with other stresses such as insects, drought, diseases and increased ultraviolet radiation. These effects are very similar to those shown in other places of the World, like Europe and East Asia.

The impact-oriented critical load approach to SO_2 *emission reduction strategy*
Critical load calculations

In 1985, the federal government of Canada and the seven easternmost provinces signed the Eastern Canadian Acid Rain Program, the purpose of which was to reduce SO_2 emissions in eastern Canada to 50% of the 1980 levels by 1994. Based on the best modeling and experimental studies of the time, it was hoped that these sulfur dioxide emission cutbacks would reduce sulfate wet deposition from the levels as high as 40 kg/ha/yr. to less than 20 kg/ha/yr. throughout eastern Canada. The specific value of 20 kg/ha/yr. was deemed a 'target critical load', which was expected to protect moderately sensitive aquatic ecosystems from acidic deposition. Highly sensitive ecosystems were left vulnerable.

The Eastern Canadian Acid Rain Program was highly successful at reducing SO_2 emissions and sulfate wet deposition in eastern Canada (see Figure 14). Sulfur emissions actually declined more than the desired 50% by 1994, and have continued to decline modestly in the present. These SO_2 emissions in the United States have also reduced dramatically, particularly since the implementation of the Canada-United States Air Quality Accord in 1991. This has been especially important to the aquatic and terrestrial ecosystems in eastern Canada, since US emissions are responsible for a large proportion of the acid deposition received in eastern Canada due to transboundary transport.

Current atmospheric transport and deposition models predict that all of the acid-sensitive areas of eastern Canada will receive less than the target critical load of 20 kg/ha/yr. of sulfate wet deposition by 2010, the year that USA emission reductions will be fully enacted. Unfortunately, there is strong evidence that even these levels of sulfate wet deposition will be too high, and large areas of eastern Canada will still be vulnerable to acid rain (Clair *et al*, 1999). To assist policy makers in developing better deposition targets for the future, scientists used the critical load concept. Here, critical load is determined as an estimate of the amount of deposition that a particular region can receive without significant damage to its ecosystems. This definition is similar to that used in Europe (see above). The critical load concept, as applied in Canada, has been a useful tool for identifying ecosystems at risk and for estimating future needs for emission abatement strategy. The concept integrates monitoring information on wet deposition and aquatic chemistry with model predictions on long-range transport of air pollutants, ecosystem response to airborne pollutants, and emission cutback scenarios.

As applied to aquatic ecosystems in eastern Canada, *critical load is defined as the amount of wet sulfate deposition that must not be exceeded in order to protect at least 95% of the lakes in a region from acidifying to a* pH *level of less than 6.0* (Ro and Vet, 1999).

Recent studies suggest that a pH of 6.0 is needed to protect most aquatic organisms (see Box 2)

Box 2. Loss of fish species and lake acidification (After Bunce, 1994)

The natural pH of surface waters is between 6.5 and 8.5 depending on type of water body, underlying geological rocks, water trophy levels, biogeochemical food web,

Table 1. Relationship between pH *and aquatic living organism losses.*

Surface water pH	Aquatic organism lost
6.0	Death of snails and crustaceans
5.5	Death of salmon, rainbow trout and whitefish
5.0	Death of perch and pike
4.5	Death of eel and brook trout

etc. It has been recognized that the declining pH less than 6.0 is accompanied with undesired changes in biodiversity and even death of many aquatic species (Table 1)

Below pH ca. 4.0, the lake becomes a suitable habitat for white moss, which prefers an acidic environment. This plant forms a 'felt mat', which may grow to 0.5 m or more thick, on the lake bottom. The mat prevents the exchange of nutrients between the water and the bottom sediments and also prevents the sediments from exerting any buffering action. The resultant lake waters are crystal clear, but this water supports very few forms of aquatic living organisms.

The loss of game fish is expected to be severe for lakes whose pH has already dropped to 5 or lower. By 1976, about 50% of the lakes in the Adirondack Mountains of New York State, USA, had no fish in them, whereas forty years ago almost all these lakes supported a population of sport fish. This observation correlates with comparisons of the alkalinity of Adirondack lakes today vs. sixty years earlier: of 274 lakes for which data were available, 80% had suffered loss of alkalinity, the median loss being 50 mol H^+/L. The loss of the fish has serious consequences for regions like upstate New York and Northern Ontario, where tourism is a mainstay of the economy.

It is worth noting that, with this definition of critical load:

a) wet sulfate deposition plays a surrogate for total (wet and dry) deposition of all acid-forming compounds that contribute to lake acidification;

b) 5% of lakes in a given region will be left vulnerable to the continued effects of acidity loading.

The critical load concept was applied to eastern Canada for two reasons, mainly:

(1) to identify the areas of Canada that receive sulfate wet deposition in excess of the calculated critical loads, and

(2) to estimate future SO_2 emission reductions necessary to ensure that all areas of eastern Canada will receive wet deposition of sulfate less than the critical loads.

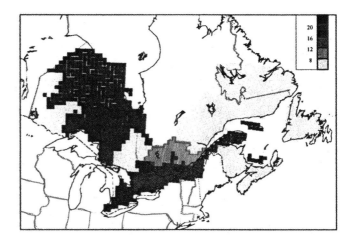

Figure 18. Regional critical loads in eastern Canada (from RMCC, 1990). Different shaded areas show the critical loads of wet sulfate deposition (kg/ha/yr.) that can be tolerated by lakes in those areas.

The first step in the application of the concept was to determine the critical load values for the different regions of eastern Canada. This was done using historical measurements of lake acidity in concert with the Integrated Assessment Model (IAM) which links atmospheric transport and deposition models with water chemistry and empirical biological response models. Details of the method are given in Jeffries and Lam (1993).

The critical loads calculated for eastern Canada are shown in Figure 18.

The lightly shaded areas in Figure 18 have the lowest critical loads and are therefore most sensitive to wet sulfate deposition and lake acidification. The dark areas have high critical loads (> 20 kg/ha/yr.) and are least sensitive. The areas with lowest critical loads occur on the Precambrian Shield region of Ontario and Quebec because of the underlying granitic bedrock and the shallow, poorly buffered soils. The Atlantic coast also has low critical loads because of lower base cation concentrations in the lakes, i.e., lower acid-neutralizing capacity (Jeffries *et al*, 1986). We can see that large areas of eastern Canada have critical loads considerably less than the old 'target critical load' of 20 kg/ha/yr. This demonstrates that, while the "target critical load" was useful as a policy-making tool in the 1980s, it was clearly insufficient to protect many lakes in eastern Canada.

Exceedances of critical load

After calculation and mapping of critical loads, the next step is to identify those areas where the critical loads were/are exceeded, i.e., where the wet deposition loading of sulfate exceeds the critical loads. In these areas of exceedance, lakes will continue to

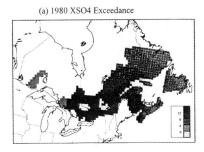 

Figure 19. Exceedances of critical loads in: (a) 1980, (b) 1995. Wet deposition exceedances of the critical loads (calculated as a difference between the annual sea-salt corrected sulfate wet deposition and the annual critical load of sulfate in each grid cell). Note that the lightest grid cells represent no exceedance of the critical load (Ro and Vet, 1999).

have pH values lower than 6.0 until such time that wet sulfate deposition decreases below the critical loads.

The critical load exceedance pattern is shown across eastern Canada for the years 1980 and 1995 (Figure 19)

Here, the darkest areas have higher exceedances of the critical loads and are therefore most vulnerable to acid deposition. The differences in the exceedance patterns of 1980 and 1995 (Figure 19a and 19b) indicate that the area of exceedance, and the amount of exceedance in most areas, declined considerably in 15 years. By way of comparison, the 1995 area of exceedance is 61% lower than that in 1980, a clear illustration that the decline in annual sulfate deposition from 1980 to 1995 resulted in a large reduction to the number of lakes vulnerable to acid rain. In spite of the decline, there still exist in 1995 large areas in eastern Canada where the critical loads are exceeded (Figure 19b). In 1995, this area was approximately $510,000 \, km^2$ and encompassed roughly 60,000 lakes.

Year-to-year changes in the area of exceedance are shown in Figure 20.

Three curves are shown, corresponding respectively to the areas where critical load exceedances are > 8, 4, and 0 kg/ha/yr. The diagram clearly indicates that the exceedance areas declined markedly from 1980 to 1995, with the areas of very high exceedances (> 8 kg/ha/yr.) almost disappearing by 1995.

Figure 21 illustrates the year-to-year change in the total amount of wet sulfate deposition (non-sea-salt) that exceeded the critical loads in eastern Canada.

The values in the diagram were calculated as the sum of all the sulfate exceedances across eastern Canada (e.g., exceedance amount equals the sum of all exceedances shown in Figures 19a and 19b for 1980 and 1995, respectively). Figure 21 shows a decline in the exceedance amounts from 1980s to 1995, with the biggest declines occurring in two periods, such as the early 1980s and in 1995. Both of these were periods of rapidly declining SO_2 emissions in the USA. It is clear from Figure 21 that the amount of wet-deposited sulfate in excess of the critical loads in eastern Canada

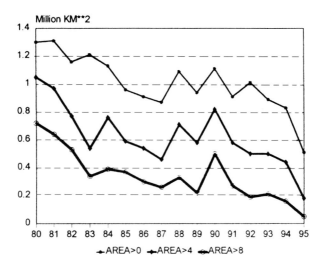

Figure 20. Temporal trends in Exceedance Area associated with critical load exceedance greater than 0, 4 and 8 kg/ha/yr. in eastern Canada. The area with > 0 exceedance declined by more than 50% by 1995 compared to 1980, and the area with > 8 kg/ha/yr. exceedances declined by roughly 90%. All three categories reached a minimum exceedance area in 1995 (Ro and Vet, 1999).

has declined markedly. Nevertheless, greater than 0.18 MT of sulfate in excess of the critical loads was wet-deposited in 1995 and continued to put lakes in eastern Canada at risk of continued acidification.

Sulfur dioxide emission abatement scenario in North America based on critical loads and their exceedances

Having demonstrated that large areas of critical load exceedance still exist, the final step in the analysis is to predict, using long-range transport models, the amount by which SO_2 emissions should be reduced in both Canada and USA so that no areas in eastern Canada receive deposition in excess of critical loads. This was done using the Canadian long-range transport model known as the Atmospheric Deposition and Oxidants Model (ADOM). Details of the model and the modeling exercise are described in Moran (1997) and are too lengthy to describe in this text. Simply stated, several different emission control scenarios were run through the model, and the predicted sulfate wet deposition results were compared to the critical loads. Two important conclusions were derived from the analysis (Moran, 1997 and Venkatesh *et al*, 2000):

(1) Hundreds of thousands of square kilometers in eastern Canada, encompassing tens of thousands of lakes, will continue to receive wet sulfate deposition above

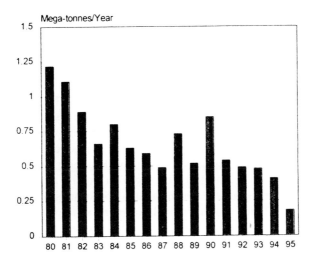

Figure 21. Annual integral Exceedance Amounts of sea-salt corrected wet sulfate in eastern Canada. Note: the markedly higher exceedance amount in 1990 is due to in part to a higher precipitation amount in this year (Ro and Vet, 1999).

the critical loads for aquatic ecosystems, even after the existing Canadian and USA SO_2 emission control programs are fully implemented in the year 2010;

(2) Further SO_2 emission reductions, estimated to be of the order of 75% beyond current reduction commitments, are required in both Canada and the United States to protect lakes in eastern Canada from sulfate deposition in excess of the critical loads.

The estimate of 75% further emission reductions contains uncertainties inherent to the entire critical loads/atmospheric modeling exercise. Nevertheless, it clearly indicates that large reductions in emissions must be initiated in the future if the remaining acid-sensitive aquatic ecosystems in eastern Canada are to be saved from the effects of acid rain. Recognizing this, the federal and provincial governments of Canada have signed a cooperative agreement on acid rain known as the Canada-Wide Acid Rain Strategy for post-2000. The long-term goal of this agreement is to develop new targets for future sulfur dioxide reductions and to conduct future research into the contribution of nitrogen to the acidity loading.

1.6. Sensitivity of East Asian ecosystem to acid deposition

There is agreement both nationally and internationally that long-range transboundary air pollution may span continents: pollutants are transferred from Europe to North America and Asia as well as in the opposite directions (Sofiev, 1998). Consequently, the calculation and mapping of critical loads as indicators of ecosystem sensitivity to acid deposition in regions outside of Europe and North America are of great scientific and political interest. Some researches have been made to calculate the acidification loading for Asia (Dianwu *et al*, 1994; Shindo *et al*, 1995; World Bank, 1994; Kuylenstierna *et al*, 1995; Kozlov *et al*, 1997; Bashkin and Kozlov, 1999).

It has been argued (Bashkin *et al*, 1996) that the best approach to the calculation and mapping of critical loads on ecosystems in Asia is to use various combinations of expert approaches and geoinformation systems, including different modern methods of expert modeling and environmental risk assessment. These systems can operate using databases and knowledge bases relative to the areas with great spatial data uncertainty. As a rule, the given systems include an analysis of the cycles of various elements in the key plots, a choice of algorithms describing these cycles, and corresponding interpretation of the data. This approach requires numerous cartographic data, such as vegetation, soil, geochemical and biogeochemical maps, information on pollution and buffering capacity of soil, water and atmosphere. This approach is the most appropriate for Russia as well as for various Asian countries such as China, India, Thailand where, at present, adequate information on the great spatial variability of natural and anthropogenic factors is either limited or absent.

The applicability of these approaches for the assessment of acidification loading on the terrestrial ecosystems in Asia is made here using the examples of Asian domain (Asian part of Russia, China, Japan, Taiwan, Korea and Thailand). In spite of the great differences in climate, soil and vegetation conditions, these regions can serve as a good test of the proposed methodology.

Characterization of soil-biogeochemical conditions in Asia

The interest in acidic deposition has resulted in the development of intensive biogeochemical investigations of a large number of ecosystems in North America, Europe, Asia and South America (Moldan & Cherny, 1994; Bashkin & Park, 1998). The biogeochemical cycling concept is designed to summarize the cycling process within various components of ecosystems such as soil, surface and ground water, bottom sediments, biota and atmosphere. Ecosystem and soil maps can serve as a basis for biogeochemical mapping (see Chapter 7).

Monitoring of acid rain in Asia

As regards the pollutants monitoring, from the measurements available so far it could be concluded that acid rain is coming to be a major problem in Asia. In many industrially developed and new developed countries such as Japan, China, Taiwan, South Korea, Thailand etc the values of pH < 5 are encountered at many sites, and they represent more than 50% of monitored rain events on a regional scale. In some developing countries of South-East Asia (Myanmar, Laos, Cambodia) most rainwater pH measurements tend to be around 5.6, the pH of 'natural' rainwater and the acid rain precipitation are mainly due to localized industrial pollution. There is some evidence that pH values below 5 at unpolluted sites may be due to the contribution of weak organic acids, such as formic and acetic acids (Radojevic 1998).

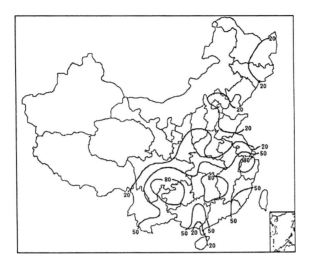

Figure 22. Contours of acid rain frequency corresponding to pH *5.6 in China (Hao et al, 1998).*

The experimental results obtained by Chinese scientists obviously show that the areas suffering from acid rain in China have extended northwards from the south of the Yangtze River in 1986 to the whole of East China at present. The statistical results from the Acid Rain Survey in 82 cities from 1991 to 1995 indicate that the annual average pH value of the precipitation was lower than pH 5.6 in nearly half of these cities or in 87% of the southern cities, which are located in the south to the Qingling Mountain and Huaihe River, and the lowest even reached pH 3.52 in the Changsha, Hunan province (Figure 22).

In addition, the frequency of acid rain was very high (higher than 60%) in one fourth of these cities (Hao *et al*, 1998). The chemical composition of acid rain in China is generally different from that in Europe, with the lower pH value and the higher sulfate, calcium and ammonium concentration. Another difference is that the concentration of calcium relative to sulfate is very high in China, while the nitrate concentration relative to other components is low. In some cases, the fluoride concentration in precipitation appears also high in China, owing possibly to the combustion of coal with high fluoride content. Besides, the alkaline fly ash and soil dust build up the capacity for acid neutralizing during washout.

The influence of acid rain on the environment is related to the various properties of different ecosystems and varies depending upon physic-chemical characteristics of soil, vegetation type, stemflow and throughfall interactions of rainwater with canopy of different botanic species. For instance, it is well known in Japan that soils close to the stems of Japanese cedar (*Cryptomeria japonica*) trees are strongly acidic. This is partly due to the leaching of hydrogen ions from the stems. Soil solutions close to a stem (10 cm) are markedly acidic (pH $\sim$ 4.5) and contain 47 μM of total Al in

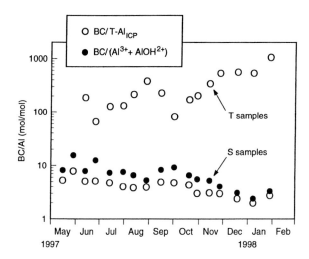

Figure 23. Seasonal variation of the molar BC/AL ratios in soils of Japanese Cedar ecosystems (Kato et al, 1998).

average. Furthermore, more than 55 % of the Al are in the form of Al^{3+}. Figure 23 shows that the BC–Al ratios in winter decline to as low as 2, which is close to 1 that is known to be used as critical value for starting soil acidification and appearance of harmful free aluminum ions (Kato *et al*, 1998).

Approximately 70% of the precipitation monitored in Taiwan is considered acid rain (exhibiting a pH lower than 5.6). Acid deposition shows large spatial variation with northern Taiwan exhibiting the highest acid deposition and southeastern Taiwan showing the lowest deposition. The spatial pattern of acid deposition is correlated with the degree of urbanization and industrialization. Continuous exposure to high levels of acid deposition could lead the forest to nutrient imbalance and thereby undermine forest health (Lin 1998). It is also known that the geochemical and biogeochemical mobility and migration of the majority of heavy metals are increasing with the decreasing soil and water pH values that are occurring due to acid deposition. Thus, it has been shown that the biological accumulation of Cd, Cu, and Zn, expect for Pb, in the leaves of vegetables was affected by the acid rain, and the rating of effectiveness on the phytoavailability of heavy metals caused by acid deposition followed the trend: Cd > Zn > Cu ≫ Pb (Chen *et al*, 1998). The harmful single and synergetic effects of acidity, SO_2, NO_x and O_3 are experimentally shown for various Asian species (Kohno *et al*, 1998).

Critical load values of acid forming compounds on ecosystems of Asia

The experimental data obtained in various countries of East Asia allow us to consider the applicability of methodology of critical loads related to an assessment of

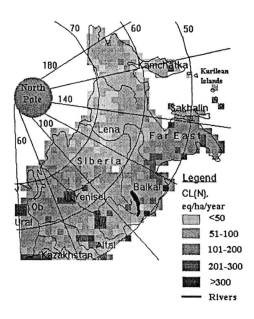

Figure 24. Critical loads of nitrogen on the ecosystems in the Russian Northern Asia (Bashkin & Kozlov, 1999).

ecosystem sensitivity to acid rains. The critical load (CL) and Environmental Risk Assessment (ERA) approaches were used for the evaluation of ecosystem sustainability to acid deposition in East Asia. The calculations of critical loads for the assessment of the sensitivity of the ecosystem to acidic deposition have been made using biogeochemical approaches including the intensity of biogeochemical cycling and periods of active temperature duration. On the basis of these coefficients the soil-biogeochemical regionalization is carried out for the whole area of Asia and the values of critical loads for acid-forming compounds are calculated using modified steady-state mass balance (SSMB) equations.

Critical loads of sulfur and nitrogen on North Asian ecosystems

The models for calculation of critical loads in North Asian ecosystems are shown in Box 3. In the ecosystems of the Asian part of Russia these values of critical loads for N, CL(N), and S, CL(S), compounds are shown to be less than in Europe due to many peculiarities of climate regime (long winter with accumulation of pollutants in snow cover) and depressed biogeochemical cycling of elements (see Chapter 7, Section 1). The minimum values of both CL(N) and CL(S) are < 50 eq/ha/yr. and the maximum ones are > 300 eq/ha/yr. (Figure 24 and Figure 25).

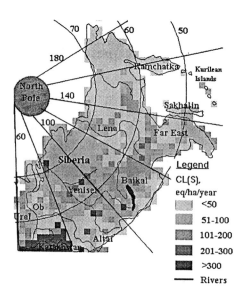

Figure 25. Critical loads of sulfur on the ecosystems in the Russian Northern Asia (Bashkin & Kozlov, 1999).

Table 2. Distribution of critical load values of sulfur, nitrogen and their exceedances for the North Asian ecosystems.

Values range, eg/ha/yr.	Percentage of area under different critical load values		Percentage of area under different critical load values	
	For nitrogen	For sulfur	For nitrogen	For sulfur
< 50	8.3	40.5	88.0	72.7
50–100	40.8	32.4	6.9	14.3
101–200	41.3	18.2	3.9	8.3
201–300	8.0	1.5	1.1	2.6
> 300	1.6	7.4	0.1	2.1

At current rate of atmospheric deposition 88.0% and 72.7 % of ecosystems in North Asia have no or small ($<$ 50 eq/ha/yr.) exceedances of CL(N) and CL(S), respectively (Table 2).

However, atmospheric deposition in excess of calculated critical loads of N and S are exceeded for 10% and 20%, respectively, of the studied ecosystems in this region.

A comparison of these critical load values with those calculated for corresponding ecosystems in Europe (see above Section 1.4) reveals lower values in the Asian areas. This can be explained by the more prolonged winter period causing an accumulation of pollutants in the snow layer and their influence on biogeochemical cycling of nutrients during the short spring and summer period. The values of active temperature coefficient, C_t are within the limits of 0.15–0.57 for the majority of the North Asian ecosystems, whereas in the European part of Russia, for example, the corresponding values are 0.25–0.87. This points to the shorter but very active periods of biogeochemical turnover in almost all North Asian ecosystems accelerating the effects of the acidification loading on the ecosystems.

Box 3. Models for calculation of S and N critical loads at North Asian Ecosystems (After Bashkin and Kozlov, 1999)

Critical loads of nitrogen

$$CL(N) = {}^*N_u + {}^*N_i + {}^*N_{de} + {}^*N_{l(crit)}, \qquad (9)$$

where * means that each of the terms refers to the values at the actual total atmospheric deposition at a side. N_u, N_i, N_{de} and $N_{l(crit)}$ are permissible nitrogen uptake, soil immobilization, denitrification and leaching, correspondingly.

Permissible atmospheric nitrogen uptake (${}^*N_u$) was given as:

$$ {}^*N_u = N_{upt} - N_u, \qquad (10)$$

where N_{upt} is annual accumulation of nitrogen in biomass and N_u is annual uptake of N from the soil.

N_{upt} was calculated accounting for the coefficients of biogeochemical turnover (see Chapter 7). Annual N_u from soil was calculated on a basis of nitrogen mineralizing capacity (NMC) of soils, which was determined experimentally or calculated using regression equations (Bashkin *et al*, 1997). So,

$$N_u = (NMC - N_i - N_{de})C_t, \qquad (11)$$

where $N_I = 0.15\,NMC$, if $C:N < 10$, $N_i = 0.25\,NMC$, if $10 < C:N < 14$, $N_i = 0.30\,NMC$, if $14 < C:N < 20$, $N_i = 0.35\,NMC$, if $C:N > 20$; $N_{de} = 0.145\,NMC + 6.447$, if $NMC > 60\,kg/ha/yr$, $N_{de} = 0.145\,NMC + 0.900$, if $NMC < 10\,kg/ha/yr$, $N_{de} = 0.145\,NMC + 2.605$, if $10 < NMC > 60\,kg/ha/yr$.

Permissible immobilization of atmospheric deposition N (${}^*N_i$) was found as:

$$
\begin{aligned}
N_i &= [(0.20NH_y + 0.10NO_x)/C_b]C_t, && \text{if } C:N < 10, \\
N_i &= [(0.30NH_y + 0.20NO_x)/C_b]C_t, && \text{if } 10 < C:N < 14, \\
N_i &= [(0.35NH_y + 0.25NO_x)/C_b]C_t, && \text{if } 14 < C:N < 20, \\
N_i &= [(0.40NH_y + 0.30NO_x)/C_b]C_t, && \text{if } C:N > 20,
\end{aligned}
\qquad (12)
$$

where NO_x and NH_y are oxidized and reduced N wet and dry deposition.

Permissible denitrification from atmospheric deposition N ($*N_{de}$) was found as:

$$*N_{de} = (N_{de}/NMC)N_{td}C_t, \tag{13}$$

where N_{de}/NMC is denitrification fraction, which depends on many features of soils and calculated on a basis of experimental data and N_{td} is total N deposition.

Finally, permissible critical leaching of atmospheric nitrogen ($*N_{l(crit)}$) was given as

$$*N_{l(crit)} = QC_{Ncrit}, \tag{14}$$

where Q is annual surplus of precipitation (runoff) and C_{Ncrit} is permissible nitrogen concentration in surface water.

Critical loads of sulfur

Since for the majority of ecosystems in North-Eastern Asia the ratio of precipitation to potential evapotranspiration (P : PE) is equal to 1.00 or slightly exceeds 1.00 (except of ecosystems with some cambisols, histosols and andosols) the values of runoff can be neglected in calculation of critical loads of acidity, CL(Ac), and they were found as

$$CL(Ac) = (BC_w \times C_t)/C_b. \tag{15}$$

And critical loads of sulfur were calculated as

$$CL(S) = S_f \times CL(Ac), \tag{16}$$

where BC_w is weathering of base cations and S_f is sulfur fraction in total sum of sulfur and nitrogen deposition.

The values of BC_w were determined on a basis of FAO soil nomenclature, soil parent material and soil texture according to UBA (1996) and values of C_b and C_t were applied to accounting biogeochemical cycling intensity and duration of active temperature period (see Chapter 7, Section 1). The root zone was assumed to be equal to 0.5 m.

Exceedances of critical loads

The values of exceedances were calculated as follows:

$$\begin{aligned} Ex(N) &= N_{td} - CL(N), \\ Ex(S) &= S_{td} - CL(S), \end{aligned} \tag{17}$$

where Ex(N), EX(S) are the values of exceedental input of nitrogen and sulfur compounds above the calculated critical loads, and N_{td}, S_{td} are nitrogen and sulfur deposition.

Regarding the uncertainty analysis of critical loads of acid forming compound at different terrestrial ecosystems of the Asian part of Russia, one can see the following. The strongest influential factor for the values of CL(N) is the parameter, nitrogen uptake by plant biomass, N_u. For the biggest part of Siberia and Far East territory this parameter has the first rank and only in the case of Humid Cambisols and Cryic Gleysols ecosystems is its rank second. Among others, nitrogen immobilization, N_i and nitrogen denitrification, N_{de}, parameters, which are closely inter-correlated, are the weakest ones. The given parameters do not practically have any influence on the values of CL(N), except Dyctric Cambisol ecosystems. The influence of nitrogen leaching, N_l, values is decreasing in the following row of ecosystem-forming soils: Regosols > Humid Cambisols = Cryic Gleysols = Gelic Podzols > Andosols > Dystric Cambisols > Eutric Cambisols > Luvic Phaeozems = Chernozems > Kashtanozems. These results reflect, to a significant extent, the geographical change of ecosystems from north to south and correspondingly an alteration of relationship of temperature and moisture constituents in hydrothermic coefficient. So, the main impact to the assessment of the influence of inputting parameters (N_u, N_i, N_{de}, N_l) on both uncertainty and sensitivity of outputting values of CL(N) belongs to N_u. It is connected firstly with a deficit of nitrogen as the main nutrient in all North Asian ecosystems as well as an existing spatial and temporal variability of this parameter that relates to a significance and correctness of experimental and computed values of N_u. In accordance with a relatively better knowledge of the hydrological picture and relatively homogenous values of critical concentration of nitrogen in surface waters, C_{Ncrit}, included in the calculation of critical nitrogen leaching, $N_{l(crit)}$, values, the input of the given parameter into the uncertainty of CL(N) is expressed in a lesser degree. Furthermore, the runoff processes are practically not significant for ecosystems of Luvic Phaeozems, Chernozems and Kashtanozems due to low P:PE ratio. During the calculations of CL(N) for ecosystems of North East Asia, the values of critical immobilization and denitrification from N depositions both in relative and absolute meanings played a subordinate role that obviously reflects their minor contribution into uncertainty and sensitivity analysis of the computed output values of ecosystem sensitivity to acidic deposition.

Thus, the ERA estimates shown in Table 3 characterize the significance of the endpoints, such as nitrogen content in plant tissues and surface waters for many ecosystems of North East Asia. These endpoints have to be taken into account during risk management step of the ERA flowchart for emission abatement strategy development (see Figure 5).

Critical loads of sulfur and acidity on Chinese ecosystems

Based on the mineralogy controlling weathering and soil development, sensitivity of ecosystem to acid deposition is assessed with a comprehensive consideration on the effect of temperature, soil texture, land use and precipitation. The results show that the most sensitive area to acid deposition in China is Podzolic soil zone in the Northeast, then followed by Latosol, Dark Brown Forest soil and Black soil zones. The less sensitive area is Ferralsol and Yellow-Brown Earth zone in the Southeast, and the least sensitive areas are mainly referred to as Xerosol zone in the Northwest,

Table 3. Percentage of various endpoints contribution to total environmental risk assessment of ecosystem sensitivity to acid deposition in Northern Asia (Bashkin, 1998).

Ecosystem forming soils	Endpoint assessment method	Endpoints			
		N content in plant issues, N_u	N content in surface waters, N_l	Denitrified N, N_{de}	N enrichment of soil, N_i
Regosols and Lithosols	SRC	50	41	2	7
	RTU	30	27	21	22
Cryic Gleysols and Humic Cambisols	SRC	37	56	2	5
	RTU	22	37	21	20
Gelic Podzols	SRC	47	45	3	5
	RTU	40	39	10	11
Andosols	SRC	40	39	10	11
	RTU	81	17	1	1
Eutric Cambisols	SRC	41	9	25	25
	RTU	63	10	14	13
Distric Cambisols	SRC	96	2	1	1
	RTU	38	32	16	14
Luvic Phaerozems and Chernozems	SRC	96	1	1	2
	RTN	65	6	12	17
Kashtanozems	SRC	95	1	4	5
	RTU	61	5	19	15

Note: SRC - Standard Regression Coefficient, RTU - RooT of the Uncertainty.

Alpine soil zone in the Tibet Plateau, and Dark Loessial soil and Chernozem soil zone in central China. These regional different soil sensitivities to acid deposition can be attributed to the differences in temperature, humidity and soil texture (Hao *et al*, 1998). It has been shown that the assessment of ecosystem sensitivity to acidic loading depends strongly on the calculation of chemical weathering of soil base cations due to an input of proton with depositions.

The critical loads of acid deposition have been mapped for the Chinese ecosystems, as shown in Figure 26.

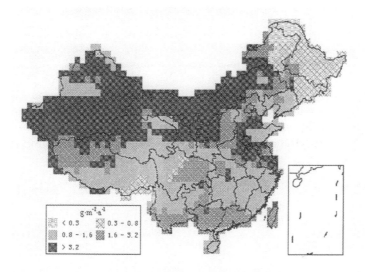

Figure 26. Critical loads of sulfur at terrestrial ecosystems of China (Hao et al, 1998).

It can be seen that most areas sensitive to acid deposition are in Southeast China, and that the insensitive is in the Northwest. The sensitive areas, including the catchment of the Changjiang (Yangtze) River and the wide areas to the southward, are warm and rain-abundant. The natural vegetation is the Tropic Rain Forest, Seasonal Rain Forest, and Subtropical Evergreen Forest ecosystems. The dominant soils in these areas are acid Ferralsols, with obvious accumulation of iron and aluminum. These soils can tolerate approximately 0.8–1.6 g/m^2/yr. of sulfur from acid deposition and belong to the intermediate sensitivity class 3. In Northwest China, the climate is semiarid or arid. The predominant ecosystems are Dry Steppe and agroecosystems, with spots of Deciduous Forest or Mixed Forest ecosystems. The soils represented by Xerosol and Alpine types are carbonate-rich and saline. Consequently, they are resistant to acidification.

The critical load class 1, the most sensitive class of ecosystems, is chiefly referred to Podzol soils of Coniferous Forest ecosystems. This class occupies the small areas in the northeast of China, mostly on the north part of Da-xing-an-ling Mountain, which is covered by coniferous forest, with annual precipitation 400–500 mm and annual mean temperature −4.9–0 °C. Podzol soils were derived from acid granite or quartz rocks and their formation was often influenced by leaching and settling of organic acid complex compounds. Hence, the soils show acid reactions and the base saturation is very low. The soil clay minerals are composed of hydrous mica and small amounts of other unweatherable minerals such as montmorillonite, kaolinite, roseite and chlorite. The low temperature and the coarse texture of soil are important to the low weathering rate of soil minerals there. Therefore, these areas must be paid great attention, even if acid deposition has not yet appeared.

Class 2 is found in Dark Brown Forest soil and Black soil zones in the Northeast, and Latosol soil zone in the south of the Taiwan area, in the north of Hainan province and near Hekou in the Yunnan province. The Dark Brown soil zone, with annual mean temperature from -1 to $-5\,°C$, is covered by coniferous forest. The coarse soil was derived from granite parent rocks and the chief clay mineral is hydrous mica. As in the Dark Brown soil zone, the temperature in Black soil areas is very low. The Black soil contains clay minerals such as hydrous mica, fulonite, gibbsite and kaolinite. It is obvious that the climate conditions and the mineral composition favor the chemical weathering of soil minerals. Contrarily, Latosol soil zone is under high temperature rain abundant conditions where the annual mean precipitation is 1900 mm and the annual mean temperature is $12\,°C$. The texture of Latosol is fine and the fraction of clay in the soil is high, which is advantageous to weathering. However, the clay minerals in this soil are dominated by kaolinite and gibbsite. Anorthite has completely decomposed and K-feldspar is rare. The weathering rate is still low as a result of the lack of weatherable minerals.

Soils of class 3 include Lateritic Red Earth in the areas southward from the Nanling Mountain, Red Earth and Yellow Earth between the Nanling Mountain and the Changjiang (Yangtzee) River, Yellow-Brown Earth in the lower reaches of the Changjiang(Yangtzee) River, Subalpine Meadow soil and Alpine Meadow soil on the Plateau of Tibet.

Class 4 is found in Paddy soil zones sporadically distributed throughout China and in the Purplish soil zone in the Sichuan River Basin.

Class 5 (the least sensitive) soil include Kashtanozem, Brown soil and Sierozem soil zones in the Plateau of Inner Mongolia and the Loess Plateau, Desert soil zones in He-xi-zou-lang and the Talimu River Basin, Subalpine Steppe soil, Alpine Steppe soil and Alpine Desert soil in the Plateau of Tibet. These kinds of soils, belonging to the soil class of Xerosol or Alpine soil, consist of easy weathering minerals such as carbonate. They show alkaline reactions, with weak leaching and sparse vegetation. Those kinds of soils are insensitive to acid deposition.

Figure 27 illustrates the percentage of areas shared by each critical load class. As we can see, the most sensitive soil (class 1) shares only 2% of the whole area in China, and the sensitive soil (class 2) is no more than 8.7%. The intermediate sensitivity and the least sensitive soils are the most extensive, which account for 35.5% and 42.4%, respectively. Class 4 covers 11.4%.

In summary, the sensitivity of an area to acid deposition and the critical load values are dependent on soil and vegetation types and meteorological conditions. In China, acid deposition often occurs in the Allite areas in the southeast of China, where soils are of intermediate sensitivity, except Latosol. The annual mean temperature in these tropic or subtropic areas is $16–25\,°C$, the accumulated temperature of $\geq 10\,°C$ is 5000–9000 $°C$, and precipitation more than 1500 mm. The characteristics of temperature high, rain abundant, and wet warm in the same season, promote reactions of soil minerals and cause the rapid weathering of soil minerals and the rapid circulation of biological materials. As a result, the weathering rates are sufficient enough for high values of corresponding critical loads. On the other hand, the areas of Podzol soil,

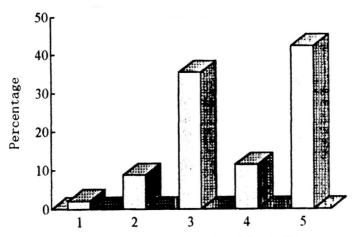

Figure 27. Percentage of areas shared by each critical load class in China (Hao et al, 1998).

Dark Brown Forest soil and Black soil are in frigid temperate conditions. The accumulated temperature of $\geq 10\,^{\circ}\text{C}$ is about 2000, the annual mean temperature is lower than $5\,^{\circ}\text{C}$, and the precipitation is low. These soils contain large sand and small clay fraction. Thus, the chemical weathering rates are low and critical loads are also low.

It can be concluded that high temperature, high humidity, wet warm in same season and high concentrations of clay and loam, are important reasons why soils and water bodies have not been found acidified in heavy acid rain areas such as the Chongqing, Guiyang and provinces of Guangdong and Guangxi. However, it should be noted that the weatherable minerals in soils in south China are scarce, and these soils are potentially at risk of acidification. Acid deposition must be strongly controlled in the Northeast area of China to protect terrestrial and aquatic ecosystems from acidification and decrease of forest production like those in Central and North Europe and North America.

Comparison of the calculated critical loads with the sulfur deposition of 1995 in China (Hao *et al*, 1996) led to the critical load exceedance map of sulfur deposition (Figure 28).

As we can see from this map, sulfur deposition exceeds critical load in a wide land area that amounts to 25% of total Chinese ecosystems, which mainly referred to the southeast of China. Among these areas, the exceedances are especially serious in the lower reaches of Changjiang (Yangtzee) River, in the Sichuan River Basin, and in the Delta of Zhujiang River.

Critical loads of sulfur in South Korea

Air pollution is still of crucial importance in South Korea. Emission of sulfur dioxide, nitrogen oxides and ammonia that have been rising in recent decades are responsible for air pollution in many places of the country. The amount of sulfur dioxide emitted

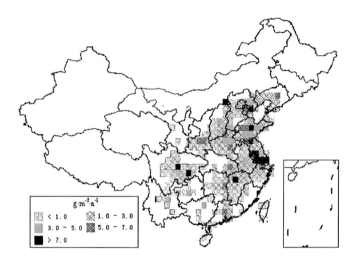

Figure 28. Sulfur critical load exceedance map of China (Hao et al, 1998).

in the area of South Korea was 1,494,000 tons in 1995 and was gradually decreasing to 1,490,000 tons in 1997.

However, this is well known that various pollutants including sulfur compounds can be transported by air from country to country in the whole Asian domain and especially in North East Asia. Thus, the model calculations have shown that in 1991–1994 about 35% of oxidized sulfur species deposited in South Korea was transported from anywhere, mainly from China (Sofiev, 1999). Accordingly, in spite of national reduction in SO_2 emission, the sulfur depositions are still very significant.

Preliminary research has shown a high sensitivity of most South Korean ecosystems to input of acid forming sulfur compounds (World Bank, 1994). This is generally related to acid geological rocks (predominantly, gneisses and granites acting as a soil parent material), acid soils with predominant light texture and low buffering capacity and sensitive Coniferous Forest (mainly pine species) ecosystems.

It has been argued that the best approach to the calculation and mapping of critical loads on ecosystems of East Asia is to use various combinations of expert approaches and geoinformation systems (EM GIS), including different modern techniques of expert modeling. However, this approach is difficult in practice due to a scarcity of data and insufficient understanding of basic controls over major pathways in the biogeochemical structure of many Asian ecosystems and their alteration under the influence of different air pollutants. Under the given conditions the EM GIS can operate using databases and knowledge basis relative to the areas with different spatial data uncertainty. As a rule the given systems include an analysis of the cycles of various elements in the key plots, a choice of algorithms describing these cycles, and corresponding interpretation of the results. This approach requires cartographic data, such as vegetation, soil and geological maps, information on buffering capacities of soils, precipitation and runoff patterns.

Indeed this approach is shown to be the most appropriate for the Asian part of Russia, China, Thailand, Taiwan, where, at present, adequate information on the great spatial variability of natural and anthropogenic factors is either limited or absent (Bashkin and Park, 1998).

During the last decades significant efforts have been applied for monitoring of main environmental compartments in South Korea. The soil, geological, and vegetation maps were created in different scales, from 1:25,000 to 1,000,000. The monitoring networks of precipitation and their chemical composition were enlarged to cover the whole area of the country. However, even now there are significant gaps in precise understanding of biogeochemical structure of different terrestrial and aquatic ecosystems that governs the sustainability of these ecosystems to acidity loading.

The Comprehensive Acid Deposition Model (CADM) has been created for calculation of dry and wet deposition of sulfur species over South Korea (Park *et al*, 1997, 1999a). This model presents the quantitative assessment of the acidity loading and alterations in deposition rates.

South Korea is in the southern part of the mountainous Korean peninsula between 126 °E–130 °E and 33 °N–38 °N. Geographically this is the northern temperate zone of the Eastern Hemisphere. The overall area of Republic of Korea (ROK) comprises 99,766 km^2.

In accordance with international approaches the following information has been collected for calculation of sulfur critical loads for all Korean ecosystems:

— general information: soil map, scale 1:1,000,000; land use map, scale 1 × 1 km; geological map, scale 1:1,000,000;

— soil chemical and physical parameters: soil texture; soil layer height; moisture content; soil bulk density; Mg + Ca + K content; log K gibbsite; pH; soil temperature; soil cation exchange capacity;

- vegetation type parameters: annual net productivity growth; net uptake of nutrients (N, Ca, Mg, K, Na, S);

- geochemical analysis: Ca, Mg, K, Na, Al, trace elements; base cation chemical weathering from soils and underlying geological parent material.

The calculation of critical loads of maximum sulfur was carried out on a scale of 11 × 14 km cells. There are 665 cells on the area of South Korea.

The calculations were carried out for the 1994–1997 period that allows us to account for the temporal variations of temperature and precipitation as well as the dynamic pattern of sulfur and base cation depositions.

Using the above-mentioned constituents, the critical loads of sulfur were calculated for the natural terrestrial ecosystems over Korea. Geographical distribution of CLmaxS values is shown in Figure 29.

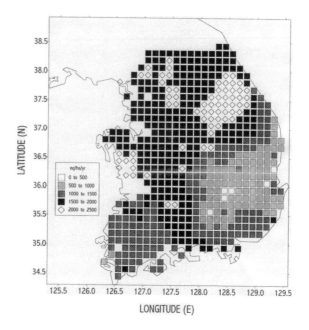

Figure 29. Critical loads of sulfur at terrestrial ecosystems of South Korea (Park and Bashkin, 2000).

We can see that about 71% of Korean ecosystems have the rank of 1000–2000 eq/ha/yr.. Very low values of critical loads (< 500 eq/ha/yr.) are characteristic for an insignificant part of ecosystems (1.0%), low values (500–1000 eq/ha/yr.) are for 15.5% of ecosystems and on the contrary high values (> 2000 eq/ha/yr.) are shown for 12.1% of considered ecosystem types.

These values are significantly different from those that have been earlier calculated by RAIN-ASIA model (World Bank, 1994). These differences might be related to the much more detailed and comprehensive national data sets on geological, soil, climate (precipitation, temperature, evapotranspiration, runoff, etc) and vegetation mapping, physico-chemical properties of soils and geological rocks. This allows the authors to calculate more precise values of all constituents used for maximum sulfur critical load calculation and mapping.

Accordingly, these CLmaxS values are generally higher than those from RAIN-ASIA model. The latter were in the range of 200–2000 eq/ha/yr, mainly 200–500 eq/ha/yr, whereas these are mainly in the limits of 1000–2000 eq/ha/yr. This is related to the peculiarities of precipitation and topography patterns in South Korea. These in turn lead to high values of surface runoff of annual mean of 6,200 m^3/ha/yr, and correspondingly to high values of acid neutralizing capacity leaching. The intensive leaching of sulfur from the soil profiles makes the local ecosystems more sustainable to high values of sulfur deposition (Park *et al*, 2001).

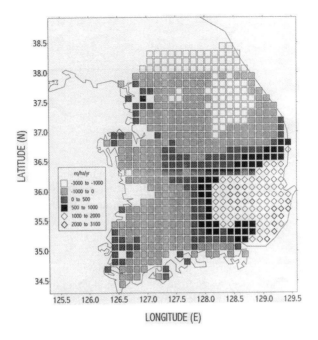

Figure 30. Exceedances of critical loads of sulfur over South Korea (Park and Bashkin, 2001).

During the 1994–1997 period the mean sulfur dry and wet deposition totally amounted to about 47 kg/ha/yr or about 3000 eq/ha/yr. (Park *et al*, 1999b). These values were maximum in the south-eastern part of the country, where the Pusan-Ulsan industrial agglomeration takes place and minimum in the north-eastern part.

Accordingly, a significant part of Korean ecosystems was subjected to an intensive input of S acid forming compounds. The values of exceedances of sulfur deposition over sulfur critical loads (ExS) are shown in Figure 30.

During 1994–1997 the Sdep values were higher than CLmaxS values at about one third of terrestrial Korean ecosystems (38%). Among them, the ExS values were in the range 176–500 eq/ha/yr. for 16.1% of total number of ecosystems, in the range of 500–1000 eq/ha/yr. were for 7.9%, in the range of 1000–2000 eq/ha/yr. were 10.7% and values even higher than 2000 eq/ha/yr. were found for 3.5% of Korean ecosystems.

The other part of Korean territory (61.8%), where the sulfur depositions were relatively less but critical load values are relatively higher (see Figure 29), was not subjected to excessive input of sulfur-induced acidity. This area can be considered as sustainable to sulfur input.

As we have mentioned above, during the 1990s up to 30–35% of sulfur deposition was due to emission of SO_2 by transboundary sources, occurring mainly in China. Thus, the emission abatement strategy in South Korea has to be developed taking into account both local and transboundary emission reduction in the whole

East Asian domain. The values of CL and their mapping presents a good possibility for the creation of ecological optimization models. At present, these CL values and corresponding mapping have been carried out by national research teams in almost all the East Asian countries, such as China, Japan, South Korea, Asian part of Russia and Taiwan (Bashkin and Park, 1998). Accordingly, this national-based mapping can be considered as a scientific basis for decreasing local and regional air pollution in the East Asian domain.

Soil acidification in Japan

Accordingly, in Japan models were developed to evaluate soil acidification caused by acidic deposition and ecosystem sensitivity to it in different spatio-temporal scales. For the prediction of catchment scale acidification, a dynamic model was made. The model took rapid chemical reactions into consideration as the quasi-equilibrium processes and processes such as chemical weathering, nutrient uptake, nitrification etc. as element flux at a constant rate to the soil system. Application of this model to the soil acidification experiments using simulated acid rain showed that changes in soil chemistry could be well expressed by the model and the acid loaded into the soil was neutralized in the top 10 cm horizon and soil acidification occurred there. Against the urgent acidification, main buffering mechanisms were cation exchange and dissolution of Al hydroxides whose relative contributions differed according to soil characteristics and its acidification stage (Shindo, 1998).

For larger scale estimation, especially for the country scale, a steady state mass balance model was to be employed. Magnitudes of acid neutralizing capacity and acid production due to chemical weathering, base cation and nitrogen uptakes and base cation deposition were estimated for total Japanese Forest ecosystems based on existent data bases of geology, soil, vegetation etc., some measured data on soil properties and parameters derived from literature. It has been additionally stressed that the mineral weathering rate has the most significant effect on the neutralizing capacity of ecosystems. An indicator was proposed to evaluate the possible ecosystem impact by acid deposition based on the steady state model under the condition of current or predicted acid deposition rates (Table 4).

Critical loads for South-East Asia

Taiwan Forest ecosystems. A RAINS-ASIA impact module is used to assess ecosystem sensitivity to acid deposition and to calculate critical load of sulfur to six forest ecosystems in Taiwan (Lin, 1998). Results indicate that forest ecosystems in Taiwan are very sensitive to acid deposition due to their low soil pH (< 5.5). Lowland subtropical forest ecosystems in Taiwan have low or moderate low critical loads for S suggesting that they are vulnerable to acid deposition (Table 5).

Yet, many forest ecosystems are exposed to acid deposition far exceeding their critical loads. Although these forest ecosystems appear healthy, there may be a sudden detrimental change once the current buffering capacity is depleted. Cation leaching

Table 4. Statistics of estimated parameters for acidity budget calculation in Japan (mol$_c$/ha/yr) (After Shindo, 1998).

Parameters	Minimum	Maximum	Average
Base cation weathering (BC$_{we(E)}$)*	96	3564	1382
Base cation weathering (BC$_{we(R)}$)*	74	5198	1729
Base cation uptake (BC$_{gu}$)**	674	2839	1663
Nitrogen uptake (N$_{gu}$)**	337	1359	819
Acid production due to uptake (BC$_{gu}$–N$_{gu}$)**	337	1480	858
Deposition rates*** H$^+$	34	1343	335
NH$_4{}^+$	151	1690	407
BC (Ca^{2+} + Mg^{2+} + K$^+$)	226	2678	908

*Statistics for all of Japan (about 360,000 grid points),
**Statistics for the secondary or planted forests (about 180,000 grid points),
***Statistics for the 29 monitoring stations.

Table 5. Data input for the calculation of critical loads (eq/ha/yr.), the calculated critical loads and current deposition rates for S at six forest ecosystems in Taiwan.

Site	Fu-shan	Pin-lin	Lien-hua-chi	Pi-lu-chi	San-pan	Tai-ma-li
ANC$_w$ (eq/ha/yr)	70	70	140	70	70	240
Q[1] (mm/yr)	2000	1600	1200	1300	2000	1250
K$_{gibb}$	150	150	150	300	150	150
CL (eq/ha/yr)	365	343	560	300	365	890
DR[2] (eq/ha.yr)	2800	2700	1400	650	1750	800

[1]Q: runoff, the difference between rainfall and evaportranspiration,
[2]DR: deposition rate.

both from the forest canopy and forest soils is observed in some forest ecosystems. Continuous exposure to high levels of acid deposition can lead the forest to a nutrient imbalance and thereby undermine forest health.

North Thailand Forest ecosystems. The input of < 1 meq/100 g soil to the forest soil in the northern part of Thailand, with the organic content of 1.33 percent, has no changes in pH value due to existing hydrogen buffering capacity. Simulated acid rain at pH 2 led to Al leaching from this soil naturally enriched by aluminum and iron, although there is no significant change of pH in the soil. The acid depositions cause also the reduction of bacteria, actinomyces and ammonifying microbes in the upper part of the soil. These could lead to low nutrient cycling rate in the ecosystem as these microorganisms play a significant role in organic matter decomposition in the soil. Due to certain buffering ability of soils to acid depositions the forest ecosystem can be sustainable in a short term period, however some negative chemical and biological changes occur in soil. This will gradually decrease the ecosystem sustainability to acidic loading (Kozlov and Towprayoon, 1998).

This and other researches carried out in North Thailand provides the results of sensitivity assessment in terms of maps for the study area, and to determine and identify the sensitive receptors and locations where abatement strategies would be implemented to reduce environmental impacts of acidification on forest and agriculture. Sensitivity of ecosystems to acid deposition was carried out using a two step procedure. Firstly, the sensitivity mapping according to revised methodology of Stockholm Environment Institute has been conducted for Thailand conditions (see Kuylenstierna *et al*, 1995 for detail). The purpose of such sensitivity mapping is to define the distribution of ecosystems with the same relative level of reactions to the given rate of acidic deposition. At the second stage the sensitivity of Thailand ecosystems to acidic deposition has been described by means of Critical Loads (CL) and exceedances, using modified the Steady-State Mass Balance model with a simplified expert-modeling approach (Figure 31).

The maps for CL of sulfur derived from this methodology have been overlaid with current (or projected future) deposition maps in order to show areas where the CL of sulfur is (or will be) exceeded. In order to manipulate with the numerous maps and data a geographical information system (GIS) was used (Kozlov and Towprayoon, 1998). As a result of both the high sensitivity of ecosystems and level of exceedances across Northern Thailand, more than 75% of the ecosystems across about 50% of this territory is at significant risk from acid deposition.

Thus, the acidity of precipitation has increased during the latest decades in many developed and developing countries of Asia. There exist clearly transboundary air pollution problems. The harmful single and synergetic effects of acidity, SO_2, NO_x and O_3 are experimentally shown for various natural and agricultural ecosystems. The critical load concept is shown to be a good guide for assessing ecosystem sensitivity to acid deposition in many East Asian countries. The calculated critical loads are exceeded at present acidity loading for many ecosystems in East Asian countries (Russia, China, Thailand, Taiwan, Japan, South Korea).

1.7. Acid deposition influence on the biogeochemical migration of heavy metals in the food web

An interesting study of acid rain effects on the biogeochemical accumulation of heavy metals (Cd, Cu, Pb, and Zn) in crops has been presented by Chen *et al*, 1998. The authors have compared the ratios of relative concentration of four heavy metals in

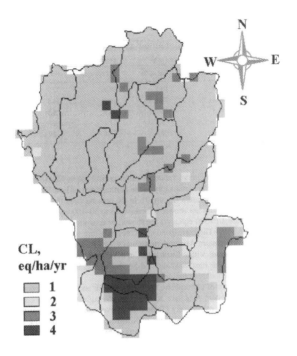

Figure 31. Critical loads of acidity on the ecosystems in South Eastern Asia (Northern Thailand).

the brown rice and leaves of vegetables sampled from an acid rain affected area and a non-affected area. The data indicated that the ratios of relative concentration of Cd, Cu, Zn in brown rice and 19 vegetable species growing in acid rain area (Lung-tang) and growing in acid rain non-affected area (Lung-luan-tang) sampled from 1996 to 1997 are almost higher than 1, or higher than 3, except for Pb (Table 6). These results suggested that biogeochemical accumulation of heavy metals in brown rice seems not affected by long-term acid rains, on the contrary to the case of vegetables species in northern Taiwan. Therefore, these accumulations are dangerous for humans eating vegetables produced in an acid rain affected area.

Table 6 also reveals that the mean concentration of Pb in the brown rice and leaves of 19 vegetable species between acid rain affected area and non-affected areas are almost the same. On the other hand, the ratio is close to 1. This result indicates that the acid rain does not influence the biological accumulation of Pb in the brown rice and leaves of vegetables species sampled in Taiwan. Some studies have indicated that concentration of Pb in the crops was only affected when the concentration of Pb in the soils is higher than 500 mg/kg (Kabata-Pendias and Pendias, 1992). Sloan *et al* (1997) also indicated that the relative bioavailability of biosolids-applied heavy metals in agricultural soils was $Cd \gg Zn > Ni > Cu \gg Cr > Pb$, for the soils sampled

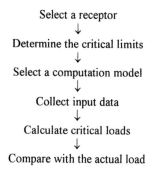

Figure 32. Flowchart for calculating critical loads of heavy metals.

15 years after biosolids application. It is quite consistent with the results of Chen *et al* (1998)'s research. Thus, the phyto-availability of heavy metals caused by acid deposition followed the trend: $Cd > Zn > Cu \gg Pb$.

2. CALCULATION OF CRITICAL LOADS FOR HEAVY METALS AT TERRESTRIAL AND AQUATIC ECOSYSTEMS

We have shown already that anthropogenic loading changed significantly the natural biogeochemical cycles of many heavy metals (HM), especially those like mercury or lead (see Chapter 8, Sections 2 and 3). In order to make adequate environmental policies concerning the reduction of the HM loading to the environment, efforts for the assessment of the effects of such pollutants and the identification of those loads of heavy metals below which harmful effects no longer occur, the so-called critical loads, is needed. Accordingly, we will discuss how these critical loads at terrestrial and aquatic ecosystems can be calculated and how to estimate the environmental standards based on biogeochemical food web estimates. Here we will also present different calculation methods to quantify the risk of inputs of several heavy metals, i.e. lead, cadmium, copper, zinc, nickel, chromium, and mercury. More details can be found in De Vries & Bakker (1988), see Further Reading.

On the contrary with critical loads for acidity, the critical loads for heavy metals refer to single metal only. Accordingly, *the critical loads equal the load causing a concentration in a compartment (soil, soil solution, groundwater, plant, animal and human organisms, etc), that does not exceed the critical limits set for heavy metals, thus preventing significant harmful effects on specified sensitive elements of the biogeochemical food web.*

Discussing the problems related to the critical load calculation, attention should be paid to (i) selection of a receptor of concern, (ii) critical limits, (iii) possible calculation methods, (iv) the necessary input data and (v) the various sources of error and uncertainty (De Vries and Bakker, 1998).

The following flowchart is useful for calculation of critical loads of heavy metals for both terrestrial and aquatic ecosystems (Figure 32)

Table 6. The ratios of relative concentration of heavy metals in brown rice and the leaves of vegetable species growing in Lung-tang area (affected by acidic rains) and Lung-luan-tang area (non-affected by acidic rains) from 1996 to 1997 in Taiwan (Chen et al, 1998).

Rice and vegetable species	acid rain/non-acid rain affecting area (sampling number)	Ratio in acid rain/non-acid rain area			
		Cd	Cu	Pb	Zn
Rice					
Rice (*Oryza sativa* Linn.)	24/15	1.25	1.05	1.09	1.03
Vegetables					
Sweet potato (*Ipomoea bataus*)	14/9	1.00	1.45	1.07	1.11
Welsh onion (*Allium fistulosum*)	10/12	0.89	1.48	3.08	2.03
Pickled cabbage (*Brassica chineniss*)	3/10	5.03	1.23	— #	1.33
Chinese chives (*Allium tuberosum*)	7/5	4.97	0.70	0.08	1.56
Mustard (*Brassica juncea*)	2/4	—	1.59	—	2.19
Lettuce (*Lactuce sativa*)	6/8	3.73	1.87	1.00	1.97
Chickweed (*Alsine media*)	3/1	—	2.40	—	0.36
Garlic (*Allium sativum*)	6/7	0.85	2.44	—	4.64
Kohlrabi (*Brassica campestris*)	1/1	2.00	2.00	—	5.50
Cabbage (Brassica oleracea)	2/1	—	1.99	—	3.06
Tassel flower (*Amaranthus caudatus*)	6/2	0.97	2.23	—	1.47
Celery (*Apium graveolens*)	2/1	—	—	—	1.55
Spinach (*Spinacia oleracea*)	2/1	—	0.75	0.80	0.42
Coriander (*Coriandrum stivum*)	1/4	8.02	5.01	—	1.80
Basil (*Ocimum basilicum*)	1/3	—	8.05	—	0.36
Radish (*Raphanus sativus*)	4/2	—	2.76	—	1.08
Pepper (*Capsicum frutescens*)	3/4	1.97	2.04	3.92	0.88
Kidney bean (*Phaseolus vulgaris*)	3/10	2.07	1.78	1.09	1.44
Water convolvulus (*Ipomoea aquatica*)	6/3	0.28	1.97	3.50	0.66

: The ratios of relative concentration can not be calculated because the heavy metal contents of rice or vegetables growing in acidic rain affecting area or in non-acidic rain affecting area is lower than that of method detection limit (MDL) of heavy metals.

2.1. Selection of receptor

In general terms, *a receptor is defined as an ecosystem of interest that is potentially polluted by a certain load of heavy metal.*

Selection of a receptor is the first step in the flowchart for calculating critical loads (see Figure 31). When selecting a receptor in view of the different effects of heavy metals, the crucial question is related to what we want to protect. With respect to risks on terrestrial ecosystems, a major distinction can be made in risks/effects on humans that use ground water for drinking or that consume crops that are grown on the soil (human ecotoxicological risks) and ecosystems (ecotoxicological risks). In order to judge the ecotoxicological risks associated with elevated heavy metal contents on terrestrial ecosystems, a further distinction should be made in the following receptors:

1. Soil microbes. Effects include reduced microbial biomass and/or species diversity, thus affecting microbial processes such as enzyme synthesis and activity, litter decomposition, associated with carbon and nitrogen mineralization, and soil respiration;

2. Soil fauna, including invertebrates, like nematodes and earthworms. Effects are connected with biodiversity, productivity and biomass changes;

3. Vascular plants including trees. Effects are related to HM toxicity, such as the reduced growth of roots and shoots, elevated concentrations of starch and total sugar and decreased nutrient contents in foliar tissues and depressed biochemical activity;

4. Terrestrial fauna, such as birds, mammals, or domestic animals. Effects are heavy metals accumulation followed by possible disturbance of physiological and biochemical reactions and metabolisms. Bioaccumulation of Cd, Hg and Cu in the food web is the most important concern.

A simplified biogeochemical food web of heavy metals in the terrestrial ecosystems, including the most important receptors (biogeochemical links) is shown in Figure 33.

With respect to soils, a receptor is thus characterized as a specific combination of land use (e.g. Forest ecosystem types, agricultural crops) and soil type. The critical loads can be calculated for both agricultural soils (grassland, arable land) with HM inputs with deposition, fertilizers, and wastes, and non-agricultural (forest and steppe) soils, where atmospheric deposition is the only input to the system.

The receptors that one may consider in both types of ecosystems are presented in Table 7.

Possible effects on soil life, plants (phytotoxicity) and on ground water are of concern in all types of ecosystems. Food quality criteria are, however, of relevance for arable land only, whereas possible secondary poisoning effects on domestic animals or terrestrial fauna are relevant in grassland and non-agricultural land. A final critical limit can be based on the most sensitive receptor. Even though effects vary for each metal, soil microbes and soil fauna are generally most sensitive.

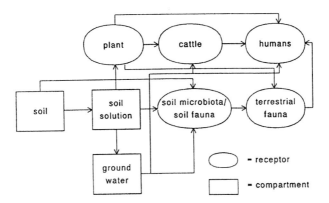

Figure 33. A simplified biogeochemical food web in the terrestrial ecosystems (de Vries and Bakker, 1998).

Table 7. Possible receptors in three main types of terrestrial ecosystems.

Receptor	Ecosystem types		
	Non-agricultural land use	Grassland and pasture	Arable land
Soil microbes	+	+	+
Plants			
phytotoxicity	+	+	+
crop quality	—	—	+
Terrestrial fauna	+	—	—
Domestic animals	—	+	—
Ground water	+	+	+

When selecting a receptor in aquatic ecosystems, this may be the aquatic organisms, the benthic organisms or the fauna (birds and mammals) or human beings that consume fish (Figure 34).

With respect to aquatic ecosystems, a receptor is thus characterized as a lake, a river or a sea. The major receptor that we will consider here is a lake, including catchment (see small catchment studies in Chapter 8). The reason is that the suggested models are all relatively simple and based on the assumption that the water compartment is homogeneously mixed. Much more complicated models are necessary to calculate critical loads for seawater or large lakes with horizontal currents and/or thermal stratification.

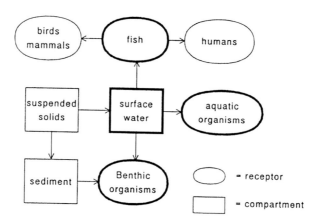

Figure 34. A simplified biogeochemical food web in the aquatic ecosystems (de Vries and Bakker, 1998).

2.2. Critical limits

The assessment of a critical limit for the receptor of concern is a second step in the flowchart for calculating critical loads of heavy metals (see Figure 31). Since critical loads of heavy metals are related to the concentration of single metal in any link of the biogeochemical food web, the correct selection of critical limits is a step of major importance in deriving the critical loads. Those critical limits, which depend on the kind of effects considered and the amount of harm accepted, constitute the basis of the critical load calculation and determine their magnitudes.

Effect-based critical limits for soils and ground water have been derived or are under development in various countries for multiple purposes (Radojevic and Bashkin, 1999). These criteria can be used to assess the environmental quality of a site or area, to set priorities in control measures and to derive emission reduction goals. Most of the values are derived from comparable starting-points such as protecting terrestrial population, water supply, food quality, and finally, animal and human health.

For most of the receptors (see Figure 33 and 34), critical limits have been defined related to ecotoxicological or human-toxicological risks, such as:

a)   Soil. Relation to direct and indirect effects on soil microbes and fauna, plants, terrestrial fauna and humans. The typical values are given in mg/kg or ppm;

b)   Ground water. Relation to direct and indirect effects on animal and human health when this water is used as a drinking water, μg/L or ppb;

c)   Surface water. Critical limits or maximum permissible concentrations (MPC) are related to direct effects on organisms in surface water or on human and animal health as drinking water, μg/L or ppb;

Table 8. Critical limits for heavy metals in soil of various countries, related to multifunctional uses.

Country	Critical limits, ppm						
	Pb	Cd	Hg	Cu	Zn	Ni	Cr
Denmark	40	0.3	0.1	30	100	10	50
Sweden	30	—	0.2	—	—	—	—
Finland	38	0.3	0.2	32	90	40	80
Netherlands	85	0.8	0.3	36	140	35	100
Germany	40	0.4	0.1	20	60	15	30
Switzerland	50	0.9	0.8	50	200	50	75
Czech Republic	70	0.4	0.4	70	150	60	130
Russia	32	2	2.1	55	100	85	90
Ireland	50	1	1	50	150	30	100
Canada	25	0.5	0.1	30	50	20	20

d) Mammals-birds-fish. Maximum permissible concentrations (MPC) in target organs are related to direct (toxic) effects, or food critical limits in human and animal food web;

e) Plants, cattle, terrestrial fauna. Critical limits in plant tissue or target organs are related to direct toxic effects, or food chemical limits related to indirect effects by animal and human consumption, mg/kg or ppm;

f) Humans. These values are connected with acceptable daily intake or ADI, in μg/kg of body weight per day. This dose is the quantity of a compound to which man can be orally exposed, on the basis of body weight, without experiencing adverse effects on human health.

Below we will consider the critical limits applied for different purposes in Europe and North America. The examples are given for multifunctional soil and water use, differentiation in land use types and bioconcentration in the food web.

Multifunctional soil use
In various countries, critical limits for soil have been derived to assure multifunctional soil use (Table 8).

Table 9. Critical limits for heavy metals in soil in surface waters of various countries, related to multifunctional uses.

Country	Critical limits, ppb						
	Pb	Cd	Hg	Cu	Zn	Ni	Cr
Sweden	1.2	0.09	—	2.1	9	9	1
Denmark	3.2	5	—	12	110	160	10
Norway	0.6	0.05	—	1.1	4.5	1.5	0.45
UK	10	—	—	5	10	5	5
Netherlands	11	0.34	0.23	1.1	6.6	1.8	8.5
Germany	5	1	0.1	—	—	—	—
Czech Republic	50	5	1	100	50	150	50
Russia	1.0	5	0.01	1	10	10	1
Canada	1.0	0.2	0.1	2	30	25	2
USA	3.2	1.1	0.01	1.2	110	160	11
WHO	10	3	1	—	—	20	50

The concentrations shown in Table 8 are in a relatively limited range, i.e., 25–100 for Pb, 0.3–2 for Cd, 0.1–1.0 for Hg, 30–70 for Cu, 50–200 for Zn, 10–85 for Ni, and 20–130 for Cr. This indicates similar ecotoxicological approaches that have been used for setting critical limits in various countries.

Multifunctional water use

Hygienic standards for dissolved concentrations of heavy metals in surface water are shown in Table 9.

The range in critical limits for dissolved species of heavy metals in surface waters is large, especially for zinc and nickel. In general, values used by Denmark and the Czech Republic are much higher than for other countries, except for Zn and Ni. For these elements, Canada and USA also use much higher values. These limits may refer to action values.

Differentiation in land use

In some countries, critical loads for heavy metals have been derived as a function of land use. An example of such a differentiation suggested in Germany (Table 10)

Table 10. Critical limits for heavy metals in Germany as a function of land use.

Land use	Critical limits, ppm						
	Pb	Cd	Hg	Cu	Zn	Ni	Cr
Multifunctional	100	1	0.5	50	150	40	50
Children's playgrounds	200	2	0.5	50	300	40	50
Domestic gardens	300	2	2	50	300	80	100
Agricultural areas	500	2	10	50	300	100	200
Recreational areas	500	4	5	200	1000	100	150
Industrial areas	1000	10	10	300	1000	200	200

indicates a strong increase in critical limits going from multifunctional land use to industrial areas.

Methods to derive effect-based critical limits

Here we will describe the methods that are used to derive critical limits for soil, based on direct ecotoxicological effects on microorganisms and plants. The indirect approaches (food web models) to derive critical limits for soil based on critical limits for terrestrial fauna such as MPC values for target organisms will be also considered.

Direct effects on soil organisms and plants

MPC's values for soils are derived using extrapolation methods that are based on ecotoxicological information. In this context, a distinction is made in acute toxicity data based on short term ecotoxicological experiments (< 1 day) and chronic toxicity data, based on long term ecotoxicological experiments (1 day–1 month). *Acute toxicity is defined by the* LC_{50} *value, which is equal to the concentration at which 50% of the considered population are dead (*LC_{50} *with LC = lethal concentration). Chronic toxicity is defined by No Observed Effect Concentrations (NOEC's), sometimes referred as No Observed Effect Levels (NOEL's), of several species in an ecosystem.* The organisms or taxonomic groups that are considered in deriving LC_{50} values or NOEC data are microbes or microbe-mediated soil processes (enzymatic activity), earthworms, arthropods and plants. If only (i) acute toxicity data or a Quantitative Structure Activity Relationships (QSAR) or (ii) chronic toxicity data for a few different taxonomic groups (microbes, enzymes, earthworms, arthropods or plants) are available, the modified Environmental Protection Agency (EPA) method is applied.

Fixed safety factors are applied in this EPA method. The safety factors used are based on arbitrary extrapolation values of 10 going from (i) the laboratory (single species) to the field (whole ecosystem) situation and (ii) acute to chronic toxicity data. A similar approach is used to derive critical limits for surface water.

Indirect effects on higher organisms

Biomagnification, which stands for the phenomenon that a chemical accumulates in species through different trophic levels in a food web, may cause toxic effects on mammals and birds as a secondary poisoning. Next to direct effects on soil organisms (see above), these indirect effects can be considered in deriving critical limits for soil, by the use of simple food web models.

To indicate the transfer of chemicals in the biogeochemical food web, both bioaccumulation factors (BAFs) and bioconcentration factors (BCFs) are used. The following definitions can be applied (de Vries and Bakker, 1998):

BAF: defined as the ratio of the test chemical concentration in (a part of) an organism (e.g., bird, mammal or fish) to the concentration in its food (e.g., laboratory fodder, plants, invertebrates, birds, mammals) at steady state. BAFs are generally used for accumulation by birds, mammals and fish and are expressed on weight basis.

BCF: defined as the ratio of the test chemical concentration in (a part of) an organism (e.g., plant, earthworm) to the concentration in a medium (e.g., water, soil) at steady state.

The food of top predators generally comprises small birds and/or mammals. Bioaccumulation of chemicals from soil to small birds and mammals takes place in at least two steps, namely a BCF from soil to food (plants and/or invertebrates), followed by a BAF to small birds and mammals.

A similar description is used for water ecosystems with fish species.

Terrestrial ecosystems. The simplest model used in terrestrial ecosystems is based on the simplified food chain:

$$\text{soil} \longrightarrow \text{soil invertebrates} \longrightarrow \text{mammals/birds.} \tag{18}$$

Assuming that the mammal or bird feeds on soil invertebrates (e.g. worm-eating birds or mammals), the simplest model to calculate an MPC based on this food web is:

$$\text{MPC}_{\text{soil}} = \text{NOEC}_{\text{species of concern}} / \text{BCF}_{\text{food species of concern}}, \tag{19}$$

where: MPC_{soil} is Maximum Permissible Concentration of a chemical in soil, ppm, $\text{NOEC}_{\text{species of concern}}$ is No Observed Effect Concentration of the food (invertebrate) corrected for the species of concern (mammals or bird, ppm), $\text{BCF}_{\text{food species of concern}}$ is Bioconcentration factor, representing the ratio between the concentration in the invertebrate, being the food of species of concern, and the concentration in soil, ppm.

Aquatic ecosystems. The simplest model used in aquatic ecosystems is based on the simplified food chain:

$$\text{water} \longrightarrow \text{fish or mussel} \longrightarrow \text{fish or mussel eating birds/mammals.} \tag{20}$$

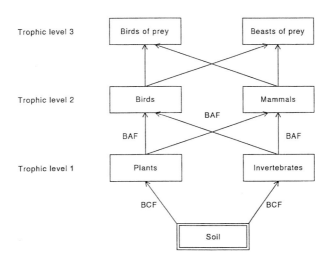

*Figure 35. Scheme of a terrestrial food web use for modeling bioaccumulation. The compart-
ments plants and invertebrates can be split up in several groups, depending on the availability
and variation among bioaccumulation data (de Vries and Bakker, 1998).*

Assuming that the mammal or bird feeds on fish or mussels, the simplest model
to calculate an MPC based on this food web is:

$$MPC_{water} = NOEC_{species\ of\ concern}/BCF_{food\ species\ of\ concern}, \qquad (21)$$

where: MPC_{water} is Maximum Permissible concentration of a chemical in water, ppb
$NOEC_{species\ of\ concern}$ is No Observed Effect Concentration of the food (invertebrate)
corrected for the species of concern (mammals or bird, ppm), $BCF_{food\ species\ of\ concern}$
is Bioconcentration factor, representing the ratio between the concentration in the
invertebrate, being the food of species of concern, and the concentration in water, ppb.

A more detailed biogeochemical food web model is shown in Figure 35.

As shown in Figure 35, there are four possible main routes going from soil to
birds or beasts of prey (soil-plant-bird, soil-invertebrate-bird, soil-plant-mammal, and
soil-invertebrate-mammal). This number increases exponentially when different plant
parts and invertebrate groups are distinguished as quantitatively important food items
for small birds and mammals. For plants a distinction can, for example, be made
between leaves, seeds, fruits and tubers. The group of invertebrates may comprise
earthworms, gastropods, larvae of insects, caterpillars, insects, isopods and spiders.
An additional distinction can be made between leaves and seeds with respect to plants
and between worms and insects with respect to invertebrates, thus leading to a total
of 16 routes going from soil to both beasts and birds of prey.

Using these routs, we can estimate the MPC values for cadmium in soil (Table 11).

Table 11. MPC values for Cd in soil, based on 16 different
exposure pathways (After Jongbloed et al, 1994).

Food chain	Critical limit of Cd in soil, ppm	
	Beast of pray	Birds of pray
Soil→ leaf → bird	37	2.3
Soil→ seed → bird	7.2	0.44
Soil→ worm → bird	1.5	0.08
Soil→ insect → bird	6.4	0.40
Soil→ leaf → mammal	48	3.6
Soil→ seed → mammal	9.4	0.68
Soil→ worm → mammal	1.9	0.12
Soil→ insect → mammal	8.3	0.61

Results show that for Cd (i) birds of prey are always more sensitive than beasts of prey and (ii) bioaccumulation is lowest in the food chain: Soil → worm → bird/mammal. The latter food chain to birds of prey is by far the most critical pathway for Cd exposure, leading to very low critical limits for soils (approximately 0.1 ppm, compare with Table 10). When one aims to protect the most sensitive species, the latter limit seems appropriate.

2.3. Calculation methods

The selection of a computation method or model is the third step in the flowchart for calculating critical loads of heavy metals (Figure 31). There are different models that can be used to calculate critical loads for terrestrial and aquatic ecosystems, based on receptor properties and on certain critical limits. Relevant aspects in relation to the selection of a calculation method are:

(i) the choice of steady-state and dynamic models in calculating critical loads,

(ii) the choice of critical limits in relation to what we want to protect,

(iii) the required model complexity (simplicity) in view of the regional applicability of the calculation method.

These aspects are briefly summarized below.

In selecting a computation model, it is important to make a clear distinction between steady-state and dynamic models. For instance, steady-state soil models, predicting metal concentrations at the soils solid phase and in the soil solution at steady-state, are particularly useful to derive critical loads based on an infinite time perspective. Dynamic models, predicting changes in heavy metal chemistry in both soil and soil solution in response to changes in metal loading, are particularly useful to predict the time period before a critical limit is exceeded (if ever) for a given input scenario. Dynamic models may, however, also be used to calculate target loads. Unlike critical loads, target loads include a definite time target (e.g., 50 or 100 years) in relation to what is considered an acceptable input for the time.

As we have discussed earlier, the critical limit to be used depends on what we want to protect. In terrestrial ecosystems this can be soil fauna, the soil vegetation or the human beings that use ground water for drinking water or that consume crops that are grown on the soil. In aquatic ecosystems, the critical limits are connected with concentration of dissolved species in water.

The choice of a model for the calculation of critical loads not only depends on the question of 'which time period' (steady-state or dynamic) and 'what to protect' (which critical limit) but also on the amount of detail required. Depending on the aim, one must chose a model with an appropriate level of detail in describing the processes in the terrestrial or aquatic ecosystems. A disadvantage of relatively complex mechanical models is that input data for their application on a regional scale is generally incomplete and values can only be roughly estimated. Even if the model structure is correct (or at least adequately representing current knowledge), the uncertainty of the output of complex models may thus still be large because of the uncertainty of input data. Simpler empirical models have the advantage of a smaller need for input data but the theoretical basis, that is needed to establish confidence in the predictions, is small, which limits the application of such models for different situations. There is thus a trade off between model complexity (reliability) and regional applicability.

When the aim of the model is to calculate critical loads on a regional scale consisting of receptors with different properties, it seems most rational to use a relatively simple model with aggregated description of processes in the total considered compartment. In choosing the model, one should be aware of the consequences of simplifications, such as ignoring certain processes (complexation, metal cycling, etc.) and making certain assumptions (steady-state, homogenous mixing, equilibrium partitioning).

In order to gain insight into the consequences of the choice of a certain model and limits, one could perform critical load calculations with different models (complex dynamic models versus simple steady-state models) using various limits, and compare the results. In this way, one also gets insight in (i) the differences in vulnerability of the various environmental compartments and related organisms and (ii) the relevance of the different processes in the systems and of the different ways of parameterizing certain processes.

As examples, the steady-state mass balance equations will be shown for terrestrial and aquatic ecosystems.

Terrestrial ecosystems. The soil is the estimated compartment for terrestrial ecosystems. Using the assumptions (see above), a complete steady-state mass balance

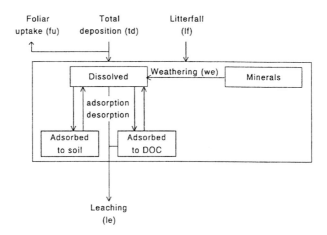

Figure 36. Schematic representation of the mass balance for heavy metals in soils of non-agricultural ecosystems (de Varies and Bakker, 1998).

of a certain heavy metal, M, in a soil layer equals (see also Figure36):

$$fM_{tl} = fM_{fu} - fM_{lf} + fM_{ru} - fM_{we} + fM_{le}, \qquad (22)$$

where all terms relate to biogeochemical fluxes of heavy metal M (in g/ha/yr. or mg/m^2/yr.): fM_{tl} is the total load by deposition and other loads (e.g., fertilizers), fM_{tl} is foliar uptake or retention, fM_{lf} is litterfall, fM_{ru} is root uptake, fM_{we} is weathering, fM_{le} is leaching.

In this equation, no distinction has been made between various oxidation states of the metals. The approach implicitly lumps all the various metal species in the soil. With respect to Pb, Cd, Cu, Zn and Ni, it is assumed that the metal only persists as a divalent cation, which is a valid assumption for these metals in soil. This assumption, however, seriously limits the application of this model to calculate critical loads for Cr and Hg.

With respect to Cr a distinction should be made between Cr(III), which is the common oxidation state in the soils, being rather immobile and so toxic, and Cr(VI), which is very mobile and very toxic. With respect to Hg, the situation is even more complex, due to the occurrence of mercuric mercury (Hg^{2+}), mercurous mercury (Hg_2^{2+}), elemental mercury (Hg^0) and organic mercury species, such as methyl mercury, $(CH_3)_2Hg$ (see Chapter 8, Section 2). Furthermore, volatilization of elemental mercury and organic mercury species is common. A description of these processes, in combination with other interactions of Hg in soil, such as reduction, absorption and complexation, is extremely difficult and the approach can only be considered as very approximate for mercury. This also holds to a lesser extent for chromium.

In soils of non-agricultural ecosystems, above ground biomass (foliar uptake) and metal cycling is considered important, due to large impact on the metal distribution in

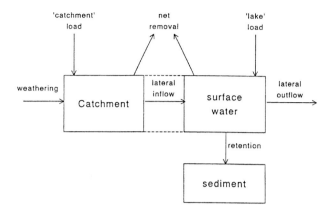

Figure 37. Schematic representation of mass balance for heavy metals in a catchment (de Vries and Bakker, 1998).

the humus layer and mineral soil profile. Especially in soils of Forest ecosystems, it may affect the accumulation in the humus layer, which is considered a very relevant compartment regarding the calculation of a critical load. In these soils, however, a steady-state element cycle is assumed, which implies that mineralization, M_{mi}, equals litterfall, M_{lf}.

In soils of agroecosystems, above ground biomass (foliar) uptake and metal cycling by mineralization and total root uptake can be lumped into a net removal term due to harvest (indicated as growth uptake, Mg_u) when the critical load is calculated for the root zone, e.g., for upper 20–30 cm. In this situation we can calculate root uptake as a function of the growth uptake, whereas the net effect of litterfall and foliar uptake is assumed to be negligible.

Thus, the mass balance of heavy metals in soils of agroecosystems is:

$$fM_{tl} = fM_{gu} - fM_{we} + fM_{le}, \qquad (23)$$

where fM_{gu} is the flux of heavy metal M by growth uptake

The critical metal leaching rate depends on critical dissolved metal concentrations, which can be derived in different ways, for instance as a critical limit for soil solution (see above).

Aquatic ecosystems. The steady-state calculation method for surface water is based on the mass balance schematically shown in Figure 37.

Figure 37 gives the simplified mass balance for heavy metals in a catchment, including both a soil compartment in the catchment area and the aquatic system with a water and sediment compartments. A complete steady-state mass balance of heavy metals for a catchment equals:

$$fM_{tl} = fM_{up} - fM_{we} + (fM_{sed} - fM_{res} + fM_{ex}) \times A_t/A_c + fM_{lo}, \qquad (24)$$

where all terms relate to biogeochemical fluxes of heavy metal M (in g/ha/yr. or mg/m²/yr.) and area (ha or m²): fM_{tl} is the total load by deposition and other loads (e.g., fertilizers), fM_{up} is metal flux by net uptake in the catchment and the lake, fM_{we} is metal flux by weathering in the catchment, fM_{lo} is metal flux by lateral outflow of water, fM_{sed} is metal flux from water compartment to the sediment compartment by sedimentation, fM_{res} is metal flux from sediment compartment to the water compartment by resuspension, fM_{ex} is metal flux from sediment compartment to the water compartment or vise versa caused by exchange processes near the sediment water interface, A_c is the surface area of the catchment, A_t is the surface area of the lake (lowest water system) in the catchment area.

Similar to terrestrial ecosystems, no distinction is made between various oxidation states of the metals. This assumption seriously limits the application of the model to calculate critical loads for mercury.

Net uptake of heavy metals is due to the removal of heavy metals in crops or trees in the catchment and/or in aquatic plants and fish in lake. Weathering relates to the release of HM from primary minerals in the catchment. Sedimentation is the result of the setting of suspended particles in the lake. As a result of this process, the pollutant absorbed to the suspended particles is transported from the water compartment to the sediment compartment. Resuspension of sediment particles is the result of the turbulence at water-sediment interface. As a result of this process, the pollutant absorbed to the sediment particles is transported from the sediment compartment to the water compartment. The exchange processes at the sediment water interface include advection or infiltration, molecular diffusion, and bioturbation and bioirrigation (the latter are the transport of HM resulting from the ventilation of tubes and burrows in the sediments by benthic organisms). To scale these processes to the catchment, the sedimentation and resuspension rates are multiplied by the ratio of the lake area and the catchment (de Vries *et al*, 1998).

2.4. Input data

Collecting input data is the forth step in the flowchart for calculating critical loads for heavy metals (see Figure 31). It includes the assessment of hydrological data, vegetation data and soil data influencing the heavy metal fluxes in the considered ecosystems. For application on a regional scale it also includes the distribution and area of receptor properties using available digitized information and GIS technologies.

Input data for the most detailed soil model include parameters describing atmospheric deposition, precipitation, evapotranspiration, litterfall, foliar uptake, root uptake, weathering, adsorption and complexation of Pb, Cd, Cu, Zn, Ni, Cr and Hg. The input data mentioned above vary as a function of location (receptor area) and receptor (the combination of land and soil type) as shown in Table 12.

The receptors of interest are soils of agricultural (arable lands, grasslands) and non-agricultural (forests, steppes, heath lands, savanna, etc) ecosystems. In non-agricultural ecosystems, the atmospheric deposition is the only input of heavy metals. Regarding the Forest ecosystems, a distinction should at least be made between Coniferous and Deciduous Forest ecosystems. When the detailed information on the

Table 12. The influence of location, land use and soil type on input data (+ effect; − no effect).

Input data	Location	Land use	Soil type
Precipitation	+	−	−
Deposition	+	+	−
Evapotranspiration	+	+	+
Litterfall	+	+	+
Foliar uptake	+	+	−
Root uptake	+	+	+
Weathering	−	−	+
Adsorption	−	−	+
Complexation	−	−	+

areal distribution of various tree species (e.g., pine, fir, spruce, oak, beech and birch) is available, this should be used since tree species influence the deposition and uptake of heavy metals and the precipitation excess. On a world scale, soil types can be best distinguished on the basis of the FAO-UNESCO Soil Map of the World, climate and ecosystem data from NASA database (1989).

In order to obtain data for all receptors within all receptor areas (grids), a first good approach is to interpret and extrapolate data by deriving relationships (transfer functions) between the data mentioned before and basic land and climate characteristics, such as land use, soil type, elevation, precipitation, temperature, etc. A summarizing overview of the data acquisition approach is given in Table 13.

3. EXAMPLES OF CRITICAL LOADS CALCULATIONS FOR HEAVY METALS

At present, the calculation and mapping of critical loads for heavy metals is only in the beginning and in Europe there are only a few examples of application of methods described in Section 2. We will refer to case studies from Germany and Russia as a most characteristic research in this direction.

3.1. Calculation and mapping of critical loads for HM in Germany

We have seen that heavy metals can cause toxic effects to living organisms when critical limits are exceeded. Present deposition rates may cause the long-term

Table 13. Data acquisition approach for input data.

Input data	Data acquisition approach
Deposition	Estimate per unit area using emission/deposition matrices, corrected for forest filtering
Precipitation	Estimate per unit area based on data of weather stations or monitoring site
Forest canopy interception	Derivation of a relationship with precipitation amount and land use
Transpiration	Calculation as a function of climate, land use and soil type
Litterfall/foliar uptake	Derivation of a relationship with deposition and land use
Root uptake	Derivation of a relationship with deposition and land use
Absorption	Derivation of a relationship with soil characteristics such as pH, organic matter content, clay content and CEC
Complexation	Literature data; derivation of a relationship with pH

accumulation of heavy metals in the soil, especially in the forest humus layers and bottom sediments. Calculations based on comprehensive models (see Chapter 8, Section 2 and 3) show the long-range transport of various heavy metals in regional and continental scale. In addition to atmospheric deposition, in agroecosystems the input of HMs is connected with phosphorus fertilizers and application of wastewater effluents. Under increasing acidification of the forest ecosystems, many heavy metals accelerate aqueous migration in the biogeochemical food web. A known example is related to Cd and Cu. The accumulation of pollutants in various terrestrial and aquatic ecosystems of Germany is almost not returnable. That is why we will consider precautionary measures based on critical load calculations of HM. The case study in Germany may give a good example of such an approach (Niegel and Schutze, 1999).

Methods of critical load calculations for heavy metals

The approach, similar to that described in Section 2 was applied for the calculation of critical loads of HMs for German soils.

Critical concentrations

According to the heavy metals' effects, the soil microbes, crops and ground waters as a source of drinking water, are the most important receptors. During migration in the food web, the heavy metals, especially Cd and Hg, can affect also higher organisms, including people. After consideration of different pathways, the most sensitive links of food webs should be chosen for establishing the critical concentration in the soil's solution (critical limit) to protect all other pathways at this concentration.

Critical concentrations with respect to soil organisms should be related to a low effect level on the most sensitive species. The effects on the processes of metabolism and other processes within the organisms should be considered and also the diversity of the species, which is most sensitive to heavy metals, has to be accounted. Critical limits must refer to the chronic or accumulated effects. For assessment of the critical concentrations in crops and in drinking water, human-toxicological information is required. In general, for establishing critical loads we should also account the additive effects of different metals and combination effect between the acidification and biogeochemical mobilization of the heavy metals in soils and bottom sediments.

The environmental standards based on total heavy metal concentration in the soil solution seem the most important criterion for the exposition of further compartments of the environment. The additional effects connected with metal speciation and complexation were not considered in the study.

A Monte Carlo-simulation is proposed to appreciate the uncertainty in the process of establishing the critical concentrations of heavy metals in the soil solution.

Models

In this case study, steady-state mass balance models are applied for critical loads calculation for the heavy metals.

$$SM_V + SM_{Dep} + SM_D SM_{Abf} = SM_E + SM_{Aw} SM_{Er} + SM_G + \Delta SM_{Vorr}, \quad (25)$$

where: SM_V is release by weathering, SM_{Dep} is input by the atmospheric deposition, SM_D is input by "usual" fertilising, SM_{Abf} is inputs by the use of waste, SM_E is output by harvest, SM_{Aw} is output by leaching, SM_{Er} is output by erosion of the soil's parts, SM_G is output by degassing, ΔSM_{Vorr} is changes of the heavy metal pool in soil.

This mass balance presents the possible links of the biogeochemical food web for various heavy metals. Some items may be neglected, like degassing of Pb, Cd, Cu and Zn metals. However, this process is of crucial importance for mercury (see Chapter 8, Section 2). The output of the heavy metals with soil erosion may also be neglected. After elimination of these processes, the simplified following equation is workable. The sum of inputs by deposition, fertilizing, and waste and rubbish as fertilizer is known as the term "Critical Load".

Thus, critical loads of any heavy metals may be calculated as follows:

$$CL_{SM} = CL_E + CL_{Aw} - \Delta CL_{Vorr}. \quad (26)$$

The mass balance model for calculation of critical loads for heavy metals includes the weathering process, net removal through the crop biomass harvest, leaching, and also leaf uptake and litterfall. Using the simple dynamic way, the distribution between adsorbed and dissolved phases was accounted.

The uncertainties in the model inputs were elaborated using statistical distribution functions for the initial parameters and also the Monte Carlo simulation.

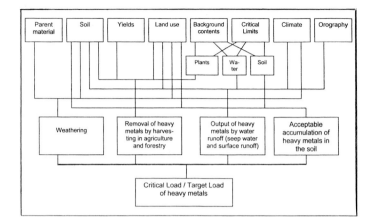

Figure 38. Database for initial information for calculation of HM critical loads in Germany (Niegel and Schutze, 1999).

The available information for calculation of critical loads of HMs in Germany is shown in Figure 38. This figure shows also the schematic algorithms for CL calculations.

The application of CL model and initial information allowed the researchers to map the critical loads of various heavy metals for different ecosystems.

3.2. Calculation and mapping of critical loads for Cd and Pb in the European part of Russia

Biogeochemistry of heavy metals has been extensively studied in the former Soviet Union due to a widespread environmental pollution. The numerous results on ecosystem sustainability or sensitivity to metal inputs have been accumulated.

The assessment of ecosystem sustainability under a heavy metal loading includes primarily the estimation of soil compartment (Solntseva, 1982; Elpatievsky, 1994; Glazovskaya, 1997). These researches as well as literature data from other countries showed that the processes of metal accumulation and transformation in soil and further migration in the biogeochemical food web, like metal uptake by plants and metal leaching from the soils, are mainly dependent on geochemical properties of the soils. The following soil parameters were shown as the most important: pH, organic matter composition (mainly on humic and fulvic acids ratio), redox reaction, and soil granulometric composition (Davies, 1980; Sanders, 1982; Kabata-Pendias, Pendias, 1984; Adriano, 1986; Balsberg-Pahlsson, 1989; Bowen, 1989; Temminghof *et al*, 1997).

M. A. Glazovskaya (1997) applied an analysis of geochemical conditions in different soils and developed principles to assess quantitatively the sustainability of ecosystems under the technogenic impact of heavy metals. The soils of the main natural zones distinguished on the East European plain area were combined into six groups by the extent of their ecological-geochemical sustainability under HM loading (from 'very sensitive' to 'insensitive'). As shown in this research, most of the soils of the East European plain area have medium or weak sustainability under metal exposure. But, quantitative parameters of HM impacts onto the soil (including the permissible levels of metal depositions) were not considered in this classification.

The quantitative assessment of biogeochemical mass-balances of the metals in various natural, urban and agricultural ecosystems were carried out in different regions of the Russian Federation (Bashkin *et al*, 1992; Uchvatov, 1995; Elpatievsky, 1994; Kasimov *et al*, 1995). Methodologically, these researches are similar to a general approach used for calculations of HM critical loads in Europe (De Vries *et al*, 1997). However, the results of these local researches could not be directly used for calculations of HM critical loads for the whole area of the East European plain due to scarcity of the data needed for computation according to the above mentioned steady-state mass balance equation of De Vries and Bakker (1998). Nevertheless, these data were used for estimating and mapping of HM critical loads for the European Russia area (Ptiputina *et al*, 1999).

Algorithm for calculation of critical loads of heavy metals in the forest ecosystems

The calculations of the critical loads for two heavy metals (Pb and Cd) were made for the terrestrial ecosystems of the European part of the Russian Federation. Using reliable experimental data, we used a simplified mass balance equation:

$$M_{tl} = M_{up} + M_{leach}, \qquad (27)$$

where M_{tl} is the total load by metal deposition, M_{up} is metal accumulation in annual wood biomass production, M_{leach} is metal leaching with total runoff.

The accumulation of heavy metal in wood biomass (metal uptake) was calculated as

$$M_{up} = G_{an} \times C_{back\,M}, \qquad (28)$$

where G_{an} is annual wood biomass production, $C_{back\,M}$ is maximum permissible (or background) level of metal concentration in tree wood.

The data on annual wood biomass production in the main forest types have been taken from (Bazilevich, 1993). We used the value of maximum permissible level of lead in the wood equal to 0.005 g/kg as an accepted in Russia (Moiseev, 1994). To estimate cadmium accumulation in the biomass, the background values equal to 0.25 mg/kg for coniferous species, 0.5 mg/kg for small leaf species and 0.6 mg/kg for broad leaf species (Uchvatov, 1995), were used. The metal accumulation in the wood biomass was not calculated for the areas without forest ecosystems, for example, tundra and steppe ecosystems. We supposed that metal conservation in the biomass of these natural ecosystems was only temporal with subsequent mineralization of organic matter and entering biogeochemical cycling.

Estimating metal leaching flux from the ecosystem, we proposed to protect the drinking groundwater and surface water. We supposed that metal concentration in the

natural water has not to be higher than the maximum permissible level accepted by
the Russian Ministry of Human Health Protection. Therefore, metal leaching from
the ecosystems was calculated as

$$M_{leach} = Q_{runoff} \times C_{water\ MPL},\qquad(29)$$

where, Q_{runoff} is annual total runoff, $C_{water\ MPL}$ is maximum permissible level of
metal concentration in the water.

The total runoff data have been taken from (Global Data Sets ..., 1988). The
value of maximum permissible level of lead concentration in the water is equal to
0.03 mg/L, the value of maximum permissible level of cadmium concentration in the
water is equal to 0.005 mg/L (MPLs Act No2932–83, 1986).

Calculation and mapping of critical loads for Pb *and* Cd

We have considered only two biogeochemical metal fluxes (accumulation in wood
biomass and leaching) that exist in the forest ecosystems. Both of these fluxes contribute to the removal of heavy metals from soil solution and, consequently, to the
ecosystem depletion in HM. The calculated values of critical loads of lead ranged
from < 50 g/ha/yr. to > 250 g/ha/yr. Cadmium critical loads are within the limits
< 5 and > 25 g/ha/yr.

For most grid cells, the values of metal leaching flux calculated in accordance
with relevant equation ($M_{leach} = Q_{runoff} \times C_{waterMPL}$) are higher than respective
values of metal accumulation in the wood biomass. This situation is typical for
the Northern and Central regions of European Russia where wet climate with an
excess of precipitation over evaporation predominates that influences the high total
runoff. Simultaneously, soil-geochemical conditions (especially, low pH) of these
zones determine the increased mobility of heavy metals accumulated in soil solution
(Perelman, 1976). We would expect from theoretical considerations the promotion of
soil self-purification due to active metal migration through the soil layers. However,
in real ecosystems, there is an imbalance between the values of metal depositions and
metal leaching related to metal adsorption by soil organic-mineral complex (Kabata-
Pendias, H. Pendias, 1984; Glazovskaya, 1997). Similar geochemical conditions are
characteristic for most landscapes of the Belorussian and Ukrainian areas (Perelman,
1976). However, we presume that specific metal accumulation in the peat soils, widely
distributed in Belorus and North Russia and partly in North Ukraine, has to be specially
taken into account to calculate the HM critical loads for these ecosystems.

Metal migration though soil layer is lower in ecosystems of the southern regions of
Russia (in Orthic Luvisols, Orthic Greyzems, Chernozems and other soils) due to both
climate features and increased metal adsorption. Simultaneously, biomass productivity of the forest ecosystems of this area is higher (Bazilevich, 1993). Accordingly,
the calculated values of Pb and Cd fluxes (leaching and wood biomass accumulation)
are similar. However, the removal of HM from eluvial landscapes is accompanied
by their accumulation in subordinate landscapes and the literary data confirms the

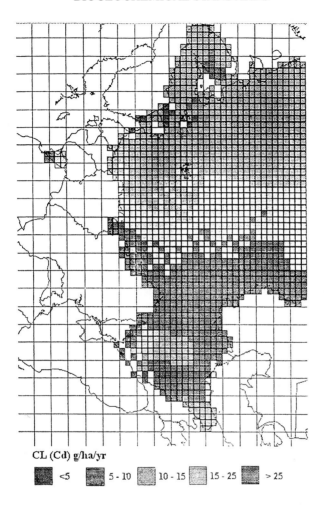

CL (Cd) g/ha/yr

■ <5   ▨ 5 - 10   ▨ 10 - 15   ▨ 15 - 25   ▨ > 25

Figure 39. Critical loads of Cd *for ecosystems of European Russia.*

active migration of chemical elements from watershed area to accumulation zones as a result of soil erosion with surface runoff (Yaroshevich, Zhilko, 1980).

The example of calculated values of critical loads of cadmium is shown in Figure 39.

FURTHER READING

Maximilian Posch, Peter A. M. de Smet, Jean-Paul Hettelingh and Robert Downing (Eds.), 1999. Calculation and Mapping of Critical Thresholds in Europe, Status Report 1999. CCE/RIVM, Bilhoven, the Netherlands, 166pp.

Wim de Vries and Dick J. Bakker, 1998. Manual for calculating critical loads of heavy metals for terrestrial ecosystems. DLO Winand Staring Centre, Report 166, The Netherlands, 144pp.

Wim de Vries W., Dick J. Bakker and Harald Scerdrup, 1998. Manual for calculationg critical loads of heavy metals for aquatic ecosystems. DLO Winand Staring Centre, Report 165, The Netherlands, 91pp.

Vladimir Bashkin and Soon-Ung Park (Eds.), (1998) Acid Deposition and Ecosystem Sensitivity in East Asia, NovaScience Publisher, 427pp.

Per Warfvinge and Harald Sverdrup, 1995. Critical loads of acidity to Swedish forest soils. Report 5:1995, Lund University, Sweden, 104pp.

QUESTIONS AND PROBLEMS

1. Discuss the political requirements for critical load concept. Present a brief review of emission reduction politics in 1980–1990 in Europe and North America.

2. Present the definition of critical loads and critical levels. Discuss the pros and cons of the critical loads concept in comparison with BAP approach to emission abatement strategy.

3. Describe the basic biogeochemical principles of the critical load concept. Characterize the main advantages of using biogeochemical cycling parameterization to set environmental standards.

4. Describe the structure of biogeochemical model "PROFILE". What biogeochemical processes does this model simulate?

5. What types of effects can you differentiate regarding the influence of acid deposition on terrestrial and aquatic ecosystems?

6. Describe the general flowchart for critical load calculation. Present a brief discussion of the main steps in CL calculation.

7. What are the main sources of uncertainties in CL calculation? Discuss these uncertainties from the position of Environmental Risk Assessment approach.

8. Discuss the application of Environmental Risk Assessment approach for estimating acid deposition impacts on ecosystems.

9. Characterize the relationships between CL and ERA approaches for impact-oriented assessment of airborne pollutant deposition on natural ecosystems.

10. Describe European models for calculation of critical loads at terrestrial ecosystems. Pay attention to discussing a critical load function.

11. Define the exceedances of critical loads. How do you calculate the exceedance values for single and combined pollutants?

12. Present examples of critical load calculations and mapping for terrestrial ecosystems in Europe.

13. Using the maps of CL and Ex for acid forming and eutrophication compounds of sulfur and nitrogen, discuss the advantages that have been achieved in Europe in emission reduction during last two decades.

14. Describe the emission inventory in North America in 1990s. Characterize the role of individual countries in emission of sulfur and nitrogen oxides.

15. Present a general review of emission abatement strategy in North America and emphasize your attention on application of impact-oriented approaches.

16. Compare the spatial distribution pattern of sulfate and nitrate deposition in North America in 1980s and 1990s. Describe the possible sources of S and N pollution.

17. Present a definition of critical load concept for assessment of permissible loading of sulfate on aquatic ecosystems in Eastern Canada.

18. Describe the ecotoxicological indices of acidity loading on aquatic ecosystems. Present the quantitative parameters of factors affecting fish death.

19. Outline the future needs of emission abatement strategy in North America regarding the acidification and eutrophication problems.

20. Present a review of the acid deposition problem in East Asia. Why this region is of the most environmental concern at present?

21. Discuss the applicability of critical load concept in various countries of East Asia. Present relevant examples of CL calculations and mapping.

22. Characterize the role of China in transboundary pollution in East Asia. Describe the CL and Ex maps on the area of China. Indicate the reasons of differentiated sensitivity of Chinese ecosystems to acid deposition.

23. Discuss acid rain problems in South East Asia. Emphasize your attention on acid deposition effects on biogeochemical cycling in Tropical Rain Forest ecosystems.

24. Discuss the interactions between acid deposition and accelerating migration of heavy metals in biogeochemical food webs. Present examples.

25. Outline the applicability of the critical load concept for setting environmental standards for heavy metals.

26. Describe the flowchart for CL calculation for heavy metals. Compare the various approaches.

27. Estimate the role of receptor in HM critical load calculations. Present a description of different receptors for terrestrial and aquatic ecosystems.

28. Present a detailed description of critical limits for various environmental compartments. Review the values of critical limits for different media.

29. Describe models for estimating critical limits. Present examples of biogeochemical food web and corresponding critical limit values.

30. Characterize the models for HM critical load calculation in terrestrial ecosystems. Highlight the relative importance of various links of biogeochemical food web in mass balance models.

31. Describe the critical load models for aquatic ecosystem. Emphasize your attention on interactions of heavy metals between water and bottom sediment compartments. Discuss the application of small catchment approach for CL calculation.

32. Review the information sources for calculation of critical loads of heavy metals and other pollutants at different ecosystems. Show the data availability for critical load calculation in your country.

33. Present case studies of HM critical load calculations in different countries. Discuss the approach for CL assessment in Germany.

34. What data are of crucial importance for calculating CL in countries like Russia or China? Discuss the application of GIS technologies for CL calculation and mapping.

35. Discuss the calculation and mapping of critical loads of any pollutants in your country. Outline the existing problems of such calculations.

36. Discuss the political and economic limitations of implying CL concept in your country. What alternative approaches do you know? Discuss pros and cons of various approaches to emission abatement strategy in your country or region.

37. Compare the biogeochemical approaches to setting environmental standards with other methods like ecotoxicological experimental testing. Discuss the advantages and drawbacks of different approaches.

CHAPTER 11

FUTURE TRENDS IN MODERN BIOGEOCHEMISTRY

We can consider the analysis of future trends in modern biogeochemistry bearing in mind the universality of biogeochemical cycles involving the mass exchange of chemical elements between living organisms and the environment of the Earth's surface. Biogeochemistry as known has been quite productive as a high priority academic and scientific discipline. In order to parameterize quantitatively various scale local, regional and global changes due to natural and anthropogenic transformation, an understanding of biogeochemical cycles in different ecosystems is the great challenge for scientific society.

In spite of numerous publications on biogeochemical cycles of various nutrients, many questions of the cycles and their limitations/enhancements connected both with natural factors and human activity remain unclear due to the tremendous diversity of global ecosystems. One of the key problems in biogeochemistry is to make quantitative predictions of interactions between cycles. The major challenge is the interaction between carbon and nitrogen cycles, especially at the global scale, to predict the consequences if the dual growth of carbon dioxide and nitrogen species in the troposphere continues in the future. One reasonable prediction is the increasing sequestration of both C and N in vegetation biomass and soil organic matter, however, many issues related to the relative openness of biogeochemical cycles and their interactions with other macro- and micronutrients are still unanswered. Large scale regional forest fires and flooding in different areas present another challenge for understanding the interactions between biogeochemical cycles.

Review of the modern literature shows that the driving mechanisms of the cycles of many elements are still uncertain. We are now only beginning to understand the qualitative and quantitative aspects of tiny active pools of many biosphere compartments, especially soil organic and inorganic matter, that play the most important role in productivity of many natural ecosystems. The interactions between small active and large 'inactive' pools are also very important. Future research should focus on mobilization and immobilization of nitrogen in soil organic matter, in both background and modified ecosystems. This will allow us to predict the dynamic trends in these pools and fluxes under increasing pollution. There are very few unpolluted background areas in the Earth's biosphere, and the quantitative parameterization of biogeochemical cycling in these areas is a great challenge for biogeochemists.

Interactions of land and coastal biogeochemical cycles present the other vast area of future research in biogeochemistry. The intersection of many scientific approaches will be necessary to provide the knowledge for much more environmentally sound management of riverine and deposition fluxes of nutrients, both macronutrients like N, P, Si, Ca, and many micronutrients, to prevent the eutrophication of coastal waters in many regions of the World. The present investigations in European and East American coastal waters might offer key insights into other regions, especially in East and South East Asia. The driving mechanisms and forces of such links of biogeochemical cycling as N fixation, denitrification and productivity need more quantitative understanding. The interactions in the systems 'land-ocean-air' are also subjects to study in more detail.

Many changes in biogeochemical cycles of different elements occur in rather narrow marginal areas between natural and anthropogenic ecosystems. These areas might be considered as the biogeochemical barriers that alter the migration fluxes and finally determine the whole biogeochemical structure in neighboring landscapes, for example, in agricultural v/v natural ecosystems, and urban v/v natural ecosystems. Known ecological idea of ecotone or 'edge as a filter' might be pointed out that these boundaries are also the interactions within biogeochemical cycles. The tremendous variability of possible marginal areas in the total global biosphere requires crucial attention from biogeochemistry and ecosystem ecology.

Human biogeochemistry is the rapidly developing branch of modern biogeochemistry dealing with the quantitative assessment of relationships between migration of chemical species in food webs in natural and technogenic biogeochemical provinces and human health. This branch of biogeochemistry will be the subject of numerous research activities in various regions of the World, both in developed and developing societies. In spite of existing knowledge on the endemic diseases in natural and anthropogenic biogeochemical provinces, these data are very limited and restricted to small areas in a few countries. Human biogeochemistry must be enlarged in various countries for understanding human illnesses, first of all cancer and cardiovascular diseases, human diet and human adaptation to transformed environment.

Biogeochemical mapping in global and regional scales is also the great challenge for future research. This mapping is a tool to synthesize biogeochemical parameters and provide the information necessary for ecological management of the regional and global landscapes. The understanding of biogeochemical peculiarities of cycling in different ecosystems will help in recognizing the safest places for industrial projects and rehabilitation of polluted sites. Many regional biogeochemical standards will be based on this mapping, such as critical loads of pollutants on terrestrial and aquatic ecosystems. The whole area of North East American lakes is one of the regional challenges where biogeochemical mapping of critical loads of acidity will be useful for predicting affects of acid rains or HM and POP deposition.

Furthermore, we can identify more trends in future research. Biogeochemistry has been defined as the study of biological controls on the chemistry of the environment and geochemical regulation of ecological structure and function. With the increasing development of biotechnology, various genetically modified strains and species are and will be one of the important driving forces in the geochemical environment and the biosphere as a whole. The generalization that in biogeochemical cycling, the active principles come from biota, which are altered only slowly by global biological and geological activity must be reevaluated. The new species, with different specific characteristics, will change the natural biogeochemical cycles of elements and prediction of these changes will be of vital concern for future biogeochemical research. On the other hand, the environment drives the evolution of living organisms and this basic biogeochemical principle will enhance the mutual interactions between biotechnology and biogeochemistry. Finally, biogeochemical technologies are already being applied in many places and we can expect that practical oriented research in this field will be very useful for the development of this sub-branch of biotechnology.

References

Aber, J. D., Nadekhoffer, K. J., Steudler, P., & Melillo, J. M. (1989). Nitrogen saturation in northern forest ecosystems. *BioScience, 39*, 378–386.

Adriano, E. D. (1986). *Biogeochemistry of trace metals*. London, Tokyo, Boca Raton, Ann Arbor: Lewis Publishers, 512 pp.

Atlas, E. L. (1975). *Phosphorus equlibria in seawater and interstitial waters*. Ph.D. Thesis, Oregon State University, Oregon.

Bailey, R. G. (1998). *Ecoregions: The ecosystem geography of the oceans and continents*. New York: Springer.

Balsberg-Pahlsson, A. M. (1989). Toxicity of heavy metals (Zn, Cu, Cd, Pb) to vascular plants. *A literature review. Water, Air and Soil Pollution, 47*, 287–319.

Barnola, J. -M., Pimienta, P., Raynaud, D., & Korotkevich, T. S. (1991). CO_2-climate relationship as deduced from Vostok ice core: a re-examination based on new measurement and on a re-evaluation of the air dating. *Tellus, 43B*, 83–90.

Bashkin, V. N. (1984). Study of landscape-agrogeochemical balance of nutrients in agricultural regions: pt.II. Potassium. *Water, Air and Soil Pollution, 21*, 97–103

Bashkin, V. N. (1987). *Nitrogen agrogeochemistry*. Pushchino: ONTI Publishing House, 268 pp.

Bashkin, V. N. (1997). The Critical load concept for emission abatement strategies in Europe: a review. *Environmental Conservation, 24*, 5–13.

Bashkin, V. N. (Ed.). (1997). *Heavy metals in the environment*. Pushchino: ONTI Publishing House, 321 pp.

Bashkin, V. N., Uchvatov, V. P., Kudeyarova, A. Yu., Vasilieva, G. K., et al. (1992). *Ecological-agrogeochemical mapping of the Moscow region*. Pushchino: ONTI Publishing house, 170 pp.

Bashkin, V. N., Evstafieva, E. V., Snakin, V. V., et al. (1993). *Biogeochemical fundamentals of ecological standardization*. Moscow: Nauka Publishing House, 321 pp.

Bashkin, V. N., Kozlov, M. Ya., & Abramychev, A. Yu. (1996)a. The application of EM GIS to quantitative assessment and mapping of acidification loading in ecosystems of the Asian part of the Russian Federation. *Asian-Pacific Remote Sensing and GIS Journal, 8*(2), 73–80.

Bashkin, V. N., Kozlov, M. Ya., Abramychev, A. Yu., & Dedlova, I.S. (1996)b. Regional and global consequences of transboundary acidification in the Northern and Northern-East Asia. In: Proceedings of International Conference on Acid Deposition in East Asia, Taipei, May 28–30, 1996, 225–231.

Bashkin, V. N., Kozlov, M. YA., & Golinets, O. M. (1996)c. Risk assessment of ecosystem sustainability to acid forming compounds in the North-Eastern Asia In: Proceedings of International Conference on Acid Deposition in East Asia, Taipei, May 28–30, 1996, 347–356.

Bashkin, V. N., Kozlov, M. Ya., Priputina, I. V., & Abramychev, A. Yu. (1997). Regional assessment of ecosystem sustainability to atmotechnogenic deposition of sulfur and nitrogen in European part of Russia. Pt.I. Quantitative assessment and mapping of critical loads of sulfur and nitrogen compounds at terrestrial and freshwater ccosystems. *Regional Ecological Problems*, No. 1, 57–78.

Bashkin, V. N., Erdman, L. K., Abramychev, A. Yu., Priputina, I. V., Kozlov, M. Ya., et al. (1997)a. *Evaluation of the relation of atmospheric deposition to riverine input of nitrogen to the Baltic sea*. Baltic Sea Environmental Proceedings, No. 68, 83 pp.

Bashkin, V. N., Erdman, L. K., Abramychev, A. Yu., Sofiev, M. A., Priputina, I. V., & Gusev, A. (1997)b. *The input of anthropogenic airborne nitrogen to the Mediterranean sea through its watershed*. MAP Technical Reports Series, No. 118, Athens: UNEP, 95 pp.

Bashkin, V. N., & Park Soon-Ung. (Eds.). (1998). *Acid deposition and ecosystem sensitivity in East Asia*. New York: Nova Science Publishers, ltd., 427 pp.

Bashkin, V. N., & Kozlov, M. Ya. (1999). Biogeochemical approaches to the assessment of East Asian ecosystem sensitivity to acid deposition. *Biogeochemistry, 47*, 147–165

Bashkin, V. N., & Gregor, H. D. (Eds.). (1999). *Calculation of critical loads of air pollutants at ecosystems of East Europe*. Pushchino-UBA, Berlin: ONTI Publishing House, 132 pp.

Bazilevich, N. I. (1974). The geochemical function of the Earth's living matter and soil formation. In: *Genesis, classification and geography of soils*. Proceedings of the 10th Congress of International Society of Soil Science. Nauka Publishing House, Moscow, Vol. VI, part 1, 17–27.

540

Bazilevich, N. I. (1993). *Biological productivity of the ecosystems of the Northern Eurasia*. Moscow: Nauka Publishing house, 293 pp.

Beaufort, W., Barber, T., & Barringer, A. R. (1975). Heavy metal release from plant to the atmosphere. *Nature, 256*, 35–37.

Beeson, K. C., & Martone, G. (1976). *The soil factor in nutrition animal and human*. New York-Basel, 152 pp.

Benjamin, M. M., & Honeyman, B. D. (1992). Trace metals. In: S. S. Butcher, R. J. Charlson, G. H. Orians, & G. V. Wolfe (Eds.), *Global biogeochemical cycles*. Academic Press, London et al., 317–352.

Berg, L. S. (1958). *Fundamentals of climatology*. Leningrad: Uchpedgiz Publishing House, 318 pp.

Bezborodov, A. N., & Eremeyev, V. N. (1984). *Physico-chemical aspects of the ocean-atmosphere interactions*. Kiev: Naukova Dumka Publishing House, 215 pp.

Billen, G. C., Lancelot, C., & Meybeck, M. (1991). N, P, and Si retention along the aquatic continuum from land to ocean. In: R. F. C. Mantoura et al. (Eds.), *Ocean margin process in global change*. Wiley & Sons.

Blain, P. G. (1992). Aspects of pesticide toxicology. *Advance Drug React. and Acute Poison. Rev., 9*(1), 37–68.

Blanck, F. C. (1955). *Handbook of food and agriculture*. Reinhold Publishing Co., New York, New York.

Bolin, B., Degens, E. T., Duvingneaud, P., & Kempe, S. (Eds.). (1979). *The global carbon cycle*. J. Wiley, New York.

Bowen, H. J. M. (1989). *Environmental chemistry of the elements*. New York. 333 pp.

Brezonok, P. L., King, S. O, & Mach, C. E. (1991). The influence of water chemistry on trace metal bioavailability and toxicity to aquatic organisms. In: Newman, M. C., & McIntosh, A. W. (Eds.) *Metal ecotoxicology*. Lewis Publishers, Chelsea, 1–31.

Brooks, R. R. (1983). *Biological methods of prospecting for minerals*. John Wiley & Sons, New York.

Brookes, P. C., Dowlson, D. S., & Jenkinson, D. S. (1982). Measurement of microbial biomass phosphorus in soil. *Soil Biology and Biochemistry, 14*, 319–329.

Bunce, N. (1994). *Environmental chemistry*, 2nd edition. Winnipeg, Canada.

Butcher, S. S., Charlson, R. J., Orians, G. H. & Wolfe, G. V. (Eds.). (1992). *Global biogeochemical cycles*. London et al: Academic Press, 377 pp.

Cabata-Pendias A., Pendias, H. (1989). *Trace elements in soils and plants*. Moscow: Mir Publishing House, 440 pp.

Caraco, N., Cole, J. J., & Likens, G. N. (1989). Evidence for sulfate-controlled phosphorus release from sediments of aquatic systems. *Nature, 341*, 316–318.

Caraco, N., Cole, J. J., & Likens, G. N. (1990). A comparison of phosphorus immobilization in sediments of freshwater and coastal marine systems. *Biogeochemistry, 9*, 277–290.

Charlson, R. J., Anderson, T. L., & McDuff, R. E. (1992). The sulfur cycle. In: S. S. Butcher, R. J. Charlson, G. H. Orians, & G. V. Wolfe (Eds.). 1992. *Global biogeochemical cycles*. Academic Press, London et al, 285–300.

Carmichael, G. R., Hong, M. S., Ueda, H., Chen, L. L., Murano, K., Park, J. K., Lee, H., Kim, Y., Kang, C., & Shim, S. (1997). Aerosol composition at Cheju island, Korea. *Journal of Geophysics Research, v. 102*, No. D5, 6047–6061.

Cha, H. J., Kim, J. Y., Koh, C. H., & Lee, C. B. (1998). Temporal and spatial variation of nutrient elements in surface seawater off the West Coast of Korea. *The Journal of the Korean Society of Oceanography, 3*(1), 25–33.

Chaklette, T. H, & Boerngen, J. C. (1984). Element Concentration in Soils and Other Surficial Materials of the Conterminous United States. *U.S. Geological Survey Prof.*, Paper No. 574-C. Washington, 1–39.

Charlson, R. J., Anderson, T. L, & McDuff, R. E. (1992). The sulfur cycle. In: Butcher, S. S., Charlson, R. J., Orians, G. H., & Wolfe, G. V. (Eds.). 1992. *Global biogeochemical cycles*. Academic Press, London et al, 285–300.

Chen, Z. S., Liu, J. C., & Cheng, C. Y. (1998). Acid deposition effects on the dynamic of heavy metals in soils and their biological accumulation in the crops and vegetables in Taiwan. In: Bashkin, V. N., & Park, S-U. (Eds.). *Acid deposition and ecosystem sensitivity in East Asia*. NovaScience Publishers, USA, 189-228.

Choi, M. S. (1998). *Distribution of trace metals in the riverine, atmospheric and marine environments of the western coast of Korea*. Ph.D. thesis, Department of Oceanography, Seoul National University, 338 pp.

Clark, F. W., & Washington, H. S. (1924). The Composition of the Earth's Crust. *U. S. Geological Survey Prof.*, Paper No. 127.

Cleveland, C. C., Townsend, A. R., Schimel, D. S., Fisher, H., Howarth, R. W., Hedin, L. O., Perakis, S. S., Latty, E. F., Von Fisher, J. C., Elseroad, A., & Wasson, M. F. (1999). Global patterns of terrestrial biological nitrogen (N_2) fixation in natural ecosystems. *Global Biogeochemical Cycles, 13*(2), 623–645.

Comly, H. H. (1945). Cyanosis of infants caused by nitrates in well water. *Journal of American Medical Association, 129*(2), 112–117.

Conner, J. J., & Chacklette, H. T. (1975). Background Geochemistry of some Rocks, Soils, Plants and Vegetables in the Conterminous United States. *U.S. Geological Survey Prof.*, Paper No. 574-F, Washington.

Cooke, J. G., & Cooper, A. B. (1988). Sources and sinks of nutrients in New Zealand hill pasture catchment. III. Nitrogen. Hydrological. *Processes, 2*, 135–149.

Cox, P. A. (1995). *The elements on earth: Inorganic chemistry in the environment*. Oxford: Oxford University Press.

Craig, P. J. (1980). Metal cycles and biological methylation. In: *The natural environment and biogeochemical cycles*. Springer-Verlag, Berlin, 105–146.

Davies, B. E. (Ed.). (1980). *Applied soil trace elements*. Chichester-New York-Brisbane-Toronto: John Wiley & Sons, 482 pp.

Davidson, E. A. (1991). Nitrous oxide emission. In: J. E. Rogers, & W. B. Whitman (Eds.). *Microbial production and consumption of greenhouse gases: Methane, Nitrogen Oxides, and Halomethanes*. American Society for Microbiology, Washington, DC.

Degens, E. T. (1989). *Perspectives on biogeochemistry*. Springer-Verlag, 392 pp.

De Duve, C. (1991). *Blueprint for the cell*. Neil Patterson, Burlington, NC.

De Kok, L. J., & Stulen, I. (Eds.). (1998). *Responses of plant metabolism to air pollution and global change*. Leiden: Backhuys Publishers, 1998, 520 pp.

De Vries, W., Posch, M., Reinds, G. J., & Kamari, J. (1993). Critical loads and their exceedances on forest soils in Europe. *The Winand Staring Centre for Integrated Land, Soil and Water Research, Rep., 58*. The Netherlands: Wageningen, 116 pp.

De Vries, W., D. J., Bakker, & Sverdrup, H. U. (1997). Effect-based approaches to assess the risks of heavy metal inputs for terrestrial ecosystems. Overview method and models. Background document for the workshop on critical limits end effect-based approaches for heavy metals and POP's, 3–7 November 1997, Bad Hazburg, Germany, Umweltbundesamt, 1225–1277.

De Vries, W., & Bakker, D. J. (1998). Manual for Calculating Critical Loads of Heavy Metal for Terrestrial Ecosystems. *Guidelines for critical limits, calculation methods and input data*. SC report 166, DLO Winand Starring Centre. 144 pp.

De Vries, W., & Bakker, D. I. (1998). *Manual for Calculating Critical Loads of Heavy Metals for Soils and Surface Waters*. DLO Winand Starring Centre. The Netherlands: Wageningen, Report 165, 91 pp.

Demina, L. L., Gordeev, V. V., & Shumilin, E. V. (1983). The bivalent system of oceanic water. In: *Ocean Biogeochemistry*. Nauka publishing House, Moscow, 90–112.

Dentener, F. J., & Crutzen, P. J. (1994). A three dimensional model of the global ammonia cycle. *Journal of Atmospheric Chemistry, 19*, 331–369.

Derevel, S. J., Fio, J. L., & Dubrovsky, N. M. (1994). Distribution and mobility of selenium in groundwater in the western San Joaquin Valley of California. In: Frankenberger, W. T., & Benson, S. Eds. *Selenium in the Environment*, Marcel Dekker, New York, 157–183.

Diakovich, M. P., Evimova, N. V., & Motorova, N. I. (1991). Prediction of children population health in connection with air pollution. In: *Ecological Influence of Air Pool*, 57–59.

Dianwu, Z., Chuyin, C., Julin, X., Xiaoshan, Z., Zhaohua, D., Jietai, M., Seip, H. M., & Vost, R. (1994). *Acid Reign 2010 in China?*, 41 pp.

Dieter, H., & Schutze, G. (1999). Determination of the critical loads for sulfur and nitrogen compounds as well as for heavy metals in Germany. In: V. N. Bashkin, & H-D. Gregor (Eds.). *Calculation of Critical*

Loads of Air Pollutants at Ecosystems of East Europe. ONTI Publishing House, Pushchino-UBA, Berlin, 47–57.

Dobrodeev, O. P., & Suetova, P. A. (1976). The living matter of the Earth. In: *Problems of General Physical Geography and Paleography.* Moscow: MSU Publishing House, 26–58.

Dobrovolsky, V. V. (1967). Landscape geochemistry and problems of public health. In: M. A. Glazovskaya, & V. V. Dobrovolsky (Eds). *Landscape Geochemistry.* Nauka Publishing House, Moscow, 40–53.

Dobrovolsky, V. V. (1994). *Biogeochemistry of the World's Land.* Boca Raton-Ann Arbor-Tokyo-London: Mir Publishers, Moscow/CRC Press, 362 pp.

Dokuchaev V. V. (1948). *Theory of Natural Zoning.* Moscow: Geography Publishing house, 365 pp.

Doremus C. (1982). *Geochemical control of dinitrogen fixation in the open ocean. Biological Oceanography, 1,* 429–435.

Driessen, P. M., Buurman, P., & Permadhy, S. (1976). *Peat and podzolic soils and their potential for agriculture in Indonesia.* In: Proceedings of ATA 106 Midterm Seminar, Tugu, Bull. 4, 198–215.

Drisscoll, C. T., Otton, J. K., & Iverfeldt, A. (1994). Trace metal speciation and cycling. In: B. Moldan, & J. Cherny (Eds). *Biogeochemistry of Small Catchments.* John Wiley and Sons, 299–322.

Drouet, M., Le Sellin, J., Bonneau, J. C., Sabbah, A. (1990). La mercure est-il un allergen des voies respiratoires. *Allerg. Immunol., 22*(3), 81–88.

Ebens, R. J, Erdman, J. A., Feder, G. H, Case, A. A., & selby, L. A. (1973). Geochemical Anomalies of a Claypit Area, Callaway Country, Missoury, and Related Metabolic Imbalance in Beef Cattle. *U.S. Geological Survey Prof.*, Paper No. 807. Washington, 24 pp.

Ebens, R. J., & Shacklette, H. T. (1982). Geochemistry of some Rocks, Mine Spoils, Stream Sediments, Soils, Plants, and Waters in the Western Energy Region of the Conterminous United States. *U.S. Geological Survey Prof.*, Paper No. 1237. Washington, 173 pp.

EEA. (1998). *Europe's Environment, the Second Assessment.* European Environment Agency, Copenhagen.

Elpatievsky, P. V. (1994). *Geochemistry of Migration Fluxes in Natural Ecosystems and Ecosystems Transformed by Technogenesis.* Moscow: Nauka Publishing House, 253 pp.

Engel, M. H., & Macko, S. A. (Eds). (1993). *Organic Geochemistry: Principles and Applications.* New York and London: Plenum Press, 890 pp.

Elmgren, R., & Larsson, U. (1997). *Himmerfjarden: Forandingar I ett naringbelastat kustekyosystem I Ostersjon.* Rapport 4565. Naturvardsverket Forlag.

Environment Canada. (1997). *Canadian Acid Rain Assessment, Volume 3: The Effects on Canada's Lakes, Rivers and Wetlands.*

Environmental Statistics Yearbook. (1998). Republic of Korea: Ministry of Environment, 581 pp.

Erisman, J. W., Brydges, T., Bull, K., Cowling, E., Grennfelt, P., Nordberg, L., Satake, K., Scheider, T., Smeulders, S., van der Hoek, K., Wisniewski, J. & Wisniewski, J. (1999). Summary statement. In: *International Nitrogen conference,* Elsevier Science.

Ermakov, V. V. (1993). Biogeochemical mapping of continents. In: V. N. Bashkin, E. V. Evstafieva, V. V. Snakin et al, *Biogeochemical Fundamentals of Ecological Standardization.* Moscow: Nauka Publishing House, 5–24.

Erswaren, H., van den Berg, E., & Reich, P. (1993). Organic carbon in soils of the world. *Soil Science Society of America Journal, 57,* 192–194.

ESCAP. (1997). *Sustainable Development of Water Resources in Asia and the Pacific: an overview.* New York, United Nations, 162 pp.

ESCAP. (1998). *Sources and Nature of Water Quality Problems in Asia and the Pacific.* New York, United Nations, 164 pp.

Evseyev, A. V. (1988). The temporal changes of background concentrations of contaminants in various natural objects. *Moscow university News, Geography serial,* No. 3, 72–78.

Evstafieva, E. V. (1996). *Physiological and Biogeochemical Peculiarities of Adaptation in Various Environments.* DrSc Thesis, Moscow: University of Volk Friendship, 32 pp.

Evstafjeva, E., Orlinsky, D., Osovsky, U., Semenov, I., Evstafjeva, I. (1999). Human health indices and criteria for CL (HM, POPs) calculation. In: V. N. Bashkin, & H. D. Gregor (Eds). *Calculation of Critical Loads of Air Pollutants at Ecosystems of East Europe.* Pushchino-UBA, Berlin: ONTI Publishing House, 118–126.

Galiulin, R. V., & Bashkin, V. N. (1996). Organochlorinated compounds (PCBs and insecticides) in irrigated agrolandscapes of Russia and Uzbekistan. *Water, Air and Soil Pollution, 89,* 247–266.

Galloway, J. N., Schelezinger, W. H., Levy, H., Michaels, A., & Schnoor, J. L. (1995). Nitrogen fixation: Anthropogenic enhancement, environmental response. *Global Biogeochemical Cycles, 9*, 235–252.

Galloway, G. N. (1998). *The global nitrogen cycle: changes and consequences.* In: Proceedings of the First International Nitrogen Conference, Elsevier Science, 15–24.

Galperin, M. V., Erdman, L. K., & Subbotin, S. P. (1994). *Modelling of Pollution of the Arctic by the S and N Compounds and Heavy Metals from the Sources in the North Hemisphere.* MSC-E Report, July 1994, 33 pp.

Gao, Y., Arimoto, R., Duce, A., Lee, D. S., & Zhou, M. Y. (1992). Input of atmospheric trace elements and mineral matter to the Yellow Sea during the spring of a low-dust year. *Journal of Geophysics Research,* v. 97, No. D4, 3767–3777.

Giblin, A. E., Hole, W., Likens, G. N., & Howarth, R. W. (1992). The importance of reduce inorganic sulfur to the sulfur cycle of lakes. In: *Interactions of Biogeochemical Cycles in Aquatic Systems*, Part 7. SCOPE/UNEP Sonderband, Hamburg, 233–244.

Glazovskaya, M. A. (1973). *Soils of the World.* MSU Publishing House, 427 pp.

Glazovskaya, M. A. (1990). Methodological Guidelines for Forecasting the Geochemical Susceptibility of Soils to Technogenic Pollution, *ISRIC Technical Report, 22*, 39 pp.

Glazovskaya, M. A. (1984). *Soils of the World.* New Delhi: American Publishing Co., 401 pp.

Glazovskaya, M. A. (1994). Criteria of soil classification by an extent of the dangerous of lead contamination. *Pochvovedenie*, No. 4, 110–120.

Glazovskaya, M. A. (1997). *Methodological approaches of an assessment of ecological-geochemical stability of soils to technogenic impacts.* Moscow: Moscow University Publishing House, 102 pp.

Global Data Sets for Land-Atmosphere Models. *ISLSP Initiative I. 1987–1988. vol. 1–5*, 1988.

Granéli, E., Wallstrom, K., Larssonn, U., Graneli, W., & Emgren, R. (1990). Nutrient limitation of primary production in the Baltic Sea area. *Ambio, 19*, 142–151.

Gryaznova, T. P., Mednikova, V. G., Vishnyakova, A. P., & Lukashev, I. E. (1989). Pesticide influence on children health. In: *Pesticides and Health*, Krasnodar, 38–42.

GSC. (1995). *The Significance of Natural Sources of Metals in the Environment.* A Report prepared by the Geological Survey of Canada (GSC) for the UN-ECE LRTAP (Heavy Metals) Convention, 29 pp.

Gundersen, P., & Bashkin, V. N. (1994). Nitrogen. In: Moldan, B., & Cherny, J. (Eds.). *Biogeochemistry of Small Catchments.* John Wiley and Sons, 255–283.

Fenchel, T., King, G. M., & Blackburn, T. H. (1998). *Bacterial Biogeochemistry.* London et al: Academic Press, 307 pp.

Fersman, A. E. (1931). *Geochemical Problems of the USSR.* Essay 1. Moscow: USSR Academy of Sciences Publishing House, 430 pp.

Field, C. & Mooney, H. A. (1986). The photosynthesis-nitrogen relationships in wild plants. In: T. J. Givnish (Ed.). *On the Economy of Plant Form and Function.* Cambridge: Cambridge University Press, 25–55.

Flett, R. J., Schindler, D. W., Hamilton, R. D., & Campbell, N. E. R. (1980). Nitrogen fixation in Canadian precambian shield lakes. *Can. J. Fish. Aquat. Sci., 37*, 494–505.

Fogg, C. E. (1987). Marine plankton cyanobacteria. In: P. Fay, & C. V. Baaalen (Eds.). *The Cyanobacteria.* Elsevier, Amsterdam, 392–414.

Fortescue, J. A. C. (1980). *Environmental Geochemistry: A Holistic Approach.* Springer-Verlag, 347 pp.

Foundation Roi Baudouin. (1992). *Nitrates et Qualite des Eaus in Agriculture et Environnement.*

Frankenberger, W. T., & Benson, S. (Eds.). (1994). *Selenium in the Environment.* New York: Marcel Dekker, 456 pp.

Fraters, D., Boumans, L. J. M., van Drecht, G., de Haan, T., & de Hoop, W. D. (1998). Mintogen monitoring in groundwater in the sandy regions of the Netherlands. In: Proceedings of the First International Nitrogen Conference, Elsevier Science, 479–485.

Friend, J. P. (1973). The global sulfur cycle. In: *Chemistry of the Lower Atmosphere*, S. I. Rasool Ed., Plenum Press, New York, 177–201.

Freney, J. R. (1996). Control of nitrogen emission from agriculture. In: Lin, H. C., et al (Eds.), Proceedings of SCOPE/ICSU Nitrogen Workshop: *The Effect of Human Disturbance on the Nitrogen Cycle in Asia*, 85–99.

544

Frolova, L. (1999). Critical meanings of parameters of lloading calculations for pollutants by a method of fuzzy sets. In: V. N. Bashkin, & H. D. Gregor (Eds). *Calculation of Critical Loads of Air Pollutants at Ecosystems of East Europe*. POLTEX, Moscow-UBA, Berlin, 75–81.

Fuller, R. D., Simone, D. M., & Driscoll, C. T. (1988). Forest clearcutting and effects on trace metal concentrations: spatial patterns in soils solution and streams. *Water, Air and Soil Pollution, 40*, 185–193.

Hall, D. O., Scurlock, J. M. O., Ojima, D. S., & Parton, W. J. (2000). Grassland and the global carbon cycle: modelling the effects of climate change. In: T. M. L. Wigley, & D. S. Schimel (Eds). *Carbon Cycle*. Cambridge University Press, 103–114.

Hao, J., Xie, S., & Lei, D. (1998). Acid deposition and ecosystem sensitivity in China. In: V. N. Bashkin, & S-U. Park (Eds.). *Acid Deposition and Ecosystem Sensitivity in East Asia*. Nova Science Publishers, ltd., 267–312.

Hellström, T. (1996). An empirical study of nitrogen dynamics in lakes. *Water Env. Res., 68*, 55–65.

Herron, M. H., Langway, C. C., Weiss, H. V., & Gragin, J. H. (1977). Atmospheric trace metals and sulfates in the Greenland Ice Sheet. *Geochemical and Cosmochemical Acta, 41*, 915–920.

Hettelingh, J-P., Sverdrup, H., & Zhao Dianwu. (1995). Deriving Critical Loads for Asia. *Water, Air, and Soil Pollution, 85*, 2565–2570.

Holland, E. A., Parton, W. J., Detling, J. K., & Coppck, D. L. (1992). Physiological responces of plant populations to herbivory and their consequences for ecosystem nutrient flows. *American Naturalist, 140*, 685–706.

Holland, E. A., Dentener, F. J., Braswell, B. H., & Sulzman, J. M. (1999). Contemporary and pre-industruial global reactive nitrogen budgets. *Biogeochemistry, 47*.

Holsten, K. (1992). The global carbon cycle. In: S. S. Butcher, R. J. Charlson, G. H. Orians, & G. V. Wolfe (Eds.). *Global Biogeochemical Cycles*. Academic Press, London et al, 239–316.

Horne, A. J. (1977). Nitrogen fixation — a review of this phenomenon as a polluting process. *Proc. Water Technol., 8*, 359–372.

Howarth, R. W. (1988). Nutrient limitation of net primary production in marine ecosystems. *Ann. Rev. Ecol., 19*, 89–110.

Howarth, R. W., Jensen, H. S., Marino, R., & Postma, H. (1995). Transport to and processing off P in near-shore and oceanic waters. In: H. Tiessen Ed. *Phosphorus in the Global Environment*. SCOPE, John Wiley & Sons, 323–356.

Howarth, R. W., Marino, R., & Cole, J. J. (1988)b. Nitrogen fixation in freshwater, estuarine, and marine ecosystems. 2. Biogeochemical controls. *Limnol. Oceanogr., 33*, 688–701.

Howarth, R. W., Marino, R., & Lane, J. (1988)a. Nitrogen fixation in freshwater, estuarine, and marine ecosystems. 1. Rates and importance. *Limnol. Oceanogr., 33*(4, part 2), 669–687.

Howarth, R. W., & Marino, R. (1990). Nitrogen-fixing cyanobacteria in the plankton of lakes and estuaries: A reply to the comment by Smith. *Limnol. Oceanogr., 35*, 1859–1863.

Howarth, R. W., Fruci, J. R., & Sherman, D. (1991). Inputs of sediment and carbon to the estuarine ecosystems: Influence of land use. *Ecological Applications, 1*(1), 27–39.

Howarth, W. W., & Stewart, J. W. B. (1992). The interaction of sulfur with other element cycles in ecosystems. In: R. W. Howarth, J. W. B. Stewart, & M. V. Ivanov. *Sulfur Cycling on the Continent*. SCOPE, John Wiley & Sons, 67–84.

Howarth, R. W., & Cole, J. J. (1985). Molybdenum avalability, nitrogen liimitation, and phytoplankton growth in natural waters. *Science, 229*, 653–655.

Howarth, R. W., Scheider, & Swaney, D. (1996). Metabolism and organic carbon fluxes in the tidal freshwater Hudson river. *Estuaries, 19*(4), 848–869.

Howarth, R. W. Ed. (1996). *Nitrogen Cycling in the North Atlantic Ocean and its Watersheds*. Dordrecht-Boston-London: Kluwer Academic Publishers, 304 pp.

Howarth, R. W., & Marino, R. (1998). A mechanistic approach to understanding why so many estuaries and brackish waters are nitrogen limited. In: *Effects of Nitrogen in the Aquatic Environment*, KVA Report 1998: 1. Stockholm: Royal Swedish Academy of Sciences, 117–136.

Howarth, R. W., Chan, F., & Marino, R. (1999). Do top-down and bottom-up controls interact to exclude nitrogen-fixing cyanobacteria from the plankton of estuaries? An exploration with a simulation model. *Biogeochemistry, 46*, 203–231.

Huber, A. L. (1986). *Nitrogen fixation by Nodularia spumigina Mertins (Cyanobacteriaceae). Hydrobiol.,* *131*, 193–203.

Ivanov, M. V. (1983). Major fluxes in the global biogeochemical cycle of sulfur. In: M. V. Ivanov, & J. R. Freeney (Eds.). *Global Biogeochemical Sulfur Cycle*. John Wiley, Chichester, 449–463.

IPCC. (1997). *Guidelines for National Greenhouse Gas Inventories*. OECD/ICDE, Paris.

Iserman, K. (1991). Share of agriculture in nitrogen and phosphorus emission into the surface waters of Western Europe against the background of their eutrophication. *Fertilizer Research, 26*, 253–269.

Izuta, T., & Totsuka, T. (1996). *Effect of soil acidification on growth of Cryptometria japonica seedlings.* In: Proceedings of the International Symposium on Acid Deposition and Its Impacts. Tsukuba, Japan, 10–12, December, 1996, 157–164.

Jahnke, R. (1992). The phosphorus cycle. In: S. S. Butcher, R. J. Charlson, G. H. Orians, & G. V. Wolfe (Eds.). (1992). *Global Biogeochemical Cycles*. Academic Press, London et al, 301–315.

Jakucs, P. (Ed.). (1985). *Ecology of the Oak Forest in Hungary. Vol. 1: Structure, Primary Production, and Mineral cycling*. Akademiaia Kiado, Budapest.

Jaworski, N. A., & Groffman, P. M., Keller, A. A. & Prager, J. C. (1992). A watershed nitrogen and phosphorus balance: the Upper Potomac River basin. *Estuaries, 15*, 83–95.

Jeffries, D. S. (1986). *Evaluation of the regional acidification of lakes in eastren Canada using ion ratios.* In: Proc. UN ECE Workshop on Acidification of Rivers and Lakes, Grafenau, FRG, 17–37.

Jeffries, D. S., & Lam, D. C. L. (1993). Assessment of the effect of acidic deposition on Canadian lakes: determination of critical loads for sulfate deposition. *Water Science Technology, 28*, 183–187.

Jie Xuan. (1999). *Vertical fluxes of dust in northern China*. In: Proceedings of Fifth International Joint Seminar on Regional Deposition Processes in the Atmosphere, 12–16 October 1999, Seoul, 43–52.

Jickells, T. D., & Rae, J. E. (Eds.). (1997). *Biogeochemistry of Intertidal sediments. Cambridge Environmental Chemistry Series, 9*, Cambridge University Press, 193 pp.

Johnson, D. W. (1992). Base cation distribution and cycling. In: D. W. Johnson, & A. E. Lindberg Eds. *Atmospheric Deposition and Forest Nutrient Cycling*. New York: Springer-Verlag, 275–340.

Jones, G. S., Blackburn, S. I., & Pareker, N. S. (1994). A toxic bloom of Nodularia spumigena Mertens in Orielton Lagoon, Tasmania. *Aust. J. Mar. Fresh. Res., 45*, 787–800.

Joosten, L. T. A., Buijze, S. T., & Jansen, D. M. (1998). Nitrate in sources of drinking water? Dutch drinking water companies aim at prevention. In: Proceedings of the First International Nitrogen Conference, Elsevier Science, 487–492.

Jorgensen, S. E., Halling-Sorensen, & Nielsen, S. N. (Eds.). (1995). *Handbook of Environmental and Ecological Modeling*. Boca Raton: Lewis Publishers, 672 pp.

Jouzel, J., Barkov, N. I., Barnola, J-M., Benzer, M., Chappellaz, J., Genthon, C., Kotlyakov, V. M., Lipenkov, V., Lorius, C., Petit, J. R., Raynaud, D., Raisbeck, G., Ritz, C., Sowers, T., Stivenard, M., Yiou, F., & Yiou, P. (1993). Extending the Vostok ice-core record of paleoclimate to the penulimate glacial period. *Nature, 364*, 407–412.

Kabata-Pendias, A., & Pendias, H. (1984). *Trace elements in soils and plants*. Boca Raton, TL. CRC Press.

Kapitsa, A. P. (2000). *Global warming?* Pers. comm.

Kasimov, N. S., Koroleva, T. V. , & Proskuryakov, Yu. V. (1995). Biogeochemistry of urban landscapes (on the example of the Tolyatti City). In: N. S. Kasimov (Ed.) *Ecogeochemistry of Urban Landscapes*. Moscow, Moscow University Publishing House, 23–282.

Keeling, C. D., & Whorf, T. P. (1994). Decadal oscillations in global temperature and atmospheric cabon dioxide. In: W. A. Sprigg (Ed.). *Natural variability of climate on decade-to-century time scales*. National Academy of sciences, Washington, DC.

Kennedy, I. R. (1994). *Acid Soil and Acid Rain* (2nd). New York-Chichester-Toronto-Brisbane-Singapore: John Wiley & Sons INC, 254 pp.

Koch, G. W., & Mooney, H. A. (Eds.). (1996). *Carbon Dioxide and Terrestrial Ecosystems*. New York-Boston: Academic Press, 443 pp.

Kohno, Y., Matsumura, H., & Kobayashi, T. (1998). Differential sensitivity of 16 tree species to simulated acid rain or sulfur dioxide in combination with ozone. In: V. N. Bashkin, & S-U. Park (Eds.). *Acid Deposition and Ecosystem Sensitivity in East Asia*. Nova Science Publishers, ltd., 143–188.

Kovalevsky, A. L. (1991). *Plant Biogeochemistry*. Novosibirsk: Nauka Publishing House, 294 pp.

Kovalsky, V. V. (Ed.). (1981). Proceedings of *Biogeochemical Laboratory, vol. 19*, 203 pp.

Kovalsky, V. V., & Suslikov, V. L. (1980). Silicon sub-regions of biosphere in the USSR. Proceedings of *Biogeochemical Laboratory*. Moscow: Nauka Publishing House, vol. 18, 3–58.

Kovalsky, V. V. (1980). Geochemical ecology and the problems of health. In: *Environmental Geochemistry and Health*. UK, London: RSC, 185–191.

Kovda, V. A. (1973). Hydromorphic soils of Mediterranean and Tropic areas. In: *Pseudoglay and Glay*. *Weinheim: Verl. Chem.*, 379–382.

Kovda, V. A. (1984). *Biogeochemistry of Soil Cover*. Moscow: Nauka Publishing House, 261 pp.

Kozlov, M. YA., Towprayoon, S., & Sirikarnjanawing, S. (1997). *Application of critical load methodology for assessment of the effects of acidic deposition in Northern Thailand*. In: Proceedings of International Workshop on Monitoring and Prediction of Acid Rain, Seoul, 29.09–1.10. 1997, 141–146.

Kozlov, M. Ya., & Towprayoon, S. (1998). Sensitivity of Thailand's ecosystems to acidic deposition. In: V. N. Bashkin, & S-U. Park (Eds.). *Acid Deposition and Ecosystem Sensitivity in East Asia*. Nova Science Publishers, ltd., 335–378.

Krapivin, V. E. (1993). Mathematical model for global ecological investigations. *Ecological Modelling, 67* (2), 103–127.

Kuo, C., Lindberg, C., & Thompton, D. J. (1990). Coherence established between atmosphere carbon dioxide and global temperature. *Nature, 343*, 709–713.

Kudeyarov, V. N., & Bashkin, V. N. (1984). Study of landscape-biogeochemical balance of nutrients in agricultural regions. III. Nitrogen. *Water, Air and soil Pollution, 23*, 141–153.

Kudeyarova, A. Yu. (1996). *Soil polyphosphates*. Moscow: Nauka Publishers, 300 pp.

Kudeyarova, A. Yu., & Bashkin, V. N. (1984). Study of landscape-agrogeochemical balance of nutrients in agricultural regions: pt.I. Phosphorus. *Water, Air and Soil Pollution, 21*, 87–95.

Kuperman, R. G. (1999). Litter decomposition and nutrient dynamics in oak-hickory forest along a historic gradient of nitrogen and sulfur deposition. *Soil Biology & Biochemistry, 31*, 237–244.

Kuylenstierna, J. C. I., Cambridge, H. M., Cinderby, S., & ChadwickK, M. J. (1995). Terrestrial Ecosystem Sensitivity to Acidic Deposition in Developing Countries. *Water, Air and Soil Pollution, 85*, 2319–2324.

Lee, K. W., & Young, J. C. (Eds). (1998). Advance Environmental Monitoring, Proceedings of the 1st K-JIST International Symposium, Kwangju, Korea, 176 pp.

Lein, A. Yu., & Ivanov, M. V. (1988). The global biogeochemical cycles of elements and human productive activity. *Geochemistry*, No. 2, 280–291.

Lewis, W. M., Melack, J. M., McDowell, W. H., McClain, M., & Richey, J. E. (2000). Nitrogen yields from undisturbed watersheds in the Americas. *Biogeochemistry*.

Likens, G. E., Bormann, F. H., Pierce, R. S., Eaton, J. S., & Johnson, N. M. (1977). *Biogeochemistry of a Forested Ecosystems*. New York: Springer-Verlag, 146 pp.

Lin, H-C., Yang, S-S., Hung, T-C., & Chou, C–H. (Eds). (1996). The Effect of Human Disturbance on the Nitrogen Cycle in Asia, Proceedings of SCOPE/ICSU Nitrogen Workshops.

Lin, T. C. (1998). Acid deposition and forest ecosystem sensitivity in Taiwan. In: V. N. Bashkin, & S-U. Park (Eds). *Acid Deposition and Ecosystem Sensitivity in East Asia*. NovaScience Publishers, USA, 379–412.

Lindahl, G., & Wallstrom, K. (1985). Nitrogen fixation (acetylene reduction) in planktonic cyanobacteria in Oregnudsgrepen. SW Bothnian Sea. *Arch Hydrobiol., 104*, 193–204.

Lisitsin, A. P., Demina, L. L., & Gordeev, V. V. (1983). *The Bivalent System of River Water and its Interaction with the Ocean*. Msocow: Nauka publishing House, 340 pp.

Logan, G. A., Hayes, J. M., Hieshima, G. B., & Summons, R. E. (1995). Terminal proterozoic reorganization of biogeochemical cycles. *Nature, 375*, 53–56.

Lowrance, R., & Leopard, R. A. (1988). Streamflow nutrient dynamics on coastal plain watersheds. *Journal of Environmental Quality, 17*, 734–740.

Mackenzie, F. T. (1998). *Our changing plante*, Secondary Edition. Upper Saddle River: Plentice Hall, 486 pp.

Manahan, S. E. (2000). *Environmental Chemistry*, 7th ed. Boca Raton etc.: Lewis publishers, 898 pp.

Marino, R., Howarth, R. W., Shamess, J., & Prepas, E. E. (1990). Molybdenum and sulfate as controls on the abundance of nitrogen-fixing cyanobacteria in saline lakes in Alberta. *Limnol. Oceanogr., 35*, 245–259

Mason, J. W., Wegner, G. D., Quinn, G. I., & Lange, E. I. (1990). Nutrient loss via groundwater discharge from small watersheds in southern and south central Wisconsin. *J. Soil and Water Conservation*, 1990, 327–331.

Maximum permissible levels (MPLs) of hazardous substances in the natural water. (1986). Act No. 2932–83. In: *Sanitary rules and norms*, 42–121–4130–86. Moscow.

McGlathery, K. J., Marino, R., & Howarth, R. W. (1994). Variable rates of phosphorus uptake by shallow marine carbon sediments: Mechanisms and ecological significance. *Biogeochemistry*, 25, 127–146.

Meybeck, M. (1982). Carbon, nitrogen, and phosphorus transport by World rivers. *American Journal of Science*, 282, 401–450.

Meybeck M., Chapman, D. V., & Helmer, R. (1989). Global Freshwater Quality: a First Assessment. *World Health Organization/United Nations Environment Programme*. Cambridge, MA: Basil Blackwell, Inc.

Moldan, B., & Cherny, J. (Eds.). (1994). *Biogeochemistry of Small Catchments*. John Wiley and Sons, 419 pp.

Mosier, A. et al. (Eds.). (1998). *Nutrient Cycling in Agroecosystems*. Kluwer Academic Publishers, 313 pp.

Moran, M. D. (1997). Evaluation of the impact of North American SO_2 emission control legislation on the attainment of $SO_4=$ critical loads in eastern Canada. Paper 97-TA 28.05, 90th AWMA Annual Meeting, Air & Waste Management Association, Pittsburgh.

Morel, F. M. M. (1983). *Principles of Aquatic Chemistry*. New York: Wiley-Interscience,

National Academy of Sciences. (1978). *An Assessment of Mercury in the Environment*. Washington, DC.: National Academy of Sciences.

National Research Council. (1993). Managing Wastewater in Coastal Urban Areas. Committee on Wastewater Management for Coastal Urban Areas. Washington, DC.: National Academy Press.

Nazarov, A. G. (1983). *History and Ecology of Biogeochemical Cycling*. Dr.Sc. Thesis, Moscow State University.

Neill, C., Piccolo, M. C., Melillo, J. M., Steudler, P. A., & Cerri, C. C. (1999). Nitrogen dynamics in Amazon forest and pasture soils measured by ^{15}N pool dilution. *Soil Biology & Biochemistry, 31*, 567–572.

Neudachin, A. P. (1999). *Agrogeochemistry of Ameliorated Peat Soils in the Northwest Part of Middle Amur low plain*. Ph.D. thesis, Biology-Soil Institute RAS, Vladivostok, 25 pp.

NIES. (1996). Proceedings of the International Symposium on Acid Deposition and Its Impacts, Tsukuba, Japan, 10–12 December 1996, 371 pp.

Nihlgard, B. J., Swank, W. T., & Mitchell, M. J. (1994). Biological processes and catchment studies. In: B. Moldan, & J. Cherny (Eds.). *Biogeochemistry of Small Catchments*. John Wiley and Sons, 133–160.

Nikitina, I. B. (1973). The geochemistry of ultra-fresh waters in the Northern Taiga Permafrost Landscapes of South Yakut area. In: *Landscape Geochemistry and Hypergenesis*. Moscow: Nauka Publishers, 24–35.

Nikolaev, A. I., Kazenovich, L. I., & Atabaev, Sh. T. (1988). *Pesticides and Immunology, Alma-Ata*. Medicine Publishing House, 118 pp.

Nilsson, I., & Grennfelt, P. (Eds.). (1988). *Critical Loads for Sulfur and Nitrogen*. Report from a Workshop Held at Stokhoster, Sweden, March 19–24, 1988. Miljo Rapport 1988: 15. Copenhagen, Denmark, Nordic Council of Ministers, 418 pp.

Nixon, S. W., Kelly, J. R., Furnas, B. N., Oviatt, C. A., & Hale, S. S. (1980). Phosphorus regeneration and the metabolism of coastal marine bottom sediments. In: K. R. tenore, & B. C. Coull (Eds.). *Marine Benthic Dynamics*. Univ. of South Carolina Press.

Nixon, S. W., Granger, S. L., & Nowicki, B. L. (1995). An assessment of the annual mass balance of carbon, nitrogen and phosphorus in Narragansett Bay. *Biogeochemistry, 31*, 15–61.

Nixon, S. W., Ammerman, J. W., Atkinson, L. P., Berounsky, V. M., Billen G., Boicourt, W. C., Boynton, W. R., Church, T. M., Ditoro, D. M., Elmgren, R., Garber, J. M., Giblin, A. E., Jahnke, R. A., Owens, N. J. P., Pilson, J. H., & Seitzinger, S. P. (1996). The fate of nitrogen and phosphorus at the land-sea margin of the North Atlantic Ocean. In: R. W. Howarth (Ed.). *Nitrogen Cycling in the North Atlantic Ocean and its Watersheds*. Kluwer Academic Publishers, 141–180.

Novak, D. & Magnussen, H. (1993). Epidemiologie des Asthma bronchiale. *Atem. Wegs und Lungenkrankh, 19*(7), 288–295.

548

Nriagu, J. O. (1979). *Copper in the Environment*, Part I: *Ecological cycling.* New York: Wiley-Interscience.
Nriagu, J. O. (1989). A global assessment of natural sources of atmospheric trace metals. *Nature, vol. 338,* 47–49.
Nriagu, J. O., & Pacina, J. M. (1989). Quantitative assessment of worldwide contamination of air, water and soils by trace metals. *Nature, 333,* 134–139.
NRC. (1993). *Managing wastewater in coastal urban areas.* Washington, DC., USA: National Scientific Council.
Odum, Yu. (1975). *Fundamentals of Ecology.* Moscow: Myr Publishing House, 740 pp.
OECD. (1991). *OECD Environmental Data,* Compendium 1991. Organisation for Economic Co-operation and development, Paris.
Oonk, H., & Kroeze, C. (1999). Nitrous oxide emissions and control. In: R. A. Meyers (Ed.). *Encyclopedia of Environmental Pollution and Cleanup.* The Wiley, 1055–1069.
Ostromogilsky, A. Kh., Anokhin, Yu. A., Vetrov, V. A., Petrukhin, V. V, & Poslovin, A. L. (1981). *Trace Elements in the Atmosphere of Background Regions of the Land and Ocean.* Obninsk: VNIIGMI-MPD Information Center.
Overgaard, K. (1994). *Nitrate pollution of groundwater in Denmark.* 20th Nordic symposium on Water Pollution, NORDFORSK.
Paerl, H. W. (1985). Microzone formation: Its role in the enhancement of aquatic N_2 fixation. *Limnol. Oceanogr., 30,* 1246–1252.
Paerl, H. W. (1990). *Physiological ecology and regulation of N_2 fixation in natural waters. Limnol. Oceanogr., 35.*
Pacyna, J. M. (1995). Emission inventory for heavy metals in the ECE. In: Heavy Metals Emissions. State-of-the-Art Report. Economic Commission for Europe. *Convention on Long-Range Transboundary Air Pollution.* Prague, June 1995, 33–50.
Pampura, T. V. (1997). The conjugate analysis of adsorption isoterms and Cu and Zn sorbed forms in Chernozem. In: V. N. Bashkin (Ed.). *Heavy Metals in the Environment.* Pushchino: ONTI Publishing House, 266–281.
Park, S-U. (Ed.). (1998). *Research and Development on Basic Technology for Atmospheric Environment in Global Scale: Development of Technology for Monitoring and Prediction of Acid Rain*(G-7 project). Republic of Korea: Ministry of Environment, 602 pp.
Park, S-U., In, H-J. & Lee, Yu-H. (1999). Parameterization of wet deposition of sulfate by precipitation rate. *Atmospheric Environment, 33,* 4469–4475 .
Park, S-U. (1996). Estimation of the Anthropogenic Emission of SO_2 and NO_x in South Korea. In: Proceedings of International Conference on Acid Deposition in East Asia, Taipei, May 28–30,1996, 30–44.
Park, S-U. (1997). *Development of Technology for Monitoring and Prediction of Acid Rain.* Seoul: SNU, 631 pp.
Parton, W. J., Morgan, J. A., Altenhofen, J. M., & Harper, L. A. (1988). Ammonium volatilization from spring wheat plants. *Agronomy Journal, 80,* 419–425.
Parton, W. J., Scurlock, J. M. O., Ojima, D. S., Gilmanov, T. G., Scholes, R. J., Schimel, D. S., Kirchner, T., Menault, H.-C., Seasteld, T., Garcia Moya, E., Kammalrut, A., & Kinyamario, J. L. (1993). Observations and modeling of biomass and soil organic matter dynamics for the grassland biome worldwide. *Global Biogeochemical Cycles, 7,* 785–809.
Perelman, A. I. (1975). *Landscape Geochemistry.* Moscow: "Vysshaya Shkola" Publishing house, 341 pp.
Perelman, A. I. (1976). *Studies of landscape geochemistry.* Moscow: "Vysshaya shkola" Publishing house, 216 pp.
Perelman, A. I., & Kasimov, N. S. (1999). *Landscape Geochemistry,* 3rd Edition. Moscow: Integration Publishers, 763 pp.
Polynov, B. B. (1946). *Landscape Geochemistry: Problems of Mineralogy, Geochemistry and Petrography.* Moscow: Moscow USSR Academy of Sciences Publishing House, 174 pp.
Posch, M., Hetteling, J-P., Sverdrup, H., et al. (1993). Guidelines for the Computation and Mapping of Critical Loads and Exceedance in Europe. In: Proceedings of 3d CCE meeting, 1993, 15–19 March, Madrid, 1–14.
Posch, M., Hettelingh, I-P., Alcamo, J., & Krol, M. (1996). Integrated scenario of acidification and climate change in Asia and Europe. *Global Environmental Change, 6*(4), 375–394.

Posch, M., Hettelingh, J-P., de Smet, P. A. M., & Downing, R. J. (Eds.). (1997). Calculation and Mapping of Critical Thresholds in Europe. Status Report 1997. *Coordination Center for Effects, RIVM Report* No. 259101007. the Netherlands: Bilthoven, 163 pp

Posch, M., de Smet, P. A. M., Hettelingh, J-P., & Downing, R. J. (Eds.). (1999). Calculation and Mapping of Critical Thresholds in Europe. Status Report 1999. *Coordination Center for Effects, RIVM Report* No. 259101009. the Netherlands: Bilthoven, 165 pp

Potter, C. S., Randerson, J. T., Field, C. B., Matson, P. A., Votousek, P. M., Mooney, H. A., & Klooser, S. A. (1993). Terrestrial ecosystem production: a process model based on global satellite and surface data. *Global Biogeochemical Cycles, 7*, 811–841.

Prasolov, L. I. (1939). *Genetic Soil Types and Soil Regions in the European Part of the USSR*. Moscow: USSR Academy of Sciences Publishing House, 164 pp.

Prospero, J. M., Barret, K., Church, T., Dentener, F., Duce, R. A., Galloway, J. N., Levy, H., Moody, J., & Quinn. (1996). Atmospheric deposition of nutrients to the North Atlantic basin. In: R. W. Howarth (Ed.). *Nitrogen Cycling in the North Atlantic Ocean and Its Watersheds*. Kluwer Academic Publishers, 27–74.

Radojevic, M. (1998). Acid rain monitoring in East and South East Asia. In: V. N. Bashkin, & S-U., Park (Eds.). *Acid Deposition and Ecosystem Sensitivity in East Asia*. Nova Science Publishers, ltd., 95–122.

Radojevic, M., & Bashkin, V. N. (1999). *Practical Environmental Analysis*. UK: RSC, 466 pp.

Rasmussen, P. E. (1994). *Current methods of estimating atmospheric mercury fluxes in remote areas. Environmental Science Technology, 28*(13), 2233–2241.

Redfield, A. C., Ketchum, B. H., & Richards, F. A. (1963). The influence of organisms on the composition of seawater. In: M. N. Hill (Ed.). *The Sea, vol. 2*. New York: Wiley, 26–77.

Resch, H. N. (1991). The balance of nitrogen composition in the FRG. In: Borght, P., & Tychon, B. (Eds). *Gestion de l'azote agricole et qualite des eaus Vander*. Cebedoc, Liege, Belgium.

Richardson, C. J. (1987). *Freshwater wetlands: transformers, filters or sinks*. In: FOREM, 3–9.

RMCC. (1990). The 1990 Canadian long-range transport of air pollutant assessment and acid deposition report: Part 4—aquatic effects. Federal/provincial Research and Monitoring Coordinating Committee, Ottawa, Ontario, Canada.

Ro, C-U., Sirois, A. & Vet, R. J. (1997). Time trends in Canadian air and prcipitation chemistry data. Paper 97-TA 28.05, 90th AWMA Annual Meeting, Air & Waste Management Association, Pittsburgh.

Rodin, L. E. et al. (1975). *Global productivity of terrestrial ecosystems*. Proceedings of the First International Congress on Ecology, Wageningen.

Rodin, L. E., & Bazilevich, N. I. (1965). *The dynamics of Organic Matter and Biological Cycling of Ash Elements and Nitrogen in Major Vegetation Type of the Earth*. Moscow-Leningrad: Nauka Publishing House.

Rodin, L. E., Bazilevich, N. I., Gradusov, B. I., & Yarilova, E. A. (1977). Arid Savanna of the Rajaputana (Tar Desert). In: *Arid Soils: Genesis, Geochemistry and Use*. Moscow: Nauka Publishers, 196–225.

Roite, A. (1991). *Fundamentals of Immunology*. Mir Publishing House, 327 pp.

Romankevich, E. E. (1988). Living Matter of the Earth. *Geochemistry, 2*, 292–306.

Ronov, A. B. (1976). Volcanisms, carbon storage and life: the trends in the global geochemistry of carbon. *Geochemistry, 8*, 1252–1277.

Ronov, A. B., & Yaroshevsky, A. A. (1976). A novel model for the chemical structure of the Earth's crust. *Geochemistry*, No. 12, 1763–1795.

Rozanov, B. G. (1984). *Lessons in Biogeochemistry*. Moscow: MSU Publishing House, 312 pp.

Ryaboshapko, A., Ilyin, I., Gusev, A., Afinogenova, O., Berg, T., & Hjellbrekke, A-G. (1999). *Monitoring and Modeling of Lead, Cadmium and Mercury transboundary Transport in the Atmosphere of Europe*. EMEP Report 3/99, 124 pp.

Salmi, M. (1963). *On relation between geology and multiple sclerosis. Acta Geographica, 17*(4).

Sanders, J. R. (1982). *The effects of pH upon the copper and cupric ion concentrations in soil solution. Journal of Soil Science, 33*, 679–689.

Satake, K., Inoue, T., Kasasaku, K., Nagafuchi, O., & Nakano, T. (1998). Monitoring of nitrogen compounds on Yakushima island, a world natural heritage site. *Environmental Pollution, 102*, S1, 107–113.

Sato, K., Wakamatsu, K., & Takahashi, A. (1998). Changes in distribution of aluminum species in soil solution due to acidification. In: Bashkin, V. N. & Park, Soon-Ung Eds. *Acid Deposition and Ecosystem Sensitivity in East Asia*. Nova Science Publishers, ltd., 125–142.

Savenko, V. S. (1976). *On the chemical composition of the oceanic atmospheric precipitation. Geochemistry*, No. 12, 1890–1893.

Scherepers, J. S., Frank, K. D., & Watts, D. G. (1983). *Influence of irrigation and nitrogen fertilization on groundwater quality. IAHS Publ., 146*, 21–29.

Schimel, D., & Sulzman, E. W. (1994). Variability of the Earth-climate system: decadal and longer timescales. In: S. P. Nelson (ed.). *The U.S. National Report (1991–1994) to the International Union of Geophysics and Geodessey.* American Geophysical Union, Washigton, DC.

Schimel, D., Enting, I. G., Heimann, M., Wigley, T. M. L., Raynaud, D., Alves, D., & Siegenthauler, U. (2000). CO$_2$ and the Carbon Cycle. In: T. M. L. Wigley, & D. S. Schimel (Eds). *Carbon Cycle*. Cambridge University Press, 7–36.

Schindler, D. W. (1977). Evolution of phosphorus limitation in lakes. *Science, 195*, 260–262.

Schlesinger, W. H. (1997). *Biogeochemistry. An Analysis of Global Changes.* Academic Press, 443 pp.

Seitzinger, S. P. (1987). Nitrogen biogeochemistry in an unpolluted estuary: The importance of benthic denitrification. *Mar. Eco. Prog. Ser., 37*, 65–73.

Seitzinger, S. P. (1988). Denitrification in freshwater in coastal marine ecosystems: ecological and geochemical significance. *Limnol. Oceanogr., 33*; 702–724.

Seitzinger, S. P., & Giblin, A. E. (1995). Estimating denitrification in north Atlantic continental shelf sediments. In: R. W. Howarth (Ed.), *Nitrogen Cycling in the North Atlantic Ocean and Its Watersheds.* Kluwer Academic Publishers, 235–260.

Seung, Y. H., & Park, Y. C. (1990). Physical and environmental character of the Yellow Sea. In: Park, C. H., Kim, D. H., & Lee, S. H. (Eds.). *The Regime of the Yellow Sea.* Institute of East and West Studies, Seoul, 9–38.

Shacklette, H. T. (1962). Biotic implications of Alaskan biogeochemical distribution patterns. *Ecology, 43*(1).

Shacklette, T. H., Sauer, H. I., & Miesch, A. T. (1970). Geochemical Environment and Cardiovascular Mortality rates in Georgia. *U.S. Geological Survey Prof.*, Paper No. 574-C, Washington, 1–39.

Shen, H., Mao, Zh., & Wu, J. (1998). Land-ocean interaction in the Changjiang estuary. In: *Geoenvironmental Changes and Biodiversity in the Northeast Asia.* The First International Symposium, Seoul, South Korea, 49–50.

Shindo, J., Bregt, A. K., & Takamata, T. (1995). Evaluation of Estimation Methods and Base Data Uncertainties for Critical Loads of Acid Deposition in Japan. *Water, Air, and Soil Pollution, 85*, 2571–2576.

Shindo, J. (1998). Model application for assessing the ecosystem sensitivity to acidic deposition based on soil chemistry changes and nutrient budgets. In: V. N. Bashkin, & S-U. Park (Eds.). *Acid Deposition and Ecosystem Sensitivity in East Asia.* Nova Science Publishers, ltd., 312–334.

Sidorenko, G. I., Kutepov, E. N., & Gedymin, M. Yu. (1991). Methodology of human health monitoring in various environments. *Issues of the USSR Academy of Medical Sciences*, No. 1, 15–18.

Siegenthaler, U., & Sarmiento, J. L. (1993). Atmospheric carbon dioxide and the ocean. *Nature, 365*, 119–125.

Simkiss, K., & Taylor, M. G. (1989). Metal fluxes across membranes of aquatic organisms. *Rev. Aquat. Sci., 1*, 174–188.

Simon, W., Huwe, B., & van der Ploeg, R. R. (1988). Die Abschatzung von Nirtatenstragen ans landwirtschaftlichen Mulzflachen mit Hilfe von Nmin-daten. *Z. Pflanzenenaehr. Bodenk., 151*, 289–294.

Skarlygina-Ufimtseva, M. D., Chernyakhov, V. B., & Berezkina, G. A. (1976). *Biogeochemistry of Pyrite Copper Deposits in the South Ural.* LGU Publisher, Leningrad, 150 pp.

Smith, D. R., Niemeyer, S., Estes, J. A., & Flegal, A. R. (1990). Stable lead isotopes evedence anthropogenic contamination in Alaskan Sea otters. *Environ. Sci. Technol., 24*, 1517–1521.

Smith, K. R., Carpenter, R. A., & Faulstich, M. S. (1988). Risk Assessment of Hazardous Chemical Systems in Developing Countries. *Occasional Report* No. 5. Honolulu: West-East Environment and Policy Institute.

Smith, W. (1999). Acid Rain. In: R. A. Meyers (Ed.). *Encyclopedia of Environmental Pollution and Cleanup.* New York: The Wiley, 9–15.

Sokolov, I. A. (1971). *The General Peculiarities of Soil Formation in East Kazakhstan, Alma-Ata.* Nauka Publishing House, 231–245.

Solntseva, N. P. (1982). Geochemical stability of natural systems to technogenic loads. In: *Mineral Extraction and Geochemistry of Natural Ecosystems*. Moscow: Nauka Publishing House, 181–216.

Soviev, M. (1998). Numerical modeling of acid deposition on Eurasian continent. In: V. N. Bashkin, & S-U. Park (Eds.). *Acid Deposition and Ecosystem Sensitivity in East Asia (monograph)*. Nova Science Publishers, ltd., 5–48.

Staaf, H., & Berg, B. (1982). Accumulation and release of plant nutrients in decomposing Scots pine needle litter. Long-term decomposition in a Scots pine forest II. *Canadian Journal of Botany, 60*, 1561–1568.

Stephens, C. J. (1971). *Laterite and silcrite in Australia. Geoderma*, No. 5, 5–52.

Sterner, R. W., Elser, J. J., & Hessen, D. O. (1992). Stoichiometric relationships among producers, consumers, and nutrient cycling in pelagic ecosystems. *Biogeochemistry, 17*, 49–67.

Stevenson, F. J. (1986). *Cycles in Soil*. New York: Wiley.

Strebel, O. J., Duynisveld, W. M. H., & Bottcher, J. (1989). Nitrate pollution of groundwater in Western Europe. *Agricultural Ecosystems and Environment, 26*, 189–214.

Sukachev, V. N. (1964). *Selected Publications*. Leningrad: Nauka Publishing House.

Tan, J. A., Wang, W. Y., Wang, D. C., & Hou, S. F. (1994). Adsorption, Volatilization and speciation of selenium in different types of soils in China. In: Frankenberger, W. T., & Benson, S. Eds. *Selenium in the Environment*. New York: Marcel Dekker, 47–51.

Tankanag, A. V. (1999). Application of GIS techniques for calculation and mapping of critical loads. In: V. N. Bashkin, & H. D. Gregor (Eds). *Calculation of Critical Loads of Air Pollutants at Ecosystems of East Europe*. Moscow: POLTEX, Berlin: UBA, 71–75.

Temminghof, E., van der Zee, S., & De Haan, F. (1997). Copper mobility in a copper-contaminated sandy soil as affected by pH and solid and dissolved organic matter // *Environmental Science Technology, 31*, 1109–1115.

Tessier, A., Campbell, P. G. C., & Bission, M. (1979). Sequential extraction procedure for the speciation of particulate trace metals. *Analytical Chemistry*, No. 51, 844–851.

Thornton, I., Ramsey, M., & Atkinson, N. (1995). *Metal in the global environment: facts and misconceptions*. International Council on Metals and Environment, 103 pp.

Trachtenberg, I. M. (1994). *Heavy Metals in the Environment*. Moscow: Nauka Publishing House, 285 pp.

Turekian, K. K. (1977). The fate of metals in the ocean. *Geochemical and Cosmochemical Acta, 41*, 1139–1144.

Turner, D. R., Whitefield, M., & Dickson, A. G. (1981). The equilibrium speciation of dissolved components in freshwater and seawater at 25°C and 1 atm pressure. *Geochemical and Cosmochemical Acta, 45*, 855–881.

Turner, R. E., & Rabalais, N. N. (1991). Changes in Mississippi River water quality this century. *BioScience, 41*, 140–147.

UBA. (1996). Manual on Methodologies and Criteria for Mapping Critical Levels/Loads and geographical areas there they are exceeded. Berlin, 142 pp + Annex.

Uchvatov, V. P. (1994). *Natural and anthropogenic fluxes of substances in the landscapes of the Russian plain*. Dr. Sci. Thesis. Pushchino, 471 pp.

Ugolini, F. C., & Spattelstein, H. (1992). Pedosphere. In : Butcher, S. S., Charlson, R. J., Orians, G. H., & Wolfe, G. V. (Eds.). 1992. *Global Biogeochemical Cycles*. London et al: Academic Press, 123–153.

Um, K. T. (1985). *Soils of Korea. Soil Survey Materials*, No. 11. Agricultural Science Institute, Rural Administration, Korea, 66 pp.

UNESCO. (1978). *World Water Balance and Water Resources of the Earth*. Leningrad: Gidrometeoizdat Publishers, 666 pp.

UNESCO. (1978). *World Water Balance and Water Resources of the Earth. Studies and Reports in Hydrology*. UNESCO

US IPA. (1992). Framework for Ecological Risk Assessment, EPA/630/R 92/001. Risk Assessment Forum, Washington, DC.

Van de Plassche, E., Bashkin, V., Guardans, R., Johansson, K., & Vrubel, J. (1997). An Overview of Critical Limits for Heavy Metals and POPs. Background document. Workshop on Critical Limits and Effect Based Approaches for Heavy Metals and Persistent Organic Pollutants, Bad Harzburg, Germany, 3–7 November 1997. 38 pp.

Vegas-Vilarrubia, T., Maass, M., Rull, V., Elias, V., Ovalle, A. R. C., Lopez, D., Scheider, G., Depetris, P. J., & Douglas, I. (1994). Small chatchment studies in tropical zone. In: B. Moldan, & J. Cherny (Eds.). *Biogeochemistry of Small Catchments*. John Wiley and Sons, 343–360.

Venkatesh, S., Gong, W., Kallaur, A., Makar, P. A., Moran, M. D., Pabla, B., Ro, C-U., Vet, R., Burrows, W. R., & Montpetti, J. (2000). Regional Quality modeling in Canada - Applications for Policy and Real-time Prediction. *Natural Hazards, 21*, 101–129.

Vernadsky, V. I. (1926). *La Biogeochemie*. Paris: Sorbonne, 167 pp.

Vernadsky, V. I. (1926). *Isotopes and living matter. Reports of the USSR AS, serial A*, 215–218.

Vernadsky, V. I. (1932). *Essay of Geochemistry*. Moscow: USSR Academy of Sciences Publishing House, 465 pp.

Vinogradov, A. P. (1938). *Biogeochemical provinces and endemic diseases. Issues of the USSR Academy of Sciences, vol. 18*, No. 4/5, 283–286.

Vinogradov, A. P. (1962). *The average content of chemical elements in the major volcanic rocks of the Earth's crust, Geochemistry*, No. 7, 555–571.

Vinogradov, A. P. (1963). Biogeochemical provinces and their role in the organic evolution. *Geochemistry*, No. 3, 199–212.

Vitousek, P. M., & Sanford, R. L. (1986). Nutrient cycling in moist tropical forest. *Annual Review of Ecology and Systematics, 17*, 137–167

Vitoushek, P. M., Fahey, T., Jounson, D. W., & Swift, M. J. (1988). Element interactions in forest ecosystems: Succession, allometry and input-output budgets. *Biogeochemistry, 5*, 7–34.

Vitoushek, P. M., & Howarth, R. W. (1991). Nitrogen limitation on land and sea: how can it occur? *Biogeochemistry, 13*, 87–115.

Vitousek, P. M., Aber, J. D., Howarth, R. W., Likens, G. E., Matson, P. A., Schinder, W. H., Schlezinger, W. H., & Tilman, D. G. (1997). Human alteration of nitrogen global cycles: Sources and consequences. *Ecol. Appl., 7*, 737–750.

Voitkevich, G. V. (1986). *Chemical Evolution of the Earth*. Moscow: Nauka Publishing House, 230 pp.

Walker, J. C. G. (1977). *Evolution of the atmosphere*. New York: Plenum Press.

Warren, W. J. (1957). *Arctic plant growth. Advancement in Science, 143*, 383–388.

Warren, H. V. (1961). Role of Geology in Diet. *J. Canadian Diet Association, 23*(1).

Webb, J. S. (1964). Geochemistry and Life. *New scientist, 23*, 504–507.

Webb, J. S., Thornton, I., & Fletcher, K. (1966). Seleniferous soils in parts of England and Wales. *Nature, 211*(5046), 327.

Winfrey, M. R., & Rudd, W. M. (1990). Environmental factors affecting the formation of methylmercury in low pH lakes. *Environmental Toxicology and Chemistry, 9*, 853–869. Wong, C. S., Chan, Y-H., Page, J. S., Smith, G. E., & Bellegay, R. D. 1994. Changes in equatorial CO_2 flux and new production estimated from CO_2 and nutrient levels in Pacific surface waters during the 1986/1987 El Nino. *Tellus, 43B*, 64–79.

World Bank. (1994). *RAINS-ASIA*. User's Manual, IISAA, Washington, 138 pp.

WRI/UNEP. (1988). World Resources 1988–1989. World Resources Institute in collaboration with United Nations Environment Programme. New York: Basic Books Inc.

Xing, G. X., & Zhu, Z. L. (2000). A primary analysis and estimation of nitrogen sources and sinks in China watershed. Pers. Comm.

Yaroshevich, L. M., & V. V., Zhilko. (1980). Modern erosion processes. In: *Problems of the Investigation of Exogenous Geological Processes*. Minsk, Belarus, 52–64.

Zakrutkin, V. E., & Shishkina, D. Yu. (1997). Distribution of copper and zinc in soils and plants of agroecosystems of the Rostov region. In: V. N. Bashkin (Ed.). *Heavy Metals in the Environment*. Pushchino: ONTI Publishing House, 101–117.

Zehnder, A. J. B., & Zinder, S. H. (1980). The sulfur cycle. In: *Natural environment and Biogeochemical Cycles*. Berlin: Springer-Verlag, 105–146.

Zhakashev, N. Zh. (1990). Monitoring of human health in industrial city. In: *Social-Hygienic Aspects of Human Health in Kazakhstan*. Alma-Ata, 72–78.

Zhang, J., Huang, W. W., Liu, S. M., & Wang, J. H. (1992). Transport of particulate heavy metals towards the China Sea: a preliminary study and comparison. *Marine Chemistry, 40*, 161–178.

Zhu, Z. L. (Ed.). (1997). *Nitrogen Balance and Cycling in Agroecosystems of China.* Kluwer Academic Publisher, 355 pp.
Zimmermann, P. H. (1998). Moguntia: A handy global tracer model. In: van Dop, H. (Ed). *Air Pollution Modeling and its Applications VI.* New York: Plenum.

INDEX

A

Accretion, 16, 21, 22, 77
Acid neutralizing capacity, ANC, 462, 471, 506
Acid rain, 137, 325, 384, 457, 477, 478, 480, 483,
    484, 487, 489–492, 501, 506–509, 533
Acid, organic, *see* Humic acid, 30, 31, 40, 41, 44,
    45, 59, 81, 90, 115, 126, 127, 129, 131,
    136, 137, 139, 144, 145, 158, 163, 174,
    182, 195–197, 202, 246, 269, 271, 295,
    308, 315–320, 323–325, 327–329, 347,
    384, 404, 413, 418, 457, 459, 461–
    463, 465–467, 470, 471, 476–481, 483,
    484, 487, 489–493, 496–511, 532, 533,
    536
Acidification, 90, 115, 252, 269, 316, 317, 319,
    329, 355, 356, 383, 459, 461, 463–
    474, 479, 483, 484, 486, 488, 490, 492,
    495, 499, 501, 506, 508, 526, 527, 533
Adenosine triphosphate, ATF, 126, 202
Aerosols, 137, 242, 244, 337, 338, 340
Agroecosystem, 166, 363, 364, 388
Algae, 53, 54, 59, 86, 113, 146, 148, 151, 161,
    202, 205, 211, 220, 221, 313, 391
Alkalinity, 30, 103, 104, 165, 187, 284, 440, 462,
    463, 479, 483, 485
Aluminum, 16, 54, 129, 148, 241, 246, 259, 271,
    301, 303, 306, 318, 322, 461, 483, 492,
    499, 508
Amazon river, 148
Amino acids, 38–40, 48, 50, 86, 110, 344
Ammonium (NH_4^+), 26, 44, 55, 80, 89, 113, 114,
    251, 290, 373, 399, 400, 491
Anoxygenic photosynthesis, 54
Apatite, 125, 129, 130, 439
Archaea, 40, 42, 45
Assimilation, 55, 106, 114, 173, 215, 220, 232,
    254, 289
Atlantic Ocean, 158, 357, 358, 368, 369, 375, 428
Atmosphere, 1–5, 8, 13, 15, 18, 21–27, 31, 32, 34,
    39, 48, 49, 53, 54, 56, 69–71, 74, 76–
    80, 82, 89, 92–97, 99, 100, 102, 103,
    106–108, 110, 111, 113, 115, 122, 123,
    135, 137, 139–144, 157–159, 167, 172,
    192, 199, 200, 206, 209, 228–231, 241,
    269, 280, 287, 303, 307, 335–341, 356,
    359, 364, 368, 369, 373, 383, 385, 394,
    397, 409, 410, 413, 415, 421, 422, 424,
    426, 427, 443, 450, 459, 478, 490

B

Background, 13, 48, 166, 169, 170, 172, 192–196,
    286, 325, 342, 418, 435, 529, 535
Bacteria, 26, 38, 40, 42, 45, 53, 54, 59, 67, 84–86,
    89, 110, 113–115, 137, 139, 141, 200,
    202, 206, 215, 220, 221, 224, 249, 282,
    302, 407, 444, 508
Balance, *see also* budget, 21, 30, 56, 93, 100, 119,
    120, 215, 229, 230, 254, 269, 298, 320,
    363, 365, 368, 392, 395, 403, 413, 422,
    429, 471, 534
Banded iron formation, BIF, 27, 54, 62
Basalt, 95, 96, 153, 167, 296
Base cation, 269, 316, 462, 465, 471, 503, 506,
    507
Bedrocks, 92
Bicarbonate, 29, 30, 59, 60, 145, 153, 154, 269
Biodiversity, 299, 340, 356, 360, 460, 464, 469,
    485, 512
Biogeochemical cycle, defined, 3, 11, 50, 56, 60,
    73, 91, 109, 115, 128, 129, 133, 134,
    139, 143, 144, 159, 160, 167, 193, 199,
    200, 217, 242, 252, 265, 267, 275, 304,
    322–325, 337, 360–362, 374, 375, 388,
    391, 392, 400, 409, 413–416, 421, 428,
    429, 459
Biogeochemical mapping, Chapter 7, 183, 297,
    307, 332, 334, 335, 341–343, 354, 432,
    433, 455, 490, 536
Biogeochemical standards, Chapter 10, 454, 536
Biogeochemistry of macroelements, Chapter 3,
    73
Biogeochemistry, defined, 1, 2, 4–11, 68, 80, 92,
    95, 110, 128, 158, 161, 162, 173, 192,
    196, 274, 282, 304–306, 335, 341, 342,
    355, 356, 368, 382, 403, 404, 416, 418,
    428, 431, 455, 457, 528, 535–537
Bioturbation, 104, 105, 524
Boron, 9, 161, 185–192, 197, 199, 285, 299, 342,
    346, 432, 440, 442, 443

C

Z